The Map Turtle and Sawback Atlas

ANIMAL NATURAL HISTORY SERIES
Victor H. Hutchison, *General Editor*

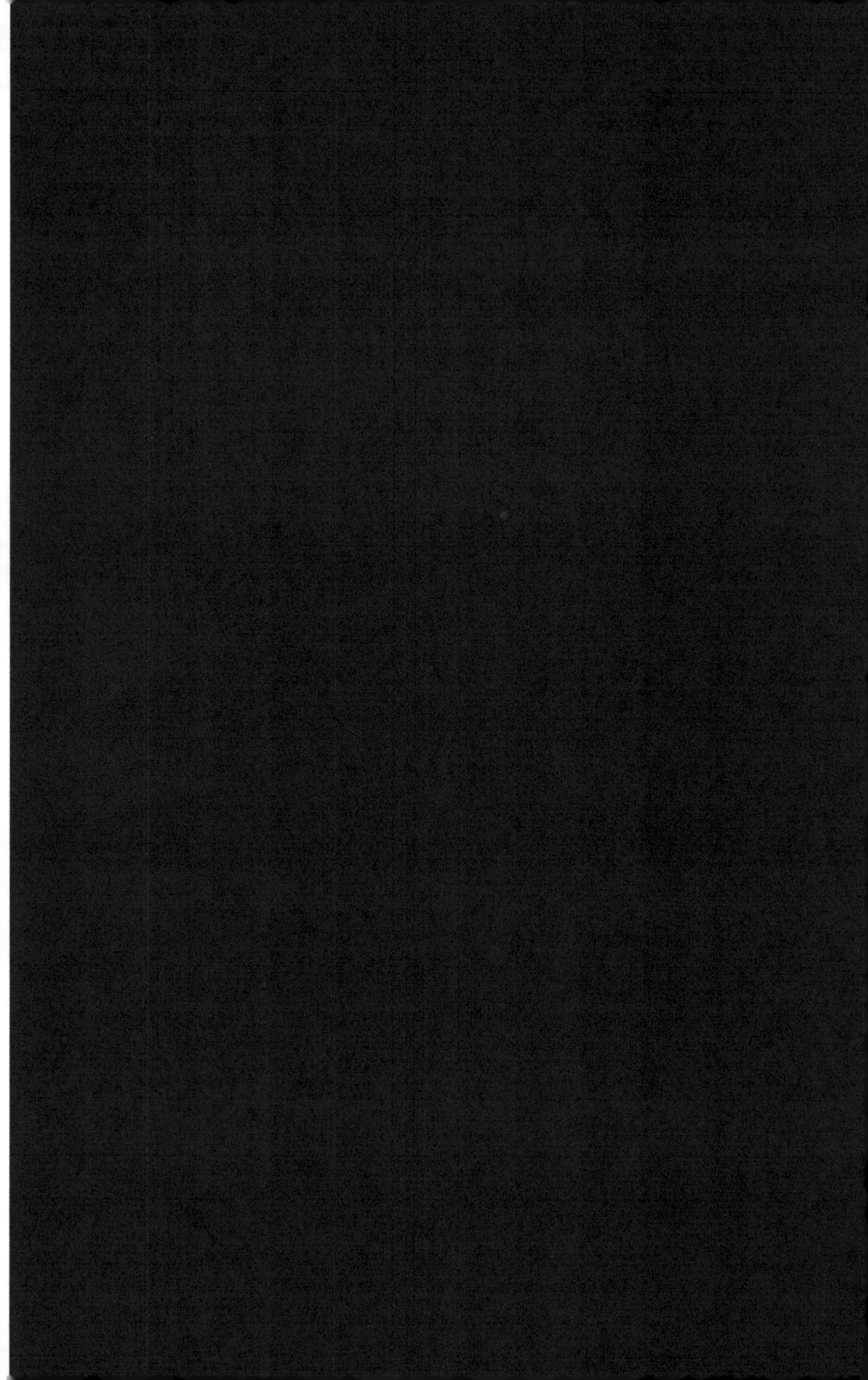

The Map Turtle and Sawback

Atlas Ecology, Evolution, Distribution, and Conservation

Peter V. Lindeman
Foreword by Anders G. J. Rhodin

University of Oklahoma Press : Norman

To my parents, Janet and Robert Lindeman, for indulging me as a child with so many books, frequent visits to the Oklahoma City Zoo, and many weekend canoeing trips in eastern Oklahoma and Arkansas, all of which helped lead me toward this career and this book

All royalties from the sales of this volume will go to the Chelonian Research Foundation, a nonprofit foundation for the conservation of turtles.

Library of Congress Cataloging-in-Publication Data
Lindeman, Peter V., 1962–
  The map turtle and sawback atlas : ecology, evolution, distribution, and conservation / Peter V. Lindeman.
    pages cm.—(Animal natural history series ; v. 12)
  Includes bibliographical references and index.
  ISBN 978-0-8061-4406-1 (cloth)
  ISBN 978-0-8061-4931-8 (paper)
1. Map turtles. I. Title.
  QL666.C547L56 2013
  597.92'59—dc23
                          2013022159

The Map Turtle and Sawback Atlas: Ecology, Evolution, Distribution, and Conservation is Volume 12 in the Animal Natural History Series.

Contents

Illustrations

Color Distribution Maps and Plates

Tables

Foreword

Map Turtles and Sawbacks in a Global Conservation Context

The beautiful and varied map turtles and sawbacks of the genus *Graptemys* endemic to North America are, in my opinion, the crown jewels of American turtle diversity. *Graptemys* is one of the most species-rich turtle genera in the world and occurs in one of the most turtle-rich hotspots of the world—emblematic of the rich turtle diversity in the southeastern United States (Buhlmann et al. 2009; Mittermeier et al., in review). With their watershed-based distribution and diversity patterns, and with distinct species more or less restricted to single or adjacent hydrobasins, these attractive turtles serve as a model of evolutionary patterns of diversification. Through them, the importance of understanding and documenting evolutionarily significant genetic and morphological patterns of speciation becomes apparent.

This thorough monograph by Peter V. Lindeman provides a detailed and comprehensive account of map turtle and sawback diversity, distribution, ecology, and conservation. It will no doubt stand as the seminal work on these turtles for years to come. Written and extensively illustrated by a focused and dedicated researcher who has spent countless hours observing and photographing map turtles and sawbacks in the wild, this book stands as a compelling testament to what one person can accomplish in significantly advancing our understanding of a complex pattern of threatened biodiversity. The author's passion for seeing and documenting these interesting turtles in their native habitats should also inspire other researchers and conservationists to take up the mantle to further pursue the unanswered questions and future directions of research and conservation efforts suggested in this volume.

Table F.1. Polytypic genera of freshwater turtles with ten or more currently recognized species ranked by percent of species endangered and threatened

	Region	Species	Percent Endangered	Percent Threatened
Cuora	Asia	12	92	100
Graptemys	North America	14	21	43
Mesoclemmys	South America	10	20	30
Chelodina	Australasia	14	14	14
Trachemys	Americas	16	6	44
Pelusios*	Africa	17	0	12
Kinosternon	Americas	18	0	11
All Chelonians	Global	330	31	47

Notes: "Endangered" refers to an IUCN Red List status of Critically Endangered or Endangered; "Threatened" refers to a status of Critically Endangered, Endangered, or Vulnerable.

*excluding *P. seychellensis*

Sources: Percentages are based on the 2013 IUCN Red List (www.iucnredlist.org) and updated Draft Red List assessments by the IUCN/SSC Tortoise and Freshwater Turtle Specialist Group as documented by the Turtle Taxonomy Working Group (2012). The few species assessed as Data Deficient may also be threatened at a similar percentage, increasing the total percentage of threatened species from 47 percent to about 50 percent.

As my colleagues and I have documented elsewhere, turtles are in terrible trouble around the world, and face a survival crisis of unprecedented proportions. Over the past few decades, the so-called Asian Turtle Crisis has laid waste to the previously rich turtle fauna of China and Southeast Asia, and increasing shockwaves of turtle population declines due to unsustainable exploitation for food, traditional medicine, and pets have spread rapidly from that epicenter to gradually affect the rest of the world (van Dijk et al. 2000; Turtle Conservation Coalition 2011). The impact of the turtle trade has reached the rest of Asia, Madagascar, and, increasingly, North America, where recent analysis has shown significant and growing effects of the globalization of the commercial turtle trade on our native turtle populations (IUCN/SSC Tortoise and Freshwater Turtle Specialist Group 2010). As a result of these increasing threats and exploitative pressures in the face of continuing habitat loss, turtles have alarmingly become one of the most endangered large groups of vertebrates in the world, currently with about 50 percent assessed as Threatened under IUCN Red List criteria (see table 1) (Turtle Conservation Coalition 2011; Turtle Taxonomy Working Group 2012). And, unfortunately, that percentage is still climbing as we continue to assess turtle species around the world—turtles as a group are now closer to 55 percent threatened after a recent IUCN Red List workshop in

Togo assessing the status of African turtles (IUCN/SSC Tortoise and Freshwater Turtle Specialist Group, unpublished data).

The genus *Graptemys,* with its 14, is one of several freshwater turtle genera with 10 or more species, as recognized by the Turtle Taxonomy Working Group (2012). The other large polytypic genera include *Kinosternon* from the Americas (18 spp.), *Pelusios* from Africa (17 spp., not including *P. seychellensis,* recently synonymized with *P. castaneus* [Stuckas et al. 2013]), *Trachemys* from the Americas (16 spp.), *Chelodina* from Australasia (14 spp.), *Cuora* from Asia (12 spp.), and *Mesoclemmys* from South America (10 spp.). Of these richly polytypic freshwater turtle genera, *Graptemys* is not nearly as severely threatened as the highly endangered *Cuora* (Asian box turtles), but the genus has a higher percentage of endangered and threatened species than most of the other polytypic freshwater turtle genera (see table 1).

Although map turtles and sawbacks are currently facing relatively lower levels of threats than many turtles in the rest of the world, those threats are gradually increasing. Unless we act in an increasingly precautionary manner, establishing appropriate regulatory frameworks and adequate protected areas and laws, these animals are destined to follow other turtle species into ever-higher levels of endangerment and possible eventual extinction. We owe it to our children and coming generations to not allow this tragedy to occur—and this book can help us to better understand and protect these amazing turtles so that they survive in the wild and into a more secure future.

Anders G. J. Rhodin, M.D.
Director, Chelonian Research Foundation
Chairman Emeritus, IUCN/SSC Tortoise and Freshwater
*   Turtle Specialist Group*

References

Buhlmann, K. A., T. S. B. Akre, J. B. Iverson, D. Karapatakis, R. A. Mittermeier, A. Georges, A. G. J. Rhodin, P. P. van Dijk, and J. W. Gibbons. 2009. A global analysis of tortoise and freshwater turtle distributions with identification of priority conservation areas. *Chelonian Conservation and Biology* 8:116–149.

IUCN/SSC Tortoise and Freshwater Turtle Specialist Group. 2010. A study of progress on conservation of and trade in CITES-listed tortoises and freshwater turtles in Asia. CITES CoP15, Inf. 22.

Mittermeier, R. A., P. P. van Dijk, A. G. J. Rhodin, and S. D. Nash. In review. Turtle hotspots—an analysis of the occurrence of tortoises and freshwater turtles in biodiversity hotspots, high-biodiversity wilderness areas, and turtle priority areas. *Chelonian Conservation and Biology.*

Stuckas, H., R. Gemel, and U. Fritz. 2013. One extinct turtle species less: *Pelusios seychellensis* is not extinct, it never existed. *Plos One* 8(4): e57116. doi:10.1371/journal.pone.0057116.

Turtle Conservation Coalition [A. G. J. Rhodin, A. D. Walde, B. D. Horne, P. P. van Dijk,

T. Blanck, and R. Hudson (Eds.)]. 2011. Turtles in Trouble: The World's 25+ Most Endangered Tortoises and Freshwater Turtles—2011. Lunenburg, MA: IUCN/SSC Tortoise and Freshwater Turtle Specialist Group, Turtle Conservation Fund, Turtle Survival Alliance, Turtle Conservancy, Chelonian Research Foundation, Conservation International, Wildlife Conservation Society, and San Diego Zoo Global.

Turtle Taxonomy Working Group [P. P. van Dijk, J. B. Iverson, H. B. Shaffer, R. Bour, and A. G. J. Rhodin]. 2012. Turtles of the world, 2012 update: annotated checklist of taxonomy, synonymy, distribution, and conservation status. Chelonian Research Monographs, no. 5:000.243–328.

Van Dijk, P. P., B. L. Stuart, and A. G. J. Rhodin (Eds.). 2000. Asian Turtle Trade: Proceedings of a Workshop on Conservation and Trade of Freshwater Turtles and Tortoises in Asia. Chelonian Research Monographs, no. 2.

Acknowledgments

I have been writing and revising this book over the past several years. During that time, many of my colleagues have responded to various and sundry inquiries about *Graptemys*, providing unpublished reports and theses as well as clarifications or additional nuggets of information. For their valuable collegiality in this regard, I particularly thank Matt Aresco, Doug Backlund, Amanda Bennett, Rob Brauman, Alvin Breisch, Kurt Buhlmann, Greg Bulté, John Carr, Gary Casper, Erica Clayton, Dave Collins, Richard Daniel, Jim Dixon, Laura Dixon, Jim Dobie, Ken Dodd, David Edds, Josh Ennen, Bryan Fedrick, Tracy Gerold, John Giles, Jim Godwin, Aaron Gregor, David Haynes, Chase Hively, Kelly Irwin, Dale Jackson, Lee Jackson, John Jensen, Bob Jones, Linda LaClaire, Chris Lechowitz, Jeff Lovich, Carl May, Lynn McCoy, Dale McGinnity, Raejeana McKinzie, Jeff Miller, Paul Moler, Hans Mueller, Erin Myers, David Ode, Chris Phillips, Bob Reynolds, Daren Riedle, Carol Rizkalla, Francis Rose, Mark Sasser, Siegbert Schulz, Will Selman, Rich Seigel, Steve Shively, Steve Smith, Patrick Stephens, Sean Sterrett, Rebecca Stowe, Eddie Sunila, Travis Taggart, Stan Trauth, Peter Paul van Dijk, and Kathryn Vaughan.

The historical aspects of this book have been the most demanding and would not have been possible without the help of many people. The staffs overseeing the Smithsonian Archives in Washington, D.C., and the Archie Carr Archives at the University of Florida gave me access to materials that were tremendously helpful. Whit Gibbons and Jeff Lovich kindly provided copies of Fred Cagle's field notes that are archived at the Savannah River Ecology Laboratory and were a resource that was invaluable to me. I particularly want to thank the people who consented to interviews concerning their role in the history of *Graptemys* research. I was fortunate to have conversations with a large number of "old-timers" from Fred Cagle's Tulane field crews of the 1940s and 1950s, including the late Allan "Case" Chaney, Dick Etheridge, Whit Gibbons, the late Ernie Liner, Clarence Smith, and Bob Webb. Notwithstanding the recent passing of two interviewees, both in their eighties, it is apparent that catching map turtles and sawbacks is one key to a long and healthy life! I also benefited greatly from my conversations and correspondence with Jody Cagle, Fred Cagle, Jr., and Mary Cagle, who shared archival photos and family lore, as well as from interviews with Dick Vogt, David Haynes, and Jeff Lovich regarding their roles in the history of studies of map turtles and sawbacks. For sharing photos, I thank Roger Bour, Valentine Cadieux, Dick Vogt, and Bob Webb.

Natural history museum curators and their staff are the absolute salt of the earth. For supplying me with Excel files full of specimen data as well as their able responses to my many inquiries and their general helpfulness, I thank Chris Raxworthy (AMNH), Ned Gilmore (ANSP), Floyd Scott (APSU), Nancy McCartney (ARK), Stan Trauth (ASUMZ), Craig Guyer (AUM), Colin McCarthy (BMNH), Steve Rogers (CM), Timothy Matson and Owen Lockhart (CMNH), Kenney Krysko (FLMNH), Alan Resetar (FMNH), Elizabeth McGee (GMNH), Chris Phillips (INHS), Kenneth Kozak and Jonathan Slaght (JFBM), Karen Butler-Clary (KU), Jimmy McGuire (LSUMZ), Jose Rosado (MCZ), Bob Jones (MMNH), Roger Bour (MNHN), Gary Casper (MPM), John Carr (NLU), Chris Wolfe (OKMNH), Ross MacCulloch (ROM), Anita Benedict and Trey Crumpton (SMBU), Kathryn Vaughan (TCWC), David Cannatella (TNHC), Harold Dundee (TU), Greg Schneider (UMMZ), Tom Giermakowski (UNM), Thomas Labedz (UNSM), George Zug and Traci Hartsell (USNM), Jonathan Campbell and Carl Franklin (UTA), Angela Dassow (UWZ), Renn Tumlinson (Henderson State University), Travis Taggart (Kansas Herp Atlas), Anthony Howell (Redpath Museum), Matthew Kwiatkowski (Stephen F. Austin University Vertebrate Museum), and Jessica Coleman (University of Texas-Tyler).

My travels to photograph turtles were aided by colleagues who graciously took time to take me to sites with good subjects. I thank Will Selman for showing me his study site on the Leaf River, Jim Godwin for a boat trip on the Coosa and Tallapoosa Rivers, and Jim Dobie for showing me sites on the Pedernales River and associated creeks.

Edinboro University of Pennsylvania students Beth Addis, Maggie Dicks, Sean Gess, and Jim Patterson assisted me in many ways during the compilation of information for this book. For their earnest endeavors on my behalf, I thank them heartily.

Several colleagues reviewed chapters and species accounts. For their invaluable comments, I thank Greg Bulté, John Carr, Jon Costanzo, Jim Dobie, Sean Doody, Jim Godwin, David Haynes, Dale Jackson, Bob Jones, Jeff Lovich, Teal Richards, Mike Seidel, Rich Seigel, Will Selman, Steve Shively, and Sean Sterrett.

Edinboro University of Pennsylvania and the Pennsylvania State System of Higher Education have supported me in the preparation of this book in many ways. The Department of Biology and Health Services provided the digital spotting scope used to capture map turtle and sawback images, and my former department chairman, Marty Mitchell, has been a great supporter and advocate of my efforts in this project. Travel for photographs and museum visits was supported by a Faculty Senate Grant and a grant from the Faculty Professional Development Fund. Finally, a sabbatical in fall 2008 was instrumental in providing the time necessary to complete some chapters and wrap up numerous loose ends in this endeavor.

I extend special thanks to Anders Rhodin, who generously provided a stipend from the Chelonian Research Foundation to help underwrite the publication of this book. All my proceeds from the sale of the book go to the Chelonian Research Foundation to support turtle conservation around the world.

Finally, I have several people at the University of Oklahoma Press who have assisted me in making this long-held dream a hard-copy reality. Karen Wieder was my first contact with OU Press, and she and Vic Hutchison gave good advice for the preparation of my initial book proposal. The late Kirk Bjornsgaard ably oversaw the completion and review of the initial draft of the book and made numerous helpful suggestions. I thank Jay Dew for stepping in after Kirk's passing to guide my revisions and see this project through to the end. Julie Rushing has been in charge of production for the many figures in the book, Emily Jerman coordinated a thorough copyediting through the very detailed reading and critique of freelance editor Darcy Wilson, Amy Hernandez oversaw production of the book jacket and marketing, and freelance cartographer Gerry Krieg did a masterful job of overhauling my original range maps to increase their resolution. To all my fellow Oklahomans at OU Press, I extend a very hearty thank you, as well as a shout of "Boomer! . . . Sooner!"

The Map Turtle and Sawback Atlas

Introduction to the Biology of Map Turtles and Sawbacks

Fittingly, a set of maps first piqued my interest in map turtles and sawbacks. My second year of teaching a heavy course load at a rural community college in western Kentucky had come to a merciful conclusion in May 1991. I had enrolled in a summer graduate-level course, the first toward a Ph.D. that was several years off, and I was thinking about a dissertation topic: something on turtles, perhaps on community ecology. I was across the state visiting relatives in Lexington and took advantage of being in a bigger city with good bookstores. I bought a copy of the newly released third edition of *A Field Guide to Reptiles and Amphibians: Eastern and Central North America* (Conant and Collins 1991; the book is part of the Peterson field guide series) to replace a first edition (Conant 1958) I had found in a used bookstore two years earlier. (I had also seen library copies of Conant's [1975] second edition.)

The revised range maps for species of the genus *Graptemys* caught my eye in a way the maps in the previous editions had not. The ranges in the revised maps became dissected at their northern limits, branching into long fingers sprawling northwestward and northeastward, undulating back and forth like snakes; the maps in the earlier editions had been more like the minimum convex polygons used in home range studies (fig. 1.1). Clearly, these turtles—of which I had known very little—were river animals. The most intriguing thing I saw that first weekend of thinking about *Graptemys* was the similarity of the range map for the Alabama map turtle, *G. pulchra* (just prior to it being split into three species), to the composite map for the three sawbacks, *G. oculifera*, *G. flavimaculata*, and *G. nigrinoda*. I read with much interest about dietary differences and differences in head width, which immediately suggested to me the topic of character displacement arising through competitive interactions.

a

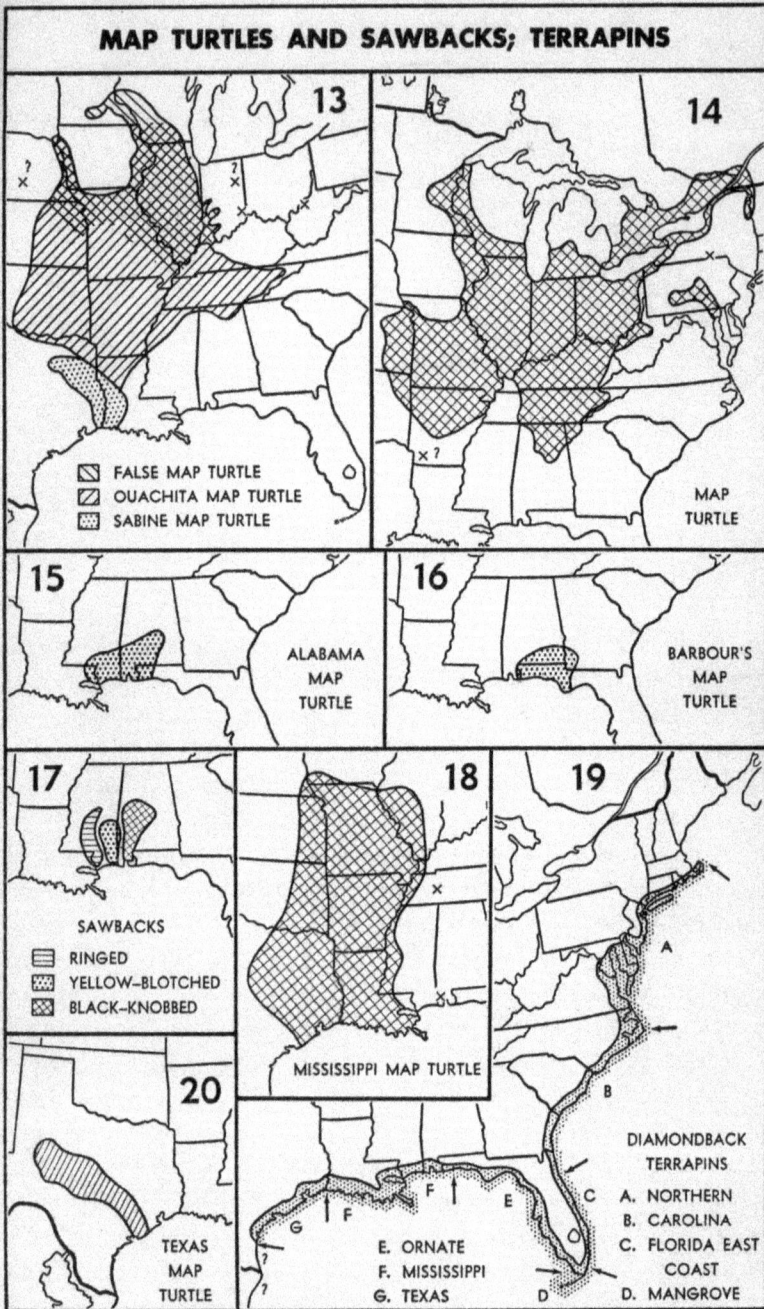

Figure 1.1. Range maps of map turtles and sawbacks from the first (a) and third (b) editions of the Peterson field guide (Conant 1958; Conant and Collins 1991). Range map from *A Field Guide to Reptiles and Amphibians of the United States and Canada East of the 100th Meridian*, 1st ed., by Roger Conant. Copyright © 1958 by Roger Conant. Range map from *A Field Guide to Reptiles and Amphibians: Eastern and Central North America*, 3rd ed., by Roger Conant and Joseph T. Collins. Illustrations copyright © 1991 by Houghton Mifflin Company. Both maps used by permission of Houghton Mifflin Harcourt Publishing Company. All rights reserved.

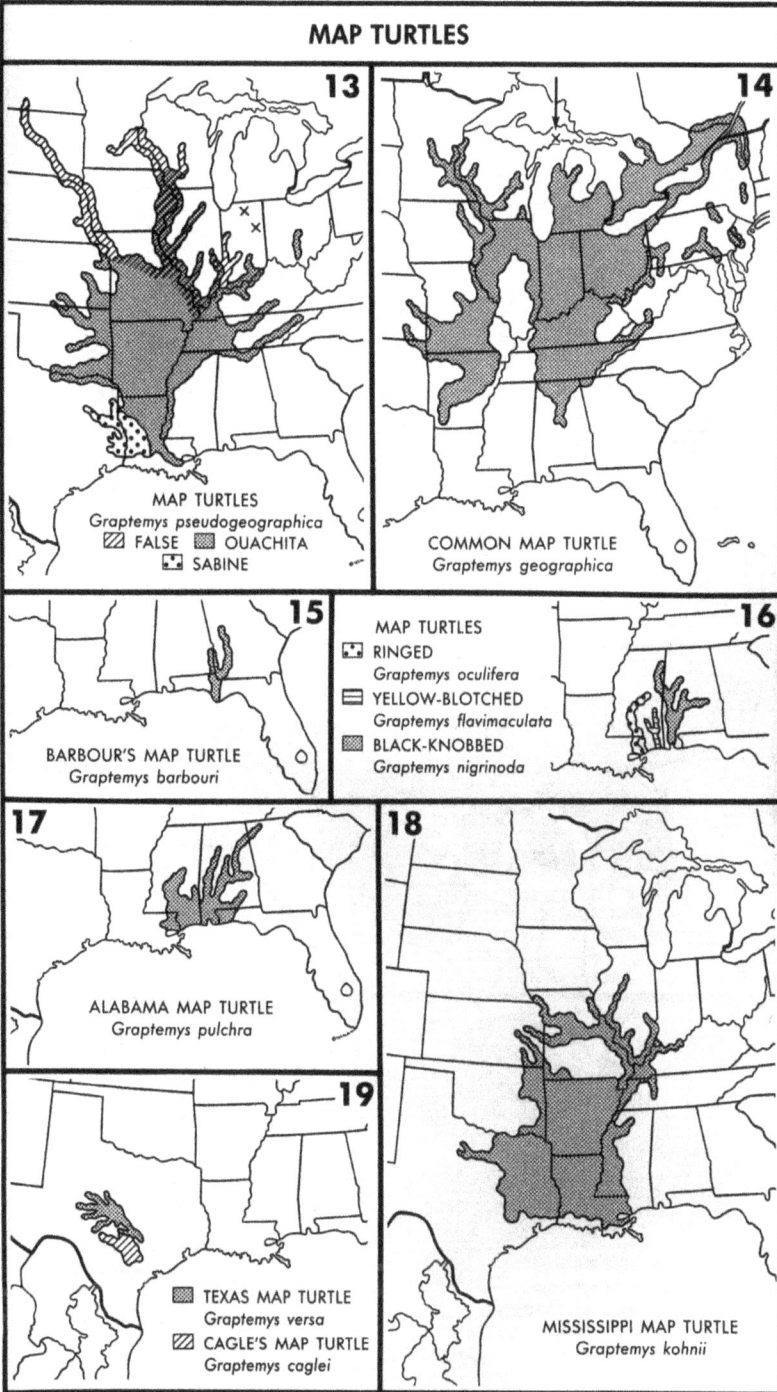

b

MAP TURTLES

13

MAP TURTLES
Graptemys pseudogeographica
▨ FALSE ▨ OUACHITA
⬚ SABINE

14

COMMON MAP TURTLE
Graptemys geographica

15

BARBOUR'S MAP TURTLE
Graptemys barbouri

16

MAP TURTLES
⬚ RINGED
  Graptemys oculifera
▤ YELLOW-BLOTCHED
  Graptemys flavimaculata
▨ BLACK-KNOBBED
  Graptemys nigrinoda

17

ALABAMA MAP TURTLE
Graptemys pulchra

18

19

▨ TEXAS MAP TURTLE
  Graptemys versa
▨ CAGLE'S MAP TURTLE
  Graptemys caglei

MISSISSIPPI MAP TURTLE
Graptemys kohnii

There are three main reasons to rank the map turtles and sawbacks exceptionally high on the list of turtle taxa that present the most intriguing suite of characteristics for ecologists and evolutionary biologists. First is the exceptional diversity of diets associated with their diverse trophic morphologies (fig. 1.2). At one end of the spectrum are the exceptionally narrow heads (it is difficult to look at a male Sabine map turtle, for example, without the phrase "pencil-necked geek" coming to mind) and, in the mouth, the narrow, keratinized alveolar surfaces that have been said to function in a scissor-like fashion during feeding (e.g., Conant and Collins 1991). Males of all species are relatively narrow headed, as are the females of some species. At the other end of the spectrum are the extraordinarily large heads of the females in some species—described by some authors as "grotesque" (e.g., Conant and Collins 1991; Lovich and McCoy 1992; C. H. Ernst et al. 1994; C. H. Ernst and Lovich 2009)—which are equipped with broad, smooth alveolar surfaces that look in some specimens as if they might have grown wider, if not for the fact that they had already collided at their medial edges in the middle of the roof of the mouth. Those with the narrower jaws and alveolar surfaces generally feed on soft-bodied invertebrates, such as insect larvae, sponges, and bryozoan colonies, and may also feed on algae. The broader-headed females typical of some species feed mainly on hard-shelled mollusks that they crush between the alveolar surfaces of their powerful jaws. A large "fathead" female, if left in a tub of water overnight following capture, can produce a prodigious pile of mollusk shell fragments.

Co-occurrence of species, generally as pairs, produces a strong pattern in which a species with narrow-headed females coexists with a species with broader-headed females, such that the diets of the females in these sympatric pairs tend to be very different. The pattern suggests an important role of competition in determining coexistence of map turtle and sawback species and is, at least superficially, consistent with ecological character displacement. The lack of strong dietary differences in the males of coexisting species, however, is every bit as puzzling as the female pattern is intriguing.

The second major characteristic of interest in map turtles and sawbacks is their degree of sexual size dimorphism (fig. 1.3). Their size difference not only exceeds anything known in the world's three-hundred-plus other turtle species (with the possible exception of a few poorly known Asian geoemydids; Gibbons and Lovich 1990), it also makes map turtles and sawbacks the champions of sexual size differences among tetrapod vertebrates (Stephens and Wiens 2009). Adult males are typically half or less the shell length of adult females. Because they are also half or less the shell width and shell height of females (also, females are relatively more highly domed), males may weigh just *one-tenth* what females weigh on average. At a field site, it is an attention-getting parlor trick for a *Graptemys* ecologist to show off a pair of captured turtles to unsuspecting members of the public and tell them that the smaller turtle—the one their children may have just excitedly called a "baby," with the requisite oohing, aahing,

Figure 1.2. Diversity of trophic morphology in sympatric map turtles. (a) A young female Pearl map turtle (on the left) with a hypertrophied head and jaw musculature adapted for mollusk crushing, basking alongside an adult female ringed sawback (at the right) with a more gracile head. Photo from the Pearl River at Highway 35 near Carthage, Leake County, Mississippi.
(b) An adult female Alabama map turtle with hypertrophied head and jaw musculature basking above an adult female black-knobbed sawback with a more gracile head. Photo from the Alabama River in Roland Cooper State Park, Wilcox County, Alabama.

Figure 1.3. Typical sexual size dimorphism of *Graptemys* as exhibited by black-knobbed sawbacks, with three adult males on the left and middle branches (note the enlarged tails, denoting maturity, visible in two) and three adult females on the middle and right branches. Photo from the Alabama River in Selma, Dallas County, Alabama.

and baby talk—is in fact a fully mature individual capable of inseminating the other animal, which dwarfs it in size, and that the male is unlikely to grow more than another inch or so in its life, if that much.

The final intriguing feature of map turtles and sawbacks is their exceptional diversification, which is all the more a wonder because much of it has occurred over a very short time span, geologically speaking. The fourteen species make the genus the species-richness champion of turtle genera in the United States and Canada. Isolation of the river drainages of the Gulf Coast has driven diversification, with most of the diversity having arisen within the last three million years (Lamb et al. 1994). While the species are diagnosable based on markings—and for the most part are easily recognizable—our ability to differentiate them genetically has been more limited as a consequence of their recent origin. Considering their dietary diversification together with their rapid speciation, the map turtles and sawbacks are justifiably viewed as an example of the fascinating process of adaptive radiation (Lindeman 2000a), a sort of southeastern U.S. rivers' answer to the finches of the Galapagos or the anoles of the Caribbean. For some animals, a river drainage is just another sort of island.

Figure 1.4. Nine basking Barbour's map turtles. Such aggregations place map turtles and sawbacks among the most conspicuous reptiles in the world. Photo from the Flint River tailwaters below Lake Blackshear Dam, Killebrew Park, Worth County, Georgia.

There are also aesthetic reasons that many turtle biologists (and turtle hobbyists) become enamored with the map turtles and sawbacks. The first is embodied in the inspirations for the two common names for these species. Markings on the shell, limbs, and head and neck make these animals among the most attractive turtles on earth; the lines really do resemble those on a topographic map. The shell architecture, with the strong lateral serrations and spiky midline keel that typify most species, calls to mind the blade tips of a circular saw. Their "sawback" morphology is shared by only a few Asian geoemydid turtle species.

The other aesthetically pleasing aspect is the opportunity these turtles afford us to observe them in their natural habitats. Few turtles show themselves off the way map turtles and sawbacks do (fig. 1.4). Basking is probably a more-developed habit in these species than in any other group of North American turtles—either that, or they are far more numerous than we know, because most basking surveys in their habitats are dominated by these species. Much has been made in field guides of their tendency to evacuate into the water at the slightest provocation, but in point of fact, it is not hard to learn how to approach close enough with binoculars or a spotting scope to get a good look at them basking.

Some will jump into the water, but often some stay; even when they all jump, with a little patience one can watch as the turtles emerge again. Their basking poses are an endless source of fascination: the improbable heights artfully scaled by the bulky creatures; the spread-eagle sunning stances they adopt; the occasional mouth-wide-open gaping; the limbs-and-head-withdrawn stances of very dry, very hot individuals that seem as though they are engaged in a kind of endurance basking; the overlapping stacking of individuals on a site as it becomes crowded; and the heads in an aggregation that swivel this way and that, warily watching the world above the waterline.

While the species of map turtles and sawbacks give ample scientific and aesthetic reasons to command our attention, paradoxically, the pace of *Graptemys* investigations in herpetology has never been particularly rapid, and large segments of the public are not acquainted with these turtles. Our knowledge of the genus was jump-started by Fred Cagle and his students, who collected enormous numbers of specimens of most of the species during the late 1940s and 1950s, but they ultimately generated more questions than they were able to answer. Two major subsequent boosts to the growing *Graptemys* literature were provided by the boom years of autecological studies carried out by a host of graduate students between the mid-1960s and mid-1980s and by work conducted on a few species after they had come to be of conservation concern, beginning later in the 1980s. Nevertheless, this group, which constitutes about a quarter of the species diversity of turtles in the United States and Canada, has been the subject of far less than a quarter of the turtle literature produced in these two countries. Sea turtles, gopher and desert tortoises, painted turtles, slider turtles, common snapping turtles, and spotted turtles are among the species that have commanded far greater research attention than the map turtles and sawbacks.

Part of the problem for *Graptemys* research has undoubtedly been the difficulty in capturing the animals. Fieldwork requires a boat and tending traps in deep water, generally while dealing with the forces of current. Most investigators report a low capture rate for map turtles and sawbacks using the traditional baited three-hoop nets, so alternative techniques have had to be sought for these species (e.g., Chaney and Smith 1950; Vogt 1980b; Browne and Hecnar 2005; Sterrett et al. 2010).

Over the years I have collected several anecdotes that demonstrate the surprising degree to which the public is unaware of map turtles and sawbacks (notwithstanding, of course, obsessive fanciers who desire to keep these species as pets and whose commerce has contributed to population declines). Lack of recognition is common even among those who spend much time on the rivers or lakes these species inhabit. In conversations at boat ramps, campers' makeshift tent sites under bridges, and other access points, the terms "map turtle" and "sawback" rarely elicit recognition. Often the general public lumps these species with other deirochelyine emydid species, such as sliders, cooters, and painted turtles, under generalist labels. In Kentucky, it seemed that many people I met at my field site saw the state's thirteen turtle species as just three: snapping turtles, box turtles, and "mud turtles," the last a potpourri of species to which they assigned all the emydids I captured. Along the rivers of southeastern Mississippi, the term "streaked-necks" (with "streak-ed" pronounced in two syllables) serves as a catchall term for sliders, cooters, and map turtles and sawbacks (W. Selman, pers. comm. 2007). Similarly, the Cajun common name *ventre jaune* ("yellow belly") is applied in Louisiana without discrimination among the same set of species (Fontenot 2004). On one telling occasion, I had captured a male Sabine map turtle on the Calcasieu River in southwestern Louisiana. Three shirtless young river rats approached to admire my catch. Speaking in Cajun accents, they indicated they did not know the species, one in particular insisting repeatedly he had "never seen a 'tuhtle' like that before." I assured them they would have seen more Sabine maps on the Calcasieu than any other turtle species, yet they continued to regard the little male turtle as a fascinating novelty.

Likewise, at a boat ramp on the Pearl River near Hopewell, Mississippi, I once spoke with a local who spent a lot of time on the river but professed to know nothing of its ringed sawbacks. After I explained that I was a college professor down from Kentucky to conduct counts of the species, which was federally listed as threatened, he seemed to suddenly recall that he had heard of the turtles: weren't they the species in decline because people were eating them? Confused, I replied that I had never heard of anyone eating ringed sawbacks and briefly outlined for him the other threats to the species. At some point later he mentioned that people he had talked to called them not "sawbacks" but "loggerheads," and it did not occur to me until I was driving off to my next survey site that he thought I had been talking about alligator snapping turtles. How very odd, he must have thought, that an egghead professor would come from out of state to study alligator snappers, yet know nothing about their use by the locals in turtle soup! More recently, I showed a black-knobbed sawback to a young woman in Selma, Alabama, who claimed she had virtually been raised on the Alabama River yet had never seen such a turtle. Selman, Qualls, et al. (2008) discussed the lack of awareness Mississippians have for yellow-blotched sawbacks, in spite of the turtles' conspicuous basking habits and status under federal law.

Recognition of map turtle species is no better farther north. My first agency-issued permit for trapping common map turtles at a site on Lake Erie allowed me to trap "mop turtles." Pennsylvanians know little about the species, perhaps one small reason being my local newspaper, which, in the first few years after I moved to Pennsylvania, three times published photographs of common map turtles (e.g., fig. 1.5)—each time with captions identifying them as painted turtles!

Map turtles and sawbacks may slowly be gaining greater recognition. Due to their attractive markings and shell shapes, they have slowly begun to gain

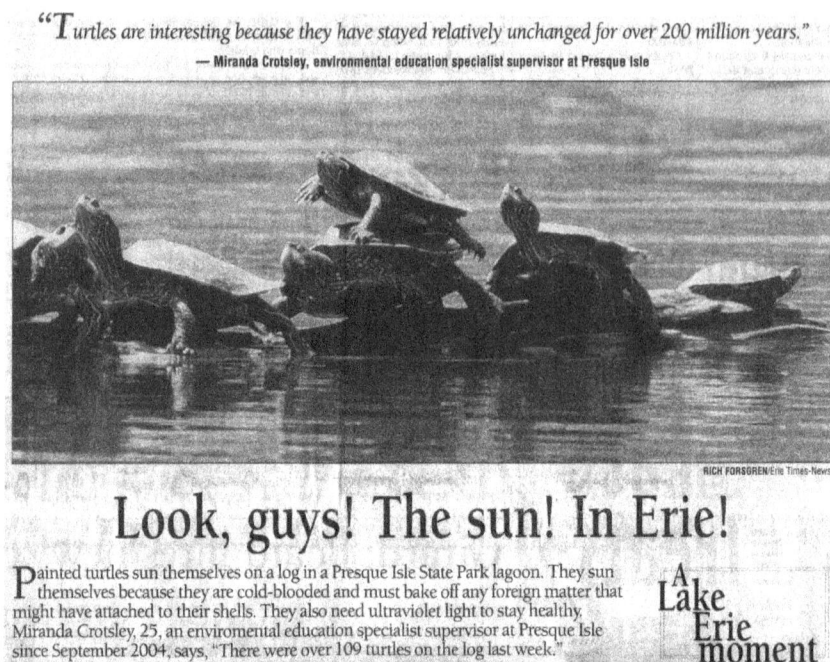

"Turtles are interesting because they have stayed relatively unchanged for over 200 million years."
— Miranda Crotsley, environmental education specialist supervisor at Presque Isle

RICH FORSGREN/Erie Times-News

Look, guys! The sun! In Erie!

Painted turtles sun themselves on a log in a Presque Isle State Park lagoon. They sun themselves because they are cold-blooded and must bake off any foreign matter that might have attached to their shells. They also need ultraviolet light to stay healthy. Miranda Crotsley, 25, an enviromental education specialist supervisor at Presque Isle since September 2004, says, "There were over 109 turtles on the log last week."

A Lake Erie moment

Figure 1.5. A newspaper photograph of six common map turtles, misidentified as painted turtles (*Erie Time-News*, 28 June 2005).

on sea turtles and tortoises in depictions on commercial clothing produced for nature lovers (e.g., fig. 1.6). Curiously, I have found Cagle's map turtle, although named and described only about four decades ago (Haynes and McKown 1974) and little studied since, depicted more often than any other species (fig. 1.7; see also T-shirt in fig. 1.6). Recently, a Cagle's map turtle graced the cover of the new edition of *Turtles of the United States and Canada* (C. H. Ernst and Lovich 2009). The term "map turtle" seems to generally elicit curiosity—any *Graptemys* biologist has been asked numerous times the meaning behind the moniker—and may generate interest in these turtles. For example, *Map Turtle: Graptemys geographica*, a sculpture by geographer Kirsten Valentine Cadieux, was installed in the Vanderbilt University Garden of Great Ideas in 2000 (fig. 1.8). Cadieux depicts a human form lying below a turtle whose carapace contains an image of the Vinland world map and maps of Tennessee, the Mississippi River drainage, the Great Lakes, and the western U.S. cartographic grid system. She called her sculpture "an interpretation of mapping the world" reflecting "the wilderness and rural life as they exist in the American cultural imagination" ("New sculptures planted in 'Garden of Great Ideas,'" *Vanderbilt News*, 27 November 2000, p. 11).

My objectives are to review the state of our knowledge of map turtle and sawback biology, to make the case that they are worthy of greater research focus

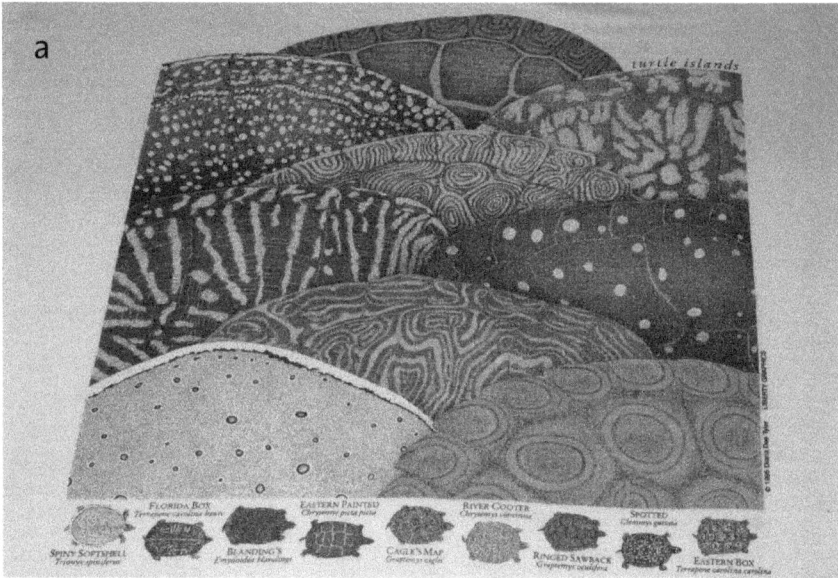

Figure 1.6. (a) Detail from a T-shirt featuring the carapaces of nine U.S. turtle species, including a ringed sawback (lower right) and a Cagle's map turtle (middle column, center). (b) A ball cap depicting a black-knobbed sawback. Both items are commercially available.

Figure 1.7. Cagle's map turtle depictions in (a) a bookmark (third turtle from top) and (b) an advertisement promoting recreational vehicles (upper right turtle).

Figure 1.8. *Map Turtle:* Graptemys geographica, a 2000 sculpture by Kirsten Valentine Cadieux, from the Garden of Great Ideas sculpture trail gifted to Vanderbilt University by the Newington-Cropsey Foundation, Hastings-on-Hudson, New York. Photo courtesy Kirsten Valentine Cadieux.

with respect to basic biology as well as applied biology regarding conservation, and to outline several promising avenues for research. From the outset, I have wanted this book to be more than just a compilation: I want it also to be an inspiration. To that end, the last three chapters are my view of what the future holds for scientific investigations of *Graptemys* biology. Many important and intriguing studies of these turtles await, far more than I or any one person could hope to conduct. It is my hope that this book will provide a sort of guidebook to future investigations.

Because I feel strongly that these species may be cute as captives but are much more engaging when exhibiting natural behaviors in their native habitats, I have eschewed the posed photographs against artificial backgrounds that are the standard fare in turtle volumes. Instead I favor "candid" shots taken at 30× to 40× magnification with spotting scopes that have built-in digital cameras. All such photographs were taken from riverbanks or lakeshores without the use (or bother) of observation blinds. The results, while novel at present, will undoubtedly one day seem crude and amateurish as the technology for digital scoping improves and better-skilled photographers try their hands at capturing these beautiful and fascinating subjects.

2

History of Studies of Map Turtle and Sawback Biology

Early North American naturalists passed over map turtles and sawbacks in their discoveries. Perhaps most notably, Meriwether Lewis and William Clark led their Corps of Discovery up the Missouri River in 1804 to explore lands acquired by the United States in the Louisiana Purchase. The only encounters with turtles they recorded took place in 1805 and 1806, by which time they had traveled beyond the extensive range of the false map turtle (*Graptemys pseudogeographica*) in the Missouri River (Benson 1978). For several decades after the first description of a species of *Graptemys*, in 1817, work on the genus was purely taxonomic and based primarily on preserved specimens, with a slow accumulation of new descriptions. By the mid-1900s, efforts to understand the biology of these species accelerated and became more closely linked to direct researcher experiences with the species in their natural habitats, primarily due to the efforts of Fred Cagle and his Tulane University field crews. The next wave of primarily autecological studies was provided in graduate student theses and dissertations and was followed more recently by research focused on conservation. Both of the more recent periods of focus continue to strongly influence the researchers who pursue map turtle and sawback studies today.

Early Taxonomic Studies

The inaugural contribution to *Graptemys* studies was made by Charles Alexandre LeSueur (fig. 2.1) almost two hundred years ago. He was born in 1778 in Le Havre, France, and died there in 1846 a year after his appointment as curator of the Muséum d'Histoire Naturelle du Havre. In between, he traveled widely, eventually settling in the United States for twenty-two years. He lived first in Philadelphia, from 1816 until 1825, and then in New Harmony, Indiana, until

Figure 2.1. Charles Alexandre LeSueur at the age of 40 in 1818, two years after his arrival in the United States and his discovery of the species now known as *Graptemys geographica*. Portrait by Charles Wilson Peale.

his return to France in 1837. He is known mainly as an ichthyologist, particularly as an illustrator.

In his first summer in the United States, LeSueur and geologist William Maclure traveled from Philadelphia to Pittsburgh and then to Erie, Pennsylvania, Buffalo, New York, the southern shore of Lake Ontario, and points east (Lindeman 2009b). In 1817, LeSueur described a turtle species he christened the Lake Erie tortoise, *Testudo geographica*, which he had caught "in a marsh, on the borders of Lake Erie" (LeSueur 1817, p. 86). No specimen was noted, nor has any ever been found, but he included a drawing (fig. 8.2) showing exceptional attention to the detail of carapacial markings that nicely match *G. geographica*. The illustration provides just enough detail of the markings on the turtle's head to confirm the identity of his find, notwithstanding the fancifully inaccurate "bulls-eye" markings encircling the tail. While LeSueur gave no further geographic information on type locality, it was apparently the Presque Isle peninsula or Presque Isle Bay, which the peninsula forms at Erie, Pennsylvania (Lindeman 2009b; see fig. 8.2). LeSueur stated in his publications on fish that he had visited Erie and Presque Isle during the 1816 trip, and these are the only sites along the Pennsylvania and New York shoreline known to have common map turtles.

Figure 2.2. A specimen bearing the handwritten name *Emys pseudogeographica* (MNHN 9136) from LeSueur's collections on the Wabash River. Photo courtesy of Roger Bour.

LeSueur later collected (and, indirectly, named) a second species, the false map turtle, *G. pseudogeographica*, which would become the center of taxonomic confusion with regard to both its correct name and its differentiation from and relationship to several taxa subsequently described. His specimens were from the Wabash River at New Harmony, where he had moved, having traveled to New Harmony on a riverboat dubbed the "Boatload of Knowledge" with other eminent naturalists taking part in a scientific and socialist commune founded by Robert Owen and Maclure. LeSueur discussed and described the specimens in print, regarding them as a new species, in his formal description of the soft-shells *Apalone mutica* and *Apalone spinifera* (LeSueur 1827). He published no name for his new hard-shelled species, however. Four years later, in his *Synopsis Reptilium*, John Gray (1831) described a new taxon, *Emys lesueuri*; under this he listed a "β" form, *Emys geographica*, which he seemed to consider synonymous with *Emys pseudogeographica*, a name he attributed to "LeSueur Mss. (Mus. Paris)" (p. 31). The reference by Gray is apparently to LeSueur's *handwriting* on the specimens' plastra (fig. 2.2; Bour and Dubois 1983), rather than a "name in manuscript" (Hay 1892b, p. 379) in the sense of preparation for publication.

In subsequent decades, most references recognized two species but varied in choice of name for the second. Some works used the specific epithets *geographica* and *pseudogeographica* (e.g., Holbrook 1842; Cope 1875), while other works (and many museum catalog entries) used *geographica* and *lesueuri* (or *lesueurii*); the latter clearly applied to what is today recognized as *pseudogeographica* (e.g., Agassiz 1857a, 1857b; Garman 1890). Hay (1892b) finally clarified the matter. He noted that J. E. Gray's (1831) diagnosis of *lesueuri* was based on the triangular postorbital mark that is a highly consistent character in *geographica* (albeit not one noted in LeSueur's description), as well as the fact that Gray himself in his subsequent works recognized *geographica* and *lesueuri* as synonymous and *pseudogeographica* as separate. Hence Gray's simple act of listing of the specific epithet *pseudogeographica* makes him the author of the taxon—notwithstanding its (accurate) description and epithet having been provided earlier by LeSueur. This was because only the description, not the epithet, had been *published* by LeSueur. Since Hay's (1892b) clarification, *pseudogeographica* has been the specific epithet used for the species, but the species has still been a further center of taxonomic confusion and debate, being alternately regarded as separate from or conspecific with *oculifera*, *kohnii*, *versa*, *sabinensis*, and *ouachitensis* (see chapter 8; today only *kohnii* is considered conspecific with *pseudogeographica*).

Further descriptions of *Graptemys* were not undertaken until the work of Georg Baur (fig. 2.3) in the 1890s. Baur, a German born in Weisswasser in 1859, had come to the United States in 1884 to work for O. C. Marsh at the Peabody Museum at Yale University. He took a position with Clark University in Worcester, Massachusetts, in 1890, then moved to the University of Chicago in 1892. His earliest work was primarily osteological and paleontological and concerned a variety of vertebrate taxa, but by the late 1880s he had become increasingly focused on turtles. In October 1890, settled into his new position at Clark University, he sketched out a proposal for a book-length treatment of turtles, *The Testudinata of North America*, a work he would never complete.[1]

In 1889 or 1890 Baur began a correspondence with Joseph Gustave Kohn, a wealthy amateur naturalist in New Orleans. Kohn loaned specimens from Louisiana that Baur (1890, 1893) used to

Figure 2.3. Georg Hermann Carl Ludwig Baur, from an 1899 obituary notice by William Morton Wheeler.

describe two new taxa. Kohn's handwritten "List of Turtles in Collection of Gustave Kohn"[2] lists *Malacoclemmys geographicus* and *M. lesueuri,* with both being from Louisiana, the former "not common" and the latter "common." Baur (1890) described them as *Malacoclemmys oculifera* (i.e., Kohn's *lesueuri*) and *Malacoclemmys kohnii* (i.e., Kohn's *geographicus*) in *Science* in a publication dated 7 November. Kohn also sent specimens of *oculifera* to the U.S. National Museum (cataloged on 10 June 1889 as "*Malaclemys LeSueuri*").

Four days later, Baur sent a letter to Leonhard Stejneger at the U.S. National Museum in which he referenced the descriptions and alerted Stejneger that two specimens (USNM 8808), which Baur apparently had previously examined, also represented a new species.[3] Three weeks later, he asked Stejneger to send him the specimens,[4] which Stejneger quickly did.[5] It did not take Baur long to confirm his suspicions. A week later, he wrote to Stejneger, "The genus *Malacoclemmys* is one of the most plastic members of the Testudinata. I have received the two Montgomery specimens [USNM 8808]. They are new. . . . This is a highly interesting group."[6] Baur considered other names for the two Montgomery specimens (first *Graptemys alabamensis,* for their collecting locality, then *Graptemys grandis,*[7] perhaps a reference to their large heads), but he settled on *Graptemys pulchra* (Alabama map turtle), the specific epithet being derived from the Latin *pulcher,* meaning "beautiful." (The choice of epithet would be apt for living specimens but is surprising in this case, given the poor state of preservation of the two types. In a 1941 letter to Archie Carr, Jr., Stejneger described them as being in poor shape, soft, and missing epidermal scutes [see fig. 8.6]; in his photographs of one specimen,[8] it looks unchanged from its present condition. The specimens had been collected on a U.S. Fish Commission expedition[9] fourteen years before Baur borrowed them and likely were already in poor condition when he saw them. He may have been influenced in the choice of epithet by a live specimen of a large female from the Pearl River that he had received from Kohn;[10] hence it is very likely the specific epithet *pulchra* was in fact inspired by a specimen of a species named *Graptemys pearlensis* more than a century later!)

Baur became enraptured by *Graptemys* during his two years at Clark University. Writing to Stejneger, he suggested that *Malacoclemmys* (i.e., *Malaclemys*), the diamondback terrapin genus under which Baur (1890) had just described *oculifera* and *kohnii,* should be split into three genera: (1) a terrapin genus, (2) a separate *Graptemys* (following Agassiz's 1857a work in recognizing a separate *Graptemys*) to contain *G. geographica* and one of the new species, and (3) "n.g." (new genus) to contain "*G. LeSueuri*" and the other new species.[11] Baur contemplated further generic splitting for his never-completed turtle opus: *Graptemys* would retain *geographica* and *kohnii*; a new genus, *Neoclemmys,* was planned for *lesueuri* (i.e., *pseudogeographica*) and *oculifera*; and another new genus, *Megaloclemmys,* was planned for *pulchra.*[12] Indeed, Baur had already noted in his description that *pulchra* might warrant placement in a separate genus.

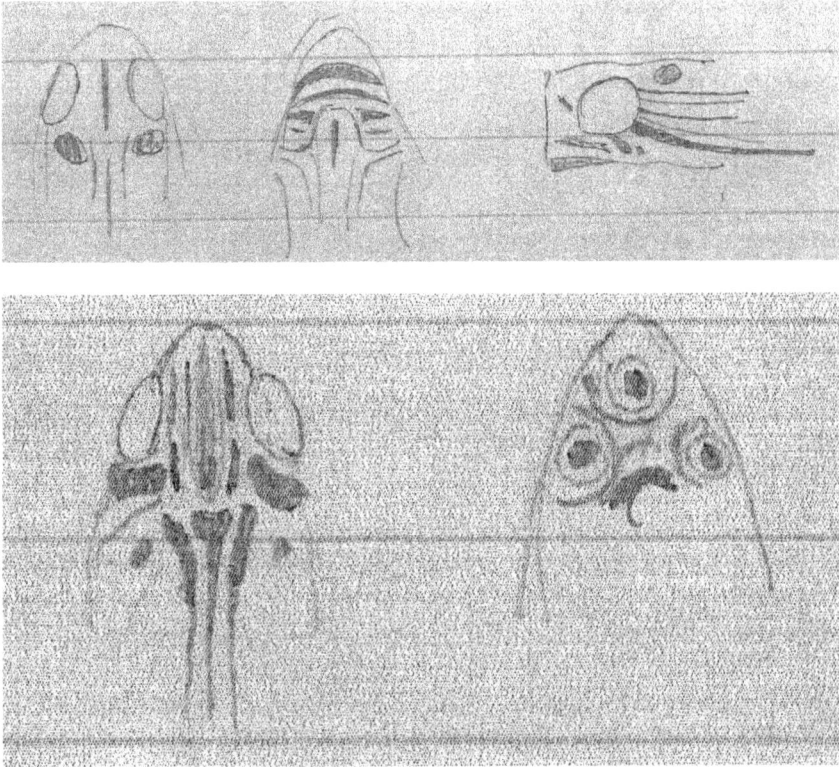

Figure 2.4 (*top*). Drawings of the head markings of "*Graptemys intermedia*," based on specimens called *Graptemys sabinensis* today, from an unpublished manuscript by Georg Baur from the 1890s (Smithsonian Archives Record Unit 7441).
Figure 2.5 (*bottom*). Drawings of the head markings of "*Graptemys* sp. n.," based on specimens called *Graptemys ouachitensis* today, from an unpublished manuscript of Georg Baur from the 1890s (Smithsonian Archives Record Unit 7441).

Baur continued to write to Kohn for more specimens, particularly of *Graptemys* (and Kohn requested specimens of actual *geographica* and "*LeSueuri*," the species he had originally thought his Louisiana specimens represented).[13] A list of specimens Baur wrote showing what he had received from Kohn includes an entry under May 1891 for three *G. oculifera* and one *G.* "*grandis*" (i.e., *G. pulchra sensu lato*),[14] which Kohn stated he had purchased in New Orleans markets.[15] Specimens from the Mermentau and Sabine Rivers in southwestern Louisiana were sent in 1893,[16] and two unpublished manuscripts show Baur was poised to christen them *Graptemys intermedia*.[17] Baur's drawing of head markings (fig. 2.4) and the Mermentau and Sabine localities[18] indicate they were of the species known as *Graptemys sabinensis* Cagle 1953; indeed, the baseplate for a specimen in the Tulane University Museum of Natural History (TU 7690) has "*Graptemys intermedia* Baur" penciled on its underside.

Baur's undated manuscript also has a page titled "*Graptemys* sp. n." with the locality "Saline River, Benton, Arkansas," with drawings (fig. 2.5) and a specimen catalog number (USNM 17818) that indicate he was examining *G. ouachitensis* (Ouachita map turtle).[19] Hence, had he completed his opus, Baur likely would have predated Fred Cagle (1953b) by more than half a century in naming two new taxa (see below). Baur also drew the head markings of a specimen of *G. nigrinoda* (black-knobbed sawback, USNM 17820; also cataloged—surely erroneously—as being from the Arkansas site[20]), but he assigned it to his *G. oculifera* rather than recognizing it as a separate species.[21]

Baur's work increasingly overwhelmed him as the 1890s progressed. Although convinced of the distinctiveness of *G. pulchra* in late 1890, he did not publish his description until 1893, following a sheepish letter to Stejneger apologizing for an interrupted correspondence and a busy schedule.[22] Kohn wrote Baur repeatedly regarding the Mermentau specimens,[23] finally writing in exasperation, "I have received as yet no information as to what you have done" regarding the "new *graptemys* [sic]" while nevertheless offering more specimens.[24] Baur scrawled "*G. intermedia*" at the bottom of the last letter, but his response to Kohn is unknown (but see below for evidence that Kohn was made aware of the planned epithet *intermedia*). By 1897, Baur, who had "a highly nervous organization" (Wheeler 1899, p. 18), was mentally exhausted and suffering from paresis, a mild partial paralysis. At the urging of friends, he traveled to Munich for a year of recuperation, but he did not improve and was ultimately placed in an asylum, where he died in 1898 at the age of 48.

Oddly, the specific epithet *intermedia* (as *Malacoclemmys intermedia*) was included in a checklist of the Louisiana herpetofauna by Beyer (1900) and attributed to Baur but was otherwise never published. Beyer (1900) acknowledged Kohn, who presumably supplied Baur's intended name for the species. Strictly speaking, one could regard *sabinensis* as a junior synonym of *intermedia*, with Beyer being the taxon's author (in much the way that Gray became the author of LeSueur's epithet *pseudogeographica*, described above); this issue is further discussed in the species account in chapter 8.

Two additional species were described in the first half of the twentieth century. *Graptemys versa* (Texas map turtle) was described as a subspecies of *pseudogeographica* by Leonhard Stejneger (1851–1943) in 1925, based on specimens collected a quarter century earlier by the brothers H. H. and C. S. Brimley. Stejneger's (1925) description, consisting of a single sentence fragment describing a single diagnostic feature, is the most sparsely worded description of any *Graptemys* taxon. *Graptemys barbouri* (Barbour's map turtle) was described by the University of Florida's Archie Carr, Jr. (1909-87), of later fame as a sea turtle biologist and conservationist, and Lewis J. Marchand, who collected their specimens the previous year (A. Carr and Marchand 1942). "Water goggling" supplied the initial captures. This technique, at which Marchand was legendarily proficient, involves diving underwater in clear, spring-fed rivers with eye goggles to search for turtles resting on the bottom. A. Carr (1952) gave an engaging account of their discovery of the species. Their description was the first for the genus to be based on a large series with attention to ontogenetic and sexual variation. Their publication was also the first coupling of a detailed habitat description with a species description.

Early Ecological Studies of *Graptemys*

The first published description of anything ecological regarding the map turtles and sawbacks was a brief account of false and Ouachita map turtles basking on tree trunks and nesting (LeSueur 1827). Nothing new on ecology was published until H. Garman (1890) and Oliver Hay (1892a, 1892b) described dietary differences between common and false map turtles. The former were reported to have wider heads, wider alveolar surfaces, and, correspondingly, a diet containing more mollusks. Hay was also the first to note the exceptional size dimorphism in the genus, based on three male and four female common map turtles he had captured in Indiana's Lake Maxinkuckee while attending a biology conference in 1891, plus museum specimens of additional females. H. H. Newman (1906a) published the first extensive natural history information, based on common map turtles in the same lake—one of five turtle species for which he stated he had developed "a really personal acquaintance" through spending "long hours, quiet days, in thoughtful observation" as the turtles pursued their "daily round of occupations" (p. 126). Much of Newman's description, which was couched in largely qualitative terms, concerns behavior and habits recognizable to anyone who has studied northern, lake-dwelling populations (although his description of long-term captives committing "suicide" by inspiring water has not been confirmed!). Barton Evermann and Howard Clark (1916) similarly provided largely qualitative descriptions of the general ecology of the species in the same lake, where they conducted biological surveys for the U.S. Fish Commission over a decade and a half. In describing Barbour's map turtle based on specimens from the Chipola River, Florida, A. Carr and Marchand (1942) also included extensive notes on habitat and behavior. Outside of these contributions, however, the study of map turtle and sawback biology beyond taxonomic matters would largely lapse until the genus was brought squarely into the spotlight by Fred Cagle and his students.

Fred Cagle's Tulane University Field Crews, 1940s and 1950s

Fred Ray Cagle was born on 9 October 1915 in Marion, Illinois, the youngest of four children of Fred and Agnes (Guiney) Cagle. As a young boy he was considered the smart one in the family and was an academic standout in school. His older sister, Mary, remembers him as a boy who "didn't waste any time—he was always doing something."[25] He took an avid interest in dissecting frogs and kept

Figure 2.6. A female specimen of *Graptemys oculifera* from the Gustave Kohn collection (TU 27).

specimens in the family garage. At one point he arranged for other children to bring him butterfly specimens, which he mounted in a display. An encounter with a turtle that he let loose inside the house is part of the family lore. As a boy, Cagle also took an avid interest in music, learning to play the violin, a pastime he enjoyed throughout his life. His father frequently had to admonish him for staying up too late on school nights to read.

Tragedy struck five days before Christmas when Cagle was twelve years old: a natural gas explosion in the Stiritz No. 1 mine killed his father and six other coal miners. The following morning his father's badly burned body was retrieved from the mine. Compensation from the mining company was meager at best, but the two oldest siblings, Clarence and Mary, worked to scrape together enough money to send their bright, youngest brother to college during the Great Depression.

Cagle earned a B.S. at Southern Illinois University in Carbondale in 1937. He then completed both an M.S. (1938) and a Ph.D. (1943) at the University of Michigan, where he studied under Norman Hartweg following the death of his original mentor, Frank Blanchard. Fieldwork for his dissertation on the growth of red-eared slider turtles was completed while he held a teaching and museum-director position at Southern Illinois University in Carbondale from 1939 to 1941. He then served three years in the U.S. Air Force during World War II, training pilots for the physiological rigors of high-altitude flight and achieving the rank of captain. According to his son, Fred Cagle, Jr., family lore had it that

Figure 2.7. Fred Cagle (standing) and Case Chaney in Cagle's
office in Dimwiddie Hall, Tulane University, ca. 1950.
Photo courtesy of Case Chaney.

his father helped train the pilots who dropped the atomic bombs on Hiroshima
and Nagasaki, although he never spoke of the mission.[26]

Cagle came to Tulane University at the start of the 1946 school year. His
field notes indicate that by 11 October of that year he was in the field collecting amphibians and reptiles.[27] The Tulane University field crews organized for
subsequent years would involve numerous students and begin the revitalization
of a collection that became the Tulane University Museum of Natural History.
The collection originally had been associated with Tulane in 1895, having been
initiated by George Beyer a few years after a letter from Gustave Kohn to Georg
Baur bemoaned the lack of a suitable facility of his exquisite taxidermic mounts
of reptile and amphibian specimens (e.g., fig. 2.6).[28] Kohn's specimens had once
been the centerpiece of a public museum, but by the time Cagle arrived on campus, the museum was closed and the specimens had been all but forgotten.

Allan "Case" Chaney enrolled at Tulane in 1946 after his discharge from
service in the U.S. Navy. He would prove to be an invaluable student field assistant to Cagle (fig. 2.7). Cagle chose Chaney to examine and report on the

old Tulane collection. Besides finding passenger pigeon and ivory-billed wood-pecker specimens, Chaney also brought back reports of Kohn's mounted speci-mens of *G. oculifera* and *G. pulchra* (the latter specimens are today regarded as *G. pearlensis*).[29] This may have occurred in the spring of 1948. Cagle wrote to Archie Carr that his examination of "several hundred" *Graptemys* specimens from various localities in Louisiana "does not indicate the validity of either *kohnii* or *oculifera*," but in a follow-up letter he noted that the Kohn specimens had "forced me to change my mind" and recognize the taxa as valid.[30]

That Cagle quickly realized the significance of the *oculifera* specimens is clear from the museum catalog he began at Tulane: specimens 1–19 are lizards and anurans collected by Cagle and his students in June 1947, but numbers 20 and 22–27 are occupied by Kohn's *G. oculifera*.[31] Between the collecting activities of Kohn and Cagle, only a single specimen of *G. oculifera* had ended up in a natu-ral history museum (CM S-7518, labeled as having been collected by Morrow J. Allen in 1930 in Lafourche Parish, Louisiana, an unlikely locality given the spe-cies' endemism in the Pearl drainage of Mississippi and Louisiana); Cagle was nonetheless unaware of that specimen. Curiously, Baur's "*pulchra*" specimens from the same localities and his Mermentau and Sabine River specimens of the taxon Cagle later named *sabinensis*, although equally significant as "lost" taxa, would not be recorded in the catalog until 1949.

Students in the Tulane field crew remodeled a truck for their excursions and nicknamed it "Sally" (fig. 2.8). Inside were a fold-out desk and benches along the edges that slept two crew members. They added a rack on the top to carry large olive cans full of formaldehyde. A Higgins trailer converted to a two-person tent and canopy and carried a boat on top and its two outboard motors under-neath. The period predated by a decade or so the development of a nationwide network of motor lodges, so the crew camped near their field sites in the trailer and canvas tents. Students traveled in a caravan of the truck and trailer and their own vehicles and ate canned food or, at times, some of the fish, turtles, frogs, or even alligators they had captured. (Ernie Liner, who many years later published a herpetological cookbook, remembered no instances of the crew consuming *Graptemys* specimens.[32]) Chaney recalled selling a forty-pound cat-fish in the nearest town on one trip and using the proceeds to purchase a large ham, a food he relished the rest of his life.[33] The students earned fifty-dollar stipends per semester or summer term for their efforts.

Beginning in 1947, Cagle's crews initially collected virtually any vertebrates they could capture at a variety of sites in Louisiana, but the emphasis was on amphibians and reptiles from the start. Cagle had planned a book on the herpe-tofauna of Louisiana and received a five-year grant for the project from the uni-versity's Council on Research, although the book was eventually scaled down to a sparsely illustrated dichotomous key (Cagle 1952a) published by the Tulane Book Store and sold to students for seventy-five cents. The turtles captured in 1947 were the basis of a publication by Cagle and Chaney (1950) in which they

Figure 2.8. The Tulane field crew, with "Sally" and the Higgins trailer, crossing the South Llano River in 1953. Left to right are Fred Cagle, Sr., Don Tinkle, and Fred Cagle, Jr. Photo courtesy of Robert Webb.

noted seeing basking *Graptemys* but having low success trapping them, with the important exception of hand collecting at Caddo Lake (see below). In his field notes for Lake Providence, an oxbow lake of the Mississippi River, Cagle noted that *Graptemys* and razorback musk turtles (*Sternotherus carinatus*) were "frequent" on basking substrates, although baited traps yielded only sliders (*Trachemys scripta*) and softshells (*Apalone* sp.). "A method must be worked out for trapping of *Graptemys* and *Sternotherus*," he added.[34]

Rediscovering the long-neglected Louisiana species *G. oculifera* may have been on Cagle's mind early on; he and his field crews made two trips in 1947 and an April 1948 trip to Mandeville, which Baur had recorded as the species' type locality. Kohn's specimens were discovered to have "Pearl River" inscribed on their plastrons (fig. 2.9), and on 3 June 1948, a field crew rediscovered *G. oculifera* at a site southwest of Varnado, Louisiana, where Big Creek joins the Pearl. John Boley, described by Chaney as a "hanger on"[35] in the early field crews, certainly caught favorable attention from Cagle that day by finding a nesting female one night along a sandy peninsula in the river. The five-day stay on the Pearl was otherwise frustrating, however, as basking *Graptemys* could be seen from a distance but not captured. Shooting them was even attempted, without success (fig. 2.10). Later in June 1948, Leslie Ellis caught the crew's first specimen of the yet-undescribed Sabine map turtle (*Graptemys sabinensis*) on the Calcasieu River. The crew went on to visit the Sabine River, the Guadalupe River near

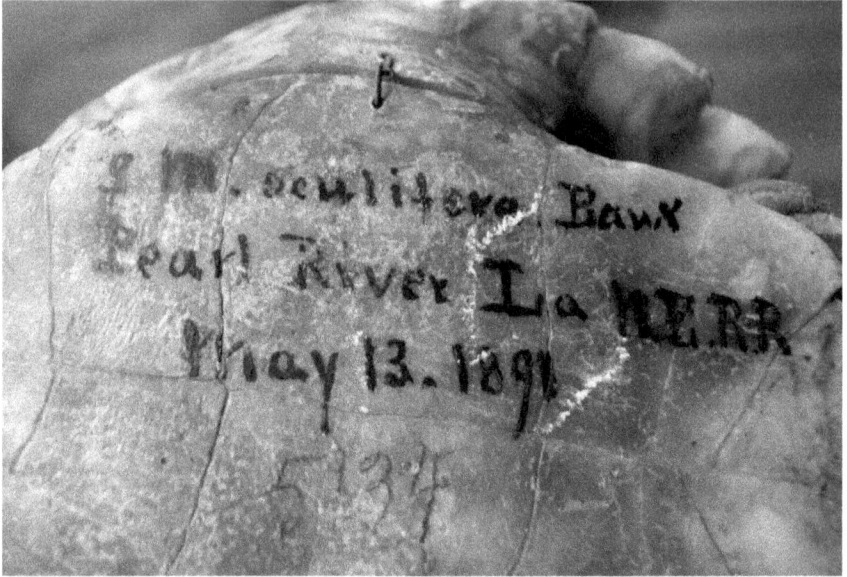

Figure 2.9 (*above*). Plastron of TU 24, a *Graptemys oculifera* originally from the Gustave Kohn collection, showing reference to the Pearl River; this led Cagle and his field crews to rediscover the species nearly sixty years later.

Figure 2.10 (*left*). Allan "Case" Chaney on the Pearl River ca. 1948, trying without success to shoot ringed sawbacks and Pearl map turtles from their basking sites. Photo courtesy of Case Chaney.

Chaney's hometown of Kerrville, Texas, and portions of the Colorado drainage in Texas. Cagle was the director of a summer Audubon Conservation Camp near Kerrville in 1948 and 1949. With no good capture methods for *Graptemys*, he missed out on getting the first specimens of a species that would later bear his name, Cagle's map turtle (*G. caglei*), as well as the then-little-known Texas map turtle (*G. versa*).

In 1949, Cagle was named head of a new doctoral program in zoology at

Tulane. "Perhaps no other spot in America is more conducive to zoological research and study than is the Gulf Coast of Louisiana," he stated in an interview regarding the new program. "Most zoologists agree that this area is one of the blind spots in zoology—a sort of zoological unexplored territory, yet abundant in zoological material."[36] Regarding *Graptemys*, the blind spots continued to clear only slowly during that summer. Seven samples totaling forty-one Mississippi map turtles (*G. pseudogeographica kohnii*) were taken from various sites in Louisiana; the method of their collection is not recorded. Cagle also captured six Texas map turtles from the South Llano River near Telegraph as a sidelight to his work at the Audubon camp, but once again he failed to encounter the turtle from the Guadalupe River that would become his namesake.

A breakthrough in the summer of 1950 was made possible by collecting conducted below the Caddo Lake Dam three years earlier, in June 1947. Basking *Graptemys* had been seen abundantly in the lake, yet only one specimen had been caught in baited traps. Sampling with headlamps at night, Chaney, Leslie Ellis, Cyrus Crites, and Cagle were able to capture large numbers of mostly juvenile turtles, of which over a third were *G. p. kohnii* and the then-undescribed *G. ouachitensis*. This was the inspiration for the technique that Chaney and a new graduate student at Tulane, Clarence Lavett Smith, would use to steer Cagle's general herpetological work toward a greater emphasis on *Graptemys* beginning in 1950.

More an ichthyologist more than a herpetologist, Clarence Smith was nonetheless welcomed to the field crew following his first year in the master's program. While steering the crew's plywood-hulled boat early in the season, he managed to hit the same submerged stump twice, poking a hole in the hull each time. Improvising, he used pieces of tin from food cans and fresh hot tar from a nearby highway to patch the holes. The hasty patch job would hold throughout a pivotal field season. Chaney and Smith worked together to develop an efficient technique for capturing the turtles, which were seen daily but rarely took the bait of their hoop traps and trot lines. Smith hung over the bow while Chaney maneuvered the boat into piles of emergent deadwood favored during the day as basking perches. Wearing a headlamp, Smith spotted turtles resting just beneath the surface on deadwood and branches of live black locust trees. With a mattress padding the bow, he quickly grabbed the turtles and whisked them into the boat. Of the submerged locust branches Cagle (1952b) would later write, "Success is often the consequence of permitting enthusiasm to override the pains of thorns penetrating the flesh" (p. 227).

On two weekend expeditions to the Pearl River near Varnado in April 1950, the crew captured 33 *G. oculifera* and 41 *G. pearlensis*. Ernie Liner remembered capturing a female on a sandbar that was the first *G. pearlensis* specimen taken by the crews, earning the favor of a delighted Cagle. In the first week of June, Chaney and Clarence Smith went out at night on the river near Angie. They captured 64 *Graptemys* and 17 other turtles, averaging 3 turtles an hour. Tulane catalog entries list 32 *G. oculifera* and 48 *G. pearlensis* for the Angie trip but may include specimens shot, caught in traps, or found nesting; Cagle (1953a) later gave totals for all nighttime collecting on the Pearl as 51 *G. oculifera*, 105 *G. pearlensis*, and 22 other turtles.

For the rest of the summer Chaney and Clarence Smith caught turtles by night and slept by day in hammocks, protected by mosquito netting. Canvas gunnysacks filled with live turtles were transported to the Tulane museum for preservation and data collection, including shell measurements Cagle took with his homemade caliper-like "cheloniometer" (fig. 2.11). In the museum, Cagle generally concentrated on processing turtle specimens while leaving student crew members to the other herpetological and fish specimens, although Chaney and Smith were often asked to assist with the turtles. Cagle included head width and alveolar width measurements; his interest in the diverse dietary habits of the genus clearly was growing.

Following their success on the Pearl River, Chaney, Clarence Smith, and their fellow crew members Richard Etheridge, Paul Anderson, Ernie Liner, Don Tinkle, Ned Lambermont, and camp cook Sam Nichols quickly returned to the field for a visit to the Ouachita River near Harrisonburg (fig. 2.12). The crew's work was almost derailed during an evening foray into town, when Etheridge and Smith were nearly arrested because they had absentmindedly forgotten to remove machetes from their belts. Hearing the story later, Cagle groused, "I hope you didn't tell them you were part of a Tulane field crew."[37] The crew grabbed 199 turtles, including 175 *Graptemys* (38 *G. ouachitensis* and 50 *G. pseudogeographica kohnii* were cataloged), in twenty-four collecting hours over the course of a week, better than 8 turtles an hour.

Figure 2.11. Cagle's homemade "cheloniometer," used to measure turtle shell dimensions.

Figure 2.12. Ernie Liner and "Sally" at the Ouachita River near Harrisonburg in 1950. Photo courtesy of Richard Etheridge.

A short trip to the Sabine River near Negreet, Louisiana, in early July yielded even better results: 325 *Graptemys* (275 *G. sabinensis* and 31 *G. pseudogeographica kohnii* were cataloged) and 11 other turtles in twenty-one hours, 16 turtles an hour. Cagle now had a big series he could use to document the two new taxa Baur had intended to describe six decades earlier. The greatest capture success was yet to come, however, in a three-day trip to the Chipola River in Florida, where Cagle sent the students in pursuit of the recently described *G. barbouri*. Initially capture success was poor, with just five specimens taken from their submerged resting spots on the first night. Carr and Marchand's water-goggling technique was tried, but their success was not replicated. Chaney and Clarence Smith noted occasional turtles at the surface and devised a technique of steering the boat back and forth from bank to bank on their way downstream. The outboard motor was mounted on a 360° swivel; thus when Chaney turned the motor 180° and revved it to bring the boat to a stop, it produced an underwater din that brought turtles to the surface, probably panicked by the reverberations off the river's limestone bottom. Suddenly it was bedlam in the twelve-foot plywood skiff, which the two students loaded down with 125 *G. barbouri* that night. They switched frantically between leaning out to make new captures and kicking turtles back into the boat that were attempting to clamber back into the river. The second night's collecting ended only when the boat was too full to continue, with 257 *G. barbouri* taken from a four-mile reach. Over thirteen hours, capture success was nearly a turtle every two minutes. For the first time, substantial numbers of adult females were part of the crew's haul. Cagle joined the crew in the field just after their success and found himself hip-deep in turtle heaven. On their way home from the Chipola, the crew tried out their new techniques on the Escambia River, but the magic was not there this time; high-water levels limited success to just 12 specimens, the first of a species that more than four decades later would be christened *G. ernsti* (Escambia map turtle).

The year 1951 was a slow one for *Graptemys* collecting: only one trip was taken, to the Bogue Chitto River in Washington Parish, which netted 14 *G. pearlensis* and 2 *G. oculifera*. Cagle was gearing up for bigger things, however. He received a two-year, $14,200 grant from the National Science Foundation (1952, p. 50) for his proposal "Speciation in the genus *Graptemys*." According to Tinkle (1957), the field crews traveled a total of forty thousand kilometers in 1952–55. The focus on turtle collecting shows in his collations: 2,693 turtles were collected, as compared with 726 anurans, 461 snakes, 414 lizards, 390 salamanders, and 3 alligators. The genus that had initially been a source of frustration—highly visible but often virtually uncatchable—now became the most abundantly collected genus, accounting for approximately 41 percent of Tinkle's turtle total for these four years, according to specimen records of the Tulane Museum.

The 1952 field season began with a trip to Texas to sample the Sabine, Trinity, and Brazos Rivers, where Cagle and his crew caught ten *G. p. kohnii*. Cagle registered surprise at the lack of *G. sabinensis* or other narrow-headed species in the latter two drainages.[38] The crew also sampled *G. versa* in the San Saba, Concho, and Llano segments of the Colorado drainage. In between the Concho and Llano stops, they visited the Pecos River in western Texas, with Cagle puzzling over the lack of *Graptemys* in apparently suitable habitat.[39] His notes on questions prompted by *G. versa*—about its comparatively small body size, overlap or lack of overlap with congeners, unknown food habits, and unusual similarity in markings to a those of sympatric species of *Pseudemys*—show the degree to which his focus had shifted from general herpetological collecting to the ecology and evolution of the map turtles and sawbacks (fig. 2.13). Bob Webb joined the field crew at Lake Texoma the summer after he finished his master's thesis at the University of Oklahoma, as the group wended its way back to New Orleans via a side trip to rivers in Oklahoma and Arkansas. Webb remembers Cagle at one point ruminating on plans to split *Graptemys* into three genera, one of which he planned to name *Rheoemys*, with only one species, *R. versa*, constituting it.[40]

In early August 1952, attempts to sample the *Graptemys* of the Tennessee River in northwestern Alabama were unsuccessful, which Cagle in his notes blamed on the lentic waters created by Wilson Dam that seemed to favor sliders (*Trachemys scripta*).[41] The northerly excursion would pay off handsomely on the next stop, however. On 7 August, the crew camped in an open field on a high bluff overlooking the Black Warrior River southwest of Tuscaloosa, Alabama. Oddly, none of the widely distributed field crew destinations of the previous five and a half years had been on the extensive Mobile Bay drainage system. The first evening, Walter Stone and Howard Suzuki captured eleven specimens

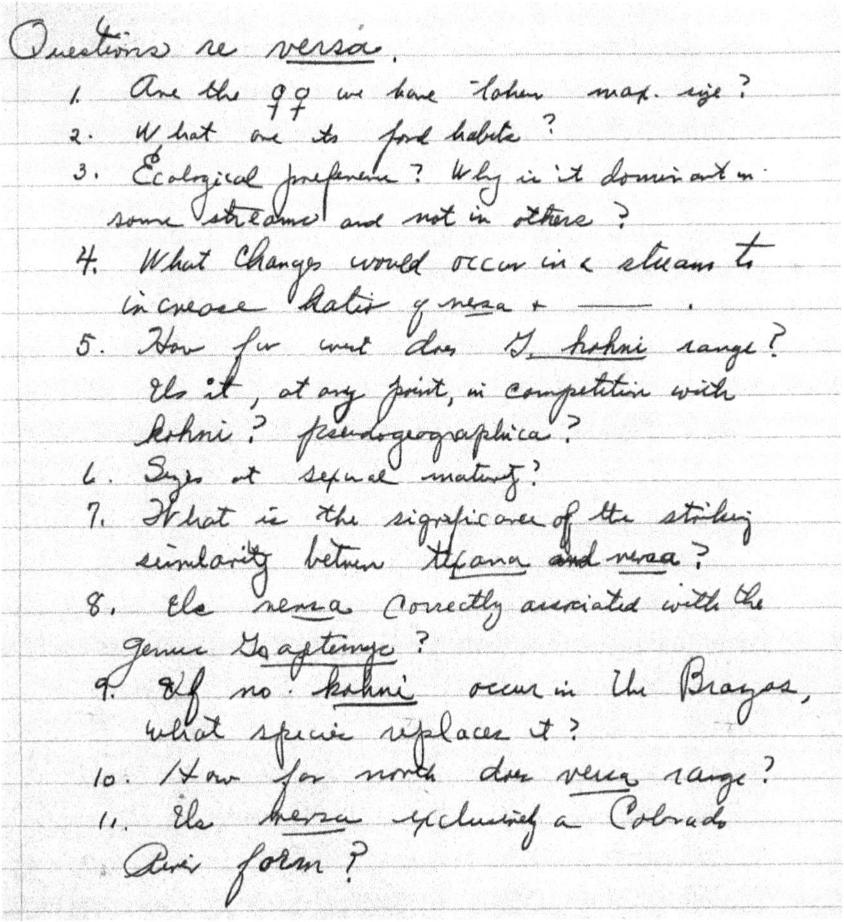

Questions re *versa*.

1. Are the ♀♀ we have taken max. size?
2. What are its food habits?
3. Ecological preference? Why is it dominant in some streams and not in others?
4. What changes would occur in a stream to increase ratio of *versa* + _____.
5. How far west does *G. kohni* range? Is it, at any point, in competition with *kohni*? *pseudogeographica*?
6. Size at sexual maturity?
7. What is the significance of the striking similarity between *tifana* and *versa*?
8. Is *versa* correctly associated with the genus *Graptemys*?
9. If no *kohni* occur in the Brazos, what species replaces it?
10. How far north does *versa* range?
11. Is *versa* exclusively a Colorado River form?

Figure 2.13. Entry dated 7 June 1952 from Cagle's field notes, made during collecting on the San Saba River in San Saba County, Texas (Cagle Archives, Savannah River Ecology Laboratory).

of a new species of sawback as well as the first "true" *G. pulchra* (*sensu stricto*) the field crews would capture and the first specimens of the species since 1876. Just after midnight, Chaney and Stone returned to the river to capture twelve more of the new species. On a trip upriver the following morning, Cagle and Suzuki counted forty-nine basking turtles; forty-six were recorded as "*Graptemys nebulus*" in Cagle's field notes.[42] Subsequent entries used the specific epithet "*nodulosus*,"[43] but by the end of the month Cagle had settled on *nigrinoda* (black-knobbed sawback) for publication.

While Chaney, Webb, Stone, Suzuki, Bob Gordon, and Cagle's thirteen-year-old son, Fred Ray, Jr., continued sampling, Cagle scrawled several questions on *Graptemys* ecology, evolution, and diversity in his field notes. He

Figure 2.14. The Pascagoula River at Old Benndale Crossing, Mississippi, where the Tulane field crew collected yellow-blotched sawbacks they shot from basking sites in August 1952. Photo courtesy of Robert Webb.

questioned whether southern Atlantic drainages such as the Savannah and Altamaha Rivers would yield new species, wondered how his new form and *G. oculifera* might contrast in diet with broad-headed forms, speculated on the role of stream piracy in producing species with such restricted ranges, and asked, "What *Graptemys* are found in the Pascagoula River? Does *nodulosus* intergrade with *oculifera* in this section?"[44] Following a stop on the Alabama River for more *nigrinoda* and *pulchra* and a return to New Orleans, the crew was dispatched to answer Cagle's questions regarding the Pascagoula on 17 August. The first afternoon, Chaney and Gordon shot and retrieved 5 specimens of a new species of sawback that Cagle, sitting out the trip, would later christen *G. flavimaculata* (yellow-blotched sawback) for its prominent orange-yellow markings. In the evening, Chaney and Suzuki took 34 specimens from the bow of the boat in three hours and fifteen minutes, following which Webb and Gordon took 26 in three and a half hours just after midnight. The following day, Chaney and Webb shot 13 more specimens (fig. 2.14) and Chaney and Gordon worked the river from 2115 to 0215, taking 76. Besides the 154 specimens of *G. flavimaculata*, the crew captured the first *G. gibbonsi* ever taken (12 specimens they regarded as *G. pulchra*). They returned a week later to collect 32 more *G. flavimaculata* and 4 more *G. gibbonsi*, and Cagle set to work describing the two new sawbacks. Two referees of Cagle's manuscript agreed that they should be recognized as full species, while the third predicted they would eventually be recognized as subspecies.[45]

The 1953 field season began with a trip to Florida. Case Chaney no longer participated in the summer field crew, which was led that year by Tinkle, Webb, the younger Cagle, and newcomers Don Blair and F. C. Johnson. Following stops to collect forty more *G. ernsti* from the Escambia River and thirteen more *G. barbouri* from the Chipola River, the crew headed farther east to sample the Wacissa and Suwannee Rivers and Silver Springs, probably in part to search for more undescribed *Graptemys*. They next took eighteen *G. barbouri* on the Flint River, although Cagle's notes indicate the real interest was in determining the eastern extent of the range of the sawback complex that had figured so prominently in the previous year.[46] On the return to New Orleans, the crew sampled the Coosa River, the Mulberry Fork of the Warrior River system, the type locality of *G. nigrinoda* on the Black Warrior River, and the lower Chickasawhay River in the Pascagoula drainage. On the Coosa River and Mulberry Fork, the common map turtle (*G. geographica*) was documented for the first time in the upper Mobile Bay drainages.

In Hattiesburg, Mississippi, a stop to pick up forwarded mail brought Blair his military draft notice.[47] Without returning to New Orleans, he boarded a bus straight to boot camp in Virginia. A subsequent western swing by the crew included a stop on the Calcasieu River to obtain fourteen *G. sabinensis* on the way into Mexico for a week and a half of sampling, topped off with a stop on the Colorado drainage in Texas, where *G. versa* was seen basking in abundance but eluded capture.[48] Johnson left the crew at the conclusion of the Mexico swing, having contracted an acute case of amoebic dysentery. A plan to sample in Victoria, Texas, was abandoned when the crew could not get their boat down the steep bluff to the Guadalupe River. Together with his work at Kerrville in 1948-49, Cagle had now unsuccessfully sought turtles at both the upstream and downstream range limits of the species that was later named for him.

In 1954, Tinkle and Webb were the two leaders in a field crew that also included Charles Albright, Don Boyer, Fred Cagle, Jr., and high-school student and camp cook Thomas Koehring. The 1954 field crew was the last to see significant participation by the senior Cagle. After they stopped at Concordia Lake, Louisiana, much of the rest of the summer was a reprise of some of the better collecting localities of previous years, including the Escambia River in Florida and the type localities of *nigrinoda* and *flavimaculata*. Cagle apparently wanted to further delineate the range of *G. barbouri*, with two fruitless stops on the Chattahoochee River and stops on the upper and lower Flint, the Ocmulgee, and the Wacissa Rivers. The crew captured 145 *Graptemys* specimens representing seven species in 1954. Cagle continued his practice of adding numerous questions to his field notes. He noted interdrainage variation in what he was calling *G. pulchra*,[49] which would be split into three species almost four decades later. He noted the absence of *G. nigrinoda* above the fall line,[50] a geographical feature that separates the coastal plain from upstream segments of the Mobile Bay drainages, despite the fact that *G. pulchra* was taken above the fall line in the Mulberry Fork and the Coosa. The distributional question was a topic he would leave to Tinkle (1959) to work on in more detail. The question of variation in *G. pulchra* was not taken up again; Walter Stone was at one point assigned to work on the megacephalic Gulf Coast species, but no thesis ever resulted.[51]

The 1955 crew was led by Tinkle, Webb having left Tulane to pursue his Ph.D. Tinkle partnered with Whit Gibbons, a junior-high student from Tuscaloosa and nephew of fellow Tulane graduate student Harriett Smith. Gibbons, then fourteen, had a strong interest in herpetology. Although his main interest as a boy was in snakes, he remembers also watching turtles from a railroad trestle over the Black Warrior River as a ten-year-old.[52] The two started out teamed with Albright, who abruptly left them without explanation at the conclusion of the first week in Arkansas. Tinkle had Gibbons keep the expedition log for the summer, complete with such Tulane conventions as notes on campsites, narratives on who collected what, and use of the twenty-four-hour military clock for recording times. During nighttime boat excursions, Tinkle often had to admonish young Gibbons for being more interested in grabbing water snakes from the branches above the water than the turtles just below the surface. Thus began a long partnership that became a mentor-student relationship between Tinkle and Gibbons.

That Cagle had essentially stopped participating in the field crews was a disappointment to Tinkle.[53] Tinkle later was exceptionally apologetic to Gibbons upon taking an administrative role with the University of Michigan Museum of Zoology, perhaps remembering how Cagle had drifted away from working in the field with him years earlier (fig. 2.15).

Cagle was a career climber, described by Ernie Liner as "always pushing to get up" in the world and as "a talker, a born administrator"[54] and by Richard Etheridge as "conscious of status."[55] Case Chaney said Cagle "wanted to be the big thing on campus."[56] Fred Cagle, Jr., sees his father's move into administration as motivated primarily by the disparity in salaries that existed between faculty and administrators in the pre-Sputnik years, before the space race with the Soviets increased funding for the sciences, but also by the greater status.[57] Already a department chairman by the mid-1950s, Cagle further reduced his teaching and research role in 1958 by taking a new position as coordinator of Tulane's research program. Although an announcement in *Copeia* in 1961 stated that "Dr. Cagle continues his studies of the systematics of *Graptemys*,"[58] his final publication on the genus was by then six years past. From 1958 to 1963, he oversaw the university's research funding, which grew from less than half a million dollars to ten million dollars. One of his interests, undoubtedly influenced by many years serving as a journal editor, was the growing difficulty researchers faced in keeping up with the literature. In a 1962 interview, Cagle

Figure 2.15. Fred Cagle (right) and Don Tinkle on the Chickasawhay River, Mississippi, in 1953. Photo courtesy of Robert Webb.

said, "information communication is so bad . . . it costs less for scientists to perform research than to find out if it has been done previously."[59]

Cagle also became active in international science relations through his affiliation with the American Institute of Biological Sciences. In 1958 and 1959 he took lengthy trips to Russia, with stopovers in other nations in Europe and Asia, for the purpose of scientific exchange; he was apparently the first U.S. scientist ever invited to tour scientific facilities of the Soviet Academy of Sciences.[60] With the Cold War at its height, Cagle was in the papers frequently, praising the Soviets' dedication to using science to set policy and calling for greater efforts by the United States, as in this quote from a 1962 *States-Item* article: "The USSR has used science effectively in this battle for world leadership. The U.S. has not. . . . Our failure is a reflection of American scientists' lack of concern with international affairs, the American public's ignorance of the importance of science, and our political leaders' failure to grasp the full import of the scientific and technological revolution."[61] Cagle and his wife, Jo, also traveled widely around the world as embassy inspectors for the U.S. State Department. In 1963 Cagle was named vice president for institutional development at Tulane, in which capacity he led the acquisition of the Belle Chasse naval bunkers where the Tulane Museum collections are housed today.

Even as he drifted away from fieldwork in his journey from professor to department chairman to administrator, Cagle continued to direct graduate students, including the doctoral works of Gordon and Tinkle in 1956, Chaney and

Boyer in 1958, Bob Shoop in 1963, and Jim Dobie in 1966. His own research career had ended abruptly shortly after the description of *G. flavimaculata* and *G. nigrinoda*, with many questions about turtle ecology and evolution left for future generations to rediscover and ponder. The same book of detailed field notes that Cagle kept in 1954 contains only sporadic field notes from early 1955, followed by diary-like entries for 1957 and 1958 (some in his son's hand) from a family trip to western national parks and brief entries from trips to the Pearl River.[62] The last major herpetological work to bear Cagle's name was the reptiles section in *Vertebrates of the United States* (Blair et al. 1957). On 8 August 1968, Cagle died at the age of 52, the victim of a dissecting aortic aneurysm.

Cagle was generally well liked and respected by his field crews. Liner, Etheridge, and Clarence Smith all remember getting along well with Cagle, who was generous in writing glowing recommendations for graduate applications.[63] In his interactions with the crew, however, Cagle could also be imperious—in his son's words, "kind of dictatorial."[64] Crew members remember often eating canned beans and wieners while watching a sizzling steak being cooked by Cagle (or, on at least one occasion, by his wife, Jo, on a visit to the campsite). During the 1950 season, Cagle explained to the students that his doctor required him to take a high-protein diet, so Clarence Smith and the others nicknamed him "HPR," for "high-protein requirement" and also "high-powered researcher."[65] Straws were drawn one night to see who would guard the campsite while the rest of the crew went into town to see what Cagle called a "morale movie." Cagle refused to draw, saying, "you work for me" and climbing into the old truck, "Sally," to attend the movie.[66] Chaney, arguably the linchpin member of Cagle's field crews, became particularly disgruntled over Cagle's threat (never carried out) to take away his scholarship and use it to attract new students, as well as a particularly rough oral exam for his Ph.D. Chaney was a professor for twenty-six years at Texas A&I University and claimed Cagle made him the teacher that he was—through his determination to *not* do as his mentor had done![67] Nevertheless, the field crews built an outstanding collection at Tulane, helped to launch many a productive herpetological career (table 2.1), and, most tellingly, created several lifelong friendships.

Studies by Graduate Students of the 1960s, 1970s, and 1980s

Almost all the research on the natural history of *Graptemys* in the 1960s, 1970s, and 1980s was done by graduate students. While much of their work unfortunately was never published in peer-reviewed publications, by and large even the unpublished work has nevertheless been deemed worthy of copious citations in the literature. Take away the research of this era and we would be greatly lacking today in even the most basic knowledge of the ecology of Barbour's, Cagle's, Escambia, false, Ouachita, and Sabine map turtles and black-knobbed sawbacks. For the most part, however, the graduate students who worked with

Table 2.1. The most prominent members of the Tulane field crews led by Fred Cagle, 1947–1955

Crew member	Years	Tulane degree	Later career
Paul K. Anderson	1950	M.S. (1951)	University of Alberta professor
Donald R. Boyer	1954	Ph.D. (1958)	Washburn University professor
Fred R. Cagle, Jr.	1952–54	—	San Diego State University professor (geology)
Allan H. Chaney	1947–52	Ph.D. (1958)	Texas A&I University professor
Richard E. Etheridge	1948–50	B.S. (1951)	San Diego State University professor
J. Whitfield Gibbons	1955	—	University of Georgia professor
Robert E. Gordon	1952	Ph.D. (1956)	University of Notre Dame vice president
Ernest A. Liner	1948–50	—	Pharmaceutical sales, E. R. Squibb and Sons
C. Lavett Smith	1950	M.S. (1951)	Curator of Ichthyology, American Museum of Natural History
Walter D. Stone	1952	—	Director, Franklin Park and Middlesex Fells Zoos, Boston
Howard Suzuki	1952	Ph.D. (1955)	Dean, University of Florida College of Health Professions
Donald W. Tinkle	1950–55	Ph.D. (1956)	University of Michigan professor
Robert G. Webb	1952–54	—	University of Texas–El Paso professor

map turtles and sawbacks during this period moved on to work on other taxa, with few choosing to return to *Graptemys* studies later in their careers.

Auburn University's Department of Zoology and Entomology was the most prolific center of graduate work on the genus, with master's theses on black-knobbed sawbacks by Jerry Waters (1974, on basking, directed by George Folkerts) and Peter Lahanas (1982, autecology, directed by Jim Dobie) and a doctoral dissertation on Escambia map turtles by Robert Shealy (autecology, directed by Bob Mount). The two master's theses were never published but have been cited numerous times; Shealy published his work as a monograph (Shealy 1976) that contains most of what is known about Escambia map turtles. Richard Timken (1968a) studied false map turtles in the Dakotas for a doctoral dissertation at the University of South Dakota. He published a short article on diagnostic markings, distribution, and population status (Timken 1968b), and his dissertation also contains reproductive data. At the University of South Florida, Roger Sanderson (1974) studied sexual dimorphism in the species with the most striking sexual differences, Barbour's map turtle, for a master's thesis under the direction of Roy McDiarmid. His unpublished thesis also contains most of what we know about the autecology of that species. Don Moll (1977), whose older brother, Ed, had made a name for himself with a dissertation on sliders in Panama, studied map turtles and other aquatic turtles along the Illinois River for his doctoral work at Illinois State University. He also found a treasure trove of old museum specimens that had been collected from his field sites decades earlier. He was able to make comparisons that showed changes in diets following environmental degradation of the river.

The first attempt to draw a phylogenetic tree for *Graptemys* was made by Ron McKown (1972) at the University of Texas, using blood proteins and karyotypes. His dissertation was never published. Norah Flaherty (1982) worked with Roger Bider at McGill University on movements and habitat use of common map turtles in Quebec, and they published the habitat-use portion of the study (Flaherty and Bider 1984). Steve Shively (1982), working with Jim Jackson at the University of Southwestern Louisiana, conducted an excellent study of environmental limits to upstream dispersal in Sabine map turtles. Their publication (Shively and Jackson 1985) remains a classic study from this era and is the lone major study of the species. At Penn State University, Thomas Pluto conducted a doctoral dissertation on movements and habitat use of common map turtles, which he published in two articles with his advisor, Edward Bellis (Pluto and Bellis 1986, 1988). At West Texas State University (now West Texas A&M), Flavius Killebrew—who had earned his doctorate studying *Graptemys* morphology at the University of Arkansas—directed a succession of master's students, Dan Porter (1990), Matt Craig (1992), and Joel Babitzke (1992). Their work on the diet, movements, and population size of Cagle's map turtles is quite good yet remains unpublished.

The most recent descriptions of new species were carried out by partnerships that included graduate students. McKown teamed with David Haynes on a description of *Graptemys caglei* (Cagle's map turtle). As a Texas undergraduate, Haynes had become interested in the diversity of *Graptemys*.[68] He traveled widely in the southeast, first simply observing and later collecting specimens from many of the old Tulane field crew sites. He met Cagle in New Orleans in 1962, and the two talked turtles, with Cagle giving Haynes reprints of his work. A year later, while traveling along the South Fork of the Guadalupe River, Haynes' wife, Michaele, spotted a turtle basking on a rock in shallow water. Although he figured it was probably just a Texas river cooter (*Pseudemys texana*), a fairly common species, Haynes waded into the river and fished it out from under the rock anyway. He knew immediately it was an undescribed map turtle species, but as an undergraduate he did not begin working on a description. He met McKown, who enrolled in the doctoral program in 1965 and traveled with him to help collect specimens for the molecular phylogenetic work. Wanting to include the undescribed species, they visited several sites on the Guadalupe looking for more specimens but finding few until McKown located a site in DeWitt County where they collected the large type series. They considered the specific epithets *guadalupensis* and *lambda* (the latter after the V-shaped marking on the top of the neck) before settling on a name honoring Cagle, who had died shortly after the type series was collected (Haynes and McKown 1974).

While a master's student at George Mason University, Jeff Lovich was asked by his major professor, Carl Ernst, to prepare an account on the Alabama map turtle for the *Catalogue of American Amphibians and Reptiles*.[69] When the account was published in 1985, Jack McCoy at the Carnegie Museum of Natural History talked with Ernst and suggested that Lovich should come to the Carnegie to look at interdrainage variation in the species. The study got off to a rough start when Lovich set out on the trip from Virginia to Pennsylvania on his Kawasaki 1000 in the fall and quickly decided that shivering all the way to Pittsburgh would not be feasible; he returned home, borrowed a car, and drove through an ice storm to reach the museum. Two later trips to the Auburn University Museum to examine their specimens and visit with Bob Mount and Jim Dobie were also important to the project, which Lovich and McCoy wrapped up after Lovich had gone on to the University of Georgia to work on his doctorate under the direction of Whit Gibbons. They restricted the Alabama map turtle to the Mobile Bay drainage and described two new species, *G. gibbonsi* (Pascagoula map turtle) from the Pearl and Pascagoula drainages to the west and *G. ernsti* (Escambia map turtle) from the Escambia Bay drainages to the east, the patronyms honoring Lovich's two graduate school mentors (Lovich and McCoy 1992). Most recently, in 2010, a group led by yet another graduate student, Josh Ennen from the University of Southern Mississippi, described the Pearl populations of *G. gibbonsi* as *G. pearlensis* (Ennen, Lovich, et al. 2010).

Vogt's Studies of the 1970s

The most prolific graduate student of the era was Dick Vogt. Vogt's dissertation was the first thorough ecological study of Ouachita and false map turtles and resolved confusion regarding conspecificity of the two predominant species of the greater Mississippi River drainage. As a sidelight, he teamed with a fellow graduate student to pioneer early ecological studies of temperature-dependent sex determination in turtles, using map turtles as principal subjects. His work with *Graptemys* continued in his postdoctoral appointment at the Carnegie Museum of Natural History.

Vogt started as a graduate student at the University of Wisconsin with two major objectives: a master's thesis on massasaugas (*Sistrurus catenatus*) and a

Figure 2.16. Dick Vogt checking a fyke net on the Rio Negro in Brazil in 1989. He learned the technique from commercial fishermen while studying map turtles of the Mississippi River in Wisconsin during the 1970s. Photo courtesy of Dick Vogt.

Figure 2.17. Jim Bull excavating map turtle eggs on an island of the Mississippi River near Stoddard, Wisconsin, for experiments on temperature-dependent sex determination in 1977. Photo courtesy of Dick Vogt.

side project, collecting information for a book on the herpetofauna of Wisconsin.[70] During deer-hunting season in the fall of 1971, ten of the forty-plus massasaugas Vogt had marked for his study were found killed. As Vogt remembers it, he realized his intended thesis would not be possible without a concurrent program of hunter eradication! Meanwhile, while incubating eggs of *Graptemys* as part of the book project that same year, he sometimes got what seemed to him to represent two to four taxa from a single clutch. His work with the *Graptemys* of the upper Mississippi River quickly grew into a dissertation project under the direction of William Reeder. Initially Vogt caught his animals on nesting beaches and by hand while wading in shallow water where they were feeding. The Wisconsin Department of Natural Resources loaned him fyke nets (fig. 2.16), and commercial fishermen on the river provided specimens caught in gill nets. The fishermen also taught Vogt about trammel nets and how to use an inverted funnel on a pole, or "carphorn," plunging it into the water's surface to drive the animals into the various nets. All were techniques that had not previously been applied to the capture of *Graptemys*.

Vogt's taxonomic conclusions sorted out a confusing mix of similar-looking taxa in the Mississippi River drainage and smaller drainages to the west. Based on extensive records of head patterns, together with analysis of courtship behaviors and protein electrophoresis, he found *kohnii* and *pseudogeographica* to be conspecific subspecies with a broad zone of intergradation but removed *ouachitensis* and *sabinensis* from their status as subspecies of *pseudogeographica*, considering them to be allopatric subspecies of a separate species.

In 1978, Vogt and another graduate student, Jim Bull, discussed a paper by Yntema (1976) on temperature-dependent sex determination (TSD), then a very new topic in herpetology. Both harbored doubts about Yntema's findings because of the low survival rates in the clutches he incubated; Bull in particular felt that TSD could not occur in nature, given the potential departure from the Fisherian 1:1 sex ratio. The two drove to Vogt's study site on the Mississippi River and unearthed fourteen incubating clutches. Twelve turned out to have only male or only female hatchlings, and the other two were highly skewed toward one sex. The following summer they completed work that would be the basis for their publication in *Science* on the subject of TSD (fig. 2.17; Bull and Vogt 1979). One of the three manuscript reviewers argued against its publication, saying it lacked sufficient new information, but Vogt protested and eventually the manuscript was accepted. Vogt and Bull thus began a partnership in TSD studies that would yield six articles over six years featuring map turtles as prominent species in an exciting new area of research in chelonian biology.

In 1968, as an undergraduate field assistant for a graduate student trapping mammals in Organ Pipe National Monument, Arizona, Vogt had met Jack McCoy, the curator for amphibians and reptiles at the Carnegie Museum in Pittsburgh. McCoy would take on the strongest mentoring role in Vogt's career. They began collecting turtles together in 1974 with money from the Netting Fund, and Vogt took a postdoctoral position at the Carnegie Museum for 1977–79 (fig. 2.18). Together they led an effort in collection building that placed the museum among the very best in the world for turtle collections. Having spotted a claim of the world's largest turtle collection in a grant application by John Legler, Vogt asked McCoy how many turtles the Carnegie had—and the two were off and running, seeking to increase their number, with collecting trips funded by the National Science Foundation and accessioning of large private collections. While many museums shy away from large turtle collections because of the large storage containers the specimens require, McCoy did not, as he dedicated museum resources to accumulating stainless steel tanks (fig. 2.19). The museum holds the world's largest collection of *Graptemys* specimens, at 10,650 (S. Rogers, pers. comm. 2006). To be sure, the number is inflated by the inclusion of many thousands of laboratory-incubated and field-collected hatchlings from Bull and Vogt and also Indiana University researcher Mike Ewert, but over 1,100 field-collected larger specimens present a treasure trove to researchers working on *Graptemys*.

Vogt's 1980 and 1981 publications, on natural history and diet, respectively, had included abundant references to his taxonomic conclusions, but the taxonomic work remained unpublished. In the fifteen years that passed between his dissertation (Vogt 1978) and publication of its systematics section (Vogt 1993), some authors of general works on turtles followed Vogt and others did not. Iverson's (1986, 1992) books of distribution maps followed Vogt and cited his dissertation as the authority, as did F. W. King and Burke (1989) in their checklist of the world's turtle species. In *Turtles of the World*, C. H. Ernst and Barbour (1989) kept *kohnii* separate from *pseudogeographica sensu stricto*, commenting that Vogt had "compiled sufficient data" for the elevation of *ouachitensis* but that the relationship of the latter two taxa needs "additional study" (pp. 220–21). (However, with the publication of their 1994 *Turtles of the United States and Canada*, coauthored by Lovich, they had accepted Vogt's conclusions in full.) In their book on the Louisiana herpetofauna, Dundee and Rossman (1989) cited Vogt's dissertation but elected to not follow its taxonomy "pending a formally published presentation by Vogt" (p. 185). The third edition of the Peterson field guide (Conant and Collins 1991) likewise retained the old taxonomy, with *ouachitensis* and *sabinensis* as subspecies of *pseudogeographica* and *kohnii* a separate species.

The new edition of the Peterson field guide—together with the fact that its coauthor, Joe Collins, had said he would not publish accounts for *pseudogeographica* and *ouachitensis* in the *Catalogue of American Amphibians and Reptiles* without prior publication of the systematic work—was too much for Vogt (never one known for holding his tongue). At a 1992 herpetology conference in El Paso, he presented his systematic work and peppered his talk with barbs aimed at the man he held responsible for the field guide slight, referring to him as "Tom" Collins throughout his talk. The talk was a rehash of a talk Vogt had

Figure 2.18. Jack McCoy using a carphorn to drive turtles into a trammel net on the Glover River in Oklahoma in 1976, with his son, Kelly, and Dick Vogt's field assistant, Lisa Mattick. Photo courtesy of Dick Vogt.

Figure 2.19. Some of the 196 stainless steel tanks at the Carnegie Museum of Natural History that are dedicated to its large turtle collections, including 10,650 specimens of *Graptemys*.

given at a conference at Auburn University in 1974 that had won him a student award; ironically, Collins had moderated his session there.[71] The highlight of the 1992 presentation was a slide of a pickup truck laden with an equine carcass, which Vogt claimed to have in the parking lot—just in case anyone was interested in further beating the dead horse! Collins, who was also under fire at the time for his suggestion to raise fifty-five disjunct, allopatric subspecies to full species status without further review, laid low at the conference, refusing to take Vogt's bait and skipping a panel discussion on the subspecies issue. Vogt's publication appeared in the *Annals of the Carnegie Museum* in February of the following year (Vogt 1993). According to Vogt, administrators at his university in Mexico had pressured him to publish work completed in Mexico following his appointment there and deemed the work in Wisconsin less important to his chances of advancement.[72] The 1993 publication thus bears a most unusual footnote on its title page, declaring its submission as a manuscript had been made on 22 March—of 1979.

Vogt's systematic work remains the only attempt to sort out the most vexing problem within the genus *Graptemys*: the status of its most similar-looking taxa. His conclusions have been upheld by molecular phylogenetic analyses (Lamb et al. 1994; Stephens and Wiens 2003) and a study of variation in markings (Lindeman 2003a). Nevertheless, in recent years some authors have remained tied to the old taxonomy (e.g., Freedberg et al. 2005) or declared themselves unable to sort out the taxa in specimens at hand (e.g., Minton 2001). Perhaps another horse carcass is in order.

The Conservation Focus

Restricted geographic ranges, reliance on rivers that attract human impacts, and attractiveness to pet fanciers have all led to the depletion of many populations of map turtles and sawbacks. The first record of concern for *Graptemys* populations concerned Barbour's map turtle. In a letter to the editor in *The Florida Naturalist*, Lee (1969) presented the "strong possibility" that Barbour's map turtle was "in danger of extinction" (p. 68). He discussed their use as food and for target practice on the Chipola River and opined that diving techniques used by collectors and the turtles' basking habits made them highly vulnerable to these threats. Because take was biased toward the large females, Lee felt sex ratios were skewed toward males in the more accessible areas. An editorial response reveals that Peter Pritchard, then a graduate student, had proposed that the Florida Game and Freshwater Fish Commission grant protected status for the species; the commission banned the shooting of basking turtles but did not otherwise protect the species at that time. Short notices in *Florida Conservation News* subsequently pronounced the species to be "nearing extinction" (Anonymous 1973, p. 2) but "increasing" on the Apalachicola River (Anonymous 1976, p. 8).

Ken Dodd, on a temporary appointment with the now-defunct Office of

Endangered Species of the U.S. Fish and Wildlife Service (USFWS), placed a short notice in the *Federal Register* (Dodd 1977b) naming twelve U.S. freshwater turtle taxa to be considered by the USFWS for federal listing under the U.S. Endangered Species Act. The act had been signed into law just four years earlier, although federal listing of species under two precursor laws dated back to 1967. Dodd's list included five of the six U.S. freshwater species that have ever achieved federal listing; only bog turtles (*Glyptemys muhlenbergii*) have been reviewed and listed since. Half his list was composed of *Graptemys* taxa, two of which would eventually be listed. In a popular article on endangered reptiles and amphibians, Dodd (1977a) discussed the case of the black-knobbed sawback, a species he felt was in steep decline and likely to be eliminated from much of its historic range in the Mobile Bay drainages due to river modifications.

The USFWS periodically published exhaustive lists of candidate species under consideration for listing (USFWS 1982, 1985, 1989, 1991, 1994), with each species listed in one of three categories: category 1, species for which the USFWS had "substantial information on hand to support . . . proposing to list," with listing anticipated; category 2, species for which "proposing to list . . . is possibly appropriate," with a call for more data; and category 3, species "no longer being considered for listing" (USFWS 1982, p. 58454) because of (A) probable extinction, (B) taxonomic revision, or (C) greater abundance or less threat than had been believed. *Graptemys* candidates between 1982 and 1996 are shown in table 2.2. Beginning with USFWS (1996), shorter lists of only the strongest candidates for listing (roughly equivalent to category 1 from previous lists) have been published.

The USFWS funded spotting-scope and trapping surveys of the three sawbacks (McCoy and Vogt 1979). The black-knobbed sawback was declared to be "doing well at most sites" (McCoy and Vogt 1979, p. 11), although fewer were seen in more heavily impacted stretches of rivers. The ringed sawback and yellow-blotched sawback were of greater concern, particularly with regard to pollution from industrial discharges and target shooting. Both would eventually be added to the U.S. Endangered Species list—the former in 1986 (Stewart 1986a, 1986b) and the latter in 1991 (Stewart 1990; Lohoefner 1991)—after USFWS biologists had completed their own surveys.

Survey data collected by the USFWS are notable for their haphazard, non-standardized, and largely unreplicated nature (table 2.3), given their implications for determining whether or not a species should be deemed imperiled and in need of conservation action. Stewart's initial basking survey for ringed sawbacks on the upper Pearl River in 1984 was of little use, as he returned to his office to find a source informing him that the species was not the only *Graptemys* native to the river.[73] In subsequent counts in 1984, he distinguished the ringed sawback from its sympatric congener, finding that the ringed sawback was a predominant species and concluding its population was "healthy and not currently threatened due to low numbers" but warning that "the degree of threat

Table 2.2. Federal status of *Graptemys* taxa reviewed for possible listing

Taxon	Review announced in 1977	USFWS status as candidate or listed species					
		1982	1985	1989	1991	1994	1996
barbouri	No	2	2	2	2	3C	—
caglei	Yes	2	2	3C	—	1	Candidate
flavimaculata	Yes	1	2	2	Threatened	Threatened	Threatened
nigrinoda	Yes	3C	3C	3C	—	—	—
oculifera	Yes	3C	1	Threatened	Threatened	Threatened	Threatened
pseudogeo-graphica[a]	No	—	—	—	—	2	—
sabinensis	Yes	3C	3C	3C	—	—	—
versa	Yes	2	2	3C	—	—	—

Note: The first reviews of freshwater turtle species in the United States were announced 13 May 1977 (Dodd 1977b). Lists published in the *Federal Register* (USFWS 1982, 1985, 1989, 1991, 1994) initially placed candidate species in categories 1–3, with numbers denoting decreasing certainty of the likelihood of their eventual listing and 3C denoting those species that were found to be more abundant than originally thought. From 1996 onward, a reduced list of candidate species has been published (e.g., USFWS 1996).
[a]Applies to populations in Indiana, Missouri, Minnesota, North Dakota, and Wisconsin.

depends upon Corps of Engineers flood control plans."[74] His status review (Stewart 1985) cited a "near obsession of various agencies to conduct flood control measures in the Pearl River basin" (p. 7) and recommended threatened status. Counts of basking ringed sawbacks were augmented with captures, mostly by the nighttime technique pioneered by the Tulane field crews.

Basking surveys for the yellow-blotched sawback began in 1985. After the second, Stewart submitted a report stating that "this species probably does not warrant listing"[75] and recommending that it be placed in category 3C (more abundant/less threat than originally believed). His field supervisor wrote "I agree" at the bottom of the report, but the species was instead moved to category 2 (proposing to list may be appropriate). Stewart returned with Bob Jones of the Mississippi Museum of Natural History in 1989 to conduct capture-mark-resight population estimates and basking surveys. Population estimates were about half those of the ringed sawback, and the status review (Stewart 1989) concluded that the species warranted being listed as threatened.

The USFWS also sent questionnaires to herpetologists with knowledge of historic trends for ringed and yellow-blotched sawbacks. Public comments were solicited regarding the proposed listing of these species, but responses were few, and the listings have not been controversial. The primary habitat impacts that supported listing were largely different for the two species: the ringed sawback was deemed threatened by the negative impacts of completed and planned activities of the U.S. Army Corps of Engineers (river impoundment, channelization, floodplain clearing, and river bar and deadwood removal) and the Soil Conservation Service (drainage ditches increasing sedimentation), whereas the yellow-blotched sawback was deemed threatened by deadwood removal, gravel mining, toxic effluents, and planned reservoirs. For both species, threats were exacerbated by wanton shooting, past overcollection for scientific and educational purposes, and ongoing collection for the pet trade.

Recovery efforts have focused on the collection of basic biological information for both listed species and on the effects of toxic substances on hormone levels in yellow-blotched sawbacks. During the late 1980s and 1990s, Bob Jones studied reproduction and growth and estimated population sizes of ringed sawbacks and also studied movements and habitat use of yellow-blotched sawbacks. Robert Brauman, Brian Horne, and Megan Moore, graduate students in Rich Seigel's lab at the University of Southeastern Louisiana, conducted studies of the yellow-blotched sawback's reproductive ecology and diet and the effects of disturbance by humans, while Mary Mendonça at Auburn University has led the studies on toxic substances and hormone levels. Studies of the yellow-blotched sawback have more recently been conducted by Will Selman, a graduate student who worked with Carl Qualls at the University of Southern Mississippi.

Cagle's map turtle (*G. caglei*) had the longest tenure as a candidate species for federal listing. A member of the 1977 group proposed for further study, it became a category 1 candidate following a 1991 petition for listing submitted by Flavius Killebrew (USFWS 1993). Killebrew and a steady succession of his students at West Texas A&M University studied Cagle's map turtle intensively from the late 1970s through the 1990s. His listing petition emphasized the threats of existing and proposed reservoir projects, collection for the pet trade, and wanton shooting. Listing was delayed by the species' low listing priority (5 on a scale of 1 to 12, indicating high but not imminent threat to the species). Late in the first term of President George W. Bush (whose administration had greatly restricted the addition of new species to the federal list), the Center for Biological Diversity submitted a petition to USFWS to list Cagle's map turtle along with 224 other species that had languished for an average of seventeen years as candidates for federal protection. The effort was unsuccessful for Cagle's map turtle. The USFWS (2006) stripped the species of its candidate status, noting an increase in population estimates by Killebrew and his students and finding that the species was safeguarded by a lack of definite plans for new reservoirs on the Guadalupe River and protected against collection and shooting due to designation of the species as threatened by the state of Texas.

Barbour's map turtle was surveyed by the USFWS in 1992, but the recommendation was against listing it (Stewart 1992). Basking density was considered good on surveyed stream reaches (table 2.3), and the biggest concern was the effect of the pet trade, with state protection (at that time not yet instituted in

Table 2.3. Summary of survey work conducted by the U.S. Fish and Wildlife Service in the 1980s and early 1990s

Date	Method	River reach (approximate length)	Results	Target species basking density
		Ringed sawback (1984–87)		
13 June 1984	Basking counts	Pearl north of Ross Barnett Reservoir (22 km)	Abundance uncertain; species not distinguished	
21 June 1984	Basking counts	Pearl from Ross Barnett Dam south to Hwy. 25	1 G. oculifera; "most common turtle" was G. pearlensis	
10 July 1984	Basking counts	Pearl near Monticello (11 km)	18 G. oculifera, 12 G. pearlensis, 20–22 Graptemys sp.	1.6 G. oculifera/km
	Night capture		Capture of 16 G. oculifera, 3 G. pearlensis	
11 Oct. 1984	Basking counts	Pearl near Columbia (1 km)	28 G. oculifera, 27 G. pearlensis, 40 Graptemys sp.	28.0 G. oculifera/km
15 May 1985	Basking counts	Pearl near Ratliff Ferry (2 km)	~3:1 ratio of G. oculifera: G. pearlensis	
	Night capture		Capture of 17 G. oculifera, 5 G. pearlensis	
12 June 1985	Basking counts	Pearl near Bogalusa (11 km)	3 G. oculifera, 9 G. pearlensis, 1 Graptemys sp.	0.3 G. oculifera/km
	Basking counts	Bogue Chitto near Sun (5 km)	4 unidentified turtles	
30 April 1987	Daytime dipnetting	Pearl near Ratliff Ferry	Capture of 19 G. oculifera, 5 G. pearlensis	
	Night capture		Capture of 34 G. oculifera, 7 G. pearlensis	

		Yellow-blotched sawback (1985–89)		
26 June 1985	Basking counts	Leaf near New Augusta (4.8 km)	7 G. flavimaculata, 19 G. gibbonsi, 3 Graptemys spp.	1.5 G. flavimaculata/km
30 July 1985	Basking counts	Pascagoula near Wade (1.6 km)	27 G. flavimaculata, 14 G. gibbonsi, 25 Graptemys spp.	16.9 G. flavimaculata/km
31 July 1985	Basking counts	Pascagoula near Poticaw (1.2 km)	64 G. flavimaculata, 7 G. gibbonsi, "numerous" Graptemys spp.	53.3 G. flavimaculata/km
1 August 1985	Basking counts	Leaf near McLain (10.4 km)	7 G. flavimaculata, 2 G. gibbonsi, 6 Graptemys spp.	0.7 G. flavimaculata/km
3 April 1989	Capture-resight	Pascagoula River and Ward Bayou	106 G. flavimaculata and 1 G. gibbonsi marked; population estimates of 337 and 262 G. flavimaculata/river km, respectively	
24–26 April 1989	Basking counts	Lower Leaf and upper Pascagoula (48 km)	83 G. flavimaculata, 23 G. gibbonsi, 22 Graptemys spp.	1.7 G. flavimaculata/km
		Barbour's map turtle (1992)		
July 1992	Basking counts	Middle Apalachicola (21.4 km)	170 G. barbouri, 101 turtles (other species and unidentified)	7.9 G. barbouri/km
July 1992	Basking counts	Chipola Cutoff (6.4 km)	51 G. barbouri, 84 turtles (other species and unidentified)	8.0 G. barbouri/km
July 1992	Basking counts	River Styx (0.8 km)	4 G. barbouri	5.0 G. barbouri/km
1992	Basking counts	Upper Apalachicola (12.9 km)	98 G. barbouri, 149 turtles (other species and unidentified)	7.6 G. barbouri/km

Georgia) considered the more appropriate route to protecting the species. The Texas map turtle was surveyed widely throughout its range in the middle and late 1980s by Norm Scott, Jim Dixon, David Kizirian, and others. The USFWS Region 2 Herpetological Advisory Team concluded that the species was abundant and widespread enough that it did not warrant further concern (J. Dixon, pers. comm. 2007). While the Sabine map turtle was dropped from candidate status after 1989, the USFWS designated it a species of concern and funded a survey of the species in Louisiana by Steve Shively (2001) of the Louisiana Department of Wildlife and Fisheries Natural Heritage Program. He found good abundance, confirmed the species' continued existence in the Mermentau River (where it had not been reported since Gustave Kohn's day more than a century earlier), and reported some new localities for the species. There was little evidence of change in population density in the Whiskey Chitto River, where Shively had conducted his master's thesis two decades earlier.

Settlement of a recent lawsuit brought by the Center for Biological Diversity against the USFWS resulted in a decision to seek information regarding the potential federal listing of 374 aquatic species in the southeastern United States. Included among the species to be considered are 5 species of *Graptemys*: Alabama, Barbour's, Escambia, and Pascagoula map turtles and black-knobbed sawbacks (USFWS 2011).

For species threatened by commercial international trade, CITES (the Convention on International Trade in Endangered Species of Wild Flora and Fauna) is an international treaty signed by 172 countries to regulate and, in some cases, prohibit trade. Species may be listed on three appendices: Appendix I species require export and import permits for international shipment and may not be traded for commercial reasons; species are added to Appendix I (or removed from it) based on votes by delegates of member nations at the biennial Conference of the Parties (CoP). Appendix II species require export permits and are also designated by votes of CoP delegates. Export "quotas" (maximum levels of annual trade) may be assigned to some species on Appendix II. Species on Appendix III also require an export permit but are added to the list at the request of the exporting nation without a vote of the CoP delegates.

The USFWS proposed to list all *Graptemys* species on Appendix II due to concerns over trade (Howe and MacBryde 1996). The proposal was later modified to exclude three widespread species that initially had been included only because of similarity of appearance (Howe and MacBryde 1997). The fate of the modified proposal at the tenth CoP, in Zimbabwe, was a case history in how the noble purposes of CITES can be derailed by political interests that supersede conservation imperatives. Canadian delegates spoke in favor of the proposal, but delegates from the Netherlands, representing the European Union, spoke in opposition, suggesting Appendix III listing instead and noting that most international trade concerned two of the widespread species dropped from the proposal.[76] Delegates voted thirty-seven to nineteen in support of the proposal, meaning it fell one vote short of the two-thirds majority needed. Rejection was thus driven by delegates of the biggest importing nations in Europe. The following year, a representative from a Scandinavian reptile hobbyist group published a rather bizarre letter to the editor in *Herpetological Review* concerning his advisement of Danish CITES delegates (Bringsøe 1998). In it, he took the USFWS to task, alleging their proposal was an attempt to shift the blame for *Graptemys* declines from habitat modification to trade. He suggested that Appendix III listing would be inappropriate, for the same reason. He argued that international trade was not a problem for most species but failed to present information supporting his assertion. The obvious inference of his screed was that because the U.S. government and its citizens were responsible for species declines through habitat degradation, Europeans and others who value the species as collectible pets should not be punished by international restrictions on their hobbies! Bringsøe appeared to not consider the notion that unregulated collection for international trade would exacerbate declines rooted in problems with habitat.

The USFWS next proposed that all *Graptemys* species be listed on Appendix III (Lieberman and VanNorman 2000). The requirement for CITES export permits would, it was suggested, allow for increased attention to the extent of the trade and ensure that international trade would be conducted in accordance with all federal, state, and local laws. The USFWS also argued that Appendix III listing would allow for greater regulation of humane conditions for live shipment. The USFWS placed all *Graptemys* species on Appendix III effective 14 June 2006, following a comment period during which the main issues raised concerned how listing would affect commercial turtle farms (Maltese 2005). One advantage to the listing is that it requires that all declarations of live map turtles and sawbacks shipped from the United States be identified at the species level, rather than lumped under the name "*Graptemys* spp.," the label used for the majority of prior exports (see chapter 5).

My Own Studies, 1991 to Present

As a boy growing up in Oklahoma City and interested in herpetology, I may have been predisposed toward my later interest in the map turtles and sawbacks by one particular experience. My family took frequent canoe excursions on the Illinois River in eastern Oklahoma. On one trip, I remember sliding downstream with my siblings in two canoes, our jaws dropping open as we passed within a few meters of a cluster of branches that held about twenty small, wide-eyed turtles that waited until the last possible minute to evacuate into the water. Only years later would I realize that they must have been *Graptemys*, most likely male Ouachita map turtles.

My first research involving *Graptemys* was my doctoral study of community ecology, an area in freshwater turtle studies that has not been particularly well

Figure 2.20. The old Highway 596 bridge on the Pascagoula River at Merrill, Mississippi, just below the confluence of the Leaf and Chickasawhay Rivers.

explored. I was an assistant professor at Madisonville Community College in western Kentucky, my summers consumed by a joint doctoral program of Murray State University and the University of Louisville. Ouachita and Mississippi map turtles were among my five focal species of common, medium-sized turtles in Nickell Cove on Kentucky Lake, together with the slider *Trachemys scripta*, the river cooter *Pseudemys concinna*, and the smooth softshell *Apalone mutica*. I compared the species for microhabitat use, diet, and diel and seasonal use of time, with an emphasis on the effects of competition, phylogeny, and morphology on resource use.

In July 1991 I took a quick trip to Mississippi to look for some of the southern *Graptemys* that had so intrigued me when I had bought my new Peterson field guide that summer (see chapter 1). Using an old pair of 7× binoculars, I spotted all three sawbacks and a few specimens I recorded as Alabama map turtles, following the taxonomy of the day. The highlight of the trip was a stop at the old Highway 596 bridge at Merrill on the Pascagoula River (fig. 2.20), which remains one of my favorite turtle spots to this day. At that time it was closed to traffic but available to pedestrians. Although it was early in the evening and I expected to see few turtles, I got up-close views from above of nine beautiful yellow-blotched sawbacks. If not already hooked on *Graptemys*, I certainly was as of that evening.

Prior to my pilot study season in 1992, I purchased a small spotting scope with a zoom lens that went up to 36×. I used it in 1992 to select Nickell Cove as my study site for my dissertation study in Kentucky, but the scope was also a boon to my southern travels. That first year, I took it on a spring break trip to the Pearl and Pascagoula drainages and then on a long, winding August tour through Mississippi, Louisiana, Arkansas, and Texas on my way to a herpetology conference in El Paso. Because the scope had only a twelve-inch tripod, I carried a tall wooden barstool with me on which to balance it, then leaned over awkwardly and tilted my head back, developing a chronic crick in my neck in the process. My older daughter, Ursula, then four, accompanied me on the spring trip, the highlight of which was standing on an inner-bend sandbar of the upper Pearl near Carthage and recording twenty ringed sawbacks and seven "Alabama" map turtles. (Later that year I learned that we had seen the newly christened G. gibbonsi—since split from that species as G. pearlensis—and that I would need to go farther east some time later to add "real" G. pulchra to my life list.)

The summer trip was an extraordinary tour of Graptemys diversity, with eight species seen. Turtles seen through a spotting scope are quite unlike those fished out of the netting of a trap or pictured in a typical field guide, outside their element. In the scope's field of view, they pose with necks extended to their full reach, turn their heads from side to side to scan for danger, gape widely, stretch out their limbs and digits (postures I call "turtle yoga"), and in general add a nice, bioaesthetic touch to any aquatic environment. Over the next several summers in western Kentucky, Ursula and her younger sister, Veronika, would sometimes feign boredom and disinterest when Dad set up the scope, but once turtles were actually in my field of view ("Hey, girls, two common map turtles and a slider"), they always jockeyed to take first turn at the eyepiece. As much as I enjoyed simply watching the turtles bask—yes, watching turtles dry is more interesting than watching paint dry—the data-collection possibilities of the spotting scope dominated my thoughts.

The summer of 1993 was my first year of formal data collection for my dissertation. Funding from Austin Peay State University bought me a new scope with a zoom up to 60× and a full-height tripod. With my focus on my dissertation project, by no means was I prepared to take a week off in June to attend an international conference on turtle conservation at State University of New York at Purchase, yet I did so anyway. I was sure that I would hear much about the conservation of the two federally listed sawbacks I had seen on my southern trips. I was surprised that the only presentation at the week-long conference having to do with Graptemys was a poster presentation by Mike Goode on captive breeding at the Columbus Zoo. Bowled over, I assumed (falsely, as it would turn out) that this meant little work was being conducted on the two listed species. I resolved to complicate my professional life for the next few years by simultaneously doing my doctoral work and completing a spotting-scope study of their range-wide abundance and its correlation with deadwood abundance. I obtained funding for travel to Mississippi through a small grant from the Chelonian Research Foundation (Lindeman 1998). Meanwhile, I discovered that Bob Jones of the Mississippi Museum of Natural History was well along in his studies of the species' life history and habitat use, but by that time I was hooked on the idea of conducting my own abundance surveys. Bob helped me with advice on selection of sites and much information about the rivers and their management.

Periodically during the next two summers, when the weather down south seemed promising, I would leave my Kentucky study site for southern Mississippi, where four days of zigzagging from river crossing to river crossing were just enough to count turtles and deadwood at the forty-one sampling points I had established. For comparison's sake, I also included my two species in Kentucky; those data could at least be taken around my trapping schedule at Nickell Cove. Sampling days were a mad scramble from nine in the morning to three in the afternoon, with no time for stops except to count turtles. (The riverbank underneath a bridge makes for a good urinal in a pinch.) While I scanned the river I read data on turtles and deadwood into a tape recorder, unfazed by the puzzled looks and comments of the locals fishing from the banks or unloading their boats. One day, I drove half an hour from the Chunky River to the upper Pearl River. Climbing out of the car, I found to my horror that my scope was missing from its spot in the backseat. I drove back to the Chunky, my knuckles white on the steering wheel, to find the scope and tripod standing in the sand next to the river. A curious passerby had removed the lens cap and eyepiece protector to have a look but, fortunately, was not so enamored of turtle watching as to declare "finders keepers."

I was initially drawn to researching *Graptemys* ecology by the intriguing patterns of species coexistence and dietary diversity within the genus. At one early point I sat down with all the old Cagle publications to write out a summary of the published information on relative head and alveolar width. A lack of conformity in data reporting among his studies and others led me to add to my dissertation with a study based on measurements of nearly twenty-three hundred museum specimens. It seemed odd to me that so little had been done regarding anatomical adaptation to diet and its implications for ecology in *Graptemys* in the four decades between Cagle's work and my own, but I was grateful to have the chance to conduct the study myself. Females fell into three nonoverlapping groups with regard to relative head and alveolar widths, and radiation within the group suggested ecological character assortment more strongly than evolutionary character displacement. A summer appointment in a program that paired research-oriented community college faculty with researchers at the University of Kentucky introduced me to Mike Sharkey, a wasp systematist, who introduced me to comparative analyses. We analyzed evolutionary changes in

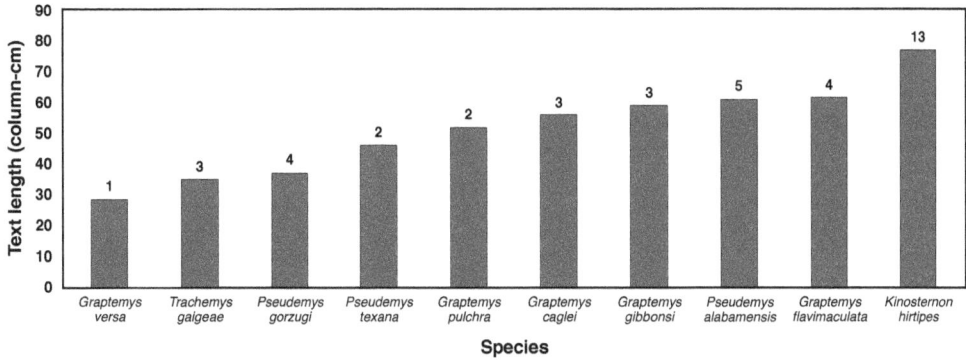

Figure 2.21. The ten shortest species accounts in C. H. Ernst, Lovich, and Barbour's (1994) *Turtles of the United States and Canada*. Above each bar is the number of published ecological fieldwork papers that they cited for that species.

relative head and alveolar width and confirmed the greater correlation of these measurements in female turtles than in males.

With my doctoral work wrapping up at Louisville in the spring of 1997, I had decided my next move would be to fill in a gap in our basic knowledge of *Graptemys* diet and life history. (By that time I was completely sold on devoting my focus to this genus, which I felt might dominate my research efforts for the rest of my career; so far, it has.) The shortest of the fifty-two species accounts in C. H. Ernst, Lovich, and Barbour's (1994) *Turtles of the United States and Canada* was the entry for the Texas map turtle, *G. versa* (fig. 2.21), so I obtained another small grant from the Chelonian Research Foundation to travel to Texas (Lindeman 2001a). Earlier travel to museums in Texas had acquainted me with a series of forty-two specimens that Bryce Brown of Baylor University and two assistants had collected by hand while wading the South Llano River one morning in 1949.[77] I reasoned that dissections would allow me to compare diet and life history before and after the invasion of Texas rivers by Asian *Corbicula* clams, which females in my Kentucky population of Mississippi map turtles were feeding upon greedily. I also wondered if hand capture at that pace was still possible five decades later.

Following the last final exam I would ever need to take, I made piles of the gear I planned to take to Texas the next day. I needed a bike ride to clear my head from the academic ordeal. I was riding faster than usual, glad to be leaving my academic world to return to fieldwork, when I ran off a sidewalk alongside a major road. I tried to steer from the grass back onto the sidewalk but hit its edge, flew over the handlebars, and landed on my right shoulder, breaking off the distal two inches of my right clavicle and shattering it into pieces. As I rolled over, sat up, and put my left hand to the shattered shoulder, my first clear thought was that the Texas fieldwork would have to be put on hold for a year. So,

Figure 2.22. Screen-walled barracks at the Texas Tech University Center in Junction, which served as field accommodations during a study of the Texas map turtle in 1999 and 2000.

in May 1998, after a short visit to the Strecker Museum at Baylor for dissections and a visit with Bryce Brown at his home, I finally arrived at the South Llano River, eager to wade the shallow, clear water to look for turtles. In a short wade that first evening I caught only a stinkpot (*Sternotherus odoratus*), but I hoped it would be a good portent of things to come. The next morning I did capture a male Texas map turtle by hand, but ultimately it was good I had brought my traps, as the decades had brought changes to the South Llano and hand captures were now few and far between.

Subsequent years spent studying Texas map turtles included my first opportunities at significant involvement of my students in my research. In 1999 I returned to the South Llano with four students from Madisonville Community College. At the Texas Tech University Center in Junction, we stayed in open-air cabins, which were former military barracks (fig. 2.22) made infamous by the football coaching legend Paul "Bear" Bryant. As the new head coach at Texas A&M in September 1954, Bryant took his team far away from the program's boosters and other distractions to the Junction campus, where he oversaw tough workouts in temperatures exceeding 100°F (38°C) each day. The brutal drought-induced weather caused all but thirty-five players to quit the team, most of them sneaking into the town bus station under cover of darkness. Two years later, the "surviving" Aggies won a conference championship.

On our third night in Junction, we narrowly escaped adding to the infamy of the barracks. As a violent storm rolled in, we quickly learned how to unroll the canvas flaps to cover the screened walls. After the wind began to howl and the flaps beat wildly against the screens, we had to run back outside into a driving rain to find the tie-downs for the flaps. The next morning we learned that an area thirty miles away had been devastated by a category 4 tornado.

The turtle catching was excellent and came with an unexpected bonus. We were collecting data on habitat in order to compare male and female trapping sites, but I was unsure where I could go with the comparison beyond simply documenting the extent to which the sexes differed. Fortune shone, however, as we captured eight young females that overlapped our sample of males in body size: here was a group of turtles with male-like body sizes but female-like dietary habits, so I could ascertain whether size-related swimming ability or diet was behind sexual differences in diet.

In 1999 I took my present position at Edinboro University of Pennsylvania and soon discovered the abundant common map turtle population at Presque

Figure 2.23. The author at a popular basking log in Graveyard Pond, Presque Isle State Park, Pennsylvania, where more than 1,500 common map turtles have been marked since 1999. Presque Isle is the type locality for the first species of *Graptemys* to be described, a coincidence that was discovered several years after the inception of the mark-release study there.

Isle on Lake Erie. My work with students has continued at Edinboro, primarily with undergraduates working on independent-study assignments. Several students assisted me in studies of the turtles' reproduction and consumption of invasive mussels. More recently, undergraduate projects have resulted in publications: Danielle Collins analyzed the relationship between trophic morphology and Asian clam shell size in Texas map turtles, Jim Patterson examined predation on invasive mussels by stinkpots at Presque Isle, and Kathleen Ryan examined patterns of reproductive allometry in common map turtles. Following her work on that project, Kathleen and I became partners more generally—we began raising our combined six children together and were even engaged for a time. The marriage did not take place and our blended family has come undone, but I still occasionally see Kathleen, who worked for several years as a park naturalist at Presque Isle, and we remain friends.

For the first six summers of trapping on Presque Isle, I had no idea of the historical significance of the site in the history of *Graptemys* studies (nor did anyone else, apparently). Early work on this volume led me to research more about who LeSueur was. I discovered his handwritten itinerary in the LeSueur collections of the Muséum d'Histoire Naturelle in Le Havre, which I was able to view online at the museum's website, and was able to put two and two together and determine that he had described the common map turtle based on collecting at Presque Isle (Lindeman 2009b). With a database now tallying over fourteen hundred animals marked and hundreds of recaptures at LeSueur's type locality (fig. 2.23), I delighted in discovering that I have been working at the site where *Graptemys* biology began nearly two centuries ago. The advent of a third century of *Graptemys* investigations now quickly approaches, powered by a strong new wave of interest in the genus in just the last few years, once again led by a cohort of graduate students. I anticipate that great strides will be made in our appreciation and understanding of this fascinating group of turtles in the coming decades.

Notes

1. Georg Baur to G. Brown Goode, 3 Oct. 1890, Leonhard Stejneger and Georg Hermann Baur Collection, ca. 1850–1943, Smithsonian Archives Record Unit 7441 (hereafter cited as Smithsonian Archives).

2. Gustave Kohn to Baur, Oct. 1890, Smithsonian Archives.

3. Baur to Stejneger, 11 Nov. 1890; Smithsonian Archives.

4. Ibid., 1 Dec. 1890.

5. Stejneger to Baur, 4 Dec. 1890, Smithsonian Archives.

6. Baur to Stejneger, 11 Dec. 1890, Smithsonian Archives.

7. Kohn to Baur, 2 Oct. 1891, 3 Nov. 1891, and 14 Nov. 1891; Baur to Stejneger, 12 Mar. 1892; undated "Catalogue of Tortoises Received from Gustave Kohn, New Orleans, La." in Baur's handwriting; all Smithsonian Archives.

8. Photographs, Smithsonian Archives.

9. U.S. National Museum specimen catalog.

10. Kohn to Baur, 14 Oct. 1891, Smithsonian Archives.

11. Baur to Stejneger, 11 Dec. 1890, Smithsonian Archives.

12. Baur, unpublished manuscript, "Preliminary List of the Testudinata of North America," Smithsonian Archives.

13. Kohn to Baur, 20 Apr. 1892, Smithsonian Archives.

14. Undated "Catalogue of Tortoises received from Gustave Kohn, New Orleans, La." in Baur's handwriting, Smithsonian Archives.

15. Kohn to Baur, 3 Nov. 1891, Smithsonian Archives.

16. Ibid., 20 Nov., 25 Nov., 8 Dec., and 14 Dec. 1893.

17. Baur, unpublished "The Testudinata of North America," 29 Oct. 1894; unpublished, undated, and untitled manuscript consisting of species accounts; both Smithsonian Archives.

18. Kohn to Baur, 19 Apr. 1894, Smithsonian Archives.

19. Baur, unpublished, undated, and untitled manuscript consisting of species accounts, Smithsonian Archives.

20. U.S. National Museum specimen catalog.

21. Baur, unpublished, undated, and untitled manuscript consisting of species accounts, Smithsonian Archives.

22. Baur to Stejneger, 6 Feb. 1893, Smithsonian Archives.

23. Kohn to Baur, 20 Nov. 1893, 25 Nov. 1893, 8 Dec. 1893, and 14 Dec. 1893, Smithsonian Archives.

24. Ibid., 19 Apr. 1894.

25. Mary Cagle Overbey, telephone interview, 27 Feb. 2008.

26. Fred Cagle, Jr., telephone interview, 9 Oct. 2007.

27. Fred Cagle field notes, Cagle Archives, Savannah River Ecology Laboratory.

28. Kohn to Baur, 20 Nov. 1990, Smithsonian Archives.

29. Allan Chaney, interview, 7 Aug. 2006, Kerrville, Tex.

30. Fred Cagle to Archie Carr, 16 Mar. 1948 and 5 May 1948, Archie Carr Archives, University of Florida.

31. Tulane University Museum of Natural History specimen catalog for herpetology.

32. Ernie Liner, interview, Jul. 2006, New Orleans, La.

33. Chaney, interview.

34. Fred Cagle field notes, Cagle Archives, Savannah River Ecology Laboratory.

35. Chaney, interview.

36. "Tulane to Start Zoology Course for Graduates," New Orleans States-Item, 23 Jun. 1949, p. 6.

37. Richard Etheridge, interview, Jul. 2006, New Orleans, La.

38. Fred Cagle field notes; Cagle Archives, Savannah River Ecology Laboratory.

39. Ibid.

40. Bob Webb, telephone interview, Sep. 2006.

41. Fred Cagle field notes; Cagle Archives, Savannah River Ecology Laboratory.

42. Ibid.

43. Ibid.

44. Ibid.

45. George Penn to Carr, 22 April 1954, Archie Carr Archives, University of Florida.

46. Fred Cagle field notes, Cagle Archives, Savannah River Ecology Laboratory.

47. Ibid.

48. Ibid.

49. Ibid.

50. Ibid.

51. Webb, telephone interview.

52. Whit Gibbons, interview, Jul. 2006, New Orleans, La.

53. Ibid.

54. Liner, interview.

55. Etheridge, interview.

56. Chaney, interview.

57. Fred Cagle, Jr., telephone interview.

58. "Editorial Notes and News," Copeia 1961, p. 259.

59. "TU Prof Charges Billions Wasted," New Orleans States-Item, 17 Aug. 1962, p. 13.

60. "Tulanian Visits USSR Institutes," New Orleans Times-Picayune, 30 Aug. 1958, p. 6.

61. Edwin Hoag, "Scientific Labs, Cold War Front Moving Closer," New Orleans States-Item, 26 Jun. 1962, p. 21.

62. Fred Cagle field notes, Cagle Archives, Savannah River Ecology Laboratory.

63. Liner and Etheridge, interviews, Jul. 2006; Clarence Smith, telephone interview, Sep. 2007.

64. Fred Cagle, Jr., interview.

65. Smith, telephone interview.

66. Chaney, interview.

67. Ibid.

68. David Haynes to author, 30 Nov. 2008.

69. Jeff Lovich to author, 3 Nov. 2008.

70. Dick Vogt, interview, Jul. 2006, New Orleans, La.

71. Ibid.

72. Ibid.

73. James Stewart to USFWS, internal memo, 25 Jun. 1984.

74. Ibid., 6 Dec. 1984.

75. Ibid., 27 Aug. 1985.

76. "Summary Report of the Committee/Meeting," CITES, accessed 28 Feb. 2013, http://www.cites.org/eng/cop/10/E10-ComI.pdf.

77. Bryce Brown, interview, 12 May 1998, Waco, Tex.

3

Evolutionary History and Relationships

The historical progression of phylogenetic studies of the map turtles and saw-backs is a microcosm of the development of general methods and philosophies of phylogenetic systematics. The first two species described predated the Darwinian revolution in systematics; they were assigned to exceptionally diverse turtle genera linked by superficial similarities. By the time more taxa were described, shortly after acceptance of evolution and its influence on classification, turtles had been split into numerous families and genera. Initial attempts at inferring relationships in *Graptemys* and their relationships to other genera show the influence of the evolutionary taxonomy school of thought; this has been replaced by the cladistic school of thought, which has guided more recent studies in *Graptemys* phylogeny. Early attempts at inferring the phylogeny of the group lack the rigor characteristic of more recent studies, rigor that has increased with the growth of the computational methodology for using larger matrices of character data to infer trees.

Fossils have played little role in elucidating *Graptemys* evolution due to the apparent recent divergence of species and a lack of taxonomic diversity among fossil material. Biogeographic hypotheses have been developed recently, but their confirmation and further development await improved knowledge of river drainage histories and improved resolution of relationships within the three clades of the genus.

Relationship of *Graptemys* to Other Genera

The development of the earliest views on the affinities of the map turtles and sawbacks is reflected in a winnowing down of their potential relatives through the generic assignment of each species as it was described. The common map turtle, the first species described, was named *Testudo geographica* (LeSueur 1817), joining the *Ur*-genus of turtles that in the early nineteenth century included taxa as diverse as tortoises, sea turtles, softshells, snapping turtles, sideneck turtles, and various species classified today as emydids and geoemydids (F. W. King and Burke 1989; Adler 2007). Say (1825) moved the species to *Emys*, to which the second species described, the false map turtle, was also assigned, as *Emys pseudogeographica* (J. E. Gray 1831). Holbrook (1842) likewise placed both in *Emys*; thus they joined a diverse genus that excluded tortoises and sea turtles but once contained species representing most major families of freshwater turtles recognized today (F. W. King and Burke 1989; Adler 2007). Modern recognition of several turtle families and scores of genera had begun to take form by the late 1800s, when the next map turtles and sawbacks were described (Adler 2007). Agassiz (1857a) first used the generic name *Graptemys* for three taxa (one of which was later deemed a junior synonym; see *G. geographica* and *G. pseudogeographica* accounts, chapter 8). Cope (1875) listed the two species in *Malacoclemmys*, a generic assignment that allied them to the diamondback terrapin. Baur (1890) described two new taxa, *kohnii* and *oculifera*, as members of *Malacoclemmys*, and Hay (1892b) also discussed the five species as congeners under their original and thus valid name, *Malaclemys*. In Baur's next description, however, he followed Agassiz in separating *Graptemys* from *Malaclemys*, naming *Graptemys pulchra* and noting differences in the skull between the genera (Baur 1893); he had apparently convinced himself of the distinctiveness of *Graptemys* within a month of his 1890 publication (see chapter 2). Hay (1908) also used the name *Graptemys* but suggested the genus had originated from *Malaclemys*. All subsequent map turtle and sawback species were placed in *Graptemys* at the time of their description, between 1925 and 2010 (F. W. King and Burke 1989; Ennen, Lovich, et al. 2010).

Genera of the family Emydidae have been a fluid group. Opinions of intergeneric relationships have changed as better evidential data matrices and more advanced forms of phylogenetic analysis have been applied. Cladistic principles of classification, which recognize only monophyletic clades as named taxa, have had substantial influence on the taxonomy of Emydidae and related families.

In the mid-1900s, Emydidae was a large group composed of freshwater and terrestrial turtles that are most diverse in North America and southeastern Asia but also found in parts of Central and South America, the Caribbean, Europe, and the Indian subcontinent. The first emydid phylogenetic tree (Loveridge and Williams 1957) arranged twenty-six genera on a tree that is antiquated in its arrangement of taxa, with several generic labels placed along branches or at ancestral nodes, rather than all being "tip" nodes as today's standards would require (fig. 3.1); presumably such groups were being depicted as potentially being paraphyletic. The tree was based on degree of development of carapacial buttresses, number of carapacial keels (zero, one, or three), presence or absence of carapacial "surface sculpture," position of the entoplastron relative to the humeropectoral sulcus, width and ridging of the alveolar surfaces of the jaws, and presence or absence of plastral hinges. From *Pseudemys*

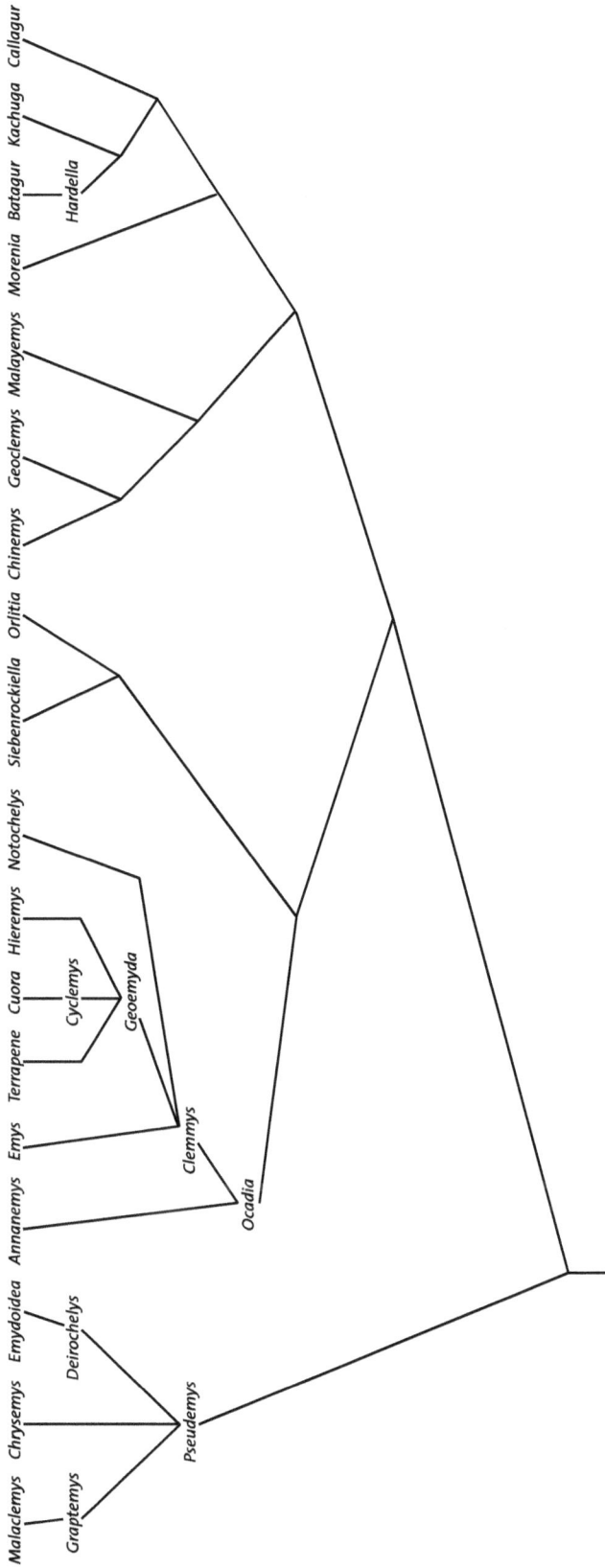

Figure 3.1. Phylogeny of genera in the Emydidae (*sensu lato*) based on seven morphological characters (redrawn from Loveridge and Williams 1957).

(then composed of cooters, redbellies, and sliders), three groups originate: a *Deirochelys* branch leading to an *Emydoidea* branch (chicken and Blanding's turtles), a *Chrysemys* branch (painted turtles), and a *Graptemys* branch leading to a *Malaclemys* branch. The other North American genera (*Clemmys*, then composed of wood, bog, spotted, and western pond turtles, plus several Old World species; and *Terrapene*, box turtles) appeared mixed among Old World genera on the other major branch of the tree.

McDowell (1964) split *Clemmys* into New World and Old World groups, with the former retaining the genus name, on the basis of morphological characters that he also used to split Emydidae into the subfamilies Emydinae (all New World genera except *Rhinoclemmys*, plus the European pond turtle, *Emys*) and Batagurinae (all Old World genera except *Emys*, plus *Rhinoclemmys*). The subfamilies were elevated to family rank by Gaffney and Meylan (1988). Their arrangement of the genera in these two families has remained stable even as other changes concerning priority of names and paraphyly have been noted (W. P. McCord et al. 2000; Spinks et al. 2004). McDowell's (1964) placement of *Graptemys* in a monophyletic clade also composed of *Emys orbicularis* from Europe and the North American genera *Chrysemys, Clemmys, Deirochelys, Emys, Emydoidea, Malaclemys, Pseudemys, Terrapene,* and *Trachemys,* plus the recently resurrected genera *Glyptemys* and *Actinemys* (Feldman and Parham 2002; Stephens and Wiens 2003), was supported in morphological analyses (Gaffney and Meylan 1988). Since the publication of Gaffney and Meylan's phylogeny, subsequent analyses of emydid relationships have generally accepted the monophyly of Emydidae *sensu stricto* and rooted trees using geoemydids as outgroups (Bickham et al. 1996; Stephens and Wiens 2003; Wiens et al. 2010). Emydid monophyly was supported in morphological and molecular analyses using three emydid species (*Graptemys pseudogeographica, Trachemys scripta,* and *Actinemys marmorata*) and representatives of every other major clade of living turtles (Shaffer et al. 1997).

In Gaffney and Meylan's (1988) morphological study, the Emydidae (*sensu stricto*) were split into two reciprocally monophyletic subfamilies, Emydinae and Deirochelyinae. Subsequent phylogenetic analyses incorporating molecular characters and an increased number of morphological characters have corroborated a basal split within the emydid clade (Bickham et al. 1996; Shaffer et al. 1997; Stephens and Wiens 2003; Wiens et al. 2010). *Graptemys* occurs in Deirochelyinae, with *Chrysemys, Deirochelys, Malaclemys, Pseudemys,* and *Trachemys.* Deirochelyinae was diagnosed by Gaffney and Meylan on the basis of five morphological synapomorphies and the larger size of females. Emydinae contains *Actinemys, Clemmys, Emys, Emydoidea, Glyptemys,* and *Terrapene* and was diagnosed on the basis of two skull synapomorphies.

The sister group to *Graptemys* has generally been acknowledged to be the diamondback terrapin, *Malaclemys terrapin,* at least as far back as the lumped genus (Cope 1875; Baur 1890; Hay 1892b). Loveridge and Williams (1957) placed *Malaclemys* as having originated from *Graptemys* (perhaps rendering *Graptemys* paraphyletic; see fig. 3.1). McDowell (1964) placed both genera back together under the name *Malaclemys*, failing to find a gap sufficient for separating them. R. C. Wood (1977) suggested that *Graptemys* might be polyphyletically derived from a widespread *Malaclemys* ancestor via separate invasions of coastal river systems from the brackish-water coastal habitats of *Malaclemys*. His principal evidence was the similarity in keeling and coloration of the carapace between *Malaclemys* populations and *Graptemys* species along the Gulf Coast, together with an inability to envision freshwater connections among drainages that would have allowed dispersal, isolation, and speciation within *Graptemys*.

Dobie (1981) examined the morphology of *Malaclemys* and all species of *Graptemys* then recognized and argued strongly against combining the genera. He found seven putative synapomorphies, absent in *Malaclemys*, that supported a monophyletic *Graptemys*: (1) a rounded symphysis of the lower jaw, (2) bordering of the foramen palatinum posteriosus by the palatine rather than the maxilla, (3) loss of a medial notch in the premaxillae, (4) double notching of some peripheral bones of the carapace, (5) extensive overlap of the nuchal bone by the first pleural scute, (6) confinement of the anterolateral border of the first vertebral scute to the nuchal bone, and (7) ridges on the inner surface of the first and fifth costal bones. Similarly, Dobie identified thirteen putative autapomorphies in *Malaclemys terrapin* not shared by any *Graptemys*. In analyzing mitochondrial DNA (mtDNA) sequences of *Graptemys* and *Malaclemys*, Lamb and Osentoski (1997) explicitly rejected R. C. Wood's (1977) hypothesis of a polyphyletic origin of the former. They found 99 percent bootstrap support for a monophyletic *Graptemys* that included all recognized species, with twenty-six synapomorphic nucleotide positions, versus 93 percent bootstrap support for a monophyletic *Malaclemys* (represented by populations of five of the seven named subspecies), with fourteen synapomorphic positions.

Gaffney and Meylan (1988) attempted to identify morphological synapomorphies to produce an emydid cladogram. They placed *Graptemys* in an outgroup position relative to four deirochelyine genera but they expressed uncertainty as to whether *Malaclemys* was sister to *Graptemys* or another basal offshoot of the deirochelyine lineage (fig. 3.2). Based on twenty-one morphological, two protein, three parasite, and two ecological characters, Seidel and Jackson (1990) produced a phylogeny of deirochelyines that had *Graptemys* sister to *Malaclemys* with an unresolved relationship to *Pseudemys* and *Trachemys* (fig. 3.3). Bickham et al. (1996) used sequence data from the 16S RNA gene of mtDNA to examine emydid phylogeny, with greater coverage of emydines (all species included) than of deirochelyines (one species per genus). Their analysis placed *G. geographica* sister to *T. scripta* with 71 percent bootstrap support, with their clade sister to *Malaclemys* with 96 percent bootstrap support (fig. 3.4).

Stephens and Wiens (2003) published an emydid phylogeny based on a more

Terrapene Emydoidea Emys Clemmys Graptemys Malaclemys Chrysemys Deirochelys Trachemys Pseudemys

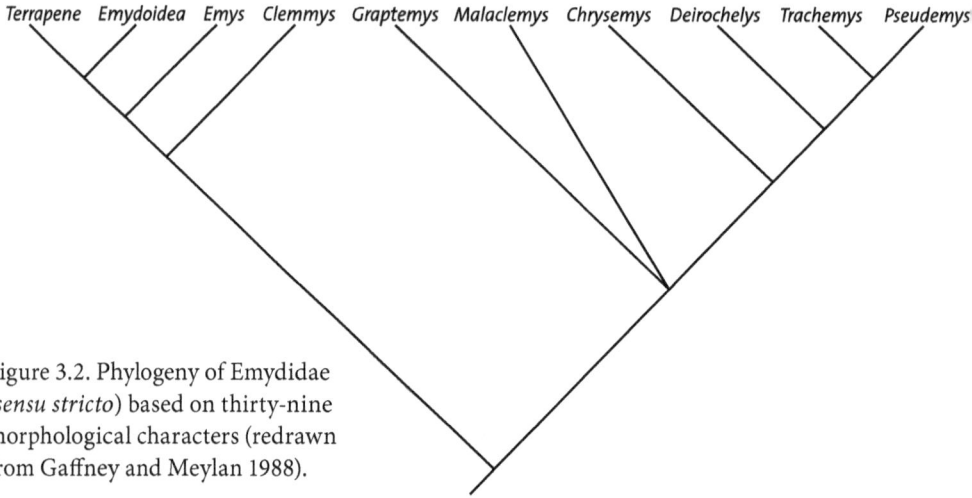

Figure 3.2. Phylogeny of Emydidae (*sensu stricto*) based on thirty-nine morphological characters (redrawn from Gaffney and Meylan 1988).

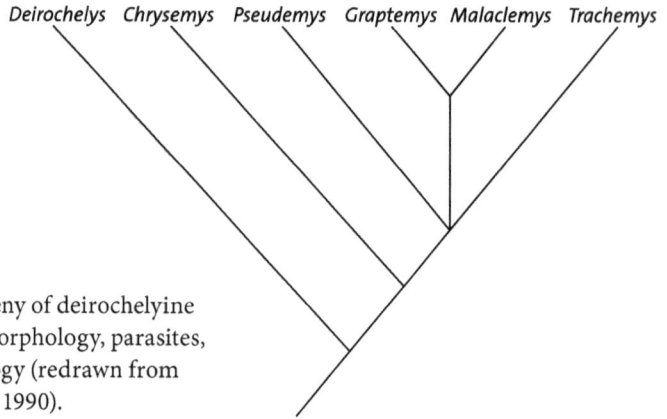

Deirochelys Chrysemys Pseudemys Graptemys Malaclemys Trachemys

Figure 3.3. Phylogeny of deirochelyine genera based on morphology, parasites, proteins, and ecology (redrawn from Seidel and Jackson 1990).

extensive data matrix that included all but one emydid species, many of which were represented by every named subspecies, with 237 morphological characters and 547 parsimony-informative mtDNA nucleotide positions. Combined analyses placed *Graptemys* and *Malaclemys* as sister taxa with the following successive outgroups: a paraphyletic *Trachemys*, then *Pseudemys*, *Chrysemys*, *Deirochelys*, and the six genera of Emydinae (fig. 3.5). Bootstrap support was above 50 percent for only four of the six successive clades containing the map turtle genus: *Graptemys* (81 percent), *Graptemys* + *Malaclemys* (75 percent), the five genera of Deirochelyinae without *Deirochelys* (82 percent), and the Deirochelyinae as a whole (73 percent; based on their fig. 7). Subsequent analysis of an expanded data set corroborates the intergeneric relationships (Wiens et al. 2010; see their online supplemental fig. 13).

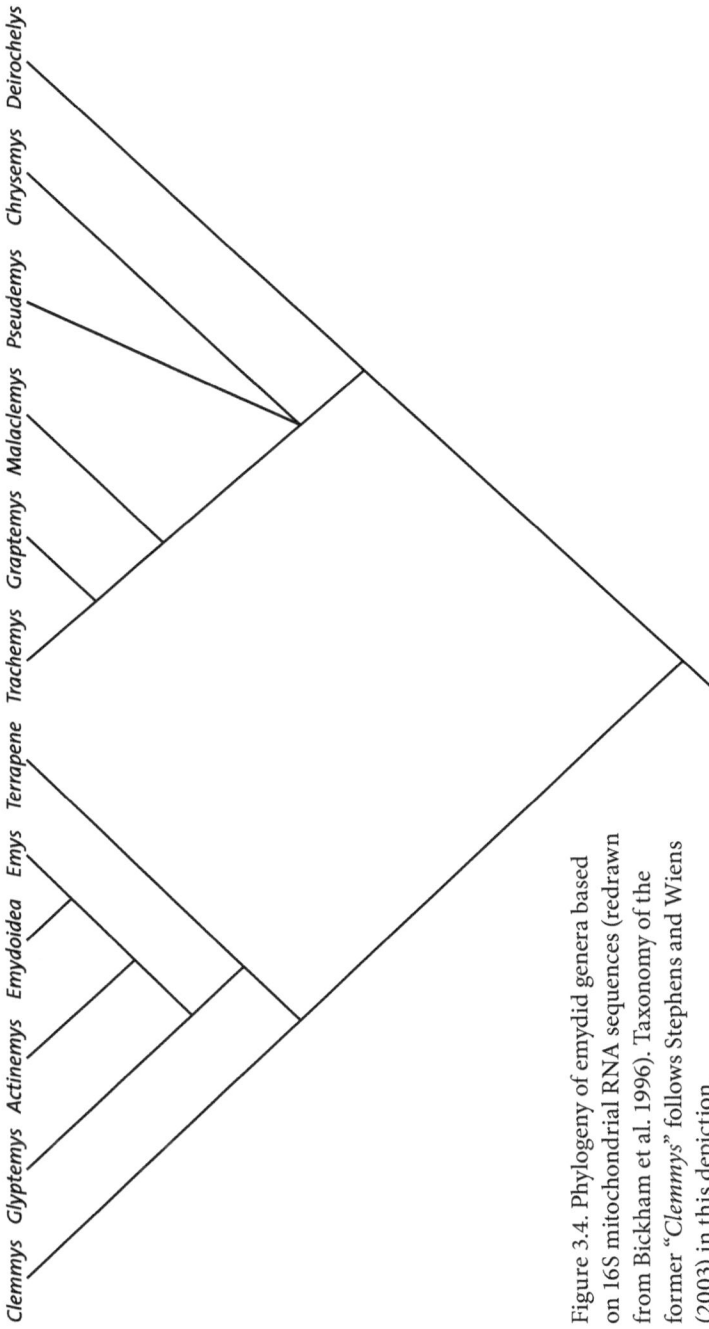

Figure 3.4. Phylogeny of emydid genera based on 16S mitochondrial RNA sequences (redrawn from Bickham et al. 1996). Taxonomy of the former "*Clemmys*" follows Stephens and Wiens (2003) in this depiction.

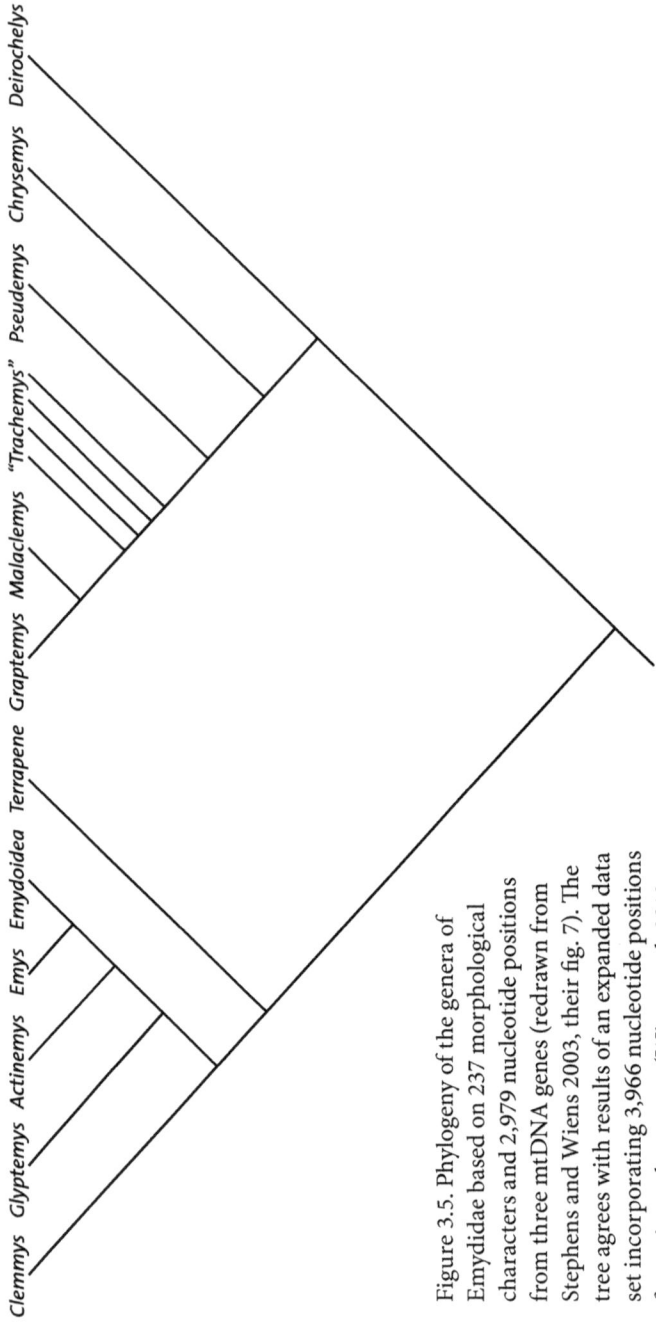

Figure 3.5. Phylogeny of the genera of Emydidae based on 237 morphological characters and 2,979 nucleotide positions from three mtDNA genes (redrawn from Stephens and Wiens 2003, their fig. 7). The tree agrees with results of an expanded data set incorporating 3,966 nucleotide positions from six nuclear genes (Wiens et al. 2010, their online supplemental fig. 13).

Intrageneric Phylogeny of *Graptemys*

As is the case with intergeneric relationships, early discussion of the intrageneric relationships of map turtles lacks the phylogenetic rigor of more recent attempts. Early attempts used small numbers of characters in an attempt to define "groups" of species within *Graptemys*, with little discussion of phylogenetic relationships within or among such groups. The first attempt to draw a phylogenetic tree for the genus was based on poorly differentiated molecular characters and a few morphological characters and included one species name placed on a branch leading to three other names (McKown 1972); since then, more rigorous attempts have produced largely dichotomous trees with mixed levels of support for the clades they include (Lamb et al. 1994; Lamb and Osentoski 1997; Stephens and Wiens 2003; Wiens et al. 2010).

Early authors noted similarities and differences among species as the number of described *Graptemys* grew, but they focused on species delimitation rather than relationships. In describing *G. barbouri*, A. Carr and Marchand (1942) seemed on the verge of placing it among the five congeners then recognized (with *kohnii* then separated from *pseudogeographica*) but fell short of doing so, finding *G. barbouri* at once "intermediate" between *G. geographica* and *G. pseudogeographica* and closest in head markings to *G. pulchra* but otherwise "very different" from that species, which was poorly known then (p. 100).

Cagle's Tulane field crews amassed large numbers of *Graptemys* specimens, which led to clarifications of species and descriptions of four new taxa. He made the first attempts to sort species into allied groups, showing the evolution of his thinking about *Graptemys* phylogeny (Cagle 1952b, 1953a, 1953b, 1954). He recognized *barbouri* and *pulchra* (*sensu lato*) as sisters based on morphological and ecological similarities and their distribution in adjacent rivers (Cagle 1952b). He commented on the uniqueness of *oculifera* when it was the only sawback species known (Cagle 1953b) and suggested that it was possibly related to *kohnii*, based on similarities of the *oculifera* skull to that of juvenile *kohnii*; he also stated that *oculifera* was not closely related to the sympatric species *pulchra* (i.e., *pearlensis*; Ennen, Lovich, et al. 2010), from which he found it to be very dissimilar.

In describing *sabinensis* and *ouachitensis*—both as subspecies of *G. pseudogeographica*—Cagle (1953a) suggested two possible arrangements of taxa then known. The first had a broad-headed group composed of *barbouri, pulchra,* and *kohnii* and a narrow-headed group composed of *oculifera, ouachitensis, pseudogeographica, sabinensis,* and *versa*. In the second arrangement *kohnii* was placed in the narrow-headed group, which was split into two subgroups, one containing *pseudogeographica* and *kohnii* and the other containing *ouachitensis, sabinensis,* and *versa* (*oculifera* was not mentioned in the second arrangement). His exclusion of *geographica* from discussion may indicate early emergence of the viewpoint that it is outgroup to all other species of the genus, a view upheld in all phylogenetic trees drawn for the genus (see below). Finally, in describing *nigrinoda* and *flavimaculata*, Cagle (1954) used morphology and their neighboring allopatric distributions to place them together with *oculifera* as a "unique complex" (perhaps meaning a monophyletic group; p. 167), of which he considered *nigrinoda* to be the most dissimilar (perhaps meaning basal) species. Killebrew (1979) echoed Cagle's conclusion in an osteological study.

In describing *caglei*, Haynes and McKown (1974) cited width and markings of the head as evidence that its closest relatives are *pseudogeographica* and *versa*, the two species nearest to it in geographic range. Vogt (1981b, 1993) discounted a close relationship between *caglei* and *versa*, however, and suggested *caglei* is more closely related to *ouachitensis* and that *versa* is a more distant relative of other *Graptemys*. Bertl and Killebrew (1983) compared the osteology of *caglei* and *versa* and placed them together in a narrow-headed species group, with *pseudogeographica*, *flavimaculata*, *nigrinoda*, and (presumably) *oculifera* and *ouachitensis*, separate from a broad-headed group that included *barbouri* and *pulchra sensu lato*. Bertl and Killebrew (1983) also recognized the presence of prefrontal processes and the absence of ridges and shelving of the supraoccipital, squamosal, parietal, and postorbital bones as diagnostic of the narrow-headed group when compared with the broad-headed group. Their split within the genus would later be recognized in reciprocally monophyletic clades by Lamb et al. (1994).

Dobie (1981) further commented on groups of related species within *Graptemys*. He recognized a group of the three sawbacks (*flavimaculata*, *nigrinoda*, and *oculifera*), the megacephalic Gulf Coast group (*barbouri* and *pulchra sensu lato*), a group centered in the Mississippi River drainage and drainages to the west (*caglei*, *ouachitensis*, *pseudogeographica*, and *versa*), and a monotypic *geographica*. He likewise did not attempt to elucidate intergroup relationships. While Dobie did not state what characteristics were diagnostic for his groups, Lovich and McCoy (1992) diagnosed the second group based on large body size and extremely broad head sizes of females, extreme sexual size dimorphism, extensive postorbital markings, spines of the carapacial keel, and a possibly unique chromosome number (see below).

McKown's (1972) attempt to recover *Graptemys* phylogeny using protein electrophoresis and karyotypes was hampered by the low levels of variation found among species. Two hemoglobin patterns were found: one shared by *M. terrapin* and *kohnii*, a mix of *ernsti* and *gibbonsi* specimens (not then recognized as distinct species), *barbouri*, *caglei*, *versa*, and *sabinensis*; the other by *geographica*, *flavimaculata*, *oculifera*, *nigrinoda*, and *ouachitensis*. The patterns for total plasma proteins and transferrins consisted of identical bands for all *Graptemys* with marked differences from *M. terrapin*. Karyotypes revealed 2n = 50 chromosomes in most species but 2n = 52 in *G. barbouri* and *G. pulchra* (*sensu stricto*; apparently no karyotyping was conducted on populations then considered conspecific). All species with 50 chromosomes had 13 pairs of

geographica   flavimaculata   oculifera   nigrinoda   pseudogeographica   sabinensis   versa   caglei   kohnii   pulchra*   barbouri

ouachitensis

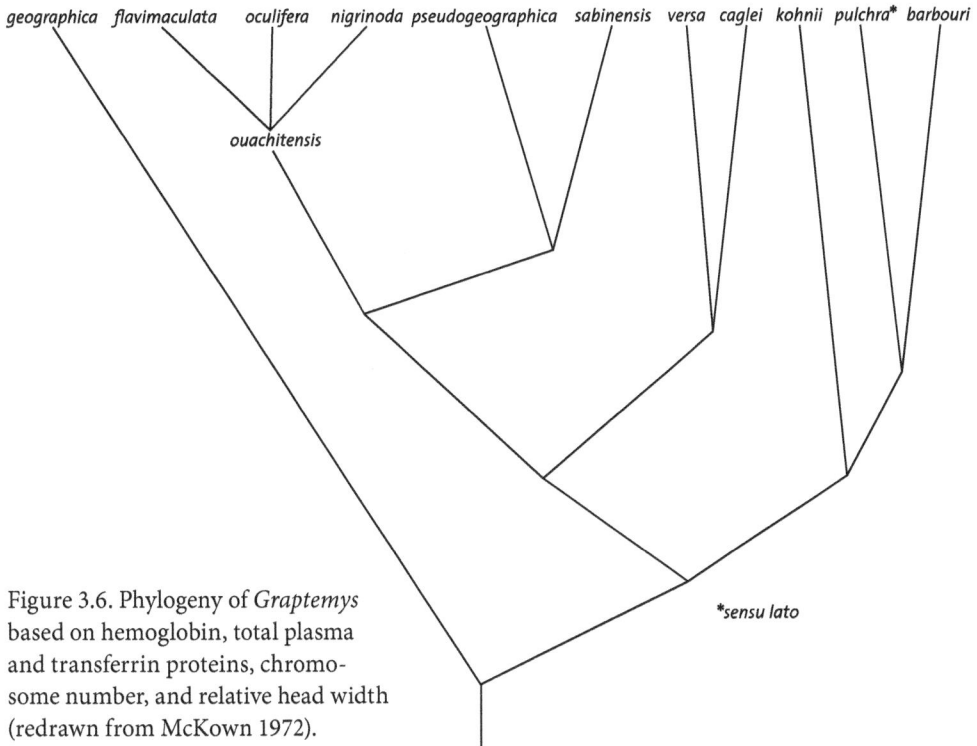

Figure 3.6. Phylogeny of *Graptemys* based on hemoglobin, total plasma and transferrin proteins, chromosome number, and relative head width (redrawn from McKown 1972).

*sensu lato

macrochromosomes and 12 pairs of microchromosomes, similar to *Malaclemys terrapin*, while *G. barbouri* and *G. pulchra* were reported to have an additional pair of microchromosomes. (McKown's findings of 2n = 52 for *G. barbouri* and *G. pulchra* were not supported by Killebrew 1977, who found the former to be 2n = 50.) Undaunted by the lack of resolution of the data, McKown combined his work with information on morphology, apparently drawing heavily on Cagle's publications, and produced a phylogeny (fig. 3.6). His tree has a basal *G. geographica*; a *kohnii* (*pulchra, barbouri*) clade; and a clade of all remaining species, in which the sawbacks form an unresolved clade depicted as derived from *ouachitensis* (the label for which appears along the branch leading to the sawbacks), *pseudogeographica* is sister to *sabinensis*, and a clade of the Texas endemics, *versa* and *caglei*, is outgroup to the other two clades.

The first computer-assisted attempt to recover *Graptemys* phylogeny from a species-by-character data matrix was a study of mtDNA restriction sites and sequences (Lamb et al. 1994). Digestion with eighteen endonucleases for fifteen taxa of *Graptemys* (all the recognized species and subspecies), plus the outgroup *Malaclemys terrapin*, revealed only nine haplotypes within *Graptemys*: *flavimaculata, nigrinoda, oculifera, pseudogeographica,* and *versa* exhibited no variation, nor did *ouachitensis* vary from *sabinensis*. Sequencing 380–84 base pairs of the cytochrome *b* gene yielded only seven nucleotide positions that varied among *Graptemys*, three of which were autapomorphic substitutions, leaving only four that were phylogenetically informative. The control region yielded more differentiation, with 37 of 344 nucleotide positions having substitutions within *Graptemys*, of which 15 were autapomorphies and 22 were phylogenetically informative. Still, the combined sequence data yielded a highly unresolved phylogeny. Lamb et al. used a combined data matrix of sequences and restriction sites to produce a tree with three well-supported clades (fig. 3.7): (1) a basal, monotypic *geographica*; (2) a clade of four species with megacephalic females (in which *pearlensis* was the outgroup while *ernsti* and *pulchra* were sister species); and (3) a clade of the remaining eight species, in which *flavimaculata* and *oculifera* were sister taxa and *ouachitensis* and *sabinensis* were sister taxa, and these two clades were sister to one another. Successive outgroups were then *versa*; *pseudogeographica* and *nigrinoda* (an unresolved polytomy); and *caglei*. Bootstrap values ranged from 91 to 100 percent in supporting the three major clades, but most other clades had considerably less support.

Lamb and Osentoski (1997) published a slightly different tree in a reanalysis of sequences that employed greater variation within the outgroup *Malaclemys terrapin*, plus an additional outgroup, *Trachemys scripta*. The same three major clades were recovered, but resolution within the largest was poor (fig. 3.8). Novel sister-group relationships placed *ernsti* with *barbouri* and *ouachitensis* with *caglei*, while *sabinensis* occurred in an unresolved trichotomy with *flavimaculata* and *oculifera*. Bootstrap support values were generally low.

Stephens (1998) analyzed thirty-four skull characters using various methods of parsimony analysis to determine the phylogeny of *Graptemys*. He analyzed male and female data separately due to sexual dimorphism in some characters, although no data were available for one or both genders of some taxa. Results varied according to gender and parsimony model, and while tree topology differed from that of Lamb et al. (1994) in some respects, most bootstrap values were again low.

Stephens and Wiens (2003) expanded the skull character data set in an analysis of combined morphological and molecular variation across the family Emydidae. In their combined analysis, they recovered an intrageneric phylogeny of *Graptemys* that was similar to that of Lamb et al. (1994) in many respects (fig. 3.9). The basal species in the genus was *geographica*, and a megacephalic clade was recovered with 81 percent bootstrap support. Within the latter, *barbouri* was basal and *ernsti* and *gibbonsi* were sister species—that is, the taxa long considered conspecific (Lovich and McCoy 1992) were recovered as a clade. Placement of *caglei* as outgroup to the megacephalic clade, rather than as allied to the remaining clade of species, was poorly supported (<50 percent bootstrap value). Placement of the remaining eight taxa occurred in a clade with less than 50 percent bootstrap support. Notable findings within the largest clade included a sawback clade, with *nigrinoda* outgroup to *flavimaculata* + *oculifera*; the novel sister-species status of *sabinensis* and *versa*; and the sister-

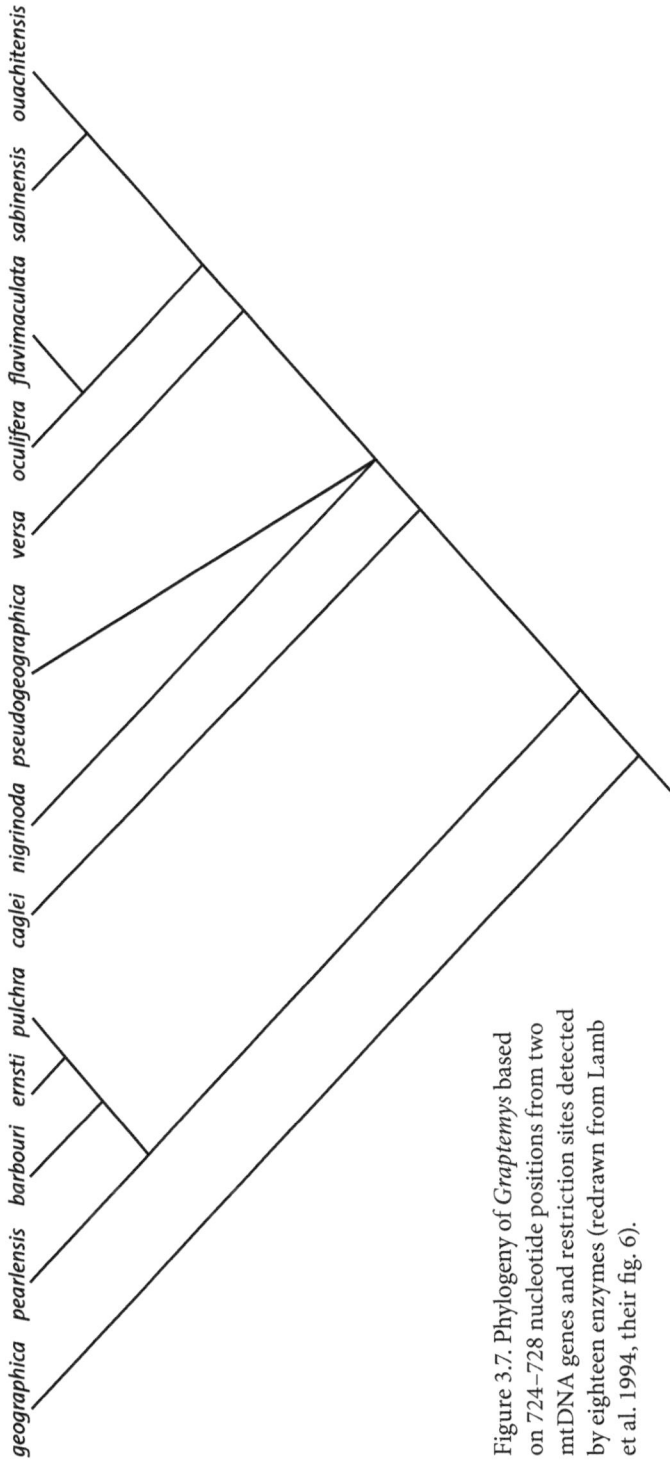

Figure 3.7. Phylogeny of *Graptemys* based on 724–728 nucleotide positions from two mtDNA genes and restriction sites detected by eighteen enzymes (redrawn from Lamb et al. 1994, their fig. 6).

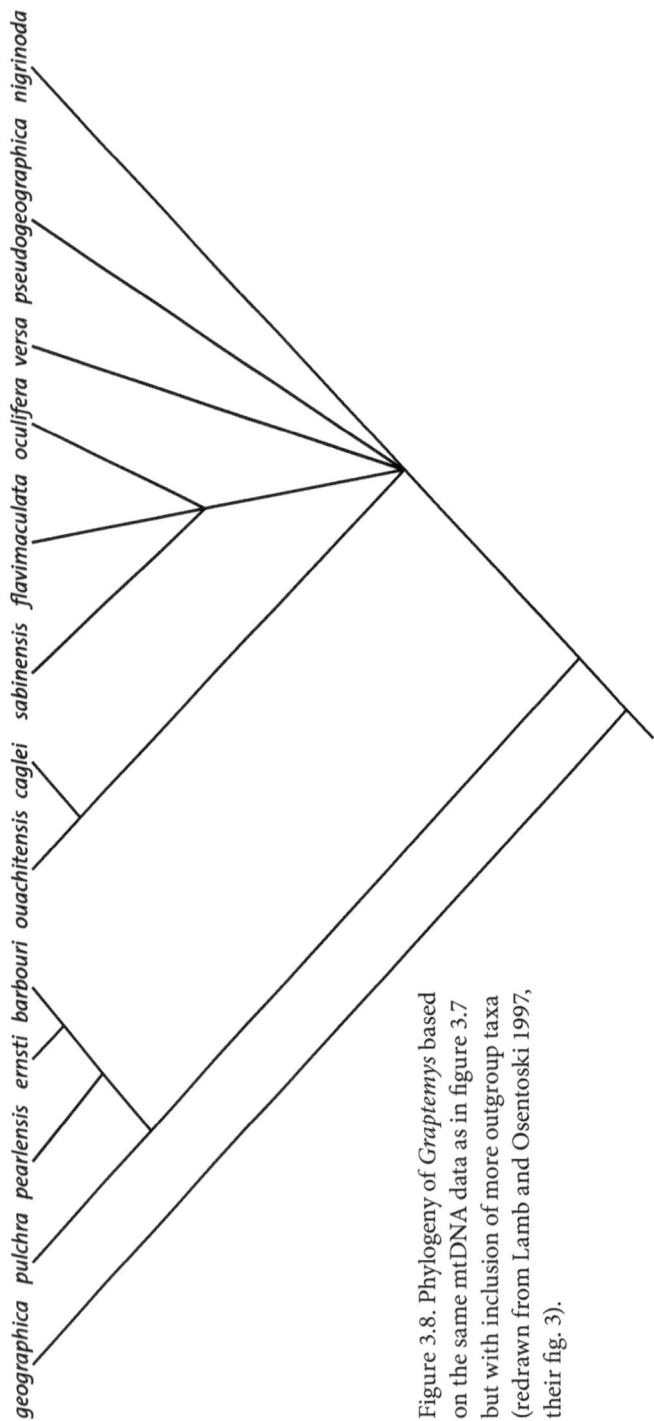

Figure 3.8. Phylogeny of *Graptemys* based on the same mtDNA data as in figure 3.7 but with inclusion of more outgroup taxa (redrawn from Lamb and Osentoski 1997, their fig. 3).

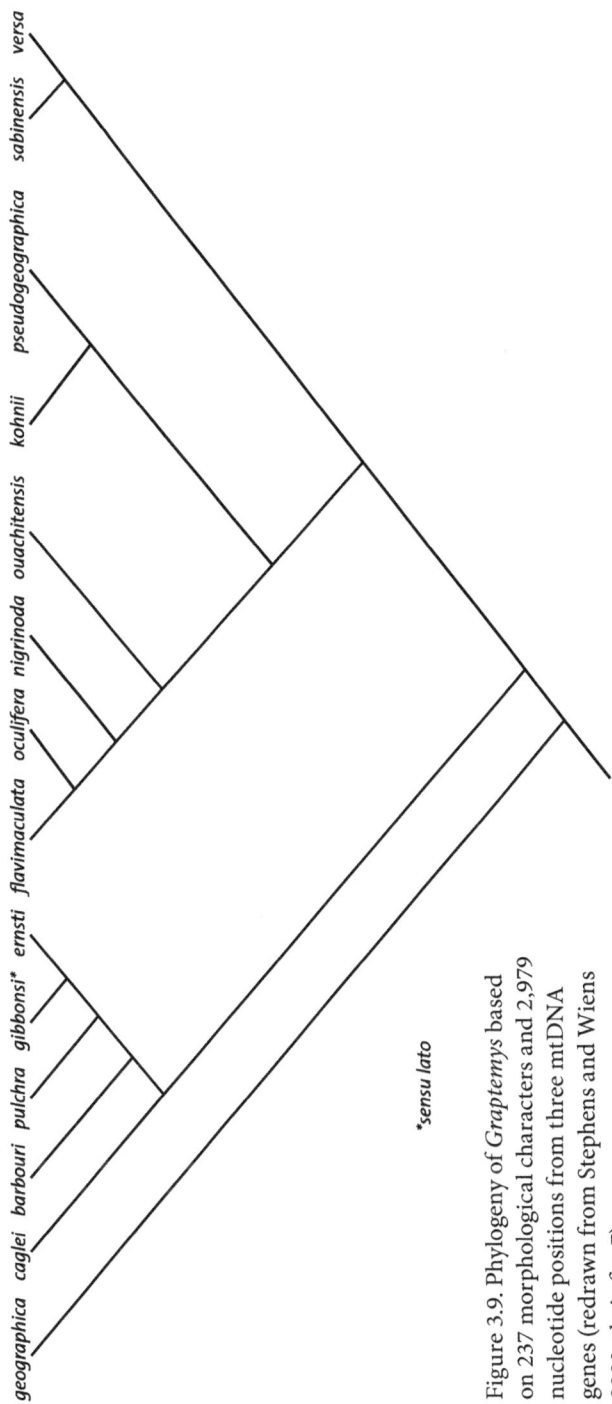

Figure 3.9. Phylogeny of *Graptemys* based on 237 morphological characters and 2,979 nucleotide positions from three mtDNA genes (redrawn from Stephens and Wiens 2003, their fig. 7).

*sensu lato

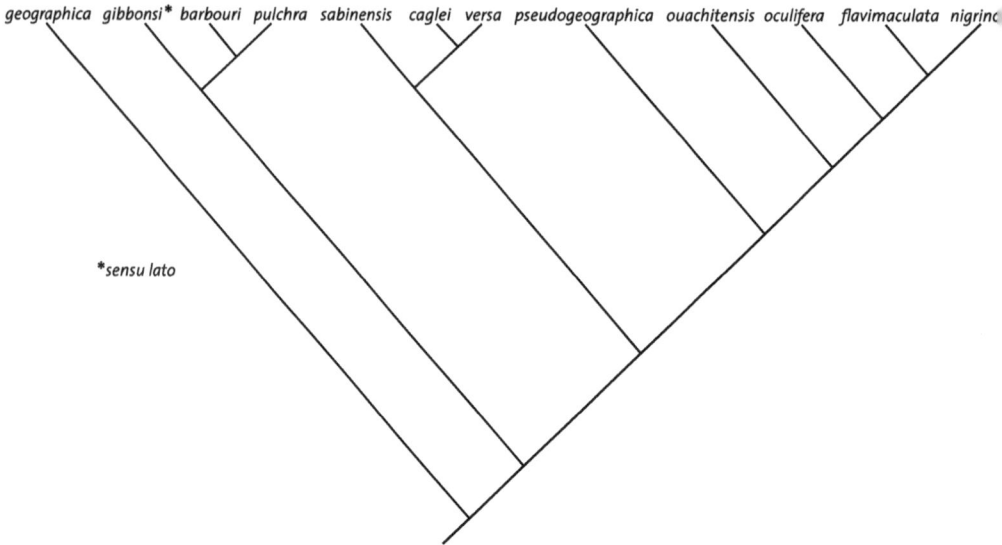

Figure 3.10. Phylogeny of *Graptemys* (excluding *ernsti*) based on 245 morphological characters, 1,264 total nucleotide positions from two mtDNA genes, and 3,966 total nucleotide positions from six nuclear genes (redrawn from Wiens et al. 2010, their online supplemental fig. 13).

taxa status of *pseudogeographica* and *kohnii*, separate from *ouachitensis* (i.e., support for the taxonomy of Vogt 1993). The only subclades within the clade of eight taxa with greater than 50 percent bootstrap support were the sawbacks (84 percent) and the *flavimaculata + oculifera* (62 percent).

A recent phylogenetic analysis of *Graptemys* appears in a dissertation by Myers (2008). She used sequences of a nuclear gene and three sequences of mtDNA to derive trees by maximum parsimony, maximum likelihood, and Bayesian analyses. All analyses confirmed the three major clades of Lamb et al. (1994), with *geographica* again basal. Bootstrap support in the largest clade was generally low, and she elected to show many of its nodes as unresolved polytomies. Within the megacephalic clade, *gibbonsi* consistently occurred as basal and *ernsti* and *pulchra* were sisters, with low to moderate bootstrap support. In the Bayesian analysis (but not in the other analyses), there was high support for a novel *caglei + versa* clade, uniting the central Texas endemics as sister species.

Most recently, Stephens and Wiens (2009) and Wiens et al. (2010) analyzed data sets incorporating new mtDNA and nuclear sequences. Their tree based on Bayesian analysis of combined morphological, mtDNA, and nuclear data (fig. 3.10) shows strong support for the three major clades of *Graptemys*, a sawback clade, and a *caglei + versa* clade. Sequences from mtDNA and nuclear DNA yielded sometimes highly discordant phylogenies, however; partitioned analyses of the six nuclear genes yielded such novelties as the sawback clade being the outgroup to all other *Graptemys* and a clade consisting of *geographica*, *pulchra*, and *sabinensis* (i.e., three species otherwise consistently found to be divided among the three major clades of the genus!). Wiens et al. (2010) concluded that "many more" (p. 457) nuclear genes would be necessary to fully resolve *Graptemys* relationships. Spinks et al. (2009) conducted simulations designed to determine how much nuclear gene sequence data might be necessary to fully resolve emydid relationships and suggested twenty-four loci, or twenty thousand base pairs, which they thought may soon be attainable.

Fossil Record of *Graptemys*

Graptemys is not well represented in the fossil record. A Tertiary fossil species originally assigned to the genus likely predates the cladogenetic origin of the genus by a considerable period. Eleven fragmentary Quaternary specimens have been found in Alabama, Florida, Indiana, Kansas, Michigan, and Texas.

Tertiary material is available from four sites in western South Dakota. A complete carapace and plastron (ACM 3607) from river deposits in Meade County were described as *Chrysemys inornata* (Loomis 1904). Hay (1908) tentatively referred the specimen to *Graptemys* on the basis of a dorsal midline keel and the elongated shape of the first suprapygal bone of the carapace. J. Clark (1937) described *Graptemys cordifera* from a crushed carapace and plastron (YPM 13838) from river deposits in Washington County, diagnosing it based on a heart-shaped entoplastron that differed from the quadrangular entoplastron in *G. inornata* and differences in the relative sizes of neural bones 7 and 8. He noted the presence of a low, rounded dorsal midline carapacial keel in both *G. cordifera* and *G. inornata*, presumably as *Graptemys*-like characteristics. The specimen measured 267 mm in carapace length (CL) and 212 mm in plastron length (PL). Bjork (1967) described two specimens (UMMG 59187) of *Graptemys* from Harding County, similar to both *G. inornata* and *G. cordifera*. The specimens were clearly river fauna, as they were collected alongside fossils of gars (*Lepidosteus*), an alligatorid, and trionychid turtles. Hutchison (1996) tentatively synonymized *cordifera* with *inornata* based on a carapace and plastron from Pennington County (SDSM 10053) with intermediate characteristics.

The ages of the Tertiary fossils appears to be around the Eocene-Oligocene boundary of circa 34 million years ago (M.Y.A.), apparently well before the origin of *Graptemys* as a separate genus. Loomis (1904), Hay (1908), and J. Clark (1937) cited early Oligocene ages for their material, but Bjork (1967) cited the arguments of G. G. Simpson (1946) for a late Eocene age for his. Lamb et al. (1994) and Lamb and Osentoski (1997) placed the date for the separation of *Graptemys* from *Malaclemys* during the middle Miocene (ca. 7–11 M.Y.A.); more recently, Near et al. (2005) used multifossil calibration of trees derived from gene sequences to estimate that *Graptemys* and *Trachemys scripta* last shared a common ancestor 15.36 (±3.16) M.Y.A. J. Clark (1937) noted Hay's (1908) uncertainty in referring Loomis's specimen to *Graptemys* and echoed that uncer-

tainty in referring his own specimen to *Graptemys*. Hutchison (1996) placed the
South Dakota material (excluding Bjork's specimens, which he apparently did
not examine) in a new genus, *Pseudograptemys*, and found even their identification as emydids suspect.

The first Quaternary fossil *Graptemys* came from terraces above the Trinity
River in Henderson County, Texas, a complete plastron (174 mm) and the anterior two-thirds of a carapace (BEGUT 30907-62). Stovall and McAnulty (1950)
stated it showed "no great difference" (p. 222) from Recent *G. geographica*, although McCoy and Vogt (1994) assigned it to *G. pseudogeographica* without
comment, perhaps because only the latter species presently occurs in the Trinity River. The specimen dates from the Wisconsinan age of the late Pleistocene.
Preston (1979) referred the specimen to *Trachemys scripta*, also without stating
his reason.

Slaughter et al. (1962) described the third, fourth, and fifth "nuchals" (i.e.,
neural bones) and right third, fourth, and fifth "pleurals" (i.e., costal bones) of
a carapace (SMUMP 60109, which was lost while on loan; Preston 1979). They
identified it as a large *G. geographica*. It was found near the Trinity in Dallas and
Denton Counties, Texas, and dated at more than thirty-seven thousand years
before present (Y.B.P.; Slaughter et al. 1962), which Preston (1979) stated was late
in the Sangamon interglacial. Although the Trinity is inhabited by *G. pseudogeographica* but not by *G. geographica* today, Slaughter et al. based their identification on the lack of pronounced midline carapacial tubercles. McCoy and Vogt
(1994) listed the specimen as *G. pseudogeographica*, again without comment.

A fossil dredged from the mouth of the Saginaw River in Michigan (UMMP
51249) was identified by R. L. Wilson and Zug (1966) as *G. pseudogeographica*.
It consists of a right hypoplastron believed to be four to six thousand years old,
which would make it the most recent fossil for the genus. The authors felt it
represented *G. pseudogeographica*, which is not found in Michigan today, rather
than *G. geographica*, which is native to Michigan, based on comparison with
five shells of the former and eight of the latter. The characteristic they used
(angle of the inguinal buttress relative to the plastron, supposedly obtuse in
G. geographica but almost perpendicular in *G. pseudogeographica*) was found
by Holman (1988) to vary within both species; he argued that the specimen was
more likely a *G. geographica* on biogeographic grounds and suggested it simply
be considered *Graptemys* sp. indet., an indeterminate species of the genus.

Holman (1972) called a left hyoplastron (UMMP V-60356) from Ellsworth
County, Kansas, identical to *G. geographica* specimens. It was found approxi-
mately 130 kilometers west of the present range of the species and is from the
Yarmouth interglacial, circa 200,000 Y.B.P.

D. R. Jackson (1975) described three fragments from the Sante Fe River of
Florida that he definitely considered to be *Graptemys*, plus a fourth he felt was
from either a *Graptemys* or a *Pseudemys*. He suggested the *Graptemys* material
was conspecific with Recent *G. barbouri* but noted its allopatric status in the Su-

wannee drainage; the specimens were later referred to a separate, now-extinct species described by Ehret and Borque (2011). A nuchal bone (UF 10572) lacked a carapacial keel and was from a specimen estimated to be 210-30 mm CL and 180-97 mm PL. A third neural bone (CFM 100) had a keel protuberance similar to that of *G. barbouri* and came from a specimen estimated to be 270-80 mm CL and 230-40 mm PL. A mandible (UF 19161) showed evidence of the hypertrophy of the head typical of *G. barbouri* and was estimated to come from a specimen measuring 280-90 mm CL. Finally, a hyoplastron (UF 19246) exhibited features common to *Graptemys* and some *Pseudemys concinna*. While D. R. Jackson (1975) considered the specimens to be Pleistocene, the nuchal bone and the hyoplastron were from deposits thought to represent the Blancan age, which is generally regarded as late Pliocene rather than Pleistocene.

Preston (1979) reported three shell fragments of *G. pseudogeographica*: (1) a portion of a nuchal bone (UMMP V-65876) from Meade County, Kansas; (2) a hyoplastron, first costa, and third and fourth peripheral bones (TMM 40099-55) from Brazos and Burleson Counties, Texas (originally reported by Hay [1923] to be a *Chrysemys picta* but reidentified by Preston); and (3) a right xiphiplastron (TMM 41606-1) from Milam County, Texas. All three were regarded as late Pleistocene, with the first two thought to come from about the Illinoian-Sangamon boundary circa 128,000 Y.B.P.

Holman et al. (1990) reported a lower mandible and three peripheral bones (RMM 6888) from a cave in Colbert County, Alabama, in an area dated at 11,820 Y.B.P. Two fragmentary specimens (IUBT-20 and IUBT-21) assigned to *Graptemys* sp. indet. were reported from 16,000 to 13,000 Y.B.P. in Daviess County, Indiana (Holman and Richards 1993).

G. E. Phillips (2006) described an epiplastron and second neural bone (MMNH VP 1788, 1686) from late-Pleistocene deposits along Catalpa Creek in the Tombigbee drainage of Lowndes County, Mississippi, and identified them as *Graptemys* cf. *pulchra*. Also found were two carapace bones (MMNH VP 1572, 1743) he tentatively identified as *Graptemys* sp.

Ehret and Borque (2011) described an exceptionally megacephalic, blunt-faced species, *Graptemys kerneri*, based on a skull (UF 239000), eight mandibles, and numerous carapace and plastral fragments collected from the Waccassasa, Suwannee-Sante Fe, and Aucilla drainages of Florida. Analysis of rare earth elements was used to date the specimens as Rancholabrean, deposited circa 15,000 Y.B.P.

Archaeological remains assigned to *G. geographica*, *G. pseudogeographica*, or *Graptemys* sp. are known from Tennessee, Illinois, Michigan, Missouri, Ohio, and Wisconsin (Adler 1968; Guilday et al. 1978). Given the taxonomic confusion surrounding *G. pseudogeographica* and *G. ouachitensis* (Vogt 1993), some fossils or remains from the southern United States assigned to the former may in fact represent the latter (Vogt 1995a, 1995b); identification of their fragments is unlikely, given their morphological similarity.

Biogeography of *Graptemys*

There has not been a formal attempt to ascertain the biogeographic history of *Graptemys*, but some authors have summarized their findings on phylogenetic relationships within the genus in light of the most striking aspect of map turtle and sawback biogeography: the isolation of Gulf Coast species that occur in neighboring drainages. Also often noted are the exceptionally recent age of differentiation among Gulf Coast forms and the greater differentiation in *Graptemys* than in other riverine turtle taxa. Recent genetic studies have allowed the use of molecular clock dating, confirming the recent differentiation of southern species.

Cagle (1954) first invoked exceptionally recent speciation within *Graptemys* regarding the sawbacks. Tinkle (1958b) compared speciation rates of *Graptemys*, the kinosternid genus *Sternotherus*, and the trionychid genus *Apalone* and found, like Cagle (1954), that the sawbacks had diverged very recently from a common ancestor; by contrast, he found speciation patterns in the broadheaded group (then considered to represent just two species, with *G. pulchra* spread over four adjacent drainages) to be very comparable to patterns within the *Sternotherus carinatus* complex and *Apalone* (particularly *A. mutica*). He also suggested that the sawbacks might be derived from *Graptemys* in the Tennessee drainage, their divergence driven by subsequent episodes of stream capture.

McKown (1972) speculated on the biogeographic influences that have driven the radiation of *Graptemys*. In his scenario, the embayment of the Mississippi River, reaching several hundred kilometers upstream of the present channel, formed an east-west barrier to *Graptemys* dispersal that was not overcome until the late Miocene or Pliocene. Since that time, repeated glaciation and its effects on climate, sea level, stream capture, and floodplain and delta formation have been the primary influences on *Graptemys* biogeography. McKown placed the split between *G. geographica* and the ancestor of the rest of the genus (which he called *G. pseudogeographica*) in the Oligocene and had these two species separated to the east and west, respectively, of the Mississippi Embayment. He echoed Tinkle (1958b) in suggesting a connection via stream capture of the Tennessee and Mobile Bay drainages, giving rise to the sawback group. He further felt that coastal floodplains, formed during interglacial periods, provided the probable dispersal routes that allowed for the east-west dispersal of *G. p. kohnii*, *G. ouachitensis*, *G. caglei*, *G. versa*, and the megacephalic group, although he could not discount further stream capture events. He also noted the greater degree of restriction to riverine habitats in *Graptemys* than in other Gulf Coast aquatic turtles and suggested this aspect of their ecology was responsible for their more rapid speciation rates; Lamb et al. (1994) would later also invoke the strongly riverine nature of map turtles and sawbacks in promoting speciation and drainage endemism.

Lovich and McCoy (1992), Lamb et al. (1994), and Lamb and Osentoski (1997) envisioned speciation in *Graptemys* driven by sea level changes coincident with glacial and interglacial periods. In these scenarios, high sea levels during interglacial periods swamped small coastal drainages and extirpated their populations while simultaneously partitioning more extensive drainages at higher elevations, promoting endemic species formation. Ensuing lowering of sea levels during glaciation allowed drainages to coalesce in emergent coastal plains, promoting dispersal. The authors drew several comparisons between the allopatric divergences in *Graptemys* and those in other riverine fauna, particularly fish but also crayfish. They did not invoke stream capture events as a cause of isolation. Likewise, Ehret and Borque (2011) stated that their dating of the fossil species *G. kerneri* was consistent with dispersal and differentiation via drainage connection at low sea level rather than via stream capture.

Studies of mtDNA allowed the first molecular clock dating of major divergences among *Graptemys* species (Lamb et al. 1994; Lamb and Osentoski 1997). Using an estimated rate of sequence divergence of 0.2–0.4 percent/million years for the cytochrome *b* gene, both studies placed the divergence of *Graptemys* and *Malaclemys* in the late Miocene, at 7–11 M.Y.A. Lamb et al. further placed the split between *G. geographica* and the rest of the genus at 6–8 M.Y.A., following a period of elevated sea level, and the split between the two large clades of the rest of the genus at 2.5–3.5 M.Y.A., likewise at the end of a period of elevated sea level.

The place of *Graptemys* as a recently diverged and monophyletic genus within the deirochelyine lineage seems well supported by molecular and morphological studies. Intrageneric relationships appear to be best known for early divergences; more recent divergences are more uncertain, with lower levels of bootstrap support and conflicting results among studies. Nuclear gene phylogenies may help resolve relationships. It seems unlikely that new fossil finds will be of service in sorting out *Graptemys* relationships; the Gulf Coastal Plain in particular has been almost devoid of pertinent fossils. A clear understanding of relationships is a prerequisite for better delineating the biogeographic history of the genus. A promising avenue may be provided by phylogeographic studies, pending identification of molecular markers that show sufficient intraspecific variation for studying the historical dispersal of species. Such markers may provide clues as to whether allopatric speciation has been mediated by changes in sea level or by stream capture. A better understanding of drainage histories would also be an important aid to better delineating the biogeography of the map turtles and sawbacks.

4

Ecology

The most extensive research on map turtles and sawbacks has concerned ecology. Sexual size dimorphism and dietary diversity (within and among species) are among the most intriguing aspects of these turtles. There has been stronger focus on habitat, diet, reproductive ecology and life history, sex determination, and basking behavior, whereas areas that have been less well studied and warrant greater attention include movements, survivorship and longevity, community ecology, and interspecific interactions.

Habitat

Prime habitat for most *Graptemys* populations is moderate current in a medium to large river with abundant sunlight penetration. Deadwood "snags" stranded against the substrate or the outer bends of a river are used for basking and are also vital subsurface components that provide cover and substrates for the growth of aquatic prey communities (J. B. Wallace and Benke 1984; Shively and Jackson 1985; Lindeman 1999c; Angradi et al. 2009). In rivers, water quality ranges from clear, spring-fed streams flowing through rocky beds such as the Guadalupe (Cagle's map turtle), Llano and San Saba (Texas map turtle), and Chipola (Barbour's map turtle) to sandy, murkier rivers such as the Neches (Sabine and Mississippi map turtles), Pearl (ringed sawback and Pearl map turtle), and Escambia (Escambia map turtle).

Flowing water is not absolutely essential for most species. Three of the Laurentian Great Lakes and numerous smaller glacial lakes are inhabited by northern populations of the outgroup species of the genus, the common map turtle, which also occupies streams of a variety of sizes. Disjunct oxbow lakes harbor individuals of many other species (see species accounts, chapter 8); for the yellow-blotched sawback, use of oxbow lakes may be seasonal (Jones 1996). Backwater floodplain habitats, ranging from connected sloughs to areas that hold water only during flooding, are used by the three widespread species of the greater Mississippi drainage, the common, false, and Ouachita map turtles (Moll 1977; Gritters and Mauldin 1994; Bodie and Semlitsch 2000b; Bodie et al. 2000; Tollefson 2004; Dreslik and Phillips 2005; Rizkalla and Swihart 2006; J. E. Wallace et al. 2007). Seasonal inundation of floodplains appears to be associated with higher catch rates of *Graptemys* and other river-dependent aquatic turtle species (J. E. Wallace et al. 2007). Most species have persisted in river impoundments as well, although their abundance relative to other turtle species may decline in impoundments (see chapter 5).

In the coastal plain of the southeastern United States, mollusk-dependent species with megacephalic females show an interesting pattern in abundance regarding substrate. Sand-bottomed river segments tend to have sparser populations, while segments underlain by limestone formations (generally, farther upstream) have much greater abundance (Shealy 1976; Godwin 2000, 2003; G. E. Wallace 2000; Enge and Wallace 2008; Sterrett 2009). The limestone segments may present more abundant molluscan prey.

The most detailed analysis of habitat features that influence occurrence concerned Sabine map turtles in the Whiskey Chitto River, a tributary of the Calcasieu River in Louisiana (Shively 1982; Shively and Jackson 1985). Path analysis supported a primary limiting role of sunlight penetration as influenced by creek width. Sunlight influenced the productivity of submerged deadwood as a substrate for prey communities, as measured in algal chlorophyll and insect density. Data for seven study stretches are shown in figure 4.1; the site farthest upstream was the last section of the stream inhabited by Sabine map turtles. The creek's mollusk-dependent species, the Mississippi map turtle, occurred farther upstream and increased in abundance above the limit of the Sabine map turtle (Shively and Jackson 1985). Those analyses were complemented by Hively's (2009) results for Sabine and Mississippi map turtles in the upper Sabine River and its former channel in Texas. Mississippi map turtles occurred in the river and the old channel, which was narrower and had less sunlight penetration and less current, but Sabine map turtles occurred only in the new river, where their abundance was highest in areas with more flow and greater width. Similarly, upstream range limits of megacephalic species (the Alabama, Pascagoula, and Pearl map turtles) typically exceed those of sympatric sawbacks, especially in smaller tributaries (see range maps in color plates section).

Sterrett (2009) related habitat features to locations of telemetered female Barbour's map turtles in Ichawaynochaway Creek, Georgia. The turtles exhibited fidelity to deep areas with large, submerged logs, underwater limestone ledges, and sandy substrates, as demonstrated by 50 percent kernel estimates that were small relative to home ranges. On two creeks, the number of Barbour's map turtles captured via snorkeling and baited traps was also positively correlated with the percentage of riparian forest, suggesting its importance for supplying instream deadwood (Sterrett et al. 2011).

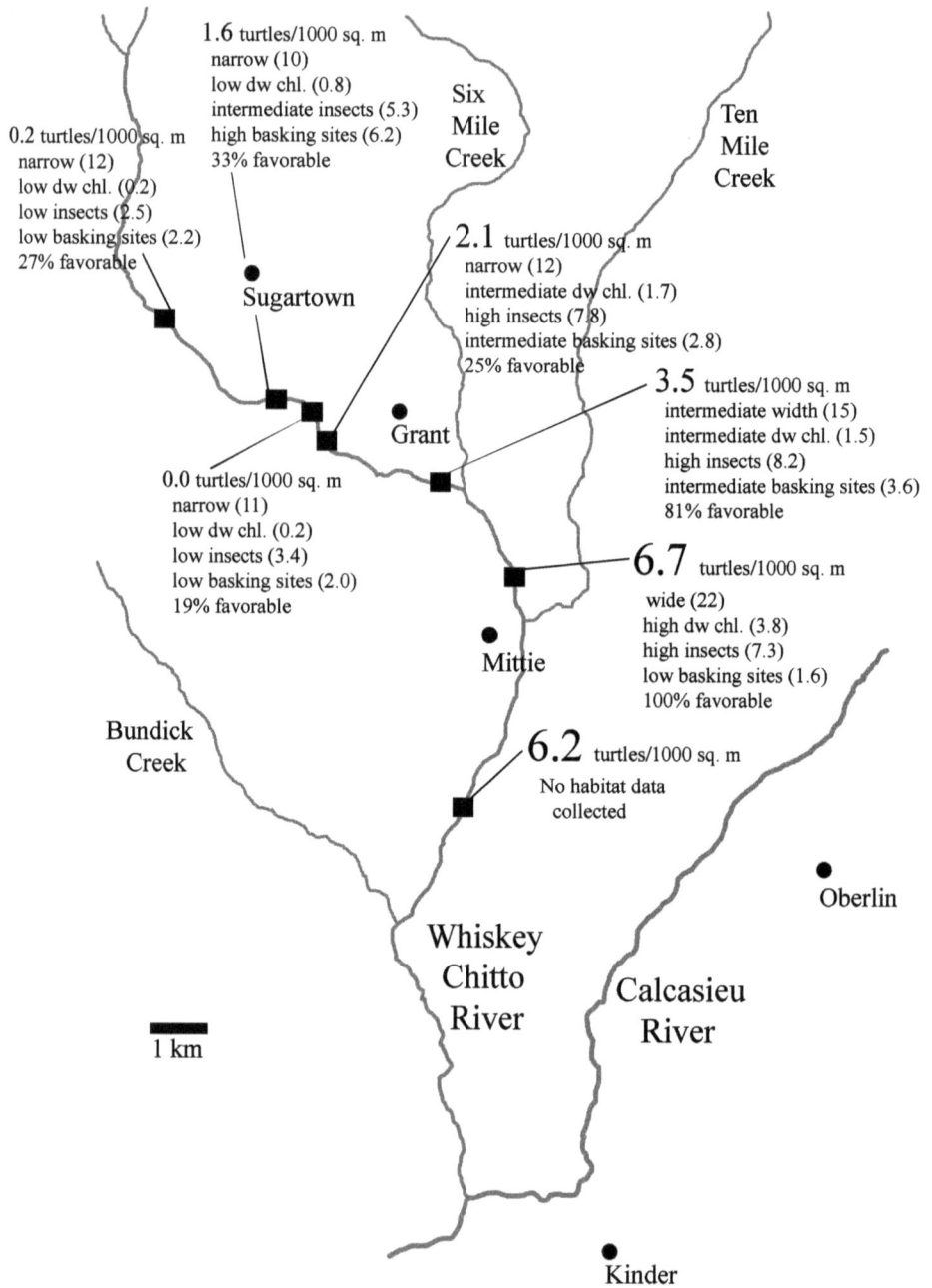

1.6 turtles/1000 sq. m
narrow (10)
low dw chl. (0.8)
intermediate insects (5.3)
high basking sites (6.2)
33% favorable

Six Mile Creek

Ten Mile Creek

0.2 turtles/1000 sq. m
narrow (12)
low dw chl. (0.2)
low insects (2.5)
low basking sites (2.2)
27% favorable

Sugartown

2.1 turtles/1000 sq. m
narrow (12)
intermediate dw chl. (1.7)
high insects (7.8)
intermediate basking sites (2.8)
25% favorable

3.5 turtles/1000 sq. m
intermediate width (15)
intermediate dw chl. (1.5)
high insects (8.2)
intermediate basking sites (3.6)
81% favorable

Grant

0.0 turtles/1000 sq. m
narrow (11)
low dw chl. (0.2)
low insects (3.4)
low basking sites (2.0)
19% favorable

6.7 turtles/1000 sq. m
wide (22)
high dw chl. (3.8)
high insects (7.3)
low basking sites (1.6)
100% favorable

Mittie

6.2 turtles/1000 sq. m
No habitat data collected

Bundick Creek

Oberlin

1 km

Whiskey Chitto River

Calcasieu River

Kinder

Sexual differences in habitat use have been described for riverine populations of common, Cagle's, and Texas map turtles and yellow-blotched sawbacks
(Pluto and Bellis 1986; Craig 1992; Jones 1996; Lindeman 2003b; Carrière and
Blouin-Demers 2010). Males tend to occupy shallower, slower water closer to
shore. Small juvenile female Texas map turtles that overlapped males in body
size had wide heads and wide alveolar surfaces that allowed a diet more similar
to that of larger females than to that of similar-sized adult males. Habitat use
by these "male-sized" females was also more similar to that of larger females,
suggesting that prey distribution plays a primary role in structuring sexual
differences in habitat use (Lindeman 2003b). In a small lake in Ontario, male,
"male-sized" juvenile female, and larger adult female common map turtles
overlapped greatly in habitat use (Bulté, Gravel, et al. 2008). None of the groups
closely tracked prey abundance in their habitat use; thus the pattern in the lake
differed from the pattern for the river in Texas.

Movements

Some data on movements made by *Graptemys* come from mark-recapture studies, which have the disadvantage of small sample sizes for locations and may be
biased regarding where traps are set. Radiotelemetric comparisons of the movements of males and females have been conducted in eleven populations of five
species (table 4.1), with analysis of home range size and sexual differences in the
distance and timing of movements. Home ranges have typically been quantified
as linear length of stream occupied but have also been estimated as areas.

Nine of thirteen comparisons found no sexual difference in home range size,
although sample sizes were generally small and variation within each gender
was extensive; in the other comparisons, home ranges in three were larger for
females and in one were larger for males (table 4.1). The most detailed sexual
comparisons of movements and home range were conducted for common map
turtles in large river and lake environments in Ontario (Carrière et al. 2009).
The most prominent differences, between and within sites, were that adult females at the river site moved greater distances (particularly in June) and had
larger home ranges. In the river, adult females may have been less restricted

Figure 4.1 (*facing page*). Density and habitat data for the Sabine map turtle (*Graptemys sabinensis*) at seven sites on Whiskey Chitto River in southwestern Louisiana
(Shively 1982, Shively and Jackson 1985). The top figure for each site is estimated
population density, followed by relative characterizations of five habitat parameters:
(1) channel width (in meters, m); (2) chlorophyll (chl.) and (3) insect density on submerged deadwood (dw; both in milligrams per 1,000 cm^2 of limb surface); (4) basking site area (in square centimeters per 1,000 m^2 of stream); and (5) percentage of
23 m stretches deemed favorable habitat based on these and other habitat measures
and their correlation with observed turtle density.

Table 4.1. Comparisons of male and female home range sizes in adult *Graptemys*, with results of statistical comparisons given between means

Species	Source	Habitat	Relocation method	Males (N)	Males Mean ± SE home range		Females Mean ± SE home range	Females (N)
barbouri	Sanderson 1974	Chipola River, FL	Recaptures	(38)	365 ± 59 m	=	273 ± 48 m	(18)
barbouri	Sterrett 2009	Ichawaynochaway Creek, GA	Telemetry	NR			839 ± 139 m; 3.1 ± 0.7 ha	(21)
caglei	Craig 1992	Guadalupe River, TX	Recaptures	(438)	985 ± 93.6 m	=	738 ± 285.8 m	(24)
			Telemetry	(6)	1300 ± 394.8 m	=	1640 ± 638.8 m	(7)
flavimaculata	Jones 1996	Pascagoula River, MS	Telemetry	(6) (6)	1861 ± 879 m; 1 ± 0.3 ha	=	1550 ± 320 m; 6 ± 2.0 ha	(7) (7)
geographica	Flaherty 1982	Lake of Two Mountains, QC	Telemetry	(6)	32 ± 8.1 ha	<	68 ± 15.4 ha	(6)
geographica	Pluto and Bellis 1988	Raystown Branch, PA	Recaptures	(46)	2115 m ± NR	>	1211 m ± NR	(14)
geographica	Carrière et al. 2009	Lake Opinicon, ON	Telemetry	(16)	192.5 ± 35.1 ha	=	154.0 ± 26.5 ha	(19)
geographica		St. Lawrence River, ON	Telemetry	(7)	121 ± 21.7 ha	<	327 ± 55.4 ha	(20)
geographica	Bennett 2009	Trent-Severn Waterway, ON	Telemetry	(5)	6180 ± 1500 m	=	8510 ± 1590 m	(9)
geographica	Ouellette and Cardille 2011	Mille-Îles River, QC	Telemetry	(5)	2734 m	=	3637 m	(5)
geographica	Richards-Dimitrie 2011	Susquehanna River, MD	Telemetry	NR			2991 ± 476 m; 121 ± 21 ha	(14)
pseudogeographica	Bodie and Semlitsch 2000b	Missouri River, MO	Telemetry	(5)	3753 ± 660.3 m	<	5152 ± 995.8 m	(10)
			Recaptures	(33)	755 ± 123.5 m	=	904 ± 163.4 m	(7)

Note: Home range has variously been reported as area (ha) or as linear home range (m) along a river. NR = not reported.

by the forces of the current than were juvenile females and adult males due to body-size differences that affect swimming ability. They may also have moved farther to reach more dispersed nesting sites: hence the large June difference in movement distance.

Few studies have examined interannual changes in individual home ranges. Jones (2006) found that some female yellow-blotched sawbacks repeated movements over two years. For female common map turtles, Carrière (2007) and Richards-Dimitrie (2011) found strong home-range fidelity in both area and sites inhabited. None of these three studies had sufficient data on males to examine their home range fidelity. Craig (1992) suggested that Cagle's map turtles may also repeat their annual patterns of habitat use, as recapture data often appeared clustered with regard to site and time of year; I have noted the same phenomenon in common map turtles in Pennsylvania, with frequent recaptures at the same site on nearly the same date across years. In contrast, however, Shealy (1976) noted that Escambia map turtles at an intensively trapped site showed a nearly complete annual turnover, as he went from rarely recording new captures one fall to rarely recording recaptures in the next three years. Fidelity to habitat is common within an active season as well; for example, over half the recaptures of common map turtles that Nickerson and Pitt (2012) recorded on Missouri's North Fork of the White River were within 100 m of original capture sites.

Despite tendencies toward site fidelity, long-distance dispersal is commonly recorded. Craig (1992) recorded sixty-four instances of Cagle's map turtles making movements that exceeded 2 km, with the six longest ranging from 10.1 to 14.2 km in both upstream and downstream directions. In the Lake of the Two Mountains, Quebec, common map turtles frequently changed embayments they occupied, with annual movements of as much as 19.5 km in females and 8.2 km in males (Flaherty 1982). Sterrett (2009) recorded movements as long as 6.4 km by female Barbour's map turtles in a Georgia creek. Nickerson and Pitt (2012) recorded movements of common map turtles as long as 3.7 km in the North Fork of the White River.

Movements by adult females during nesting season can be extensive. Vogt (1980a) reported dispersal from hibernation sites by females of three species in May, when they aggregated offshore of islands used for nesting in the Mississippi River. Peak short-term distances moved by common map turtle females occurred during nesting seasons in Pennsylvania and Ontario (Pluto and Bellis 1988; Carrière 2007), and three false map turtles in Missouri moved more than 1 km to nest on levees (Bodie and Semlitsch 2000b). Yellow-blotched sawback females also moved out of a floodplain pond and back into the lower Pascagoula during the summer months, probably to use its sandbars for nesting (Jones 1996). In Ontario, Bennett (2009) recorded longest daily distances moved by female common map turtles just after the nesting season and speculated they might be dispersing from nesting areas to good foraging areas, with less delay in their daily movements for basking. In the Susquehanna River in Maryland, common map turtle females moved the greatest distances during the nesting season, the only time of the year they were observed to leave a complex of midstream islands (Richards-Dimitrie 2011).

Map turtles in northern populations make sometimes lengthy seasonal movements associated with aggregation at and dispersal from distinct hibernacula. In a narrow slough of an impounded pool of the Mississippi, Vogt (1980a) captured 131 map turtles on 13 October and then returned to capture 113 at the same site on 11 November, with only 2 recaptures. Females marked at nesting beaches traveled 4 km to the slough for hibernation, and 1 displaced female traveled 8 km to reach the slough. In Quebec, Flaherty (1982) tracked common map turtles during late August and September as they moved to hibernacula in a fast-flowing stream inlet and a turbulent site at a bridge on a lake embayment. In Pennsylvania, common map turtles moved into a communal hibernaculum in a deep, slow-moving stream reach around September and dispersed by June or July (Pluto and Bellis 1988). Common map turtles in Vermont moved minimum annual distances of 3.1–15.4 km upon leaving a hibernaculum in the Lamoille River in late March or April, with some entering adjoining Lake Champlain, before returning to the hibernaculum by mid-October (Graham et al. 2000). Bodie and Semlitsch (2000b) recorded spring and fall movements by false map turtles into and out of floodplain scour habitats used for hibernation in the Missouri. In an urban canal in Indiana, differences between summer activity centers and hibernation sites averaged 1.3 km for female common map turtles, which used woodlot-bordered sites for hibernation (T. J. Ryan et al. 2008).

Homing ability was tested in the Escambia map turtle by Shealy (1976), who displaced turtles 24 river km upstream or downstream of capture sites. Return to original capture sites over one to three years was recorded for four of thirty-four displaced females (two moved back upstream and two downstream), and five others were recaptured at their sites of displacement, while none of thirteen displaced males was recaptured.

Diet

Diet may be the most intriguing aspect of the ecology of *Graptemys*, given the morphological adaptations associated with molluscivory, the pronounced gender differences in diet that are strongest among the more mollusk-dependent species, and the co-occurrence of differing morphologies in sympatric species pairs. Some dietary patterns in the genus have been obscured, however, by changes wrought by superabundant, invasive bivalves that have become important prey and by changes in prey base associated with habitat degradation.

Dietary studies reporting both frequency of occurrence and a measure of percentage of total biomass eaten enable calculation of the index of relative importance (IRI; Bjorndal et al. 1997; see tables 4.2–4.17). Values sum to 100 for a set of samples, facilitating comparisons among data sets. Comparisons should be made with caution, however, for three reasons: First, the second set of percentages used in IRI calculations has been variously reported as percentage of total mass or percentage of total volume (determined by displacement)—either of which weights samples according to how much total material is recovered in each—or mean percentage of mass or volume of individual samples, which weights all samples equally regardless of size (Lindeman 2006a). Second, some reports come from material in stomachs via dissection or flushing of live specimens, while others come from fecal samples. Stomach flushing may yield unreliable results (e.g., Lindeman 2006a, 2007). The change in prey composition with varying degrees of digestion and relative rates of movement through the digestive tract have also not been studied and may produce systematic biases. Third, various studies have lumped males, females, and juveniles together or divided them among different sex, size, or age categories in ways that complicate comparisons. In spite of these caveats, IRI values remain the most efficient way to make qualitative, if not wholly quantitative, intra- and interspecific comparisons.

Sexual and Ontogenetic Patterns

Adult females comprise three morphological groups: microcephalic (five species), mesocephalic (four species), and megacephalic (five species: see chapter 7; Lindeman 2000a). Roughly speaking, these classes differ with respect to mollusk consumption. Mollusks are of low importance in the diets of microcephalic females (summed IRI values for molluscan prey 0–49, mean 7.5; tables 4.2–4.8), generally moderate to high importance in the diets of mesocephalic females (0–99, mean 57.9; tables 4.8–4.16), and very high importance in the diets of megacephalic females (99.7 in a single report, table 4.17; but see also the results of Shealy 1976, for which molluscan IRI could not be computed but would undoubtedly also have been high). Additional major prey taxa of females include insects, sponges, bryozoan colonies, algae, fish carrion, and mosses.

Males show little in the way of important interspecific variation in diet. To wit, Vogt (1981a) found the diets of males of three sympatric species so similar that he simply listed major prey taxa without quantification or comparison. Males have most frequently been reported to specialize on eating caddis fly larvae (e.g., IRI values ranging from 40 to 98 in tables 4.4, 4.7, 4.9, 4.13, 4.16, and 4.17; Moll [1977] also reported nearly exclusive feeding on caddis fly larvae by false and Ouachita map turtles in the Illinois River). Males of some populations rarely take caddis fly larvae or other aquatic insects, however (e.g., tables 4.2, 4.3). Sponges, algae, and various other insects are often important parts of male

Table 4.2. Diet of the yellow-blotched sawback from the lower Pascagoula River and various upstream locations

| Taxon[a] | Lower Pascagoula River | | | | | | | | | Various upstream localities | | | | | | | | |
| | Juveniles | | | Males | | | Females | | | Juveniles | | | Males | | | Females | | |
	%F	%V	IRI	%F	%V	IRI	%F	%V	IRI	%F	%V	IRI	%F	%V	IRI	%F	%V	IRI
Mussels and clams				9	1	0.1	61	55	49									
Sponges	75	84	92	86	91	98	82	43	51	39	89	86	50	55	67	43	65	58
Bryozoans	25	11	4	5	0.1	0.01	7	0.1	0.01	9	0.1	0.02	3	4	0.3	29	0.1	0.1
Insects	51	5	4	46	2	1	25	1	0.3	61	9	14	53	4	5	57	35	42
Algae				9	6	0.7	7	0.6	0.1	9	1	0.3	31	38	28			

Note: Index of relative importance (IRI) calculated based on percentage frequency (%F) and percentage of total volume (%V) for each prey taxon for each group. Data from the lower Pascagoula were obtained via collection of feces, while those from upstream localities were obtained via dissection. In some cases a taxon was present in one or more samples but with negligible volume, in which case IRI is 0. Data from Seigel and Brauman (1994).

[a]An "unidentified" category had IRI values <3 for all classes.

Table 4.3. Diet of the black-knobbed sawback from the Mobile-Tensaw Delta

Taxon[a]	Males			Females		
	%F	%V	IRI	%F	%V	IRI
Mollusks	20	10	7	24	18	16
Sponges	47	37	61	35	28	37
Bryozoans	20	12	8	29	24	26
Algae						
Cladophora	13	12	6	24	8	7
Ulothrix	13	9	4	6	3	0.7
Spirogyra	20	20	14	18	17	12

Note: Index of relative importance (IRI) calculated for each prey taxon for each group calculated based on percentage frequency of occurrence (%F) and mean percentage volume across samples (%V). Data were obtained via dissection (Lahanas 1982).
[a]Taxa that had IRI values <3 for both classes included insects, crabs, and barnacles.

Table 4.4. Diet of the ringed sawback from the Pearl drainage

Taxon[a]	%F	%N	IRI
Caddis flies	79	27	65
Dipteran flies	14	35	15
Mayflies	10	23	7
Beetles	45	6	8
Vegetation	28	6	5

Note: Data based on combined samples from males and females (Kofron 1991), with index of relative importance (IRI) for each prey taxon calculated based on percentage frequency of occurrence (%F) and percentage of total prey items across samples (%N). Data were obtained via dissection.
[a]Taxa with IRI values <3 included dragonfly larvae and earthworms.

diets, with much feeding apparently done by grazing the prey communities of submerged deadwood. Males of two sympatric species in a reservoir fed heavily on small spiral egg cases of midges (tables 4.6 and 4.15), which they apparently scraped from submerged substrates as well.

Mollusks are generally of low importance in male diets. There is little in the way of predictable variation for males among species whose females are microcephalic (range of summed IRI values for mollusks 0–7, mean 1.1; tables 4.2–4.7), mesocephalic (range 0–98, mean 24.9; tables 4.9–4.11 and 4.13–4.16), or megacephalic (2; table 4.17). One exception to the sexual contrast in diet is the nearly exclusive reliance by both male and female common map turtles on small snails in Missouri's Niangua River, where IRI for snails was 99 in a com-

Table 4.5. Diet of the Ouachita map turtle in Meredosia Lake, floodplain of the Illinois River

Taxon[a]	Juveniles		Males		Females	
	1899, 1931	1975–76	1899, 1931	1975–76	1899, 1931	1975–76
Coontail (*Ceratophyllum*)			14		11	
Waterweed (*Elodea*)			2		4	
Water nymph (*Najas*)			8		42	
Pondweed (*Potamogeton*)			12		34	
Eelgrass (*Vallisneria*)					3	
Muskgrass (*Chara*)					3	
Filamentous algae			15		0.7	
Fly larvae/ detritus	23	99	7	98		95
Caddis fly larvae/detritus			5			
Mayfly nymphs	29		5			
Dragonfly nymphs	23		18			0.2
Wasps and ants	11	0.24	4	0.4		1
Beetles	6	0.7	2	0.2		2
Homopterans	1	0.011	3	0.08		0.3

Note: Data expressed as index of relative importance calculated for each prey taxon based on percentage frequency of occurrence and percentage of total volume across samples, showing historical change in predominant prey via comparison with dissected museum specimens (Moll 1977). [a]Taxa with index of relative importance values <3 for all classes in all time periods included knotweed (*Polygonum*), stonewort (*Nitella*), terrestrial vegetation, unidentified plants, adult flies, adult dragonflies, orthopterans, unidentified insects, oligochaete worms, and unidentified animals.

bined sample of nearly equal numbers of each sex (table 4.12). Historical data from dissected museum specimens suggest a similar high reliance decades ago on snails and small sphaeriid clams by male common map turtles in Illinois (tables 4.10 and 4.11). Bulté, Gravel, et al. (2008) and Bulté, Irschick, et al. (2008) likewise reported that both male and female common map turtles preyed heavily upon snails in a lake in Ontario, although males and small females of similar body sizes ate more caddis fly larvae and fewer zebra mussels than did adult females. Richards-Dimitrie et al. (2013) fount that both male and female com-

Table 4.6. Diet of the Ouachita map turtle in Kentucky Lake, a reservoir on the Tennessee River

Taxon[a]	Juveniles %AFDM	%F	IRI	Adult males %AFDM	%F	IRI	Small females %AFDM	%F	IRI	Large females %AFDM	%F	IRI
Asian clams							0.04	6	0.006	0.5	11	0.3
Dead Asian clams							6	6	0.9			
Snails	21	64	44							0.04	11	0.02
Scuds	10	36	12									
Midge egg cases	2	27	1	36	55	43	0.7	17	0.3			
Dipteran fly larvae	12	18	7	11	82	20	4	61	8	0.4	22	0.4
Dipteran fly adults	5	27	5	1	9	0.3				10	22	11
Beetles	7	27	7	6	18	2	0.4	11	0.1	6	44	14
Mayflies				24	36	20	10	22	6	16	33	26
Bryozoa				5	9	1	24	50	34	13	22	14
Sponges				1	9	0.3	10	22	6	0.08	11	0.05
Fish										9	11	5
Filamentous algae	14	18	8	14	36	12	33	44	42	16	22	17
Leaves							1	6	0.2	7	22	7
Water milfoil	13	27	12									

Notes: Index of relative importance (IRI) calculated from mean percentage ash-free dry mass (%AFDM) and percentage frequency of occurrence (Lindeman 1997c, 2000b). Small females were <160 mm midline plastron length, the size of the smallest known mature female. Data were obtained via stomach flushing.
[a]Taxa with IRI values <3 for all classes included oligochaete segmented worms, leeches, spiders, sow bugs, ostracods, copepods, centipedes, caddis fly larvae and adults, water boatmen, wasps, ants, homopteran and hemipteran bugs, stonefly nymphs, ectoproct colonies, seeds, and grasses.

mon map turtles in the lower Susquehanna River in Maryland were mollusk specialists but differed in taxa of mollusks taken. Females ate large pleurocerid snails almost exclusively, while males ate smaller snails (physids, hydrobiids, and planorbids) and corbiculid Asian clams, but no pleurocerid snails, and overall relied less heavily on mollusks than females did.

The diets of unsexed juveniles have been somewhat less well quantified. While large mollusks are avoided, small snails are a large component of ju-

Table 4.7. Diet of Sabine map turtles from the Whiskey Chitto River in Louisiana

Taxon[a]	Juveniles			Adult males			Females		
	%F	%M	IRI	%F	%M	IRI	%F	%M	IRI
Asian clams	29	6	2	20	5	1	12	4	0.5
Caddis fly larvae	100	78	90	100	81	90	94	28	28
Unidentified insects	43	6	3	10	2	0.2	12	0.01	0.001
Moss	14	0.06	0.01	60	0.9	0.6	94	8	8
Algae	57	6	4	80	9	8	100	60	64

Note: Calculation of index of relative importance (IRI) based on percentage frequency of occurrence (%F) and mean percentage dry mass across samples (%M; data from Shively 1982). Data were obtained via dissection.
[a]Taxa with IRI values <3 for all classes included dragonfly nymphs, mayfly nymphs, dipteran fly larvae, dobsonfly larvae, beetles, and vascular plants.

venile diets in three cases, with IRI values exceeding those of all other larger-sized classes (tables 4.6, 4.13, and 4.15). Otherwise, juvenile diets mirror those of larger individuals: caddis fly larvae are often important (tables 4.7, 4.13, and 4.17), or sponges may predominate (table 4.2).

Invasive Mussels, Habitat Alteration, and Dietary Change

The genera *Corbicula* (Asian clams) and *Dreissena* (zebra and quagga mussels) have introduced sometimes dramatic changes in the dietary ecology of meso- and megacephalic females. *Corbicula* mussels were introduced as food sources by Asian immigrants in the northwestern United States in the 1920s and spread to the rivers inhabited by map turtles and sawbacks by the 1950s (McMahon 1982). They were first reported as predominant prey in female diets in Escambia and common map turtles (Shealy 1976; Moll 1977). Shively and Vidrine (1984) reported nearly exclusive reliance on Asian clams by most female Mississippi map turtles in two Louisiana streams. Heavy consumption of *Corbicula* has also been reported in female Cagle's (IRI = 94; table 4.9) and Texas map turtles (IRI = 94–99; table 4.16). In the latter species, a series of specimens collected in 1949, prior to invasion of its habitat, allowed characterization of their change in diet. In the 1949 sample, the most important prey of females were caddis fly larvae, native fingernail clams, sponges, bryozoans, and algae, but in samples collected in 1998–99, on average 95 percent of fecal samples from adult females were Asian clams. Asian clams were also important in the diet of Mississippi map turtles in a large reservoir on the Tennessee River in Kentucky, albeit not as predominantly (IRI = 44; table 4.15), and in Maryland, they were the third most important component of the diet of male common map turtles but nearly absent from female diets (Richards-Dimitrie et al. 2013).

Table 4.8. Diets of adult female common, false, and Ouachita map turtles from the Mississippi River in Wisconsin

Taxon	Common map turtle %F	%V	IRI	False map turtle %F	%V	IRI	Ouachita map turtle %F	%V	IRI
Mollusks	81	66	90	55	19	22	39	3	2
Fish	19	12	4	26	11	6	20	5	2
Caddis fly larvae	33	1	0.6	39	3	2	57	8	9
Mayfly nymphs	14	10	2	29	18	11	37	42	32
Damselfly nymphs	24	4	2	21	1	0.4	11	1	0.2
Vegetation	5	4	0.3	63	42	55	67	32	44
Miscellaneous	24	3	1	32	6	4	50	9	9

Note: Calculation of index of relative importance (IRI) based on percentage frequency of occurrence (%F) and percentage of overall volume (%V) of prey recovered. Data were obtained via dissection and are approximated here from bar graphs in Vogt (1981a).

Table 4.9. Diet of Cagle's map turtles from the Guadalupe River

Taxon[a]	Males %F	%M	IRI	Females %F	%M	IRI
Asian clams	1	0.2	0.005	50	88	94
Fingernail clams	1	0.04	0.001	14	0.8	0.2
Snails	45	16	13	43	0.7	0.6
Caddis fly larvae	82	46	70	29	1	0.9
Beetles	30	5	3			
Mayflies	26	10	5	21	0.3	0.1
Damselflies and dragonflies	37	8	5	8	0.6	0.4
Grapes	1	0.4	0.01	21	8	4

Note: Index of relative importance (IRI) for each prey taxon for each group calculated from percentage frequency of occurrence (%F) and percentage of total dry mass across samples (%M). Data were obtained via dissection (Porter 1990).
[a]Taxa with IRI values <3 for all classes included dipteran flies, hemipteran and homopteran bugs, bees, wasps, ants, dobsonfly larvae, crickets, grasshoppers, stoneflies, unidentified insects, sow bugs, other noninsect arthropods, American elm, willow seeds, willow leaves, and pondweed.

Table 4.10. Diet of the common map turtle in Meredosia Lake, floodplain of the Illinois River

Taxon[a]	Juveniles		Males		Females	
	1899, 1931	1975–76	1899, 1931	1975–76	1899, 1931	1975–76
Fly larvae/detritus	31	98	1	98	2	80
Mayfly nymphs	14				0.2	
Dragonfly nymphs	19		0.4		0.5	
Wasps and ants	5	0.5	0.2	1		4
Beetles	3	0.4	0.04	0.5		3
Snails	22		93		80	0.5
Fingernail clams			5	0.05	18	4
Fish						6
Unidentified animals	4	0.06				

Note: Data expressed as index of relative importance calculated for each prey taxon based on percentage frequency of occurrence and percentage of total volume across samples, showing historical change in predominant prey via comparison with dissected museum specimens (data from Moll 1977).
[a]Taxa with index of relative importance values <3 for all classes in all time periods included terrestrial vegetation, adult flies, adult dragonflies, homopterans, unidentified insects, crayfish, oligochaete worms, and river mussels.

Eurasian zebra and quagga mussels were introduced to the Great Lakes as larvae stowed away in the ballast of large ships in the 1980s (Hebert et al. 1989; May and Marsden 1992), but their place in the diet of the common map turtle was not established until about two decades later. Serrouya et al. (1995) studied molluscan prey preferences in juvenile common map turtle females kept in laboratory conditions. The turtles would consume zebra mussels but preferred a native snail that was calculated to yield 72 percent more energy per unit of time spent feeding. Females in a Lake Erie population of common map turtles fed very heavily on zebra and quagga mussels (IRI = 73 in large juvenile females and 98 in adult females; table 4.13), however, suggesting that prey preferences were probably secondary to prey availability. Bulté and Blouin-Demers (2008) also found zebra and quagga mussels predominating in the diet of female common map turtles in Ontario and emphasized the change that consumption of dreissenid mussels has produced in the energetics of north temperate lakes. Prior to invasion, common map turtle females fed nearly exclusively on prey that are part of the littoral food web, but zebra and quagga mussels feed primarily on pelagic organisms; thus as much as one-third of the standing crop of the common map turtle in Lake Opinicon now originates from pelagic trophic pathways. Zebra and quagga mussels now overlap with the ranges of two species of map turtles in which females feed on mollusks, not only in the Great Lakes

Table 4.11. Diet of the common map turtle in the Illinois River and a reservoir

Taxon	Illinois River (1890s)			Peoria Lake (1975–76)		
	Juveniles	Males	Females	Juveniles	Males	Females
Fly larvae/ detritus	15					
Other flies	0.1			11	13	
Mayfly nymphs	23					
Dragonfly nymphs	19					
Dragonfly adults	0.03			12	6	
Wasps and ants	6			46	25	
Beetles	4			18	21	
Homopterans	1.4			5	1.3	
Unidentified insects	2.1			5	4	
Crayfish	1.5			0.02	14	13
Snails	13	53	20		12	
Fingernail clams	10	41	80	2	3	25
River mussels		7				
Asian clams				0.3		51
Fish	3			0.5		10
Unidentified animals	0.7			0.2	1	1

Note: Specimens from the Illinois River were collected during the 1890s, and those from Peoria Lake in 1975–76. Data expressed as index of relative importance calculated for each prey taxon based on percentage frequency of occurrence and percentage of total volume across samples (data from Moll 1977).

but also in the greater Mississippi drainage (fig. 4.2), and may thus be important prey on a wider geographic scale than has been appreciated.

Accurately characterizing the contrast in diet between meso- and mega-cephalic females is complicated by inclusion of invasive mollusks in the diet. The historic change in the diet of mesocephalic female Texas map turtles indicates a shift from very moderate feeding on native mollusks (summed IRI values = 24) to nearly exclusive feeding on invasive clams (IRI = 94 for stomach contents, 99 for feces; table 4.16). Because female Cagle's map turtles have similar trophic morphology and feed almost exclusively on the same invasive species in comparable riverine habitat in Texas, a parallel historic change is presumed to have occurred in this species as well (Lindeman 2006a). In Lake Erie, mesocephalic common map turtle females fed almost exclusively on dreissenid

Table 4.12. Diet of the common map turtle
from the Niangua River, Missouri

Taxon	%F	%V	IRI
Snails	99	94	99
Crayfish	19	3	0.6
Beetles	3	2	0.1
Flies	3	0.4	0.01
Vegetation	3	0.9	0.03

Note: Data based on combined samples from
males and females, with index of relative importance (IRI) for each prey taxon calculated based
on percentage frequency of occurrence (%F) and
percentage of total volume across samples (%V).
Data were obtained via dissection (White and
Moll 1992).

Figure 4.2. A juvenile false map turtle basking on an exposed chunk of concrete cov-
ered with desiccated dreissenid mussels, invasive species that likely are now a major
food source of adult female false and common map turtles in the upper Mississippi
River. Photo from Headquarters Marina in Harpers Ferry, Allamakee County, Iowa.

Table 4.13. Diet of the common map turtle at Presque Isle, Lake Erie

Taxon[a]	Juveniles			Adult males			Small juvenile females			Large juvenile females			Adult females		
	%F	%V	IRI	%F	%V	IRI	%F	%V	IRI	%F	%V	IRI	%F	%V	IRI
Zebra mussels with shells	33	3	2	20	4	1.3	44	28	28	92	62	73	100	95	98
Zebra mussel soft parts				6	6	0.5	11	11	3						
Snails	67	33	32	37	18	10	22	13	6	50	5	3	33	1	0.4
Caddis fly larvae	83	48	59	86	66	87	67	41	60	58	31	23	33	4	1
Monocot leaves	17	4	1	11	0.4	0.07	22	6	3	25	1	0.4	19	0.09	0.02
Seeds	50	5	3	3	0.05	0.002	11	0.1	0.03						

Note: Index of relative importance (IRI) for each prey taxon for each class of turtles calculated based on percentage frequency of occurrence (%F) and mean percentage volume across samples (%V). Small juvenile females had the same range in body sizes as adult males. Data were obtained via collection of feces (Lindeman 2006b).

[a]Taxa with IRI values <3 for all classes included mayfly nymphs, beetles, stonefly nymphs, damselflies, unidentified insects, fish carrion, vegetative stalks, duckweed, dicot leaves, water milfoil, stalked algae, and filamentous algae.

Table 4.14. Diet of the false map turtle in Meredosia Lake, floodplain of the Illinois River

	Juveniles		Males		Females	
Taxon[a]	1899, 1931	1975–76	1899, 1931	1975–76	1899, 1931	1975–76
Coontail (Ceratophyllum)			13		29	
Waterweed (Elodea)					10	
Water nymph (Najas)			4		13	
Knotweed (Polygonum)			18		2	
Pondweed (Potamogeton)					36	
Eelgrass (Vallisneria)					7	
Filamentous algae			13			
Fly larvae/detritus	16	98	4	97		95
Other flies	3	0.3	0.9	0.4		0.2
Mayfly nymphs	6		12			
Dragonfly nymphs	61		28	0.006		
Wasps and ants	5	0.7	3	0.3		0.9
Beetles	3	0.3	2	0.4		0.7
Unidentified animals	3			0.1		

Note: Data expressed as index of relative importance calculated for each prey taxon based on percentage frequency of occurrence and percentage of total volume across samples, showing historical change in predominant prey. Data from Moll (1977).
[a]Taxa with index of relative importance values <3 for all classes in all time periods included muskgrass (Chara), stonewort (Nitella), terrestrial vegetation, unidentified plants, adult dragonflies, orthopterans, homopterans, and unidentified insects.

mussels, but other studies of diet indicate the species' females often have a more varied diet (Lindeman 2006b).

Not all species show a shift to such heavy reliance on invasive mussels. For adult female common map turtles in an Illinois reservoir, Asian clams had an IRI of 51 (table 4.11), while for adult female Mississippi map turtles in a Kentucky reservoir, Asian clams had an IRI of 44 in stomach contents (table 4.15). Dreissenid mussels accounted for only 28 percent of the diet in female common map turtles in Lake Opinicon (Bulté and Blouin-Demers 2008). In the limestone-rich Chipola River, native snails had IRI values of 14 for male-sized juvenile females and 99 for larger females in Barbour's map turtle, while Asian

Table 4.15. Diet of the Mississippi map turtle in Kentucky Lake, a reservoir on the Tennessee River

Taxon[a]	Juveniles			Adult males			Small females			Large females		
	%AFDM	%F	IRI	%AFDM	%F	IRI	%AFDM	%F	IRI	%AFDM	%F	IRI
Asian clams	65	78	88	15	40	23	27	56	33	26	54	44
Dead Asian clams										7	8	2
Mussel gills				4	5	0.7						
Unionid mussels										3	8	0.6
Snails	12	22	5	1	20	1	29	56	35	3	15	1
Caddis fly larvae	5	11	0.9	2	20	2	2	22	0.8	1	8	0.2
Midge egg cases	1	11	0.1	23	40	36				1	8	0.3
Dipteran fly larvae	16	22	6	2	35	3	1	44	0.9	0.2	15	0.1
Dipteran fly adults							0.2	11	0.1			
Mayflies				5	5	1				20	31	19
Bryozoan colonies				8	10	3	12	22	6	2	15	1
Sponges				13	15	7	1	22	0.6			
Fish				0.3	5	0.07	7	11	2	24	31	23
Filamentous algae				20	30	23	22	44	22	12	23	9

Note: Index of relative importance (IRI) calculated from mean percent ash-free dry mass (%AFDM) and percent frequency of occurrence (Lindeman 1997c, 2000b). Small females were <170 mm midline plastron length, the size of the smallest known mature female. Data were obtained via stomach flushing.

[a] Taxa with IRI values <3 for all classes included water mites, spiders, beetles, ants, stoneflies, moths, planarian flatworms, and seeds.

Table 4.16. Diet of the Texas map turtle in the South Llano River

Taxon[a]	Adult males			Small females		Large females		
	1949 colons	1998–99 stomachs	1998–99 feces	1998–99 stomachs	1988–99 feces	1949 colons	1998–99 stomachs	1998–99 feces
Asian clams			0.3		43		94	99
Fingernail clams						23		
Snails		8	14	33.3	28	1	0.6	0.8
Caddis fly larvae	78	40	51		7	30	1	0.001
Mayfly nymphs		43	0.01	25				
Beetles	5	3	1	8		1	2	0.001
Hemipteran adults		4	0.004					
Unidentified insects	15	0.3	29	8	19	1		0.1
Bryozoan colonies		0.1				17	0.3	
Sponges			0.09			21		
Filamentous algae		0.2	0.4	25	0.08	5		0.001

Note: Data expressed as index of relative importance calculated for each prey taxon for each class based on percentage frequency of occurrence and mean percentage volume across samples. Small females were those that overlapped with adult males in body size. Data from 1949 were obtained via dissection, while data from 1998 to 1999 were obtained via stomach flushing and collection of feces (Lindeman 2006b).
[a]Taxa with IRI values <3 for all classes included crayfish, psephenid beetle larvae, damselfly nymphs and adults, dragonfly nymphs, dobsonfly larvae, moths, grasshoppers, springtails, sow bugs, oligochaete segmented worms, leeches, fish carrion, dicot leaves and stems, monocot leaves, grass seeds, grass inflorescences and other inflorescences.

clams were a negligible component of the diet (table 4.17). Ewert et al. (2006) suggested the situation for the species was similar in a sandy reach of the lower Apalachicola, based on cursory examination of feces.

Moll (1977) dissected specimens from the 1890s and early 1900s and collected dietary samples in the 1970s to examine the effects of habitat degradation in a floodplain lake of the Illinois River on the diets of sympatric common, false, and Ouachita map turtles (tables 4.5, 4.10, and 4.14). By the 1970s, all three were subsisting on little more than fly larvae (primarily chironomid midges), which had become superabundant in the lake bed due to sedimentation and nutrient enrichment. The museum specimens had more diverse diets, including a typically heavy reliance on mollusks by the common map turtle

Table 4.17. Diet of Barbour's map turtle from the Chipola River

Taxon[a]	Juveniles			Males			Small females			Large females		
	%F	%V	IRI	%F	%V	IRI	%F	%V[b]	IRI	%F	%V	IRI
Snails	20	0.9	0.2	38	6	2	67	22	14	100	97	99
Asian clams	20	0.7	0.2				33	0.1	0.3	20	3	0.7
Caddis fly larvae	100	73	78	100	94	98	100	88	86			
Beetles	80	25	21	13	0.1	0.01	67	0.1	0.6			

Note: Index of relative importance (IRI) for each prey taxon for each group calculated based on percentage frequency (%F) and percentage of total volume across samples (%V). Small females are those that overlap with adult males in body size. Data were obtained via dissection (Sanderson 1974).
[a]Taxa with IRI values <3 for all classes included mayfly and dragonfly nymphs, moth caterpillars, dipteran fly larvae, and algae.
[b]Sanderson's percentage volume data for small females sum to 110, presumably due to 10 extra percentage points tallied erroneously for either snails or caddis fly larvae.

and high importance values for aquatic plants and filamentous algae for the other two species. Similarly, a reservoir population of common map turtles was more insectivorous and less molluscivorous than a population in lotic habitat downstream (table 4.11).

Patterns in the Diets of Sympatric Species

Co-occurrence of map turtle and sawback species is typically characterized by differences in the relative head width of adult females (table 4.18) that relate to differences in consumption of mollusks. The pattern evokes the ecological phenomenon of character displacement, by which divergence in trophic morphology between species that come into sympatry with one another would be selectively advantageous because it would reduce interspecific competition for food. However, parsimony analysis suggests there have been few major shifts in the radiation of *Graptemys* regarding female trophic morphology, such that displacement would be limited to the early history of the genus, if it has occurred at all (Lindeman 2000a). The lack of coexisting species with similar trophic morphology is thus better suggestive of an alternate process, "character assortment," in which the lack of sympatry of species that are morphologically similar is attributed to competitive exclusion (Lindeman 2000a; the term is an inadvertent derivation of "size assortment," used by Losos [1990] for anole species in the Lesser Antilles, and the concept is also called "species assortment"; see Dayan and Simberloff 2005). While assortment is a purely ecological process, without evolved changes in morphology, character displacement involves evolutionary divergence (Roughgarden 1995; Dayan and Simberloff 2005).

It is not clear biogeographically how often species of similar trophic morphology might have come into contact with one another. Nor is it clear why

Table 4.18. Co-occurrence of *Graptemys* species in Gulf Coast river drainages

Drainage	Microcephalic	Mesocephalic	Megacephalic
Apalachicola			G. barbouri
Choctawhatchee			G. barbouri and G. ernsti
Escambia-Yellow			G. ernsti
Mobile Bay			
Upper drainages		G. geographica	G. pulchra
Middle drainages	G. nigrinoda	G. geographica	G. pulchra
Lower drainages	G. nigrinoda		G. pulchra
Pascagoula	G. flavimaculata		G. gibbonsi
Pearl	G. oculifera		G. pearlensis
Upper Mississippi	G. ouachitensis	G. pseudogeographica and G. geographica	
Lower Mississippi	G. ouachitensis	G. pseudogeographica	
Mermentau	G. sabinensis	G. pseudogeographica	
Calcasieu	G. sabinensis	G. pseudogeographica	
Sabine-Neches	G. sabinensis	G. pseudogeographica	
Trinity		G. pseudogeographica	
Brazos		G. pseudogeographica	
San Bernard		G. pseudogeographica	
Colorado		G. versa	
Guadalupe		G. caglei	

Note: Species arranged according to the trophic morphology of females.

displacement or assortment might occur for females but not be important for males, which have not been shown to have highly divergent diets in sympatry.

Ecological Implications of Sexual Differences in Diet

Intrasexual differences in diet could conceivably result from selection for intra-specific niche divergence to reduce competition between the sexes or from sexual differences in reproductive roles that favor ingestion of different prey (Bulté, Irschick, et al. 2008). Bulté, Irschick, et al. (2008) showed that females with wide heads relative to their body size had high body condition index values (chapter 7) and produced large hatchlings. They argued that their data supported the reproductive-roles hypothesis for niche divergence. Also, the size range of in-gested mollusks increased with increased body size in both sexes, which does not support the hypothesis that the sexes have diverged in diet due to intraspecific competition (Bulté, Gravel, et al. 2008). On this point, it is also of note that intraspecific niche divergence is mediated primarily by differences in body size and secondarily by divergence in the trophic apparatus (see chapter 7), but the divergence in body size arises via considerable intersexual differences in age at maturity (Lindeman 1999b). The fact that females mature so much later than males argues against selection for simple intersexual competitive divergence, as females incur a considerable fitness cost by delaying maturation; that cost is presumably more than offset by the increase in fitness secured by consumption of mollusks to fuel reproduction in large females and by the advantages of producing large eggs.

Feeding Behavior

Only anecdotal information has been published on foraging behavior, chiefly regarding males. Male Barbour's and Cagle's map turtles scrape prey from submerged limestone formations and small rocks, respectively (Lee et al. 1975; Porter 1990). The three sawbacks and false and Sabine map turtles "graze" on prey communities that grow on submerged deadwood (Waters 1974; Moll 1976b; Shively and Jackson 1985; Kofron 1991; Seigel and Brauman 1994; Selman and Qualls 2008a). On a Kentucky reservoir, I was consistently successful in dipnetting Ouachita map turtles (mostly males) from the boat wells of a dock, where they appeared to be attracted by the *aufwuchs* (encrusting growth of biological communities) that blanketed submerged supports; an exceptional aggregation of Ouachita map turtles also occurs at the floating docks of Paris Landing State Park in Tennessee (see "Thermoregulatory Behavior," below). Male Cagle's map turtles also apparently graze on live willow roots that house their preferred prey, caddis fly larvae (Killebrew et al. 2002). Consistent with grazing on deadwood, Kofron (1991) reported wood fragments in two-thirds of the stomachs of ringed sawbacks he examined. Mississippi map turtles and ringed sawbacks have also been observed to leave the water briefly to feed, in shoreline vegetation and on insect prey taken from logs, respectively (J. L. Carr 2008; Lindeman 2009a), and surface feeding on superabundant periodical cicadas has also been described (Powell and Powell 2011).

Growth

Map turtles and sawbacks show the typical turtle growth pattern: rapid juvenile growth that slows with age. Growth has been studied in eight species using mean sizes of turtles that were aged based on plastral annuli (including recaptures, to generate growth intervals) and nonlinear modeling based on turtles of known age or recapture intervals. In older turtles, annuli become harder to read because they become more closely spaced and the plastron eventually wears so smooth that no annuli can be seen; these findings are also common in other deirochelyines (D. S. Wilson et al. 2003). The prolonged rapid juvenile growth of females makes annulus counts possible into later ages. For males and females, respectively, annuli could be read to ages four and seven for common, false, and Ouachita map turtles (Moll 1976a, 1977); to ages five and nine in black-knobbed sawbacks (Lahanas 1982); to ages seven and ten in common map turtles (Iverson 1988); to ages five and six in ringed sawbacks (Jones and Hartfield 1995); to ages one and four in Mississippi map turtles (Lindeman 1999b); to ages two and four in Ouachita map turtles (Lindeman 1999b); to ages five and eight in Escambia map turtles (Lindeman 1999b); and to ages six and ten in common map turtles (P. Lindeman, unpublished data). The larger the size in females, the more growth annuli are readable; thus annuli were readable only to age six in Texas map turtles (maximum carapace length 183 mm; Lindeman 2005) but to age twelve in Ouachita map turtles in Wisconsin (maximum carapace length 242 mm; Vogt 1980a).

Based on mean sizes at known ages, growth trajectories are similar for juvenile males and females of the same age (Webb 1961; Moll 1976a, 1977; Gordon and MacCulloch 1980; Lahanas 1982; Iverson 1988; Jones and Hartfield 1995). However, these data—which cover only the first few year classes in a population—have not been critically examined for differences; in particular, the time of year when juveniles are measured strongly influences the size recorded (Lindeman 1997b), so averages of data collected over the course of a year do not address whether sexual differences occur in juvenile growth rate. Using Sergeev's (1937) formula to back-calculate plastron length from the lengths of the medial edge of plastral annuli in some studies also may be misleading, as the ratio changes with age (Moll 1976a; Iverson 1988).

Growth rates may vary with prey availability. In a floodplain lake of the Illinois River, where map turtles fed almost exclusively on fly larvae and benthic detritus, two of three species showed higher growth rates in a summer when larvae were approximately ten times as abundant as they had been the previous year, and false map turtles grew more rapidly in the lake than at a Mississippi River site where they were more herbivorous (Moll 1976a, 1977).

Nonlinear models used to study growth of ectotherms include the von Bertalanffy model, in which growth rate slows progressively with age; the logistic model, in which growth rate is maximal at half of asymptotic body size (the size at which growth slows to be negligible); and an intermediate model, the Gompertz model, in which juvenile growth can start out slower than maximal. As is true in most freshwater turtles, the von Bertalanffy model fits the data best for *Graptemys*, except in the case of female Escambia map turtles, for which data better fit the Gompertz model (Jones and Hartfield 1995; Lindeman 1999b; Bulté and Blouin-Demers 2009). Shealy (1976) suggested that the pattern of most rapid growth in large juvenile and small adult female Escambia map turtles might occur as a result of increased inclusion of an abundant exotic clam in their diet. Males approach their smaller asymptotic sizes more rapidly than females, a result explained by their earlier maturation (Jones and Hartfield 1995; Lindeman 1999b, 2005; Bulté and Blouin-Demers 2009).

A seasonally sinusoidal aspect to growth—that is, accelerating growth in the early growing season and decelerating growth later (Lindeman 1997b)—was indicated for both sexes of Mississippi map turtles but not for five other data sets on three other species (Lindeman 1999b). More data taken from dates throughout the active season will be necessary to critically evaluate this pattern. If confirmed, the sinusoidal pattern may arise from slow growth in the late spring and early autumn, which is related to recovery from and preparation for winter torpor (Lindeman 1997b). Lahanas (1982) found that growth of black-knobbed sawbacks essentially stopped two months prior to the cessation of feeding in the fall, as stored body fat increased.

Age at Maturity

Adult female map turtles and sawbacks range from about two to two and a half times the shell length of conspecific adult males (Gibbons and Lovich 1990) and are thus several times heavier. The exceptional degree of sexual size dimorphism that characterizes all fourteen species is apparently driven almost exclusively by strong bimaturism: an exceptional difference in age at maturity. Males mature as early as their second or third growing season, while females may not reach maturity until they are six to eighteen years old (Shealy 1976; Vogt 1980a; Lahanas 1982; Jones and Hartfield 1995; Lindeman 1999b; Lindeman 2005; Bulté and Blouin-Demers 2009). Female age and size at maturity are clearly correlated. While Texas map turtles matured at 115 mm midline plastron length in about their seventh season (Lindeman 2005), Escambia map turtles matured at 190 mm in their fourteenth to nineteenth season (Shealy 1976; Lindeman 1999b).

Body Size

Hatchling Body Size

Despite variation in adult body size, hatchling size shows no clear interspecific variation. For hatchlings from nests in the field, the range reported for plastron length across five species is just 23–35 mm and means range from 27.5 to 32.8 mm; the ranges for species with large samples span nearly the entire range of all reported sizes (table 4.19). Hatchlings grow little if at all prior to hibernation, as hatchlings in the spring and early summer are not much larger at fall emergence; indeed, specimens less than 40 mm in midline plastron length in spring or early summer are known for every species except Cagle's map turtle (P. Lindeman, unpublished data). In Kentucky, in May or June I captured nineteen hatchling Ouachita map turtles that averaged 31.8 mm in plastron length

Table 4.19. Hatchling plastron lengths at emergence from the nest

Species	Plastron length (mm)			N	Source
	Minimum	Mean	Maximum		
flavimaculata	28	29.9	31	24	Seigel and Brauman 1995
geographica	23	27.5	30	350	Nagle et al. 2004
nigrinoda	NR	32.8	NR	21	Lahanas 1982
oculifera	24	30.8	35	227	Jones 2006
ouachitensis	24	28.3	32	127	J. L. Carr 2001
pseudogeographica	27	29.7	34	25	J. L. Carr 2001

Note: NR = not reported.

(range 26–40 mm) and twenty-three hatchling Mississippi map turtles that averaged 34.4 mm (range 30–43 mm). Fall hatchling Cagle's map turtles averaged 37.8 mm (range 34–43 mm, $N = 9$), while the smallest of four hatchlings from the early half of the year measured 43 mm on 6 May, indicating that a short fall growth increment may occur in hatchlings of this southernmost species (see Lindeman 1999b).

Effect of Latitude on Body Size

Size at maturity, mean size, and maximum size all correlated positively with latitude in map turtles and sawbacks as well as in temperate-zone deirochelyines in general (Lindeman 2008a). Hypotheses that seek to explain the relationship of size to latitude have been proposed for widespread species such as the painted turtle (*Chrysemys picta*) but likely also apply to *Graptemys*: (1) larger size promotes retention of body heat in cooler northern climates, (2) larger size allows greater storage of energy for the longer northern winters, (3) larger size results from reduced interspecific competition because turtle diversity is lower at higher latitudes, or (4) large size is necessary for producing larger clutches to compensate for increased egg or juvenile mortality in the north (Iverson and Smith 1993; Lindeman 2008a).

Intraspecific Variation in Body Size

Body size may show substantial intraspecific variation on a local scale. The Texas map turtle is considerably smaller in the South Llano than it is in many other tributaries of the Colorado River drainage to which it is endemic (Lindeman 2005). Its maximum sizes are reached in the nearby San Saba (Lindeman 2005), even though it provides habitat superficially similar to the South Llano.

Two sawbacks show increased size at their downstream range limits. The yellow-blotched sawback is substantially larger in the lower Pascagoula than at

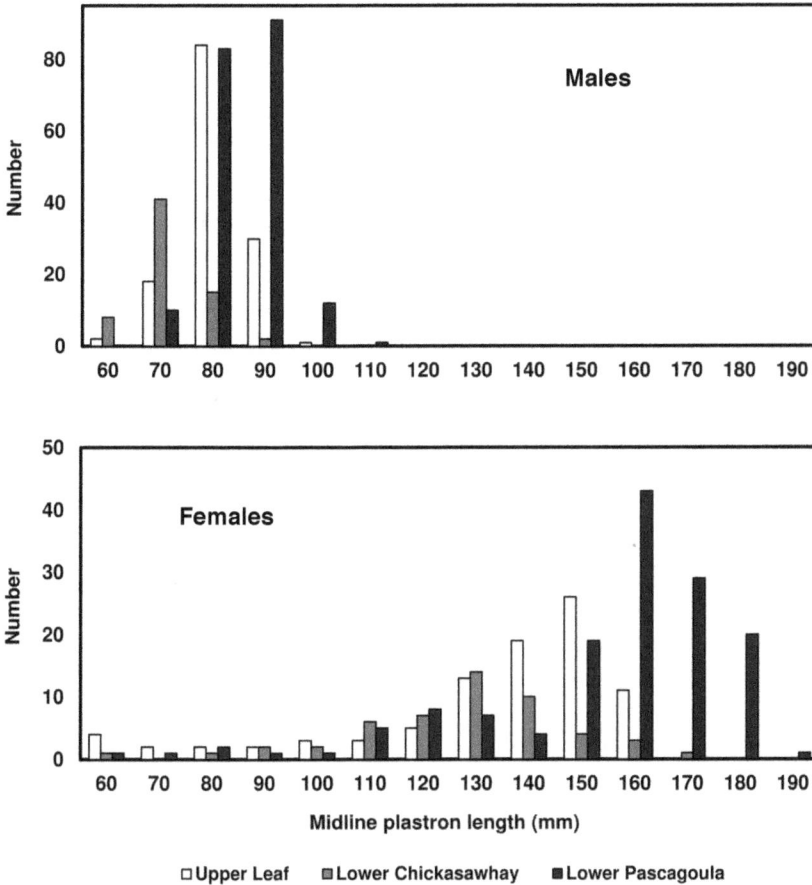

Figure 4.3. Size-frequency histograms for yellow-blotched sawbacks at three study sites (W. Selman, unpublished data 2005–2007; see also Selman 2012).

two upstream localities (fig. 4.3). Mean and maximum midline plastron lengths of females are 132 and 170 mm, respectively, on the lower Chickasawhay; 141 and 168 mm on the upper Leaf; and 159 and 189 mm on the lower Pascagoula. Corresponding figures for males are 77 and 93 mm on the Chickasawhay, 77 and 93 mm on the Leaf, and 90 and 118 mm on the Pascagoula (Selman 2012). The size contrast is also evident in data from previous workers. Museum specimens taken broadly throughout all segments of the species' range upstream of the lower Pascagoula had midline plastron lengths from 61 to 102 mm for adult males and up to 156 mm in adult females (Lindeman 2008a). On the lower Pascagoula, females mature at larger sizes than they do elsewhere: Cagle (1954) reported that the smallest of 3 adult females from the middle Pascagoula was mature at 140 mm maximum plastron length, while the smallest of a sample of 134 adult females from the lower Pascagoula measured 157 mm maximum plastron length (Horne et al. 2003). Also for the lower Pascagoula, Jones (1993a) recorded a female that was 184 mm midline plastron length and Seigel and Brauman (1995) reported a female that was 250 mm maximum carapace length, which equates to approximately 212 mm maximum plastron length (~200 mm midline) based on ratios of means (Jones 1993a).

Male black-knobbed sawbacks have a maximum midline plastron length of 97 mm for all areas upstream of the Mobile-Tensaw Delta but can be as large as 102 mm in the delta; corresponding maximum sizes for females are 180 mm upstream and 202 mm in the delta (Lindeman 2008a and unpublished data). There have been no similar comparisons for the ringed sawback at its downstream range limit in the Pearl.

The downstream changes in body sizes of the yellow-blotched and black-knobbed sawbacks are coincident with the near disappearance of sympatric congeners (Pascagoula and Alabama map turtles, respectively; Lahanas 1982; Jones 1993a; Godwin 2001, 2003). Factors that limit downstream range limits of sympatric species pairs are unknown. Salinity tolerance may differ between the sawbacks and their sympatric congeners, although a close relative of the latter, Barbour's map turtle, appears highly tolerant of brackish habitats on the lower Apalachicola (Ruhl 1991; Ewert and Jackson 1994). The changes in size suggest competitive release of sawbacks at downstream locations, where they occur almost in the absence of their broader-headed congeners, and that the increase in their body size may be an adaptation for increased intake of mollusks (Lindeman 2008a; Selman 2012).

Interspecific Patterns in Body-Size Variation

Based on the fact that broader-headed species have larger female body sizes than do sympatric conspecifics, body size might be subject to character displacement (Lindeman 2000b), which would promote increased dietary divergence within sympatric species pairs. Were this true, one would expect relaxed selection on body size in the four allopatric species with meso- or megacephalic species (Barbour's, Cagle's, Escambia, and Texas map turtles), but a comparative analysis incorporating the effect of latitude on body size did not support this hypothesis (Lindeman 2008a). Instead, the large body sizes of megacephalic females from the southeastern Gulf Coast—comparable to those exhibited at far northern latitudes by three congeners—may simply be a biomechanical necessity for supporting their large heads. Consider the group of female Barbour's map turtles in figure 4.4: if each female had its same head size but was reduced in shell length, width, and height by 20 percent, then body sizes would be comparable to those of southern, microcephalic females—but support of those massive, mollusk-crushing heads would certainly be problematic! The large female body size of the five megacephalic species is not a trivial matter, as it is likely achieved via exceptionally delayed maturation (Lindeman 1999b). Hence selec-

Figure 4.4. Five female Barbour's map turtles, showing the exceptionally enlarged heads that likely are linked to delayed maturity and large body size. Photo from Veteran's Park on the Flint River in downtown Albany, Dougherty County, Georgia.

tion for dietary biomechanics may be enmeshed in an important trade-off with reproductive maturity.

Sexual Differences in Body Size

Perhaps the most interesting question regarding *Graptemys* body size concerns the evolutionary origin of their extreme sexual size dimorphism. Their sexual size difference is not only the greatest known for any well-studied turtles (albeit possibly rivaled in some species of the geoemydid genera *Kachuga* and *Pangshura* in Asia; Gibbons and Lovich 1990), it is the greatest known among all tetrapod vertebrates (Stephens and Wiens 2009). Comparative analyses involving other temperate-zone deirochelyine turtles suggest that change in body size has been substantially greater in males, which are considerably smaller than expected after correction for latitude, than in females, which have not evolved to be appreciably larger (Lindeman 2008a). One hypothesis for why males have become so much smaller holds that they can mature early in life at small sizes because their strict use of aquatic habitats frees them from mammalian predators that males of other deirochelyines may encounter on land; females would remain tied to their larger body sizes because of selection for high fecundity and because they may encounter mammalian predators while nesting (Lindeman 2008a).

Survival and Longevity

Annual survival has not been quantified for hatchling or young juvenile stages of any species of *Graptemys*, and for older turtles, estimates have been made for only one population of common map turtles over three years, with females having slightly higher annual survival than males (means 87 vs. 83 percent; Bulté and Blouin-Demers 2009). Historically, there has been a dearth of the long-term studies necessary for such estimates; this may be changing, as several mark-recapture programs conducted over multiple years are under way. The fluid nature of large, dispersed *Graptemys* populations in rivers and large lakes (e.g., Shealy 1976; Vogt 1980a; Flaherty 1982; Pluto and Bellis 1988; Craig 1992; Jones and Hartfield 1995; Jones 1996; J. L. Carr 2001) will make estimating annual survival difficult, however, in that distinguishing between dead and dispersed individuals will not be possible.

The dearth of long-term studies of *Graptemys* populations also results in few longevity records. Snider and Bowler (1992) reported longevity records in captivity of 35 years and 5 months for a male false map turtle in the Columbus (Ohio) Zoo and 31 years and 8 months for a female Barbour's map turtle in the National Zoo in Washington, D.C. More recently, in the Columbus Zoo, two Barbour's map turtle females of unknown origin have been kept 43 and 37 years and females of all three sawbacks, plus male ringed sawbacks, have been maintained 21–23 years (T. Gerold, pers. comm. 2007). To date, the Tennessee Aquarium has maintained Barbour's map turtles for 21 years and Pearl, Ouachita, and false map turtles and ringed sawbacks for 15–16 years (D. Collins, pers. comm. 2008). In the wild, survival past 20 years of age is known in common map turtles (C. H. Ernst et al. 1994; C. H. Ernst and Lovich 2009) and ringed sawbacks (females at least 30 years; Jones and Selman 2009), but maximum longevity may well stretch decades further, based on records for other emydids (C. H. Ernst et al. 1994).

Reproduction

Seasonal Reproductive Cycles

The gonadal cycles of all *Graptemys* are probably similar. Testes grow during late summer, becoming largest by September (Timken 1968a; Shealy 1976; Vogt 1980a; Lahanas 1982; Kofron 1991). In the black-knobbed sawback, spermatogenesis begins in the testes in June and peaks in September, when mature sperm begin entering the epididymides, leaving the testes almost empty of sperm by November (Lahanas 1982); however, Shealy (1976) reported sperm in the epididymides of Escambia map turtles from April through December. The events

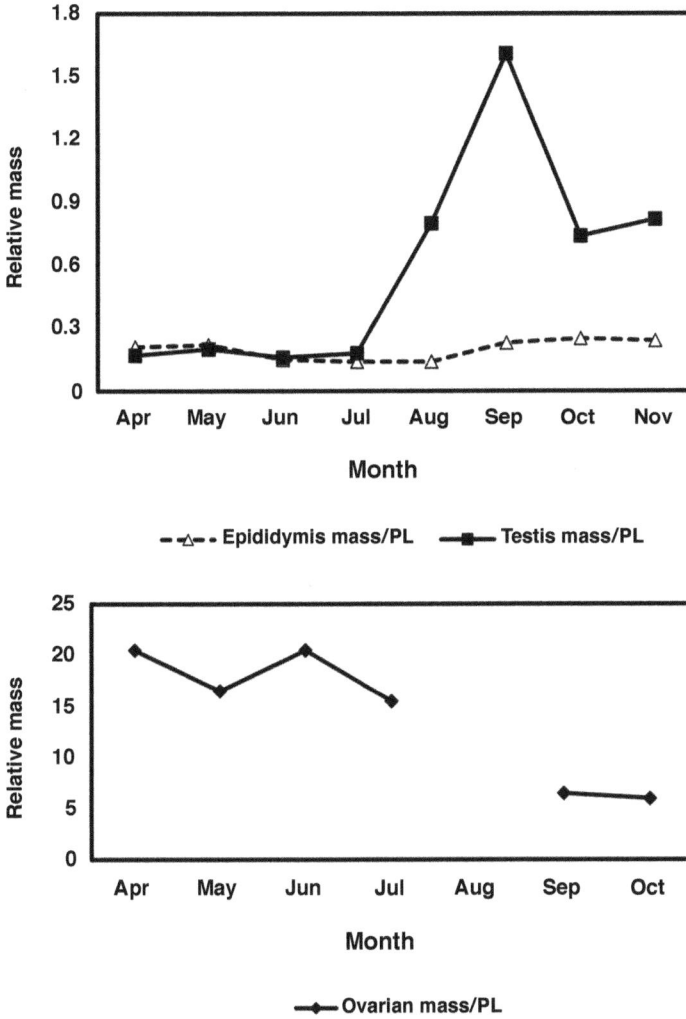

Figure 4.5. Seasonal cycles in the mass of the gonads as a function of body size (plastron length, PL) for *Graptemys nigrinoda* males (top) and females (bottom; based on data in Lahanas 1982).

of the spermatogenic cycle coincide with maximum relative mass of testes and epididymides (fig. 4.5; Lahanas 1982). Ovarian follicle proliferation and growth also begin in late summer and fall, with follicles reaching maximum size in spring (fig. 4.5; Timken 1968a; Shealy 1976; Vogt 1980a; Lahanas 1982; Kofron 1991). Females are gravid by late spring, and egg laying extends well into summer, with nesting beginning earlier and ending later at southern latitudes (table 4.20). Hatchlings emerge in the late summer or early fall (P. K. Anderson 1958; Wahlquist and Folkerts 1973; Haynes and McKown 1974; Shealy 1976; Lahanas

Table 4.20. Reports of female body size, clutch size, egg dimensions, and dates known to be gravid for species of *Graptemys*

Species	State or province	Source	Mean adult female size (mm)	CL or PL	Range
barbouri	FL	Cagle 1952b		PL	176–271
	FL	Ewert and Jackson 1994; Ewert et al. 2006			
	GA	Wahlquist and Folkerts 1973			
caglei	TX	Killebrew and Babitzke 1996			
ernsti	AL	Shealy 1976	244.7	CL	212–285
flavimaculata	MS	Horne et al. 2003			
geographica	MO	White and Moll 1991			
	PA	K. M. Ryan and Lindeman 2007	203.8	PL	169–228
	PA	Nagle et al. 2004			
	NY	Kiviat and Buso 1977			
	ON	Gordon and MacCulloch 1980			
	ON	Gillingwater 2001	234	CL?	186–261
	IN	Baker et al. 2010			
gibbonsi	MS	Lovich et al. 2009[b]	224	CL?	179–?
nigrinoda	AL	Lahanas 1982			
	AL	P. Lindeman, unpublished data	159.1	PL	135–180
oculifera	MS	Jones 2006	164.6	PL	130–215
ouachitensis	IL	Moll 1977			
	LA	Rosenzweig 2003	157.3	PL	137–182
	OK	Webb 1961	150.9	PL	154–170
	WI	Vogt 1980a	205	CL	163–242
	WI	Janzen et al. 1995			
pearlensis	MS	Lovich et al. 2009[b]			
pseudogeo-graphica	SD	Timken 1968a	206.5	PL	180–231
	SD	L. A. Dixon 2009			
	SD	A. Gregor, pers. comm. 2008	202	PL	177–238
	WI	Vogt 1980a	225	CL	193–274
	WI	Janzen et al. 1995			
	LA	J. L. Carr 2001	168.0	PL	153–190
pulchra	AL	P. Lindeman, unpublished data	199.6	PL	188–207
sabinensis	LA/TX	Ewert, Doody, et al. 2004			
	LA	P. Lindeman, unpublished data	171.5	PL	152–185
versa	TX	Lindeman 2005	133.8	PL	115–163

Note: CL = carapace length, PL = plastron length.

[a] Ranges for egg dimensions previously reported as ranges of intraclutch means.

[b] Data on female body size, egg size, and end of the nesting season pooled for *G. gibbonsi* and *G. pearlensis*.

Mean clutch size	Range	Mean egg width (mm)	Range	Mean egg length (mm)	Range	Mean egg mass (g)	Range	Dates gravid
5.2	4–11	25.9	22.2–29.3	37.1	31.0–40.4			
8.8	4–14							29 Apr–7 Jul
		29.5	27.6–30.8	39.4	38.3–41.6			
3.8	1–8	22.8		38.3		12.5		late Mar–late Aug
7.2		26.0		38.0				28 Apr–30 Jul
4.7	3–9							20 May–12 Aug
10.1	6–15							
11.9	3–21	22.3	19.2–24.9[a]	35.2	27.9–40.6[a]	10.3	6.7–14.5[a]	28 May–7 Jul
10.6	6–15							2 Jun–14 Jul
15.0	10–18							
12.7	9–17							
13.2	7–22			34.0	32.0–41.0			
9.0	6–14							
7.5		26.0		38.0			13.9–15.5	3 Apr–Aug
5.5	3–7	23.8	20.9–26.0	37.0	32.9–41.8			20 May–5 Aug
3.8	2–5	22.1	19.3–24.4					
3.6	1–10	22.7	17.6–26.8	38.9	28.3–47.0	11.9	5.0–17.4	30 Apr–15 Jul
14.0	11–18							
6.4	3–12	21.6	16.6–25.3	35.5	26.4–42.3	10.2	4.7–14.3	23 May–26 Jul
6.0	3–10							
10.5	6–16	22.0		34.0		10.2		18 May–26 Jul
11.3	8–17					11.4	7.4–14.3	
6.4		26.0		38.0			13.9–15.5	3 May–Aug
12.3	6–18	23.0		35.0				10 Jun–30 Jun
10.8	7–14	23.2	19.6–25.8	34.4	26.6–39.6			28 May–2 Jul
10.9	6–14	23.9	20.9–27.0	36.1	25.0–43.0			
14.1	8–19	25.1		34.2		9.9		18 May–26 Jul
12.8	8–19					10.9	7.6–13.7	
6.6	4–11	22.4	19.1–25.2	36.4	29.9–41.5	10.3	7.2–15.6	12 May–21 Jul
5.4	4–7	24.5	22.4–27.2					
2.3	1–4					9.9		13 Jun–16 Jul
5.3	3–7	23.5	21.3–26.5					
5.6	4–9	20.7	18.4–22.2	35.1	30.7–38.4			30 Apr–22 Jun

1982; Vogt and Bull 1984; Jones 2006), except in common map turtles, whose hatchlings overwinter in northern populations (see chapter 7).

Courtship and Mating

Face-to-face behaviors considered to represent courtship have been described for all *Graptemys* except the Alabama map turtle, mostly on the basis of captives (Cagle 1955; Wahlquist 1970; C. H. Ernst 1974; Jenkins 1978; Tryon 1978; Lahanas 1982; Fritz 1991; Vogt 1993; Artner 2001a, 2001b; Hertwig 2001). Field observations of courtship and mating are rare and anecdotal and fail to resolve the question of when mating typically occurs. Shealy (1976) reported courtship from September to November in Escambia map turtles but not in other months. Carrière et al. (2009) observed mating behavior by common map turtles at communal hibernacula in fall and spring. Courting pairs have been observed in April, July, October, and November for Texas map turtles (J. Dobie, pers. comm. 2008) and in August for black-knobbed sawbacks and Alabama map turtles (pers. observation 2009).

Males sometimes do not engage in courtship and simply mount females (Vogt 1993). More commonly, the male places his nose at the cloaca of the female and then adopts a face-to-face posture, bobbing his head vertically (common, Cagle's, Escambia, Pascagoula *sensu lato*, and Texas map turtles), vibrating the tops of his foreclaws against the female's ocular regions ("titillation"; Barbour's, Ouachita, and Sabine map turtles, ringed sawbacks), or engaging in both behaviors (false map turtles, black-knobbed and yellow-blotched sawbacks). Vogt (1993) described differences in titillation between sympatric species in which titillation bouts were similar in duration (about half a second), but the mean number of strokes per bout in false map turtles was twice that in Ouachita map turtles (10.3 vs. 5.2), possibly an important part of species recognition. Both head bobbing and foreclaw titillation are widespread in the Deirochelyinae; hence the absence of one or the other implies evolutionary loss of the behavior (Fritz 1991). Male foreclaws are elongated in some species but not in others (C. H. Ernst et al. 1994). Their use in display by some but not all species presumably relates to whether or not a species' foreclaws are sexually dimorphic.

In captivity, head bobbing and titillation occur between juveniles or same-sex pairs (Cagle 1955; Lahanas 1982; Vogt 1993). Recent studies of foreclaw displays in another deirochelyine, the slider (*Trachemys scripta*), found high rates of foreclaw displays between pairs of females (Thomas and Altig 2006) and frequent copulation without foreclaw display (Thomas 2002); hence calling such behaviors "courtship" may at best be overly restrictive labeling or at worst simply wrong, as they may have other social functions (Cagle 1955; Thomas and Altig 2006).

Nesting

Females typically nest on sandbars, islands, and lake shores, not far from the water (P. K. Anderson 1958; Timken 1968a; Wahlquist and Folkerts 1973; Shealy 1976; Gordon and MacCulloch 1980; Lahanas 1982; Vogt and Bull 1984; Killebrew and Babitzke 1996; Moulis 1997; Perkins and Backlund 2000; Horne et al. 2003; Rosenzweig 2003; Jones 2006; Barrett Beehler 2007; Carrière 2007; L. A. Dixon 2009; Geller 2012a). Nest sites are generally in large open areas, but the sun-shade ecotone is also used, often heavily, influencing hatchling sex ratios (Vogt and Bull 1984; see chapter 7). Small clearings or shaded forest sites may also be used (Shealy 1976; Ewert and Jackson 1994; Horne et al. 2003). In lakes lacking open sand at the shoreline, common map turtles move farther inland to find sandy nest sites (Newman 1906a; K. M. Ryan and Lindeman 2007).

Not all nesting occurs in natural sandy soils. Vogt (1980a) found aborted nesting attempts in new, dry, vegetation-free mounds of dredge spoil and Barbour's map turtle nests on spoil mounds created decades ago (Ewert and Jackson 1994; Ewert et al. 2006). On Presque Isle, Lake Erie, a compacted parking area with gravel mixed with sand and a 4 m tall dirt mound draw about as many nesting attempts by common map turtles (and more frequent aborted attempts) as adjacent natural sandy habitat (K. M. Ryan and Lindeman 2007; P. Lindeman, unpublished data). Some nests of the species in an Indiana lake were in plowed soil (Newman 1906a), and Cahn (1937) described nesting by a false map turtle in a dirt road.

The two Texas endemics likely have alternative nesting microhabitats as well. Texas map turtle nests have not been described, but the large sandbars typically used by species farther east are not available along most of the Colorado River and its tributaries. Killebrew and Babitzke (1996) described four nests of Cagle's map turtles from beaches on the lower Guadalupe with sand, sand loam, or clay loam substrates, but sandbars are absent from many other segments of the drainage (Haynes and McKown 1974).

Sandbars are also rare in impoundments, so reservoir populations must also use alternative nest sites. I never observed nesting of two species that I studied in Kentucky Lake (Lindeman 1998, 1999c, 2003a), an impoundment of the Tennessee River that lacks sandbars, but both were abundant and hatchlings were common every year. In June of 1994 I found a female Ouachita map turtle about 75 m from the lake and 15 m above it on a steep, grassy bank built to support U.S. Highway 68. She was oriented toward the water, and palpation revealed no eggs; thus she likely was returning from nesting in a nearby gravel roadside pull-off or on the forested hillside just beyond. While smooth softshells (*Apalone mutica*) dug nest cavities in shoreline gravel overlying their preferred sand at Kentucky Lake (Lindeman 2001b), I did not observe either map turtle species doing the same, and I suspect they nested farther inland in loamier soils.

Most nesting takes place in mid- to late morning (Shealy 1976; Vogt 1980a;

Vogt and Bull 1984; Jones 2006; K. M. Ryan and Lindeman 2007; Geller 2012a), although Moore and Seigel (2006) reported a midafternoon peak in nesting by yellow-blotched sawbacks, and nesting occurred exclusively in late evening in the black-knobbed sawbacks studied by Lahanas (1982). The most detailed descriptions of nesting behavior are for common, Ouachita, false, Escambia, and Alabama map turtles and black-knobbed and yellow-blotched sawbacks (Newman 1906a; Cahn 1937; Shealy 1976; Vogt 1980a; Lahanas 1982; Moore and Seigel 2006).

The following is a general description of common nesting behaviors. Females may aggregate at nesting beaches in the hours before emergence. The path taken from the water is a highly meandering one compared with the direct return path afterward. The female frequently presses her snout to the ground and may scratch at the sand with her forelimbs. Black-knobbed sawbacks cover their carapaces with sand with swings of the forelimbs, possibly as camouflage (Lahanas 1982). The female may begin and abandon nest holes, often because of an obvious obstruction such as a rock or plant root or sometimes when a storm begins. Aborted "desperation nests" with one to two eggs were also noted by Shealy (1976). Voiding of the cloaca may moisten the substrate and contribute to packing the sand so that the cavity remains intact. The cavity is dug by alternating motions of the hind feet and is flask shaped, with its neck slanted toward the female's head. She gradually becomes oriented on a slant with her head up and the rear of her carapace in a broad pit slightly below ground level. Deposition of eggs, spaced by one to two minutes, can be recognized by spasmodic vertical bobbing of the female's extended head. As eggs are laid, generally in two to three layers, she positions them with alternating movements of her hind feet. She then uses her hind feet to pack soil around the eggs, smoothes the soil at the surface, and returns to the water without delay; the nest is usually completed in under an hour.

Females may show nesting-beach fidelity for multiple clutches within a year or in consecutive years (Vogt and Bull 1982; P. Lindeman, unpublished data for common map turtles in Pennsylvania). Freedberg et al. (2005) tested the hypothesis that females return to their own sites of hatching by cross-transplanting female Mississippi map turtles in Tennessee that nested at two sites 6 km apart. Recapture data showed that 54 percent returned to their original nesting area to construct a second nest the same year and that 97 percent did so in subsequent years. Within one nesting area divided into 150 m wide transects, 82 percent returned to nest within the same or an adjacent transect within years and 90 percent did so across years. The authors also analyzed microsatellite and mitochondrial DNA and detected population structuring that separated the nesting areas; they even found genetic structuring within the area divided into 150 m transects, supporting the hypothesis that females return to nest near their natal nest sites.

Incubation, Hatching, and Emergence

Incubation lasted 74–79 days in artificial nests of Escambia map turtles maintained outdoors (Shealy 1976), 61–68 days in a mix of natural and artificial nests of black-knobbed sawbacks (Lahanas 1982), 52–85 days in natural nests of three species in Wisconsin (Vogt and Bull 1984), 62–75 days in natural nests of yellow-blotched sawbacks (Seigel and Brauman 1995), and 55–81 days in natural nests of ringed sawbacks (Jones 2006). Hatching is not synchronous within a nest, and hatchlings do not emerge until a few to several days after hatching, when they have resorbed their yolk sacs (Shealy 1976; Vogt 1980a; Lahanas 1982; Seigel and Brauman 1995; Jones 2006). Reported dates of fall emergence range from 26 August to 20 October in black-knobbed sawbacks (Lahanas 1982) and 15 September (possibly as early as 29 July) to 19 October in yellow-blotched sawbacks (Jones 1993a), although for the latter species Seigel and Brauman (1995) also reported a nest in which hatchlings remained for 56 days after hatching, not emerging until 2 December. Ringed sawbacks emerged as early as 27 July, and the length of time reported between hatching and emergence suggests emergence was not completed until October (Jones 2006). Vogt and Bull (1984) excavated hatchlings from their nests between 3 August and 22 September. Emergence occurs shortly after nightfall, possibly cued by falling sand surface temperatures (P. K. Anderson 1958; Lahanas 1982). Overwintering common map turtle hatchlings in Indiana emerged in April and May (Baker et al. 2010). While hatchling Escambia map turtles moved directly to the nearest water (Shealy 1976), black-knobbed sawback hatchlings dispersed in random directions, nevertheless always finding water within twenty-four hours (Lahanas 1982).

Clutch Size and Egg Size Variation

The largest mean clutch sizes (more than ten eggs) occur in northern, large-bodied populations of the three widely distributed *Graptemys* species (table 4.20). Five southern species with similarly large body sizes—Alabama, Barbour's, Escambia, Pascagoula, and Pearl map turtles—are associated with only intermediate mean clutch sizes (five to nine). Smaller southern species tend to have small mean clutch sizes (fewer than seven eggs in all three sawbacks; Ouachita map turtles in Louisiana and Oklahoma; and Sabine, Texas, and Cagle's map turtles). Egg sizes show little variation among species and hence little influence of female body size across species. Mean egg mass ranges only from 9.9 to 12.5 g. The two largest figures have been reported from relatively small-bodied species, Ouachita and Cagle's map turtles, which both have moderate egg widths but among the greatest reported egg lengths.

Within species, there is a strong tendency for clutch size and egg width to both increase with size of the female (Vogt 1980a; Lindeman 2005; Jones 2006;

K. M. Ryan and Lindeman 2007; but see Rosenzweig [2003], who reported only clutch size to be correlated with body size). Similarly, hatchling and female size are correlated (Bulté and Blouin-Demers 2008; Bulté, Irschick, et al. 2008). While correlation of clutch size with body size in turtles is common and suggestive of body-size constraints on total reproductive output, correlation of egg or hatchling size with female body size is inconsistent with optimal egg-size theory (Congdon and Gibbons 1987). Body size, specifically the diameter of the pelvic opening or the gap between the plastron and carapace, may limit egg size (Tucker et al. 1978; Congdon and Gibbons 1987; K. M. Ryan and Lindeman 2007).

Regression analyses of log-transformed data have begun to show that within a population, a partitioning of the increased reproductive effort is enabled as mature females grow larger (Lindeman 2005; K. M. Ryan and Lindeman 2007). An isometric relationship between a reproductive parameter and female body size means that as females increase in size, that reproductive parameter also increases at the same rate. The expected log-log slope under isometry is one if both variables are linear (e.g., for regressing egg width on female plastron length) or if both are three-dimensional (e.g., egg mass on female mass). The expected log-log slope is three if a three-dimensional variable is regressed on a linear measurement (e.g., egg mass on female plastron length). In common map turtles, significant negative allometry ("hypoallometry") was found for the positive correlations of clutch size and egg width and mass with female size, while clutch mass increased with female size in an approximately isometric fashion (K. M. Ryan and Lindeman 2007). Simply put, as females increased in size, total reproductive effort increased at a similar rate but clutch size and egg size increased at slower rates, apparently because they represent competing sinks for the increasing energetic output. Further support for the hypothesized anatomical constraint on egg size comes in the fact that egg length generally shows no relationship with female body size in *Graptemys* and other freshwater turtles, as any physical constraint on egg size should operate on its lesser dimension, egg width (Lindeman 2005; K. M. Ryan and Lindeman 2007).

In the context of optimal egg-size hypotheses and how they apply to *Graptemys*, it is interesting that in a laboratory study of the effects of substrate moisture on incubation, initial egg mass was positively related to the probability of hatching in two syntopic species (Janzen et al. 1995). In Ouachita map turtles, hatching probability rose from 73 percent for the smallest eggs to 81 percent for the largest, while in false map turtles, the increase in probability of hatching was very sharp among the smaller eggs, rising from 28 percent for the smallest eggs to above 90 percent for eggs of average and above-average mass. Hence the initial advantage of increased egg size may accrue during incubation, with further advantages also possible for emergent hatchlings.

(facing) Adult female Pascagoula map turtle (*Graptemys gibbonsi*) and adult female yellow-blotched sawback (*Graptemys flavimaculata*), Leaf River at Wingate Road near New Augusta, Perry County, Mississippi.

Graptemys geographica

Common Map Turtle

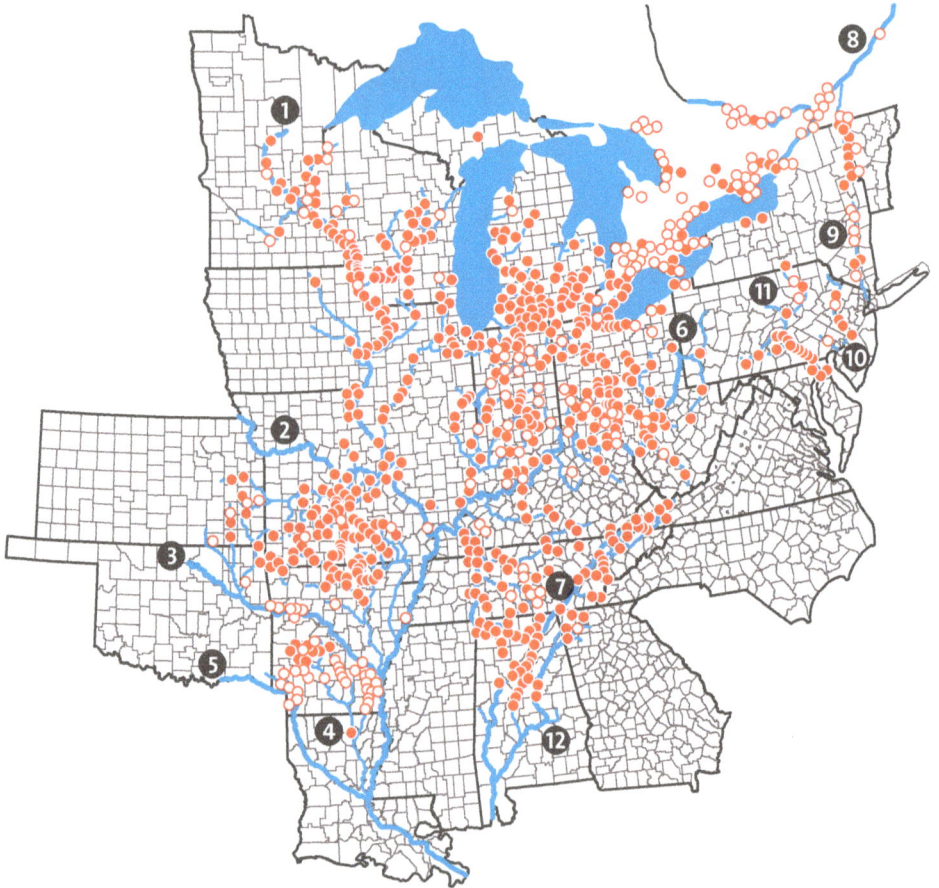

Map A.1. Distribution. Major drainage basins shown are (1) the Mississippi, (2) the Missouri, (3) the Arkansas, (4) the Ouachita, (5) the Red, (6) the Ohio, (7) the Tennessee, (8) the St. Lawrence, (9) the Hudson, (10) the Delaware, (11) the Susquehanna, and (12) Mobile Bay. Closed dots are specimen records, and open dots are reliable literature records and sightings.

Plate A.1.   Adult female, East Fork of White River at Buddha Bypass Bridge near Buddha, Lawrence County, Indiana.

Plate A.2.   Mississippi River, upstream view from Highway 54, Winona, Winona County, Minnesota.

Plate A.3.   Green River, upstream view from Houchins Ferry Road in Mammoth Cave National Park, Edmonson County, Kentucky.

Plate A.4.   Adult male, Mississippi River, Riverview Park on Schmitt Island in Dubuque, Dubuque County, Iowa.

Plate A.5.   Juvenile, backwater of Mississippi River, Pettibone Park on Barron Island in La Crosse, La Crosse County, Wisconsin.

A.1

A.2

A.3

A.4

A.5

125

Graptemys barbouri

Barbour's Map Turtle

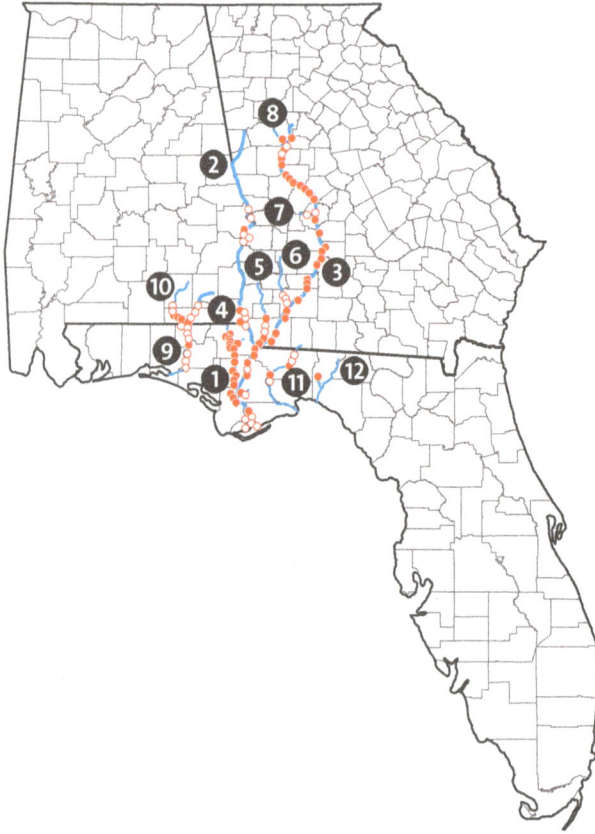

Map B.1. Distribution. Rivers shown are (1) the Apalachicola and its tributary, the Styx, as well as the side channels of its delta, the East and the St. Marks; (2) the Chattahoochee; (3) the Flint; (4) the Chipola, with Spring Creek; (5) a second Spring Creek; (6) Ichawaynochaway Creek; (7) Buck Creek; (8) Line Creek; (9) the Choctawhatchee; (10) the Pea; (11) the Ochlockonee; and (12) the Aucilla and its tributary, the Wacissa. Closed dots are specimen records, and open dots are reliable literature records and sightings.

Plate B.1. Three adult females and one adult male, Apalachicola River at Torreya State Park, Liberty County, Florida.

Plate B.2. Apalachicola River, downstream view from Torreya State Park, Liberty County, Florida.

Plate B.3. Chipola River, downstream view north of Highway 90 near Marianna, Jackson County, Florida.

Plate B.4. Three juveniles, Flint River, Earle May Boat Basin near Bainbridge, Decatur County, Georgia.

Plate B.5. Adult male, Flint River, Killebrew Park at the tailwaters of Lake Black-shear Reservoir near Warwick, Lee and Worth Counties, Georgia.

B.1

B.2

B.3

B.4

B.5

Graptemys pulchra

Alabama Map Turtle

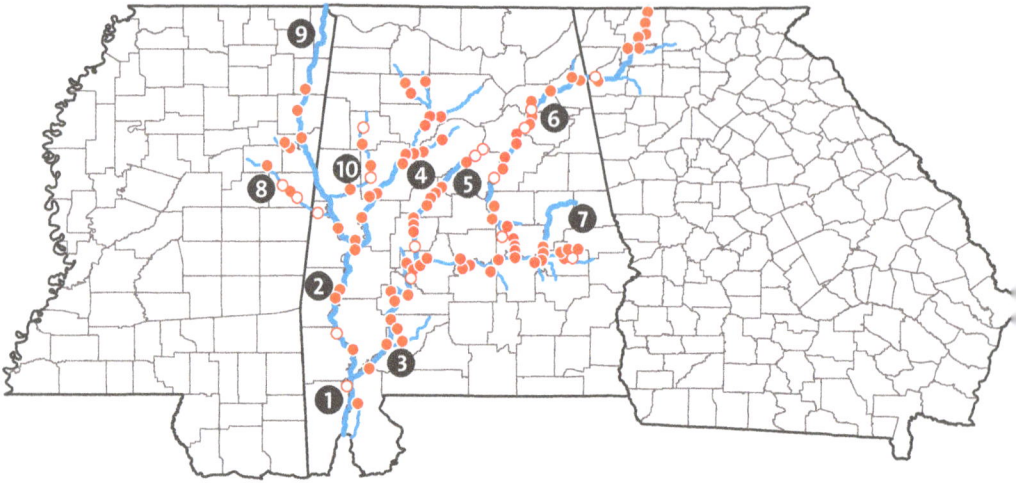

Map C.1. Distribution. Rivers shown are (1) the Mobile-Tensaw Delta; (2) the Tombigbee; (3) the Alabama, with Big Flat, Bogue Chitto, and Pintlala Creeks; (4) the Black Warrior, with Locust, Mulberry, and Sipsey Forks and Brushy Creek; (5) the Cahaba; (6) the Coosa, with Weoka Creek and the Little, with the Coosa formed in Georgia at the confluence of the Oostanaula and Etowah, the former formed by confluence of the Conasauga and Coosawattee; (7) the Tallapoosa, with Line, Cubahatchee, and Uphapee Creeks, the last formed by Opintlocco and Chewacla Creeks; (8) the Noxubee; (9) the upper Tombigbee/Tenn-Tom Waterway, with Tibbee Creek; and (10) the Sipsey. Closed dots are specimen records, and open dots are reliable literature records and sightings.

Plate C.1. Adult male, backwater of Alabama River at Roland Cooper State Park, Wilcox County, Alabama.

Plate C.2. Cahaba River, upstream view from downstream of Highway 280 crossing in Birmingham, Jefferson County, Alabama.

Plate C.3. Tallapoosa River, downstream view near Ft. Toulouse State Historical Park, Elmore County, Alabama.

Plate C.4. Adult female, Chewacla Creek at Highway 80, Macon County, Alabama.

Plate C.5. Adult female and adult male, Alabama River downstream of Steele's Landing on County Road 1, Autauga County, Alabama.

Graptemys ernsti

Escambia Map Turtle

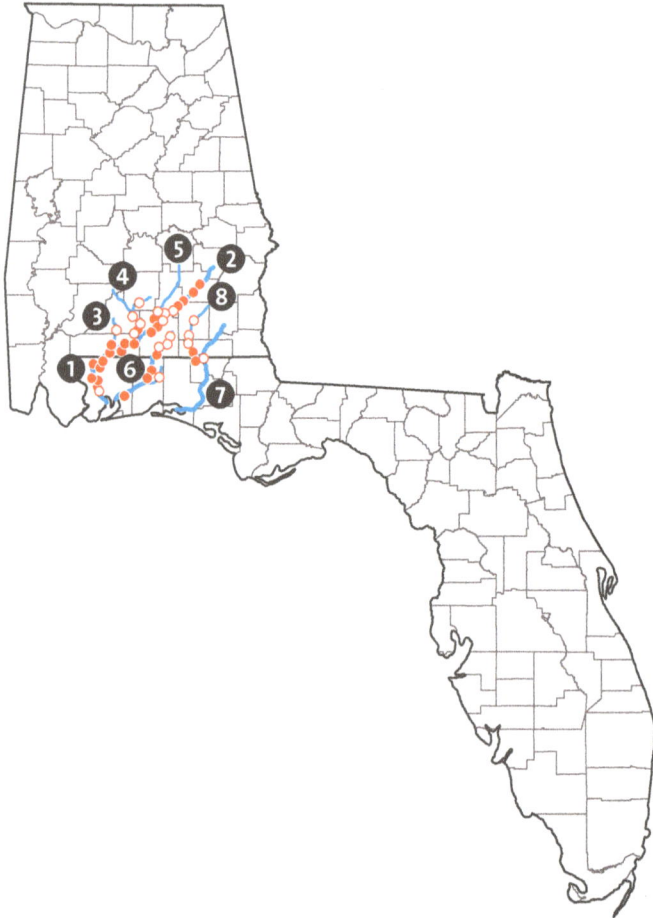

Map D.1. Distribution. Rivers shown are (1) the Escambia; (2) the Conecuh; (3) Murder Creek; (4) the Sepulga, with Pigeon Creek; (5) Patsaliga Creek; (6) the Yellow, with its tributary, the Shoal; (7) the Choctawhatchee; and (8) the Pea. Closed dots are specimen records, and open dots are reliable literature records and sightings.

Plate D.1. Adult male, Conecuh River at new County Road 4 near Roberts, Escambia County, Alabama.

Plate D.2. Conecuh River, upstream view at County Road 19 crossing near Andalusia, Covington County, Alabama.

Plate D.3. Yellow River, upstream view at County Road 4 crossing near Crestview, Okaloosa County, Florida.

Plate D.4. Hatchling and adult male, Conecuh River, Highway 29/10/15 west of Troy, Pike County, Alabama.

Plate D.5. Adult female and two hatchlings, Conecuh River, Highway 29/331/9/15 south of Brantley, Crenshaw County, Alabama.

D.1

D.2

D.3

D.4

D.5

131

Graptemys gibbonsi

Pascagoula Map Turtle

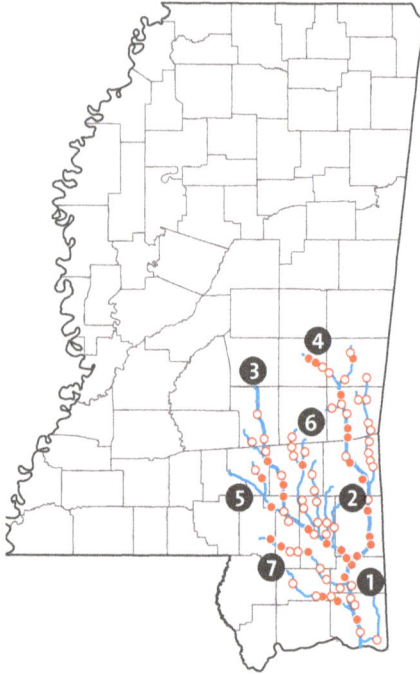

Map E.1. Distribution. Rivers shown are (1) the Pascagoula, with its tributary, the Escatawpa; (2) the Chickasawhay, with Buckatunna, Long, and Souinlovey Creeks; (3) the Leaf, with Oakohay Creek; (4) the Chunky, with Okatibbee Creek; (5) the Bowie, with Okatoma Creek; (6) the Bogue Homa and Tallahala, Tallahoma, Thompson, and Gaines Creeks; and (7) Black and Red Creeks. Closed dots are specimen records, and open dots are reliable literature records and sightings.

Plate E.1. Adult female, Pascagoula River at Highway 596 near Merrill, George County, Mississippi.

Plate E.2. Chunky River, upstream view from Highway 80 crossing near Chunky, Newton County, Mississippi.

Plate E.3. Leaf River, upstream view about 1 km north of Highway 11/42 crossing near Petal, Forrest County, Mississippi.

Plate E.4. Adult male, Leaf River, about 1 km north of Highway 11/42 crossing near Petal, Forrest County, Mississippi.

Plate E.5. Juvenile, Leaf River, about 1 km north of Highway 11/42 crossing near Petal, Forrest County, Mississippi.

Graptemys pearlensis

Pearl Map Turtle

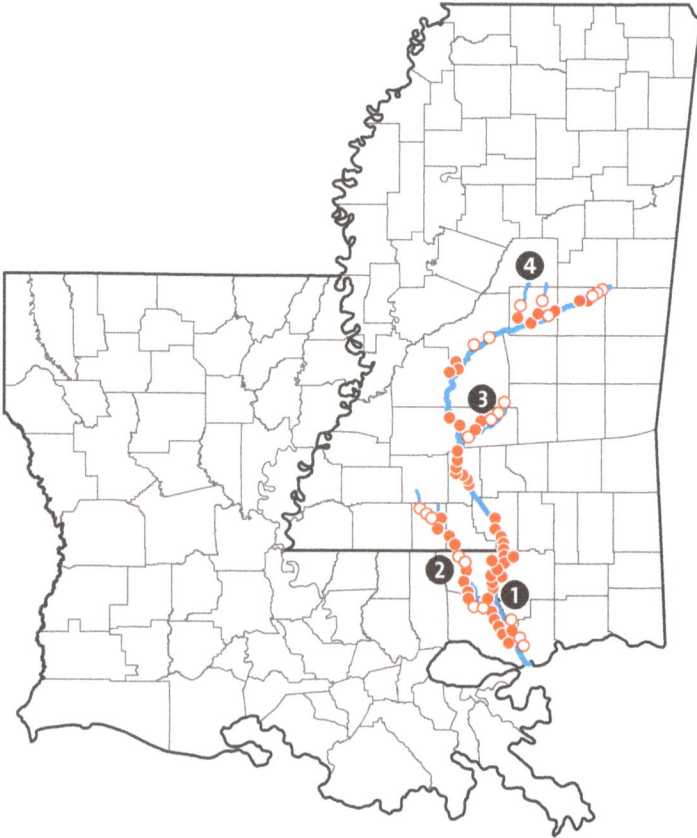

Map F.1. Distribution. Rivers shown are (1) the Pearl, with Lobutcha and Pushepatapa Creeks; (2) the Bogue Chitto, with Topisaw Creek; (3) the Strong; and (4) the Yockanookany. Closed dots are specimen records, and open dots are reliable literature records and sightings.

Plate F.1. Adult females, Lobutcha Creek at Highway 16 near Sunrise, Leake County, Mississippi.

Plate F.2. Pearl River, upstream view at Highway 35 crossing near Carthage, Leake County, Mississippi.

Plate F.3. Strong River, upstream view at Highway 18 crossing near Puckett, Rankin County, Mississippi.

Plate F.4. Juvenile female, Topisaw Creek at Leatherwood Road east of Holmesville, Pike County, Mississippi.

Plate F.5. Juvenile, Yockanookany River at Highway 16 near Ofahoma, Leake County, Mississippi.

F.1

F.2

F.3

F.4

F.5

Graptemys caglei

Cagle's Map Turtle

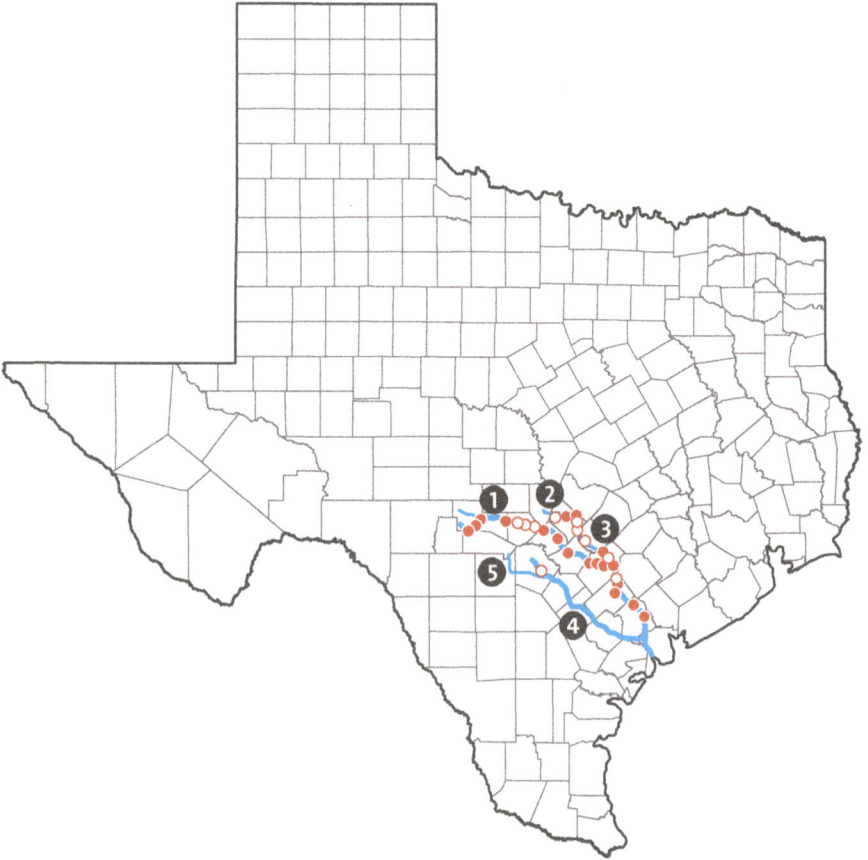

Map G.1. Distribution. Rivers shown are (1) the Guadalupe, with its North Fork and South Fork tributaries; (2) the Blanco; (3) the San Marcos; (4) the San Antonio; and (5) the Medina. Closed dots are specimen records, and open dots are reliable literature records and sightings.

Plate G.1. Adult female, San Marcos River from roadside overlook in Palmetto State Park, Gonzales County, Texas.

Plate G.2. Guadalupe River, upstream view at Highway 236, Cuero, DeWitt County, Texas.

Plate G.3. San Marcos River, downstream view from roadside overlook in Palmetto State Park, Gonzales County, Texas.

Plate G.4. Adult male, Guadalupe River, Highway 236, Cuero, DeWitt County, Texas.

Plate G.5. Adult female and adult male, Guadalupe River, Highway 183, DeWitt County, Texas.

G.1

G.2

G.3

G.4

G.5

Graptemys versa

Texas Map Turtle

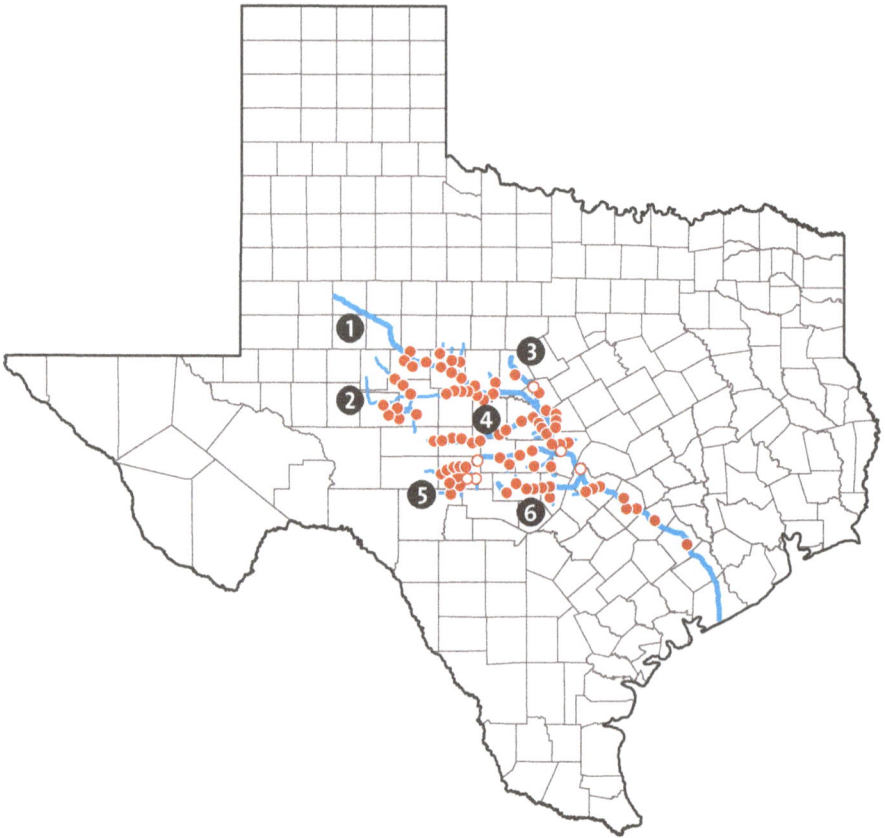

Map H.1. Distribution. Rivers shown are (1) the Colorado, with Valley, Coyote, Elm, Home, Rough, Morgan, Legion, Walnut, and Barton Creeks; (2) the Concho, with North, Middle, and South Forks and Spring and Dove Creeks; (3) Pecan Bayou; (4) the San Saba; (5) the Llano, with North and South tributaries, Paint Creek, East Johnson Fork, and Beaver Creek; and (6) the Pedernales, with White Oak, Live Oak, and Rocky Creeks. Closed dots are specimen records, and open dots are reliable literature records and sightings.

Plate H.1. Adult female, South Llano River below dam in city park in Junction, Kimble County, Texas.

Plate H.2. Llano River, upstream view at Highway 1871 crossing southwest of Mason, Mason County, Texas.

Plate H.3. San Saba River, downstream view at Highway 87 crossing south of Brady, McCulloch County, Texas.

Plate H.4. Adult males, Live Oak Creek, Vista Loop Nature Trail in Ladybird Johnson Municipal Park near Fredericksburg, Gillespie County, Texas.

Plate H.5. Adult males, Colorado River, Bastrop Riverwalk, Bastrop County, Texas.

Graptemys sabinensis

Sabine Map Turtle

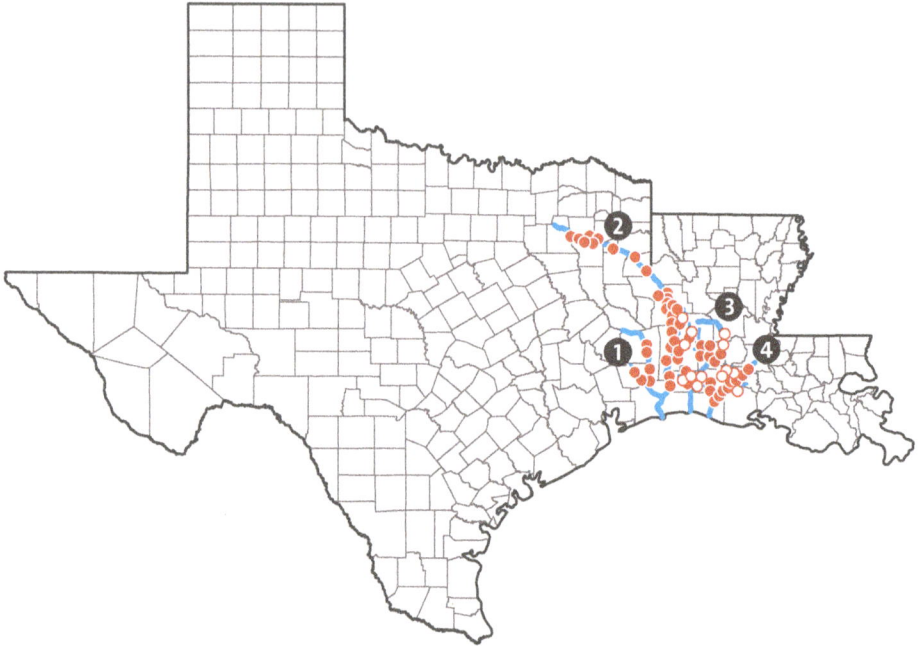

Map I.1. Distribution. Rivers shown are (1) Neches, with Village Creek; (2) the Sabine, with Bayou Toro and Anacoco Bayou; (3) the Calcasieu, with Whiskey Chitto River and the West Fork Calcasieu and the Houston; and (4) the Mermentau, with Lacassine Bayou and bayous Nezpique, des Cannes, Plaquemine Brule, and Queue de Tortue. Closed dots are specimen records, and open dots are reliable literature records and sightings.

Plate I.1.   Adult female and adult male, Sabine River at Palmer Lake Road near Junction, Beauregard Parish, Louisiana, and Newton County, Texas.

Plate I.2.   Sabine River, upstream view from Palmer Lake Road near Junction, Beauregard Parish, Louisiana, and Newton County, Texas.

Plate I.3.   Whiskey Chitto River, upstream view at Highway 26 east of Mittie, Allen Parish, Louisiana.

Plate I.4.   Adult female, Bayou Plaquemine Brule, Highway 91 north of Estherwood, Acadia Parish, Louisiana.

Plate I.5.   Adult female, West Fork Calcasieu River, Sam Houston Jones State Park, Calcasieu Parish, Louisiana.

I.1

I.2

I.3

I.4

I.5

Graptemys pseudogeographica

False Map Turtle

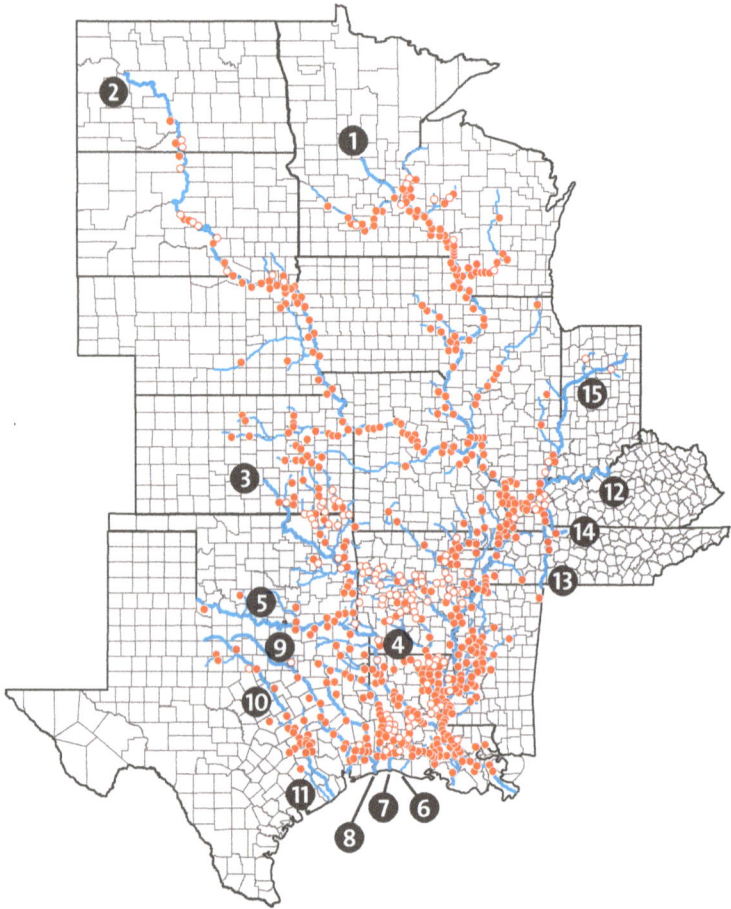

Map J.1. Distribution. Major drainage basins shown are (1) the Mississippi, (2) the Missouri, (3) the Arkansas, (4) the Ouachita, (5) the Red, (6) the Mermentau, (7) the Calcasieu, (8) the Sabine-Neches, (9) the Trinity, (10) the Brazos, (11) the San Bernard, (12) the Ohio, (13) the Tennessee, (14) the Cumberland, and (15) the Wabash. Closed dots are specimen records, and open dots are reliable literature records and sightings.

Plate J.1. Juvenile female, Anacoco Bayou at Harvey Crossing east of Knight, Vernon Parish, Louisiana.

Plate J.2. Wabash River, upstream view at Harmonie State Park, Posey County, Indiana.

Plate J.3. Neches River, downstream view at Highway 96 near Evadale, Hardin, and Jasper Counties, Texas.

Plate J.4. Adult female, backwater of Mississippi River, Pettibone Park on Barron Island in La Crosse, La Crosse County, Wisconsin.

Plate J.5. Two adult females, Village Creek, Highway 96 near Lumberton, Hardin County, Texas.

J.1

J.2

J.3

J.4

J.5

Graptemys ouachitensis

Ouachita Map Turtle

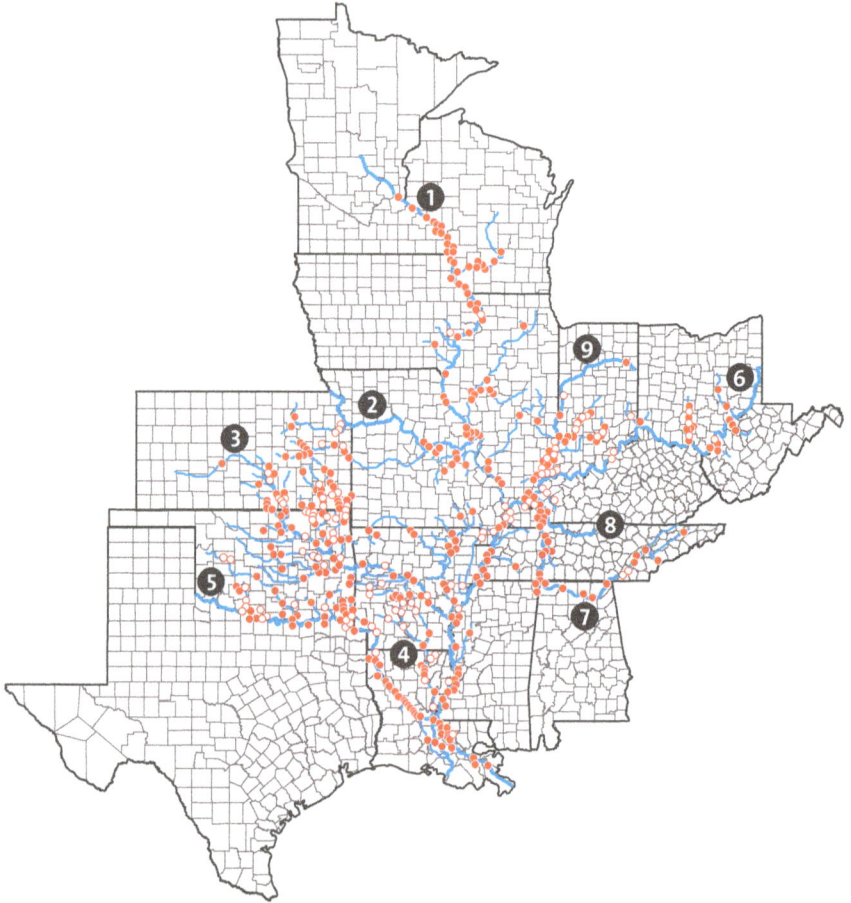

Map K.1. Distribution. Major drainage basins shown are (1) the Mississippi, (2) the Missouri, (3) the Arkansas, (4) the Ouachita, (5) the Red, (6) the Ohio, (7) the Tennessee, (8) the Cumberland, and (9) the Wabash. Closed dots are specimen records, and open dots are reliable literature records and sightings.

Plate K.1. Adult male, Poplar Creek Embayment of Lake Barkley (the impounded Cumberland River), Old Kuttawa Road near Kuttawa, Lyon County, Kentucky.

Plate K.2. Spring River, upstream view at Moccasin Bend, Highway 10 near Miami, Ottawa County, Oklahoma.

Plate K.3. Scioto River, upstream view from Highway 348 crossing near Lucasville, Scioto County, Ohio.

Plate K.4. Adult female and four adult males, Mississippi River, Pettibone Park on Barron Island in La Crosse, La Crosse County, Wisconsin.

Plate K.5. Adult female, Paris Landing State Park near Buchanan, Henry County, Tennessee.

K.1

K.2

K.3

K.4

K.5

Graptemys oculifera

Ringed Sawback

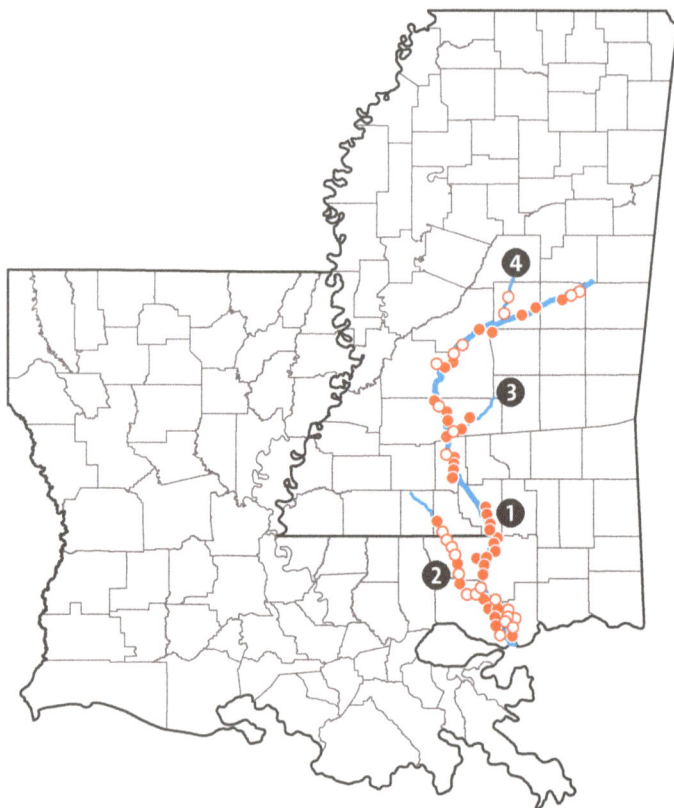

Map L.1. Distribution. Rivers shown are (1) the Pearl, (2) the Bogue Chitto, (3) the Strong, and (4) the Yockanookany. Closed dots are specimen records, and open dots are reliable literature records and sightings.

Plate L.1. Juvenile female, Pearl River, along River Road south of Edinburg, Leake County, Mississippi.

Plate L.2. West Pearl River, downstream view from Interstate 59 crossing near Pearl River, St. Tammany Parish, Louisiana.

Plate L.3. Bogue Chitto River, downstream view from Highway 25 crossing near Franklinton, Washington Parish, Louisiana.

Plate L.4. Adult male and adult female, Pearl River, River Bend on the Natchez Trace Parkway near Shoccoe, Madison County, Mississippi.

Plate L.5. Adult male, Pearl River, along River Road south of Edinburg, Leake County, Mississippi.

L.1

L.2

L.3

L.4

L.5

Graptemys flavimaculata

Yellow-Blotched Sawback

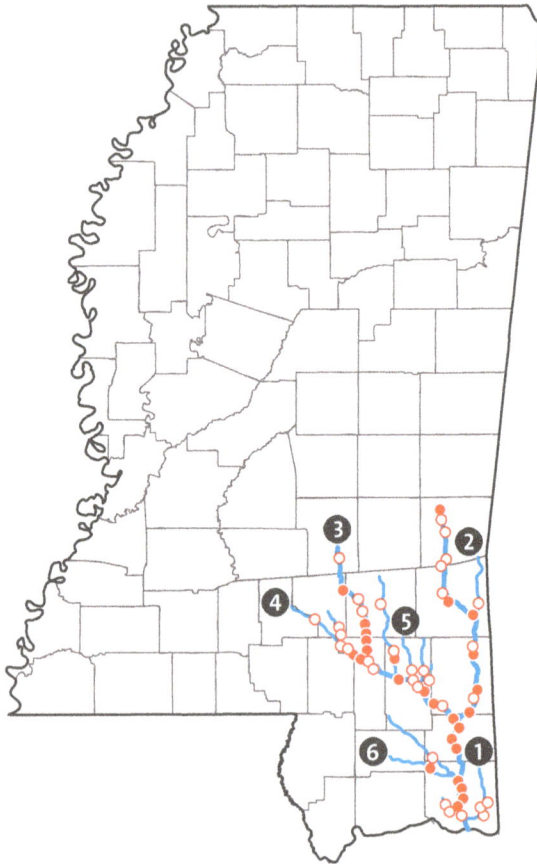

Map M.1. Distribution. Rivers shown are (1) the Pascagoula, with Bluff Creek and the Escatawpa; (2) the Chickasawhay, with Buckatunna Creek; (3) the Leaf; (4) the Bowie, with Bowie and Okatoma Creeks; (5) Tallahala Creek, the Bogue Homa, and Thompson and Gaines Creeks; and (6) Black and Red Creeks. Closed dots are specimen records, and open dots are reliable literature records and sightings.

Plate M.1. Adult female, Leaf River about 2 km north of Hattiesburg and Petal, Forrest County, Mississippi.

Plate M.2. Chickasawhay River, downstream view at Highway 57 crossing in Leakesville, Greene County, Mississippi.

Plate M.3. Pascagoula River, upstream view toward the confluence of the Leaf and Chickasawhay Rivers from the Highway 596 bridge at Merrill, George County, Mississippi.

Plate M.4. Adult male, Leaf River, Highway 15 crossing near Beaumont, Perry County, Mississippi.

Plate M.5. Adult female, Pascagoula River at Highway 596 bridge, Merrill, George County, Mississippi.

M.1

M.2

M.3

M.4

M.5

Graptemys nigrinoda

Black-Knobbed Sawback

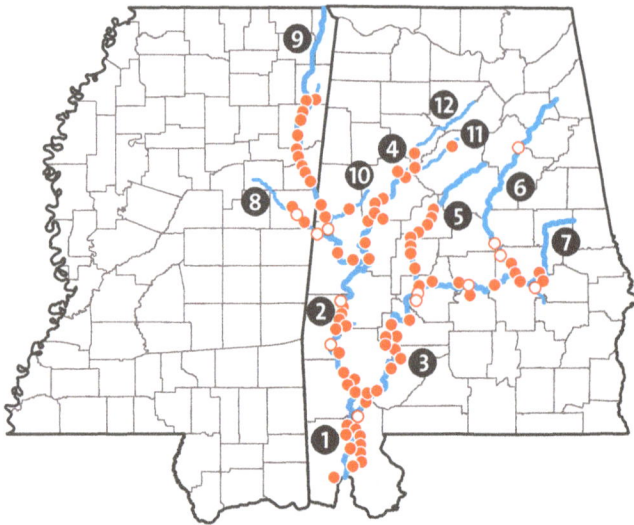

Map N.1. Distribution. Rivers shown are (1) the Mobile-Tensaw Delta; (2) the Tombigbee, with Horse Creek; (3) the Alabama; (4) the Black Warrior; (5) the Cahaba; (6) the Coosa; (7) the Tallapoosa, with Line Creek; (8) the Noxubee; (9) the upper Tombigbee/Tenn-Tom Waterway, with Bull Mountain Creek; (10) the Sipsey; (11) Locust Fork; and (12) Mulberry Fork. Closed dots are specimen records, and open dots are reliable literature records and sightings.

Plate N.1. Adult male, backwater of Alabama River at Roland Cooper State Park, Wilcox County, Alabama.

Plate N.2. Noxubee River, upstream view from Highway 17 crossing north of Geiger, Sumter County, Alabama.

Plate N.3. Alabama River, upstream view from the Edmund Pettus Bridge in Selma, Dallas County, Alabama.

Plate N.4. Adult female, Alabama River at Selma, Dallas County, Alabama.

Plate N.5. Adult female and two adult males, Tensaw River, County Road 86 near Crossroads, Baldwin County, Alabama.

Plate O.1. Adult male and adult female Cagle's map turtles (*Graptemys caglei*), Guadalupe River at Highway 183 near Hochheim, DeWitt County, Texas.

Plate O.2. Adult female Sabine map turtle (*Graptemys sabinensis*), West Fork Calcasieu River in Sam Houston Jones State Park, Calcasieu Parish, Louisiana.

Plate P.1. Four adult female Mississippi map turtles (*Graptemys pseudogeographica*) and adult male and adult female Sabine map turtles (*Graptemys sabinensis*), Sabine River at Palmer Lake Road near Junction, Beauregard Parish, Louisiana, and Newton County, Texas.

Plate P.2. Adult female Pearl map turtle (*Graptemys pearlensis*), Yockanookany River at Highway 16 near Ofahoma, Leake County, Mississippi.

Plate Q.1. Juvenile female Alabama map turtle (*Graptemys pulchra*), Cahaba River at Highway 280 in Birmingham, Jefferson County, Alabama.

Plate Q.2. Juvenile and adult female Escambia map turtles (*Graptemys ernsti*), Conecuh River at County Road 28 near Goshen, Pike County, Alabama.

Clutch Frequency

Production of more than one clutch per year has been recorded for at least some females in all studies with large samples, with internesting intervals of two to three weeks (Vogt and Bull 1982; Horne et al. 2003; Rosenzweig 2003; Jones 2006; K. M. Ryan and Lindeman 2007). Average clutch frequency in populations has proven to be a difficult life-history parameter to determine with confidence, however, and so interspecific variation, latitudinal trends, and interannual variation in clutch frequency all remain poorly known. Unlike the case for pond-dwelling populations of some turtles, it can be extremely difficult to obtain high incidence of recaptures during nesting, as most populations nest on strings of sandbars and islands along river corridors (Horne et al. 2003; Rosenzweig 2003; Jones 2006).

Recent intensive mark-recapture radiographic studies of the southern yellow-blotched and ringed sawbacks, based primarily on basking trap captures, suggest only rare production of more than one clutch per season in these species (table 4.21; Horne et al. 2003; Jones 2006). However, Tucker (2001) discussed the difficulty of producing accurate estimates of clutch frequency using mark-recapture techniques without consideration of the probability that capture occurs when oviductal eggs are present. Jones (2006) did not find multiple peaks over the course of the nesting season in proportions of females that were gravid, supporting his finding that multiple clutches are rare. Horne et al. (2003) supplemented their results with ultrasonography one season, which allowed them to detect females with both enlarged follicles and oviductal eggs. They calculated virtually the same mean annual clutch frequency as the radiographic data had indicated, but both sets of their results still may be biased toward low values for females that had nested one or more times prior to their first capture. The conclusion that production of multiple clutches was rare in yellow-blotched sawbacks was supported by concurrent hormone-level studies, which failed to show the multiple seasonal peaks in estradiol-17β that might typify production of more than one clutch (Shelby et al. 2000). Most recently, Shelby-Walker et al. (2009) used ultrasonography on yellow-blotched sawbacks to show that at least 17 percent produced two clutches in the same population Horne et al. had studied, while at least 50 percent did so in another population, possibly due to differences in the two populations' historical exposures to contaminants.

In contrast to the above results, dissection studies suggest annual clutch frequencies of four or more in southern populations of other species (table 4.21), although it is not known whether all enlarged follicles found in dissection prior to or during the nesting season really represent clutches for that season (i.e., some could be follicles for the next season or those that would be lost to atresia in the fall; Vogt 1980a; Lindeman 2005). Only one report of a nesting study, of Mississippi map turtles in Tennessee, claimed production of up to four clutches

Table 4.21. Estimates of annual clutch frequency in populations of *Graptemys*

Method and Species	State	Source	N	Range	Mean
Mark-recapture studies					
flavimaculata	MS	Horne et al. 2003	126	1–2	1.2
oculifera	MS	Jones 2006	124	1–4	1.1
Ultrasonography					
flavimaculata	MS	Horne et al. 2003	97	1–3	1.2
Dissection studies					
barbouri	FL	Cagle 1952b	7		4.2[a]
ernsti	AL	Shealy 1976	31		4.0[b]
geographica	MO	White and Moll 1991	18	2–3	2.3
nigrinoda	AL	Lahanas 1982	7	3–4	
ouachitensis	IL	Moll 1977	6	1–3	2.2
	LA	Rosenzweig 2003	10	1–3	2.1
	OK	Webb 1961	3	3–3	3
	WI	Vogt 1980a	65	2–4	
pseudogeographica	LA	J. L. Carr 2001	4	1–3	2.3
	SD	Timken 1968a	12	1–3	1.3[c]
	WI	Vogt 1980a	50	2–3	
versa	TX	Lindeman 2005	10	2–4	3.3[c]

[a]Based on mean annual reproductive potential of 22 eggs and mean clutch size of 5.2 eggs.
[b]Based on mean annual reproductive potential of 29 eggs and mean clutch size of 7.2 eggs.
[c]No corpora lutea were detected, hence the mean may be an underestimate if some females had already laid eggs at the time of capture.

annually, but unfortunately it lacked details (Freedberg et al. 2005). Hence mark-recapture and ultrasonography yield an estimate of minimum clutch frequency per female, whereas dissections yield an estimate of maximum clutch frequency per female (Jones 2006).

Body size seems to influence clutch frequency, with larger females being more likely to lay multiple clutches (Timken 1968a; Shealy 1976; Jones 2006). Some females may skip reproducing for a year (Horne et al. 2003; Jones 2006), which might explain the occasional observation of a lack of oviductal eggs and enlarged follicles in dissected specimens with mature body size (e.g., Rosenzweig 2003; Lindeman 2005).

Sex Ratio

If primary sex ratio is the ratio of males to females at hatching and functional sex ratio is the sex ratio of mature adults (Lovich and Gibbons 1990), secondary sex ratio may be defined as the ratio at some stage after hatching: for example, at the point of maturation of the sex that matures first, when sexing individuals in the field becomes possible. For *Graptemys* populations, secondary sex ratio has been most frequently reported; both secondary and functional sex ratios vary widely (table 4.22). Variation in the functional sex ratio may stem not only from variation in the primary sex ratio, which can occur in these turtles because of the influence of incubation temperature on gender (temperature-dependent sex determination, or TSD), but also from the difference in age at maturity (males mature earlier, creating a male bias) and potential differences in adult survival between the sexes (Lovich and Gibbons 1990). In addition, gender bias in trapping has not been rigorously examined in *Graptemys*, but it occurs in trapping other aquatic turtles (Ream and Ream 1966; Gibbons 1990; Koper and Brooks 1998). Gender bias in basking trap success was likely a major cause of a seasonal shift in functional sex ratios observed in ringed sawbacks, for which decreased capture of females occurred after the nesting season (Jones and Hartfield 1995). Gender bias in trapping may also explain why significantly more male ringed sawbacks were captured in the most intensively sampled population, whereas mark-recapture estimates suggested a significant female bias (table 4.22). The difficulty of trapping female *Graptemys* has been noted by some authors (e.g., Tinkle 1958a; Webb 1961).

A tendency toward a female bias may also occur in basking surveys: while small body size alone does not mean an individual is male (and thus tail size must be observed before gender can be recorded), large body size is enough to determine that an individual is female. Thus the more female-biased sex ratios reported by Godwin (2000, 2001, 2003) could stem in part from his recording of turtles as female, male, of minimum male size but uncertain gender, or juvenile (smaller than male size). Most records in the third category (often a substantial portion of Godwin's counts) would be expected to be adult males, biasing ratios that were reported based solely on the first two categories.

The only primary sex ratios reported are for nests of three species in Wisconsin and were skewed toward females by approximately 3:1 (Vogt and Bull 1984). Significant male skews are most common for secondary and functional sex ratios, but balanced sex ratios and significant female skews also occur (table 4.22). Secondary sex ratios are more male skewed in basking counts than in trapping results when both have been reported for the same populations (McCoy and Vogt 1979; Lindeman 1997c). Misidentification of some smaller juvenile females likely accounts for some of the discrepancy, but there may also be gender biases in basking frequency or capture probability.

Table 4.22. Sex ratios reported for species of *Graptemys* (minimum $N = 30$)

Species (location)	Data[a]	Source	Proportion male	N
Primary sex ratios of hatchlings				
geographica (WI)	Field nests	Vogt and Bull 1984	.21***	61
ouachitensis (WI)	Field nests	Vogt and Bull 1984	.28***	1,540
pseudogeographica (WI)	Field nests	Vogt and Bull 1984	.25***	489
Secondary sex ratios (all sexed individuals)				
barbouri (FL)	HC	Sanderson 1974	.57*	312
caglei (TX)	BT, HC	Killebrew and Porter 1989	.86***	712
ernsti (AL)	HC	Shealy 1976	.3***	357
ernsti (AL)	Basking counts	Godwin 2000	.14***	1,040
flavimaculata (MS)	FN	McCoy and Vogt 1979	.38	34
	Basking counts		.69***	221
geographica (QC)	BT, HC	Gordon and MacCulloch 1980;	.63***	312
		Flaherty 1982		
geographica (PA)	BT, HC	Pluto and Bellis 1986[b]	.55/.68***	168/885
geographica (WI)	BHN	DonnerWright et al. 1999	.54	159
geographica (IL)	BHN	R. A. Anderson et al. 2002	.58	36
geographica (IN)	BHN, DN, HC	Conner et al. 2005	.52	305
geographica (ON)	BT	Barrett Beehler 2007[b]	.43/.44	112/141
geographica (ON)	BHN, BT, HC	Browne and Hecnar 2007	.28***	163
geographica (ON)	BT, DN, HC	Bennett et al. 2009[c]	.36*/.59	59/54
geographica (MO)	HN	Pitt and Nickerson 2012[d]	.48/.42/.60	83/66/43
nigrinoda (AL)	FN	McCoy and Vogt 1979	.37***	126
	Basking counts		.66***	188
nigrinoda (AL)	FN	Lahanas 1982	.40	77
nigrinoda (AL)	Basking counts	Godwin 2001, 2003	.51	867
oculifera (MS)	FN	McCoy and Vogt 1979	.29***	99
	Basking counts		.69***	166
oculifera (MS)	BT, HC	Jones and Hartfield 1995[b]	.56***/.57***	2,587/2,984
	Mark-recapture		.33***	5423
ouachitensis (KY)	Basking counts	Lindeman 1997c	.62***	345
	BT, FN	Lindeman 1999b	.43	61

Species (location)	Data[a]	Source	Proportion male	N
ouachitensis (LA)	FN	J. L. Carr 2001	.27***	333
pulchra (AL)	Basking counts	Godwin 2003	.23***	273
pseudogeographica (KY)	Basking counts	Lindeman 1997c	.67***	890
	BT, FN	Lindeman 1999b	.45	49
pseudogeographica (WI)	BHN	DonnerWright et al. 1999	.73***	40
pseudogeographica (SD)	BHN	Timken 1968a	.21***	274
pseudogeographica (SD)	BHN, FN	A. Gregor, pers. comm. 2008	.27***	988
sabinensis (LA)	BT	Shively and Jackson 1985	.36***	132
versa (TX)	BT, FN, HC	Lindeman 2003b and unpublished data	.50	70
Functional sex ratios (adults only)				
barbouri (FL)	HC	Sanderson 1974	.87***	206
caglei (TX)	BT, HC	Porter 1990	.88***	82
ernsti (AL)	HC	Shealy 1976	.53	200
geographica (WI)	FN, GN, HC, TN	Vogt 1980a	.75***	60
geographica (OH)	BT	Tran et al. 2007[b]	.16***/.19***	98/141
geographica (ON)	Mark-recapture	Bulté and Blouin-Demers 2009	.51	652
nigrinoda (AL)	FN	Lahanas 1982	.52	60
oculifera (MS)	BT, HC	Jones and Hartfield 1995[b]	.71***	1,975
ouachitensis (WI)	FN, GN, HC, TN	Vogt 1980a	.20***	333
ouachitensis (LA)	FN	J. L. Carr 2001	.28***	316
pseudogeographica (SD)	BHN	Timken 1968a	.51	112
pseudogeographica (SD)	BHN, FN	A. Gregor, pers. comm. 2008	.47	567
pseudogeographica (WI)	FN, GN, HC, TN	Vogt 1980a	.38***	177
versa (TX)	BT, FN, HC	Lindeman 2005	.64**	55

Note: Asterisks indicate significant deviation from 1:1 in chi-square goodness-of-fit tests (*$p < .05$; **$p < .01$, *** $p < .005$).
[a]Abbreviations for trapping studies: BHN = baited hoop nets, BT = basking traps, DN = dip nets, FN = fyke nets, GN = gill nets, HC = hand captures (including dipnetting and snorkeling), TN = trammel nets.
[b]Ratios are for first captures only/all captures including recaptures.
[c]Two populations were sampled.
[d]Population sampled in three different decades.

Sex Determination

Eight species of *Graptemys* have been studied for the influence of constant temperature on embryonic development (Bull and Vogt 1979; Bull, Vogt, and McCoy 1982; Ewert and Nelson 1991; Wibbels et al. 1991; Ewert et al. 1994). All show the TSD pattern termed "Ia" (Ewert and Nelson 1991), in which cool temperatures produce males and warm temperatures produce females (fig. 4.6). Because this pattern is ubiquitous in the Deirochelyinae (Ewert and Nelson 1991; Ewert et al. 1994), it is likely the pattern throughout the genus, although TSD patterns are quite labile evolutionarily (Janzen and Krenz 2004). As with other turtles, the middle part of incubation is the temperature-sensitive stage for sex determination; Bull and Vogt (1981) used switch experiments to show that at constant temperatures, sex is determined between embryonic stages 16 and 22 (of 27 total stages; fig. 4.7).

Bull, Vogt, and Bulmer (1982) incubated eggs of Ouachita map turtles at 29.2°C–29.3°C to examine sex determination at a threshold temperature. Significant variation was found among females and heritability of sex ratio at the threshold temperature was high, but the authors argued that in nature the genetic influence on sex ratio determination was likely not strong, given the fluctuating temperatures in nests and the possible effect of site choice on a nest's thermal environment.

One of the more promising hypotheses for explaining the existence of TSD holds that the fitness of males and females is determined by differing selective pressures, such that each gender is differently advantaged by incubation at cooler or warmer temperatures (Charnov and Bull 1977; Ewert and Nelson 1991; Shine 1999). There is a pronounced female bias in early-emerging nests of *Graptemys* in the Mississippi River in Wisconsin that lessens as more and more male-producing nests emerge later in the season (fig. 4.8; Vogt and Bull 1984). The scenario of a "head start" on growth for the larger sex would favor pattern Ia in fall-emerging deirochelyines, which all have larger females (Gibbons and Lovich 1990). In addition, in the diamondback terrapin, *Malaclemys terrapin*, females carrying larger eggs laid them in open microhabitats that favored development of females, while females carrying smaller eggs laid them in shaded microhabitats that favored development of males (Roosenburg 1996). This behavior would give hatchling females a head start on attaining their characteristic larger body sizes. No similar analyses of oviposition behavior have yet been conducted for *Graptemys*.

Freedberg et al. (2001) tested a model that suggests that in animals with TSD, the sex produced under environmental conditions that result in lower lifetime fitness will predominate in the hatchling sex ratio, as the numbers of that sex would decline more steeply than the numbers of the other sex during growth to maturation, thus yielding an adult sex ratio of about 1:1. Hatchling Ouachita map turtles incubated at five constant temperatures (24°C–31°C) did

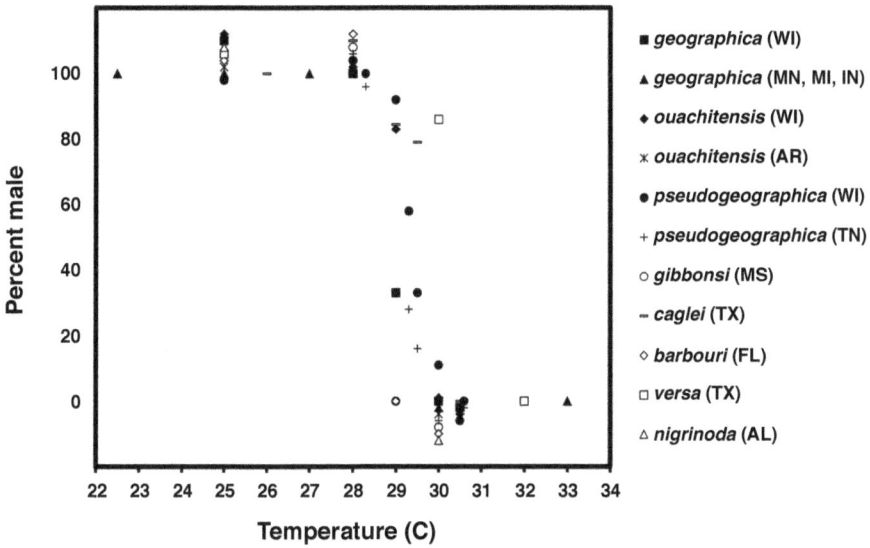

Figure 4.6. Experimental demonstrations of temperature-dependent sex determination in eight species of *Graptemys* using incubation at constant temperatures (Bull and Vogt 1979; Bull, Vogt, and McCoy 1982; Ewert and Nelson 1991; Wibbels et al. 1991; Ewert et al. 1994). Northern populations are denoted with solid symbols.

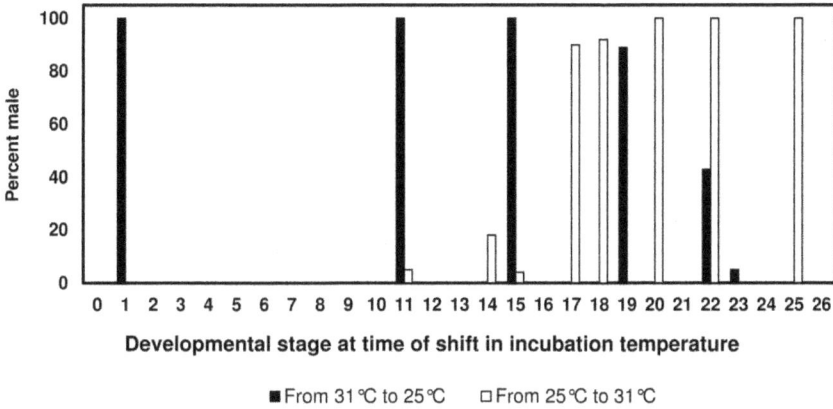

Figure 4.7. Sex ratios produced by switching *Graptemys ouachitensis* eggs between 25°C (a male-producing temperature under constant incubation) and 31°C (a female-producing temperature under constant incubation) at various stages of development, in which 0 = oviposition and 26 = hatching (redrawn from Bull and Vogt 1981).

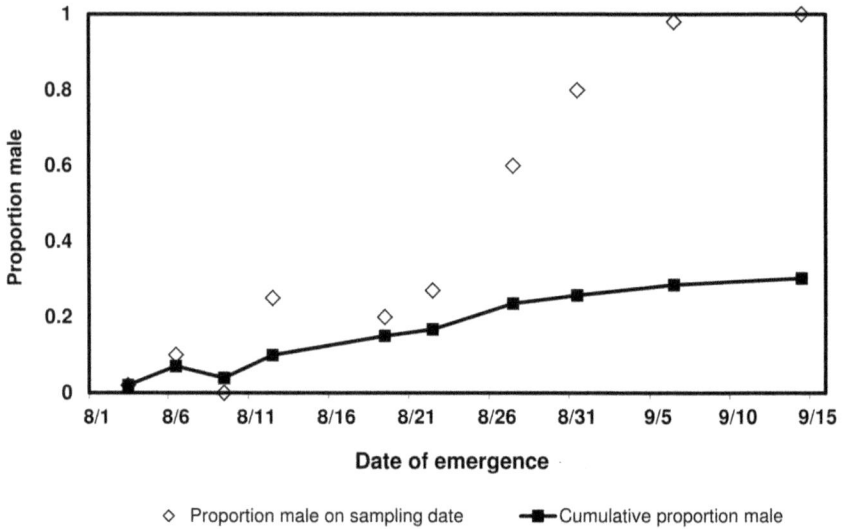

Figure 4.8. Sex ratios of hatchlings emerging from natural nests in Wisconsin in 1980, showing proportion of hatchlings on each individual sampling date that were male and cumulative percentage of males over the season (data from Vogt and Bull 1984).

not support the predictions of the model, however. Over nine months following hatching, hatchlings from the cooler, male-producing temperatures had slower growth rates and righting responses (see also Freedberg et al. 2004) and lower survival rates, but hatchling sex ratios the authors cited from field studies favored females by almost 2:1, not males as the model would predict. (The authors did not quantify the hatchling sex ratio in their Indiana source population, however.) They argued that female-skewed sex ratios could indicate selection for homing by daughters to natal nest sites. Higher-quality sites would produce larger numbers of female offspring that would return to them, biasing hatchling sex ratio over time.

Freedberg et al. (2005) concluded that the close genetic relationship of Mississippi map turtle females found nesting in the same area, combined with transplanted females' ability to return to the nest sites where they had first been captured, constituted support for the hypothesis that natal homing produces female-biased hatchling sex ratios. Under this hypothesis, females returning to natal nest sites would tend to be nesting in "proven" sites that produce predominantly female offspring, enforcing the bias toward females.

The possibility of geographic trends in the pivotal temperature (i.e., the point of transition from all-male to all-female clutches) was examined by Bull, Vogt, and McCoy (1982) for four species of *Graptemys*; the painted turtle, *Chrysemys picta*; and the slider turtle, *Trachemys scripta*. They expected southern turtles to produce all-male clutches at higher temperatures, given their higher

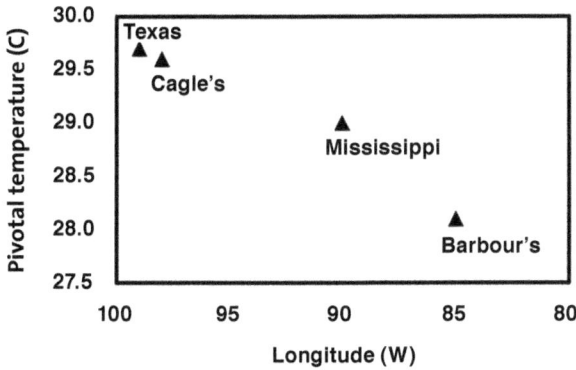

Figure 4.9. Purported longitudinal trend in the pivotal temperature of sex determination at constant temperatures for four map turtle species (Ewert et al. 1994, with modifications of the data from Ewert, Etchberger, et al. 2004).

environmental temperatures, but found less than the expected trend. Wibbels et al. (1991) examined the threshold temperature for Cagle's map turtle, the southernmost species of *Graptemys*, finding that it was slightly higher than for more northern species. They concluded, as Bull, Vogt, and McCoy (1982) had previously suggested, that the difference was insufficient to compensate for climatic differences among localities; if equal sex ratios are produced by species and populations at differing latitudes, varying the seasonal timing of nesting and nest-site choice (sunny vs. shaded) must be the primary mechanism. Ewert et al. (1994) and Ewert, Etchberger, et al. (2004) reported a longitudinal trend in pivotal temperature for four southern *Graptemys*, with higher pivotal temperatures in western Cagle's and Texas map turtles than in the more eastern Mississippi and Barbour's map turtles (fig. 4.9). They attributed the trend to the drier, less vegetated (i.e., less shaded and hence warmer) nesting habitats of the two former species. In general, in *Graptemys* the transition from all-male to all-female clutches occurs between 28°C and 30°C, with only a slight tendency for more southern species to produce males at higher temperatures (fig. 4.6).

In the field, nests of three species of *Graptemys* tended to produce all-male clutches in the shade and all-female clutches in open areas, with less than one-quarter of nests producing mixed genders (Vogt and Bull 1984). Compared to all-male nests, all-female nests had more hours per day over 30°C and 32°C, higher mean temperatures, and more variable temperatures (Bull 1985). Georges (1989) subsequently reanalyzed the same data with a model taking temperature-dependent variation in developmental rate into account. The *Graptemys* data were a good fit to the hypothesis that sex is determined by proportion of development, not proportion of time, spent above or below the pivotal temperature.

Bull et al. (1990) analyzed the potency of two warm temperatures by switching stage 18 embryos from 26.0°C to either 31.0°C or 32.5°C. Most eggs switched to 32.5°C (78 percent) produced females, but the switch to 31.0°C failed to produce females, suggesting a quantitative aspect to sex determination.

Carbon dioxide concentration also can influence sex ratio in *Graptemys* and other turtles (Etchberger et al. 2002). Higher concentrations of CO_2 (which might occur for eggs in relatively impervious soils, as the by-products of embryonic respiration build up, or in nest cavities with high quantities of decomposing organic matter) were associated with increased yield of female offspring at the intermediate incubation temperature of 28.5°C. Etchberger et al. speculated that elevated CO_2 should lower embryonic blood plasma pH, which might be linked to the change in sex determination potency.

Community Ecology

The role of map turtles and sawbacks in structuring and dynamics of the community has received little study. They are nonetheless important components of the ecosystems they inhabit as they are consistently high in relative abundance in most turtle assemblages, often outnumbering all other species combined in basking counts or trapping results (see "Population" sections in chapter 8).

The role of competition in an assemblage of turtle species that included Ouachita and Mississippi map turtles was assessed in a cove on a reservoir on the Tennessee River, an old creek inlet inundated by Kentucky Dam (Lindeman 2000b). Microhabitat use, diet, diel use of time, and seasonal use of time were studied for the two map turtles and three abundant syntopic species of similar body sizes: the slider turtle (*Trachemys scripta*), river cooter (*Pseudemys concinna*), and smooth softshell (*Apalone mutica*). Data were subjected to pseudo-community analysis (Winemiller and Pianka 1990; Pianka 1992) to test for statistically significant resource partitioning and guild structure against a null hypothesis of random structuring of species' use of resources. Little evidence of partitioning or guild structure was found, suggesting competition had not been important in structuring resource use (Lindeman 2000b). The two map turtles were the most similar pair of species for all four categories of resource use. Results were interpreted as having much more to do with phylogenetic constraints on resource use than structuring by interspecific competition. In particular, the concept of niche complementarity (Schoener 1974) would predict that two species highly similar in one category of resource use (e.g., microhabitat) would be highly dissimilar in another (e.g., diet); however, the high similarity in resource use of the two map turtle species, plus their generally high similarity in all four categories to the closely related slider turtle, was evidence that niche complementarity did not occur. Moll (1977) described a situation in which sympatric common, false, and Ouachita map turtles inhabiting a sediment-filled floodplain lake all fed nearly exclusively on midge larvae and bottom detritus (tables

4.5, 4.10, and 4.14), which also dominated the diets of other sympatric species, demonstrating the unimportance of interspecific competition when a food resource is superabundant.

Thermoregulatory Behavior

The most conspicuous behavior related to thermoregulation of *Graptemys* is basking on emergent deadwood. Although the importance of abundance, type, and location of deadwood to habitat choice has been well established (Waters 1974; Flaherty and Bider 1984; Shively and Jackson 1985; Pluto and Bellis 1986; Fuselier and Edds 1994; Jones 1996; Lindeman 1998, 1999c; DonnerWright et al. 1999; Godwin 2000, 2001, 2003; Killebrew et al. 2002; Tollefson 2004; T. J. Ryan et al. 2008; Richards-Dimitrie 2011; Sterrett et al. 2011), little focused research has been conducted to empirically test hypotheses about basking behaviors as they relate to the physiology or ecology of map turtles and sawbacks. Basking is universal among species and size and sex classes; even hatchlings, which are essentially cryptic in some other aquatic emydids (e.g., Lovich et al. 1991; Pappas and Brecke 1992), are readily seen basking in *Graptemys* populations, often among larger con- and heterospecifics but also in shallower, sheltered habitats (Waters 1974; Shealy 1976; Janzen et al. 1992; Lindeman 1993; Selman and Qualls 2011).

Basking occurs on any day of the year that is warm enough and from early morning well into the evening, but maximum numbers are observed under sunny conditions in late spring and early summer and during late morning and early afternoon (Waters 1974; Shealy 1976; Craig 1992; Lindeman 2000b; Barrett Beehler 2007; Coleman and Gutberlet 2008; Peterman and Ryan 2009; Selman and Qualls 2011), when the turtles are most active (G. R. Smith and Iverson 2004). Autumnal peaks in basking may occur (e.g., Craig 1992; Coleman and Gutberlet 2008; Selman and Qualls 2011; fig. 4.10), although such peaks may at least partially be sampling artifacts, as basking may be stimulated following cool or rainy days (Sanderson 1974; Waters 1974; Craig 1992), which are more frequent in fall. There appears to be a seasonal component that differs between the sexes such that basking late in the season (i.e., following nesting and during spermiogenesis) is male-biased (Craig 1992; Jones and Hartfield 1995; Barrett Beehler 2007; Coleman and Gutberlet 2008; fig. 4.10).

Anecdotally, turtles remain close to favored basking substrates and use them repeatedly (Chaney and Smith 1950; Waters 1974; Shively and Jackson 1985); however, telemetry may also reveal frequent use of areas not used for basking (Tran et al. 2007). In one remarkable instance of site fidelity at Presque Isle on Lake Erie, a few hours after I released an adult male common map turtle about 50 m from where I had captured it (in a basking trap set under an angled branch of a fallen tree), I captured it in the same trap. Favored perches are securely anchored over deep water and not in contact with the riverbank or shoreline

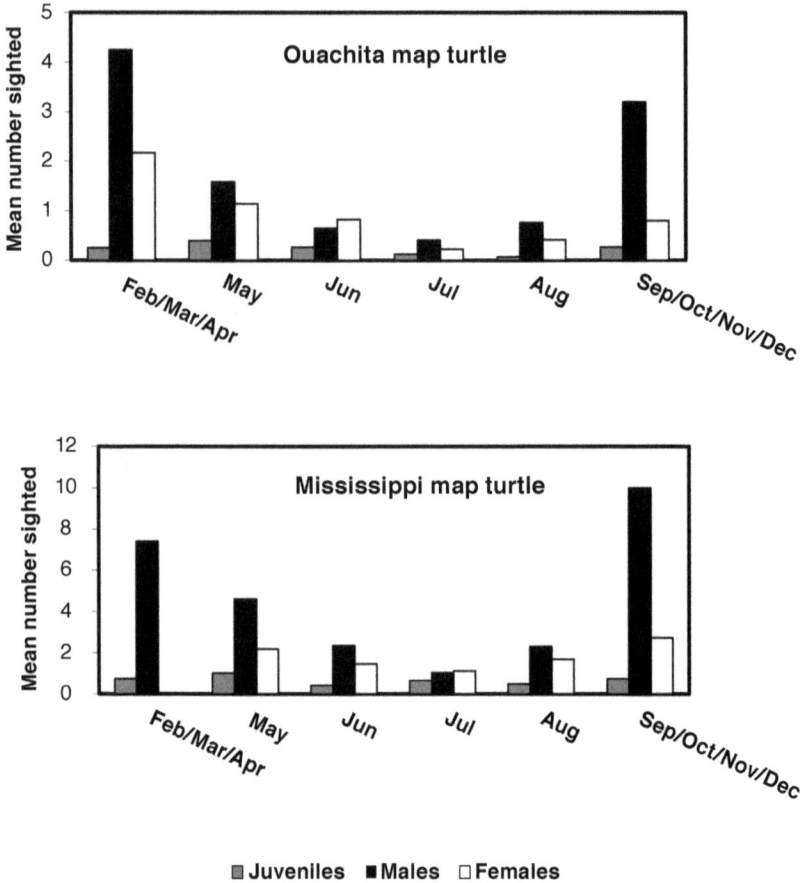

Figure 4.10. Mean numbers of basking turtles sighted in a cove of a reservoir for 182 observations made over three years (from data summarized more generally in Lindeman 1997b, 2000a), showing seasonal and sexual patterns of basking frequency for two syntopic species.

(Waters 1974; Shealy 1976; Flaherty and Bider 1984; Coleman 2006). In basking surveys dominated by *Graptemys* in three drainages, both percentage occupancy and mean number of turtles per substrate showed the same hierarchy of favored sites: tree crowns, then logs, then branches, and then tangles of woody debris washed together and emergent root masses, then stumps (Lindeman 1999c). In a canal in which few deepwater substrates not touching the shoreline were available, however, basking occurred on shoreline riprap (Peterman and Ryan 2009).

Basking map turtles and sawbacks are distinctly gregarious, crowding together on logs and branches with other turtles (fig. 4.11). In bays of a lake in Ontario, Flaherty and Bider (1984; see also Flaherty 1982) found no statistical

Figure 4.11. Basking aggregations of *Graptemys*. (a) Approximately equal numbers of Barbour's map turtles and river cooters on the Flint River, Veteran's Park in downtown Albany, Dougherty County, Georgia. (b) Eleven black-knobbed saw-backs on one branch on the Alabama River in Selma, Dallas County, Alabama; other portions of the same sunken tree crown also held several conspecifics at the time the picture was taken. (c) Five Texas map turtles and seven Texas river cooters, South Llano River in Junction, Kimble County, Texas. (d) Part of a large aggregation of common map turtles at Graveyard Pond in Presque Isle State Park, Erie County, Pennsylvania. (e) Eleven Ouachita map turtles in the Mississippi River about 300 m below the mouth of the Bad Axe River, Vernon Co., Wisconsin.

differences between basking sites that were used by large numbers of common map turtles and unused sites in bays that had no turtles. They suggested that aggregation during basking may enhance group detection of predators. Peterman and Ryan (2009) likewise reported aggregative tendencies: large numbers of turtles were frequently observed on some sites, whereas nearby sites remained unoccupied.

Basking map turtles and sawbacks often splay out their hind limbs and spread their hind feet toes, a posture they may hold for several minutes at a time (Shealy 1976; fig. 4.12). Such "turtle yoga" obviously increases insolation; besides increasing the rate of heating, it may also promote increased synthesis of vitamin D (Pritchard and Greenhood 1968). Individuals may also bask with their heads withdrawn. The head-in posture is especially commonly observed in individuals that appear to have been basking for a long duration, based on the dryness of their shells and the large number of turtles that may accumulate below them on the same basking substrate (pers. observations); hence the posture may be a strategy to keep basking but avoid overheating. Basking individuals often exhibit kicking motions and scratching of the head and neck, which may indicate heat stress or reaction to swarming insects (Waters 1974; Moll 1976c; Shealy 1976; pers. observations). Waters (1974) and Selman (2011) reported repeated submergence and reemergence on basking sites, possibly for avoidance of overheating; Ben-Ezra et al. (2008) made similar observations in captives.

Aggregations are typified by low levels of aggression among turtles, but displacement into the water can occur. In two studies that quantified the outcome of large numbers of interactions among basking map turtles, relative size was the main predictor. Displacement in common map turtles was usually of a smaller turtle by a larger turtle and often occurred as one emerged from the water (Pluto and Bellis 1986). Two map turtles were among the four emydid species in a study of the roles of relative size and status as initiator or recipient of an aggressive act (Lindeman 1999a), as described below under "Interactions with Other Species." Potential displacement may be the reason small turtles (juveniles and males) often climb high onto angled substrates (Pluto and Bellis 1986).

Basking individuals are wary, periodically moving their heads to scan in various directions. There is a more marked tendency among *Graptemys* than in most other aquatic turtles to climb to great heights while basking (Boyer 1965; Hill 2008; fig. 4.13). For example, most of Coleman's (2006) observations of basking Sabine and Mississippi map turtles were on substrates with climbing angles of about 45°. Climbing may provide greater detection of potential danger, ease crowding, and allow individuals to avoid agonistic encounters with other turtles or change the thermal environment experienced. The strong preference for deadwood not in contact with the shoreline, especially by females (e.g., Haynes and McKown 1974; Waters 1974; Shealy 1976; Shively and Jackson

Figure 4.12. Typical basking stances with splayed-out hind feet as exhibited by
(a) a male yellow-blotched sawback (Highway 596 crossing of the Pascagoula River
at Merrill in George County, Mississippi) and (b) a female Alabama map turtle
(County Road 21 crossing of the Sipsey River east of Echola in Tuscaloosa County,
Alabama).

Figure 4.13. Demonstration of the climbing proclivities of *Graptemys*. (a) Note the female and male (partially hidden) Alabama map turtles occupying the highest point of a snag (Coosa River downstream of Wetumpka, Elmore County, Alabama). (b) A female Cagle's map turtle that climbed a nearly vertical section of a large snag (Guadalupe River in Independence Park, Gonzales, Gonzales County, Texas). (c) A male Texas map turtle that climbed high onto a narrow branch (North Concho River at Highway 2034 in Water Valley, Tom Green County, Texas). (d) Common map turtles high on an angled branch in Graveyard Pond of Presque Isle State Park, Erie County, Pennsylvania.

1985; Coleman 2006; Enge and Wallace 2008; Hill 2008; Selman and Qualls 2011; pers. observations), suggests that terrestrial predators pose a significant threat during basking. With low deadwood abundance in the streams they occupy, Texas map turtles may often be seen basking on rocks instead and seem to prefer rocks midstream in the river (fig. 4.14). Common map turtles in the Susquehanna River also frequently use midstream rocks for basking (Richards-Dimitrie 2011). Evacuation from basking sites may occur at the slightest hint of danger, including boats more than 100 m removed from turtles, flying birds, swimming beavers or alligators, jumping frogs, and floating debris (Newman 1906a; Shealy 1976; Craig 1992; Moore and Seigel 2006; Selman, Qualls, and Mendonça 2008; Selman, Strong, et al. 2009; Selman and Qualls 2011).

Basking turtles in some populations can become quite inured to high-density human use of their habitats. Waters (1974) noted that recreational boaters who frequented the Alabama River were not perceived as threats by basking

Figure 4.14. Three Texas map turtles basking on a midstream rock in the San Saba River, upstream of the Highway 1311 crossing, Menard County, Texas.

black-knobbed sawbacks, which were markedly warier along the lower Cahaba River, where the shooting of basking turtles was said to be especially common. Jones and Hartfield (1995), who compared numbers sighted with population estimates, suggested that ringed sawbacks were less prone to abandon basking sites at Ratliff Ferry than at other river stretches because the turtles there have become accustomed to the high recreational use (see fig. 5.12). On average, 29 percent of the estimated population was seen on basking surveys there, compared with just 10 percent of the estimated summed populations at four other sites.

At Graveyard Pond on Presque Isle, Lake Erie, recreational kayakers and canoeists, a sightseeing pontoon boat, and powerboats traversing a no-wake zone regularly approach within 10–20 m of a large aggregation of common map turtles without eliciting mass abandonment of basking perches (fig. 4.15). In perhaps the most unusual instance of basking habituation I have seen, I observed approximately six dozen Ouachita map turtles basking on dock supports and lakeside sidewalks of a busy marina on Kentucky Lake (fig. 4.16). Large aggregations basked on the dock supports without being disturbed by people on the dock passing over them or by a party of young children taking fishing lessons just 20 m away—indeed, when the children were called away to take a short drink break, it took just a few minutes for the turtles to reclaim the

Figure 4.15. (a) A powerboat at no-wake speed and (b) a sightseeing pontoon boat tour and recreational kayakers passing near large basking aggregations of habituated common map turtles at Graveyard Pond, Presque Isle State Park, Erie County, Pennsylvania.

Figure 4.16. Basking habituation of Ouachita map turtles at the marina of Paris Landing State Park on Kentucky Lake in Henry County, Tennessee, (a) on dock supports within a meter of a gangplank leading to the boat slips and (b) during a break taken by a children's group fishing lesson.

sidewalk where the children had dropped their poles! High frequency of disturbance by boaters and other human users of *Graptemys* habitats can ultimately
have a negative impact on basking behavior, however (Moore and Seigel 2006;
Selman 2010; see chapter 5).

Detection of potential threats during basking appears to occur solely by sight.
Waters (1974) swam toward basking individuals and shouted at them from a
distance of 1–2 m without effect, until they saw him and hastily dove into the
water. I once watched a yellow-blotched sawback that was not scared from its
basking perch underneath a railroad trestle despite the prolonged, ear-splitting
screech of a train braking to a halt directly overhead. Turtles blinded in one or
both eyes may be easily taken while basking. In Kentucky, I hand-captured the
same blind female Mississippi map turtle three summers in a row while she was
basking or swimming and I was wading in chest-deep water. In each instance
she made no attempt at escape until I made contact, at which point she resisted
vigorously.

Hibernation

For *Graptemys* at southern latitudes, winter is mild, and movement and basking
may occur on warm days (Waters 1974; Shealy 1976; Craig 1992; Jones 1996).
Farther north, map turtles aggregate at communal hibernacula under ice; they
do not dig into the substrate, instead wedging themselves under rocks and logs
with necks and limbs extended on the bottom of their aquatic habitats, often
moving about the bottom under the ice (Newman 1906a; Evermann and Clark
1916; Vogt 1980a; Graham and Graham 1992; Carrière et al. 2009). A. Carr
(1952) and Shealy (1976) described similar behavior for two southern species,
Barbour's and Escambia map turtles, during short periods of cool weather.

Hibernating common map turtles seek out deep, well-oxygenated areas, and
large numbers of turtles winter in close proximity (Pluto and Bellis 1988; Graham and Graham 1992; Ultsch et al. 2000; Graham et al. 1997, 2000; Crocker et
al. 2000; Richards-Dimitrie 2011). Reese et al. (2001) speculated that because of
oxygen demand during hibernation (see chapter 7), common map turtles may
be restricted to areas that do not become hypoxic in winter, thus eliminating
shallow marshes, small ponds, and so forth as potential habitats. However, map
turtles do not follow the pattern of another turtle that relies on well-oxygenated
water at northern latitudes for hibernation—the stinkpot, *Sternotherus odoratus*, which Reese et al. (2001) stated is found in a wider variety of habitats in the
southern portion of its range, where it does not hibernate. Male common map
turtles were less aggregated and more prone to locomotion than females during hibernation (Graham and Graham 1992). Dispersal from the hibernaculum
occurs in the spring after ice breakup (Graham and Graham 1992; Pluto and
Bellis 1988), although the exact phenology and its environmental cues have not
been recorded.

Emergence of hatchlings from their nests occurs in late summer or early fall in Escambia and Cagle's map turtles and black-knobbed and ringed sawbacks (Shealy 1976; Lahanas 1982; Jones 1993a; Killebrew and Babitzke 1996); fall emergence is presumably the rule at all southern latitudes. However, common map turtle hatchlings in northern portions of the species' range overwinter in their nest chambers, with few nests producing fall emergers (Newman 1906a; Pappas et al. 2000; Gillingwater 2001; Baker et al. 2003; McCallum 2003; Nagle et al. 2004; Parren and Rice 2004). Conflicting reports have false and Ouachita map turtles at northern latitudes emerging in fall (Vogt 1980a) or spring (Christiansen and Gallaway 1984). Overwintering in the nest is common in hatchlings of many other emydids (Gibbons and Nelson 1978), including some northern populations of the closest relative of *Graptemys*, the diamondback terrapin, *Malaclemys terrapin* (Auger and Giovannone 1979; Baker et al. 2006). Spring emergence has been recorded in Indiana and Pennsylvania primarily during May, with dates ranging from 9 April to 3 June (Baker et al. 2003; Nagle et al. 2004), but did not occur until between 27 May and 20 July in an Ontario population on Lake Erie (Gillingwater 2001). During the winter, hatchlings are oriented with their heads upward and their shells overlapping one another (Nagle et al. 2004). Spring emergence may be synchronous, with all hatchlings emerging the same day, or asynchronous, spread over several weeks, but most emergence follows recent rains, when air temperature is at least 12°C (Nagle et al. 2004).

Parasites and Pathogens

Studies of the parasites of map turtles and sawbacks (table 4.23) have focused mainly on leeches (Hirudinea: *Placobdella*), coccidian parasites (Apicomplexa: *Eimeria*), digenean flukes (Platyhelminthes: Trematoda), and spiny-headed worms (Acanthocephala: *Neoechinorhynchus*). Tapeworms, flukes, round-worms, a cnidarian, a cloacal mite, and a trypanosome have also been reported, and there are sparse reports of bacterial and viral pathogens. To date, studies have been almost entirely casual observations of rates of infestation in the wild, with little analysis of routes of transmission or effects on fitness or population dynamics.

Leeches and Red Blood Cell Parasites

Two species of *Placobdella* leeches parasitize *Graptemys*. Graham et al. (1997) reported *P. parasitica* and *P. ornata* on overwintering common map turtles in Vermont. Females had higher incidence and intensity of infestation than males and were more likely to have broods of young leeches, which the authors attributed to size dimorphism and reduced fall movements by females. A higher frequency of basking by males in the fall (see above) would potentially also play a role in the difference. Saumure and Livingston (1994) reported the same

Table 4.23 Eukaryote parasites reported from species of *Graptemys*

Parasite taxon	Site infected	*barbouri*[a]	*caglei*[b]	*ernsti*[c]	*flavimaculata*[d]
Hirudinea					
Placobdella ornata	Skin				
Placobdella parasitica	Skin				
Placobdella sp.	Skin			+	
Apicomplexa					
Eimeria chrysemydis	Red blood cells		+		
Eimeria graptemydos	Red blood cells		+		
Eimeria juniataensis	Red blood cells				
Eimeria lutotestudinis	Red blood cells		+		
Eimeria marginata	Red blood cells				
Eimeria mitraria	Red blood cells				
Eimeria pseudogeographica	Red blood cells		+		
Eimeria trachemydis	Red blood cells		+		
Eimeria sp.	Red blood cells	+			
Haemogregarina sp.	Red blood cells	+			
Haemoproteus degiustii	Red blood cells	+			
Euglenozoa					
Trypanosoma chrysemydis	Red blood cells				
Cnidaria					
Myxidium chelonarum	Bile duct, gall bladder				
Trematoda					
Digenea					
Allossostomoides chelydrae	Rectum				
Cephalogonimus vesicaudus	Intestine				
Dictyangium chelydrae	Intestine				
Heronimus chelydrae	Lungs				
Heronimus mollis	Lungs				
Macravestibulum obtusicaudum	Intestine				
Microphallus opacus	Not stated				
Spirorchis artericola	Bloodstream				
Spirorchis innominatus	Heart				
Spirorchis scripta	Trachea, bloodstream				

Source notes for Table 4.23 are on page 180.

geographica[e]	gibbonsi[f]	nigrinoda[g]	ouachitensis[h]	pseudogeographica[i]	pulchra[j]	versa[k]
+						
+			+	+	+	
	+	+				
+						+
+						
+				+		
+				+		+
				+		
+						
+						
+						
				+	+	
				+		
				+		
+						
+						
				+		
+				+		
+						
+				+		
+						
				+		

Table 4.23 Eukaryote parasites reported from species of *Graptemys (continued)*

Parasite taxon	Site infected	*barbouri*[a]	*caglei*[b]	*ernsti*[c]	*flavimaculata*[d]
Telorchis angustus	Intestine				+
Telorchis corti	Intestine				+
Telorchis gutturosi	Intestine				
Telorchis necturi	Intestine				
Telorchis singularis	Intestine				+
Telorchis sp.	Intestine			+	
Unicaecum ruszkowskii	Bloodstream				
Unidentified pronocephalids	Intestine				+
Aspidogastrea					
Cotylaspis cokeri	Intestine				+
Unidentified aspidogastreans	Intestine				
Unidentified trematodes	Intestine				
Monogenea					
Polystomoides coronatum	Buccal cavity				
Cestoda					
Proteocephalus testudo	Intestine				
Unidentified cestodes	Intestine				
Acanthocephala					
Neoechinorhynchus emydis	Intestine				
Neoechinorhynchus stunkardi	Intestine				+
Neoechinorhynchus sp.	Intestine			+	
Nematoda					
Camallanus microcephalus	Intestine				
Cosmocercoides dukae	Intestine				
Gnathostoma procyonis	Muscle				
Oswaldocruzia leidyi	Intestine				
Spironoura affinis	Not stated				
Spironoura concinnae	Intestine				
Spironoura procera	Intestine				+
Spironoura wardi	Large intestine				
Spiroxys contortus	Stomach				
Spiroxys sp.	Intestine				+
Unidentified nematodes	Intestine				
Acarina					
Caminacarus dawsoni	Cloaca				

Note: No parasites have been reported for *G. oculifera, G. pearlensis,* or *G. sabinensis.* Parasite synonymies for sources through 1976 follow E. M. Ernst and Ernst (1977).

geographica[e]	gibbonsi[f]	nigrinoda[g]	ouachitensis[h]	pseudogeographica[i]	pulchra[j]	versa[k]
+				+		+
				+		
				+		
				+		
+				+		
						+
			+			
+				+		
				+		
			+			
+				+		+
				+		
		+				
+				+		
				+		
				+		
				+		
+						
+						
+						
+				+		
+				+		
		+	+			+
				+		

Table 4.23 Eukaryote parasites reported from species of *Graptemys (continued)*

[a]Wacha and Christiansen (1976); Telford (2009).
[b]McAllister et al. (1991).
[c]Shealy (1976).
[d]Steinauer and Horne (2002).
[e]Van Cleave (1913); Stunkard (1915, 1919); E. C. Harrah (1922); Horsfall (1935); Hsü (1937); Hughes et al. (1942); Hopp (1946, 1954); Rausch (1946, 1947); DeGiusti and Batten (1951); Cable and Hopp (1954); Fisher (1960); Woo (1969b); Pluto and Rothenbacher (1976); Wacha and Christiansen (1976); Saumure and Livingston (1994).
[f]Selman, Strong, et al. (2008).
[g]Lahanas (1982).
[h]Webb (1961); Vogt (1979).
[i]Barker and Parsons (1914); Stunkard (1915, 1916, 1917, 1919, 1923); Magath (1919); H. B. Ward (1921); Horsfall (1935); Byrd (1939); McKnight (1959); Fisher (1960); Cable and Fisher (1961); Ash (1962); Acholonu (1969a, 1969b); Acholonu and Arny (1970); Brooks (1975); Brooks and Mayes (1975, 1976); Wacha and Christiansen (1976); Vogt (1979); Dyer and Wilson (1997); Bochkov and OConnor (2008).
[j]C. H. Ernst and Barbour (1972); Dodd (1988b).
[k]McAllister et al. (1991); Lindeman and Barger (2005).

two species on common map turtles in Ontario. T. J. Ryan and Lambert (2005) found that 51 percent of the common map turtles in an Indiana population carried *Placobdella* sp. during the summer, with a mean of 1.7 leeches/turtle, but no significant sexual differences in infestation frequency and intensity. Shealy (1976) found abundant *Placobdella* sp. with frequent broods on Escambia map turtles during the spring in Alabama. Lahanas (1982) found *Placobdella* sp. on 24 percent of the black-knobbed sawbacks he examined throughout the season in Alabama.

Placobdella infestation is thought to be low in *Graptemys* and other deirochelyines due to their higher basking affinities relative to other turtles, particularly bottom-walking species (McAuliffe 1977; MacCulloch 1981). This untested assertion was referred to by T. J. Ryan and Lambert (2005) as the "desiccating-leech" hypothesis. They showed that colonization rates were higher in the stinkpot (*Sternotherus odoratus*), a kinosternid, than in common map turtles, even when neither species was allowed to bask; this suggests that leeches may adaptively prefer to settle on the former, which basks only infrequently (Lindeman 1996). They found that stinkpots were nearly twice as likely to be infested in the field and had an average of more than twenty times as many leeches as common map turtles. Leeches on basking map turtles and sawbacks may assume a spherical shape, which is thought to slow desiccation by minimizing their surface-to-volume ratio (Shealy 1976), and Selman, Strong, et al. (2008) and Selman and Qualls (2009b) described voluntary abandonment by leeches that had assumed such positions on a Pascagoula map turtle (whose body temperature was estimated to have reached 40°C) and on a yellow-blotched sawback, respec-

Figure 4.17. Leeches (probably *Placobdella* sp.) balled into spheres on (a) a female Sabine map turtle (left rear carapace; photo from Whiskey Chitto River at Highway 26, Allen Parish, Louisiana) and (b) a juvenile Barbour's map turtle (photo from the Choctawhatchee River at Highway 90, Washington County, Florida).

tively. In figure 4.17, two map turtles are shown with leeches in the spherical shape; when the first was startled into the water and then re-emerged minutes later to bask, its leech had flattened itself out again.

In general in reptiles, leeches appear to cause little in the way of disease symptoms, but they may be harmful as vectors of protozoan pathogens (Lane and Mader 1996). *Placobdella* sp. can transmit unicellular hemogregarines (Apicomplexa: Coccidiasina) and trypanosomes (Euglenozoa: Kinetoplastea) from turtle to turtle within or across species (Woo 1969a; McAuliffe 1977; Siddall and Desser 2001; Telford 2009). In the host, the unicellular parasites occur in the liver and are visible in the blood, with degree of infection being expressed as infected cells per one thousand or ten thousand red blood cells (McAuliffe 1977; Siddall and Desser 2001).

The only study of trypanosome prevalence in *Graptemys* found them in three of thirteen common map turtles in Ontario (Woo 1969b). The first report of hemogregarines in a *Graptemys* reported them from two of eleven Barbour's map turtles (Telford 2009). Given the widespread use of map turtles and sawbacks as hosts by *Placobdella*, it is likely that other species are also hosts of trypanosomes and hemogregarines.

The vector is unknown for another protozoan blood parasite, *Haemoproteus degiustii* (Apicomplexa: Haemosporidia), known to occur in common map turtles in Wisconsin and Barbour's map turtles in Georgia (DeGiusti and Batten 1951; Telford 2009), but it may be a dipteran fly (DeGiusti et al. 1973). Moll (1976c) observed common map turtles responding spasmodically to flies that swarmed about them at their basking sites, but there have been no studies of the dipterans that feed on basking *Graptemys* or the parasites they may transmit.

Eimerians are another group of unicellular apicomplexans that have been detected in several species of *Graptemys*. They infect their hosts' intestinal epithelial cells or, sometimes, liver or kidneys (Roberts and Janovy 2005). Their presence is determined by sampling of feces or intestinal contents for reproductive structures called oocysts via flotation in a sugar solution (McAllister and Upton 1989). Nine species of *Eimeria* have been noted in five host species (Pluto and Rothenbacher 1976; Wacha and Christiansen 1976; McAllister et al. 1991). Host specificity is apparently very low for *Eimeria* in turtles, as many species have been found in multiple genera or families of hosts (McAllister and Upton 1989; McAllister et al. 1991). Prevalence of *Eimeria* was 71 percent in seven false map turtles (Wacha and Christiansen 1976) and 69 percent in sixteen Cagle's map turtles (McAllister et al. 1991). In captive reptiles, clinical symptoms associated with these pathogens may include fibrosis, ulcers, and septicemia in the intestinal epithelium as well as anorexia, diarrhea, regurgitation, dehydration, and hemorrhagic enteritis (Lane and Mader 1996), but effects on the vigor of wild *Graptemys* have not been studied.

Endoparasites

A diversity of digenean flukes—seventeen species from nine genera—have been reported from *Graptemys* (Stunkard 1915, 1919, 1923; Harrah 1922; H. B. Ward 1921; Horsfall 1935; Hsü 1937; Byrd 1939; Rausch 1946, 1947; Brooks 1975; Brooks and Mayes 1975, 1976; Steinauer and Horne 2002; Lindeman and Barger 2005). Sites of infection include the gastrointestinal tract, the heart and major arteries, and the lungs and trachea (table 4.23). Prevalence ranges from 6 percent (for *Heronimus chelydrae* in common map turtles; Harrah 1922) to 53 percent (for *Macravestibulum obtusicaudum* in common map turtles; Rausch 1947), while infection intensity is generally low, at fewer than 25 flukes per host, except for a female yellow-blotched sawback that held 143 *Telorchis corti* (Barker and Parsons 1914; Stunkard 1915; Rausch 1947; Brooks and Mayes 1976; Steinauer and Horne 2002; Lindeman and Barger 2005). Only two studies have examined sexual differences. Male Texas map turtles had a lower prevalence of infection by *T. corti* (14 percent) than females (54 percent; Lindeman and Barger 2005), and intensity of infection by trematodes in yellow-blotched sawbacks was higher in females (Steinauer and Horne 2002). Intermediate hosts have not always been identified, but snails appear to be of primary importance (e.g., McMullen 1934; Horsfall 1935; Hsü 1937; Steinauer and Horne 2002). Pathologies associated with digenean flukes have not been reported for *Graptemys* or for reptiles in general (Lane and Mader 1996).

   Spiny-headed worms (*Neoechinorhynchus* sp.) have been reported in six species via dissections of gut contents and examination of feces for eggs (Leidy 1851; Van Cleave 1919; Rausch 1947; Hopp 1954; Fisher 1960; Cable and Fisher 1961; Acholonu 1969a; Shealy 1976; Lahanas 1982; Steinauer and Horne 2002; Barger 2004; Lindeman and Barger 2005). In Texas map turtles, females had greater prevalence and intensity of *N. emydis* (to 186 worms/host) than males (Lindeman and Barger 2005). Intensity of *N. stunkardi* also appeared to be higher in female yellow-blotched sawbacks than in males (means 16.5 and 2.7 worms/host, respectively; Steinauer and Horne 2002). Sexual differences may relate to sexual size dimorphism, differences in prey taken, or overall dietary diversity (Steinauer and Horne 2002; Lindeman and Barger 2005). Prevalence of *N. emydis* in two populations of common map turtles and of *Neoechino-rhynchus* sp. in black-knobbed sawbacks was uniformly high, however (14 of 17 hosts, 29 of 32 hosts, and 36 of 38 hosts, respectively; Rausch 1947; Hopp 1954; Lahanas 1982). Hopp (1954) recorded an average of 73 worms/host with a high of 243, and Rausch (1947) stated that infections of 200–300 worms/host were "not uncommon" (p. 441).

Rausch (1947) described a pathology in a turtle infected with *N. emydis.* The small intestine was thickened with connective tissue to a diameter of 3 cm around a reddened, hollow growth in the mucosa, in which several worms were found embedded in a "caseous brown mass" (p. 441). In general, however, acanthocephalan infections cause little damage in reptiles other than localized granulomas or ulcerations of the intestines (Lane and Mader 1996).

Transmission routes of *Neoechinorhynchus* are unclear. Ostracod crusta-ceans serve as intermediate hosts and may be ingested inadvertently by the tur-tles (Esch et al. 1979). Snails have been reported to serve as paratenic hosts (also known as transport hosts) for *Neoechinorhynchus* (Whitlock 1939; Lincicome and Whitt 1947; Lincicome 1948; Hopp 1954) and might therefore also serve as a potential route of transmission. In the Texas map turtle, dietary evidence (from museum specimens collected in 1949 and from stomach flushes and fecal samples collected in 1998–99) supported the ostracod route of transmission: snails were rare in the diets of both sexes in 1949, when prevalence was 57 per-cent in males and 100 percent in females, whereas snails were somewhat more common but equally prevalent in male and female diets in 1998–99 (Lindeman and Barger 2005; Lindeman 2005). If inadvertent ingestion of ostracods occurs during feeding on algae, sponges, and bryozoan colonies—all of which are eaten more frequently by females—it might help to explain the greater prevalence and intensity of infection in females (Lindeman and Barger 2005).

Additional parasites recorded from *Graptemys* include a cnidarian (C. H. Ernst and Barbour 1972); a tapeworm (McKnight 1959); a variety of enteric roundworms (Magath 1919; Rausch 1947; McKnight 1959; Acholonu and Arny 1970; Dyer and Wilson 1997; Steinauer and Horne 2002; Lindeman and Barger 2005); encysted juveniles of a roundworm whose adult stage was first reported from raccoons (Ash 1962); a monogenean fluke found in its host's esophagus or buccal cavity (Stunkard 1916; Acholonu 1969b); at least one species of en-teric aspidogastrean fluke (Barker and Parsons 1914; Rausch 1947; Steinauer and Horne 2002; Lindeman and Barger 2005); and a cloacal mite (Bochkov and OConnor 2008). Of these taxa, only certain enteric roundworms are known to cause significant pathology or disease in reptiles (Lane and Mader 1996).

Host Specificity

Overall, specificity for hosts appears to be low: of the forty-five parasite species recorded (table 4.23), twenty-three (51 percent) have been reported in the common map turtle and twenty-seven (60 percent) in the false map turtle, the two species that have received the most attention from parasitologists (table 4.23). Furthermore, of the seventeen parasite species recorded in the other species, only seven (41 percent) have not also been recorded in either common or false map turtles. More extensive surveys of the other species would undoubtedly turn up several new host records, many of them of known parasites of the genus. Lack of host specificity is also extrageneric: at least thirty-five of the forty-five species recorded from *Graptemys* (78 percent) also parasitize other turtles (DeGiusti and Batten 1951; Woo 1969b; C. H. Ernst and Barbour 1972; E. M. Ernst and Ernst 1977; McAuliffe 1977; McAllister and Upton 1989; Barger 2004; Telford 2009).

Many of the records in table 4.23 for false map turtles may in fact come from Ouachita or Sabine map turtles or a mix of the three species, given the long run of confusion concerning these species even after description of the latter two by Cagle (1953b; see Vogt 1993). In particular, the host species is questionable in the reports of Barker and Parsons (1914), Stunkard (1915, 1917, 1923), Magath (1919), H. B. Ward (1921), Horsfall (1935), Byrd (1939), Fisher (1960), and Ash (1962).

Bacterial and Viral Pathogens

Jacobson et al. (1989) reported the bacterium *Flavobacterium meningosepticum* from a Barbour's map turtle that died in captivity and also in specimens of seven other map turtle and sawback species, some of which had been recently taken from the wild. Specimens exhibited lethargy and cutaneous edema prior to death, and necropsies revealed infection of the liver, heart muscle, spleen, and thymus. Aerobic and anaerobic bacteria were cultured from shell lesions of captive specimens of various species of *Graptemys* (Hernandez-Divers et al. 2009). Jacobson et al. (1982) reported the deaths of captive adult Barbour's and false map turtles that were found to have subcutaneous edema, swollen livers, minimal fat tissue, muscular atrophy, and necrotic lesions of the liver, pancreas, and kidneys consistent with herpesvirus infections, which may have been the cause of death.

Interactions with Other Species

Interactions of map turtles with other species have seldom been explored in detail but include competition for basking space; predation on eggs, hatchlings, and older turtles; mutualisms associated with basking activity; and commensalisms. Much of the published literature on these topics is anecdotal or descriptive, and hypotheses regarding impacts on population dynamics are at best speculative.

Competition for Basking Space

Map turtles and sawbacks occasionally engage in mildly agonistic interactions with other basking turtles. Vogt (1980a) observed male *Graptemys* being bitten by male spiny softshell turtles (*Apalone spinifera*). In a study of the basking interactions of Ouachita and Mississippi map turtles in syntopy with slider turtles (*Trachemys scripta*) and river cooters (*Pseudemys concinna*), pushing with the shell, forelimbs, or hind limbs was recorded at low frequencies, with turtles falling or jumping into the water, turning away, yielding space, or simply ignoring the initiator (Lindeman 1999a). In interspecific interactions, species identity did not predict which turtle "won" the interaction. Larger turtles were more than twice as likely to win as smaller turtles, and initiators won four of every five interactions. Smaller turtles showed evidence of possible assessment and avoidance of larger turtles by initiating fewer interactions and often turning away from them or voluntarily displacing themselves into the water. Selman and Qualls (2008b, 2008d) also described interspecific interactions at basking sites that further support the hypothesis that displacement and avoidance are mediated by relative size.

Predators

Predation affects mainly the egg and hatchling stages in *Graptemys*. Fish crows (*Corvus ossifragus*) are the most common nest predator along Gulf Coast rivers (Shealy 1976; Lahanas 1982; U.S. Fish and Wildlife Service 1993; Horne et al. 2003; Jones 2006). During peak nesting times, fish crows begin searching in small flocks around dawn and can seem quite brazen, walking behind females searching for a nesting site and driving females away before a nest is complete (Shealy 1976; Lahanas 1982; Horne et al. 2003). Nests are apparently identified by visual cues associated with disturbed sand, and eggs are usually carried well away from the nest to be consumed at roost sites in trees (Lahanas 1982). Nests are also raided by raccoons (*Procyon lotor*), red foxes (*Vulpes fulva*), gray foxes (*Urocyon cinereoargenteus*), coyotes (*Canis latrans*), river otters (*Lutra canadensis*), feral pigs (*Sus scrofa*), bobcats (*Lynx rufus*), armadillos (*Dasypus novemcinctus*), common crows (*C. brachyrhynchos*), scarlet snakes (*Cemophora coccinea*), speckled kingsnakes (*Lampropeltis getula*), and native and non-native fire ants (*Solenopsis molesta* and *S. invicta*; Neill 1951; Timken 1968a; Vogt 1980a; Jones 1993a, 2006; Brauman and Fiorillo 1995; Moulis 1997; Godwin 2002b; Horne et al. 2003; Rosenzweig 2003; Geller 2012b). Fish crows depredated 82 percent of black-knobbed sawback nests in the Mobile River delta, where they were the only predator noted (Lahanas 1982; but see Godwin 2002b). Fish crows were also the chief predator responsible for annual losses of 42 to 100 percent of yellow-blotched sawback nests on the lower Pascagoula (Horne et al. 2003). They were only the third most prevalent predator of ringed sawback nests farther inland on the upper Pearl, taking 15 percent of unprotected nests, as compared with 23 percent taken by raccoons and 21 percent taken by armadillos (Jones 2006). In Wisconsin, raccoons depredated 81–100 percent of map turtle nests monitored over four years (Geller 2012b).

Nests that are not depredated within a few days of deposition subsequently experience reduced predation (Shealy 1976; Vogt 1980a; Lahanas 1982; Rosenzweig 2003; Geller 2012b). Both fish crows and raccoons will excavate decoy turtle nests constructed by researchers (Lahanas 1982), so visual cues are probably of greatest importance to nest predators, but olfaction may also play a role. Lahanas (1982) noted that female black-knobbed sawbacks do not urinate prior to nesting, which would reduce olfactory cues left at a nest site, but female common, false, and Ouachita map turtles do urinate during nest construction (Vogt 1980a; pers. observations).

Fly maggots feed on hatchlings, but whether they usually feed on dead or live embryos is not well established. Hall and Parmenter (2008) suggested that platystomatid and sarcophagid maggots may feed only on dead embryos in Australian sea turtle nests. Maggots of the sarcophagid *Tripanurga importuna* were reported by Vogt (1981d), however, to feed on live hatchling false and Ouachita map turtles prior to their emergence from the nest cavity. Live ringed sawback hatchlings have also been found being consumed by *Tripanurga* maggots in the nest cavity (R. Jones, pers. comm. 2009). Female flies deposited eggs on sand, and the larvae burrowed below to where the eggs were located, entering live hatchlings via their plastral yolk plugs. Not all fly larvae in nests are feeding on the embryos, as Saumure et al. (2006) reported anthomyiid seed corn maggots (*Delia platura*), a saprophagous species not known to feed on carrion, from a failed nest of the common map turtle in Ontario. The maggots may have fed on plant roots in the nest cavity or fungi that grew on the dead eggs.

Substantial annual variation in sarcophagid infestation has been reported in three studies. In a wetter-than-normal nesting season, Vogt (1981d) reported 36 percent of nests containing *T. importuna* larvae, with up to 100 percent of the hatchlings in a nest being affected; prior to the wet year, the fly larvae affected approximately 15 percent of nests. Gillingwater (2001) recorded 16 percent and 68 percent infestations of common map turtle nests in consecutive years. Similarly, Jones (2006) recorded 14 percent loss of ringed sawback eggs to *T. importuna* in one year, with all but one of the eleven lost hatchlings killed prior to pipping, followed by 5 percent loss the following year, when all losses occurred after pipping.

Predation on hatchling and small juvenile *Graptemys* has not been intensively studied. Vogt (1979) reported observing several bird species that took "large numbers" (p. 609; see also Vogt 1980a) of hatchling map turtles upon emergence from nests along the Mississippi River in Wisconsin (primarily common grackles, *Quiscalus quiscala*, red-winged blackbirds, *Agelaius phoeni-ceus*, and American crows, *Corvus brachyrhynchos*, but possibly also including ring-billed gulls, *Larus delawarensis*, and great blue herons, *Ardea herodias*, both also said to be present in abundance). Rice rats (*Oryzomys palustris*) fed heavily on hatchling common and false map turtles around Reelfoot Lake, Tennessee (Goodpaster and Hoffmeister 1952). Part of the authors' impression that turtles are "chief items of food" (p. 368) for rice rats may stem from their observations that rats entered the live-storage containers of collectors during the once-flourishing trade in pet turtles at Reelfoot; however, abundant de-predated carcasses of hatchlings were also found along the lake margins. Lastly, non-native fire ants (*Solenopsis invicta*) were reported attacking a one-year-old *G. barbouri* (Mount 1981).

The extent of predation on young map turtles by aquatic predators is not clear. While their small size and soft, pliable shell would seem to leave them exceptionally vulnerable to predatory fish and bird and mammalian predators that forage along shallow shorelines, few records of predation upon small ju-veniles exist. Bullfrogs (*Lithobates catesbeianus*) took hatchling map turtles in Missouri and Indiana, and a hatchling Pearl map turtle was found in a spot-ted bass (*Micropterus punctulatus*) in Louisiana (Graham 1984; J. L. Carr and Messinger 2002).

Experiments with largemouth bass (*Micropterus salmoides*) presented with hatchling slider (*Trachemys scripta*) and painted turtles (*Chrysemys picta*) are probably relevant to fish predation on hatchling *Graptemys*. Semlitsch and Gib-bons (1989) and Britson and Gutzke (1993) both noted the lack of turtle hatch-lings in the diets of *M. salmoides* and other predatory fish (see also F. Jordan and Arrington 2001). They conducted experiments that demonstrated that whereas dead or anesthetized hatchlings were readily taken by *M. salmoides*, live hatch-lings were egested unharmed. Fish in the experiments attacked hatchlings up to six times before ignoring them. Because the fish consumed dead and anesthe-tized hatchlings, behavior of the hatchlings inside the mouth (perhaps raking the gills with their claws) was assumed to be the reason for rejection. Britson and Gutzke (1993) speculated that hatchlings' brightly colored and vividly con-trasting plastral patterns function as aposematic, or warning, coloration, simi-lar to the bright and vivid markings of many toxic or venomous animals; they further noted that aposematic coloration had not previously been associated with noxious defensive behavior in any animal species. The hypothesis is sup-ported by the observation that in many turtles, plastral patterning fades and colors become more muted as growth occurs (i.e., perhaps the coloration be-comes less advantageous as individuals grow to sizes less prone to predation). A follow-up experiment, again with largemouth bass, demonstrated greater avoidance of hatchlings of a brightly colored species (the painted turtle) than of a cryptically colored species (the common snapping turtle, *Chelydra serpentina*; Britson 1998). Because an ontogenetic loss of patterning occurs in map turtles

Figure 4.18. Two potential aquatic predators of *Graptemys*. (a) A large alligator snapping turtle surfacing for air in the Apalachicola River at Torreya State Park, Liberty County, Florida, a site inhabited by several Barbour's map turtles. (b) An American alligator in the Chickasawhay River at Highway 57 in Leake County, Mississippi, where numerous yellow-blotched sawbacks and Pascagoula map turtles were also seen.

and sawbacks (e.g., Folkerts and Mount 1969), the hypothesis that plastral coloration functions as a warning to predatory fish may well also apply to these turtles (Britson and Gutzke 1993).

The only quantifications of predation on older map turtles concerns bald eagles (*Haliaeetus leucocephalus*) nesting near water and taking Mississippi map turtles in Texas and Barbour's map turtles in Florida (Mabie et al. 1995; Means and Harvey 1999). Mississippi map turtles constituted 0.9 percent of all prey and 3 percent of all turtle prey found under twenty-seven nest trees over seven years; softshell turtles (*Apalone* spp.) were the predominant prey. Barbour's map turtles were the predominant prey recorded for the nest tree of a single pair of eagles over five nesting seasons, constituting 84 percent of all turtles and 74 percent of all vertebrates found beneath the nest tree. The eagle pair took a narrow range of body sizes (108–157 mm carapace length), which were believed to include adult males and a larger number (over half the sample) of juvenile females.

Alligators (*Alligator mississippiensis*) and alligator snapping turtles (*Macrochelys temminckii*) are possible predators on adults (Shealy 1976; Lahanas 1982). Both are significant predators of adults of other Gulf Coast turtle species (Giles and Childs 1949; Valentine et al. 1972; Delany and Abercrombie 1986; D. Taylor 1986; Sloan et al. 1996; Barr 1997; Iverson and Hudson 2005; Elsey 2006). Nevertheless, their depredation of *Graptemys* has been confirmed for only two specimens of *Graptemys* sp., one taken by an alligator in Louisiana and the other by an alligator snapping turtle in Arkansas (Wolfe et al. 1987; Elsey 2006), although Neill (1971) stated, without details, that alligators consume Barbour's map turtles. Many of the dietary studies cited above have been performed outside the range of *Graptemys* or in lentic habitats with lower map turtle and sawback densities; also, turtle remains are often unidentified. Shealy (1976) examined the feces of "several" (p. 86) alligator snapping turtles from the Conecuh River and failed to find anything but mussel shells. Because both predators can be found in the rivers inhabited by map turtles and sawbacks, however (fig. 4.18), their importance as predators of these species remains a distinct possibility.

Mammalian nest predators such as raccoons, coyotes, and foxes sometimes also consume nesting female *Graptemys*. Oldfield and Moriarty (1994) reported that adult common map turtles in the Mississippi River in Minnesota frequently showed evidence—shell damage or missing toes or limbs—of predator attacks. P. A. Cochran (1987) reported thirteen adult female common map turtle carcasses in Minnesota that may have been depredations, as heads and limbs were missing. Sterrett (2009) reported carcasses of female Barbour's map turtles he believed had been taken by a raccoon (at a nesting beach) and a river otter (on an island). Carcasses of adult females of all three sawback species have been found on nesting beaches with their limbs and head missing or having been eviscerated through the cloaca, presumably by raccoons (pers. observation 2001; R. Jones, pers. comm. 2009).

Mutualistic and Commensal Interactions

Two cases of possible mutualism, both involving basking behavior, have been reported for *Graptemys*. Vogt (1979) observed common grackles (Passeriformes: Icteridae: *Quiscalus quiscala*) feeding on leeches on basking common, false, and Ouachita map turtles. The turtles appeared exceptionally tolerant of the approach and tugging behavior of the grackles. Vogt's (1979) observations were anecdotal, so the frequency of the interaction and the extent of any mutual benefits derived from it are unknown. His observations were made at one site over a relatively short period (six days, the only such observations in a five-year study) and may therefore have involved a single bird. While grackles are common both at Nickell Cove on Kentucky Lake, where I studied Ouachita and Mississippi map turtles from 1992 to 1999 (Lindeman 1999a, 1999c, 2000b), and at Presque Isle State Park, Pennsylvania, where I have studied common map turtles since 1999 (Lindeman 2006b; K. M. Ryan and Lindeman 2007), I have observed no interactions like those Vogt (1979) reported, in spite of frequent and often lengthy observations of turtles with a spotting scope or binoculars.

Shealy (1976) discussed the mixed-species assemblages in which Escambia map turtles bask. He suggested that they and other species provide a "sentinel" service, alerting one another to possible danger by plunging into the water. Softshells (*Apalone* sp.) were reported to be the wariest species, while large river cooters (*Pseudemys concinna*) and large female Escambia map turtles were thought to send good signals to other turtles by the obvious splashes they produced. Basking Pascagoula map turtles did not react to wood ducks (*Aix sponsa*) disturbed from a log, however, suggesting that they may not use ducks as sentinels (Selman and Qualls 2008c).

Commensal interactions with epizoic organisms appear to involve only small proportions of *Graptemys* populations. J. R. Dixon (1960) reported a colonial ectoproct (*Plumatella* sp.) growing on Texas specimens he identified as false map turtles (possibly Ouachita or Sabine map turtles). A stalked protozoan (*Vorticella* sp.) was recorded on two of nine common map turtles from the Mississippi River in Illinois, and an acinetid ciliate (*Squalorophrya macrostyla*) was found on common map turtles in central Tennessee (Cooper and Sharp 1982; Ziglar and Anderson 2002). Edgren et al. (1953) reported the green algae *Basicladia chelonum* and *B. crassa* growing on the carapace and plastron of a common map turtle, with the former also found growing on a false map turtle; forty-two other specimens of *Graptemys* sp. were free of *Basicladia* sp. Vinyard (1953) reported *Basicladia* sp. and the encrusting green alga *Dermatophyton radians* growing on Oklahoma specimens he identified as *G. pseudogeographica* (probably Ouachita map turtles). Proctor (1958) found *Basicladia* sp. on "less than 50%" (p. 637) of a large series of what he termed *G. pseudogeographica pseudogeographica* (probably Ouachita map turtles, given the collecting localities—primarily south-central Oklahoma and possibly Texas and Missouri). Both species of *Basicladia* also occurred on J. R. Dixon's (1960) Texas specimens, along with the green alga *Rhizoclonium hieroglyphicum* and the cyanobacterium *Lyngbya* sp., with half of eight specimens having one or more species of algae. Belusz and Reed (1969) recorded *B. chelonum* on four of five common map turtles, with *D. radians* and *Gongrosira debaryana* each on two specimens. Waters (1974) found that *Basicladia* sp. grew in "substantial" (p. 40) amounts on eleven of fifty-nine black-knobbed sawbacks, with 40 percent of males but no females having colonies. Waters speculated that males' smaller body sizes made it possible for them to heat more rapidly and reenter the water before fully desiccating the algae. Neill and Allen (1954) also reported the development of algal growth on Barbour's map turtles kept in outdoor enclosures, but they did not record the algal species.

Basicladia sp. typically grows on turtle shells, although the species can also occasionally be found growing on inanimate objects such as stones (Proctor 1958). Incidence of these algae in map turtles and sawbacks is low compared with other deirochelyine emydids such as painted turtles (*Chrysemys picta*) and slider turtles (*Trachemys scripta*), and emydids in general have low incidence relative to kinosternids and chelydrids (Edgren et al. 1953; Neill and Allen 1954; Proctor 1958). Proctor (1958) suggested that the rougher carapaces of kinosternids and chelydrids were better attachment points for the algae and that the loss of each of the shell laminae in one piece during ecdysis in deirochelyines made them more likely to be captured and examined in an algae-free state. Algal growths on black-knobbed sawbacks were thickest in the spring, before ecdysis, and captives allowed to bask developed little algae, whereas those denied basking opportunities developed thick colonies (Waters 1974); this suggests another reason for the denser growths on species such as the kinosternids and chelydrids, which bask less frequently than do deirochelyines. As for the difference between the slider turtles and the Ouachita map turtles, Proctor (1958) suggested that the former's denser growth of *Basicladia* sp. occurred because *T. scripta* provides a more advantageous growth substrate for the algae, spending more time in shallow, sunlit water and floating at the surface.

Ultsch et al. (2000) described an apparent commensal interaction of northern leopard frogs (*Lithobates pipiens*) with common map turtles during winter. Several dozen frogs were found in a deep (6–8 m), well-oxygenated depression in a river among approximately one hundred turtles. The frogs were found both in rock crevices and underneath turtles, possibly protecting them from fish predation.

5

Conservation Status

The major threats to species diversity around the world are overexploitation, habitat alteration, and the introduction of exotic species. All three, but particularly the first two, have played a role in species declines in *Graptemys*. Two species are federally protected in the United States, and one is federally protected at its northern range periphery in Canada, while an international list of imperiled species includes nine of the fourteen species of the genus under various levels of endangerment. State laws further restrict exploitation of most species. The primary documented threats are anthropogenic engineering of rivers, which alters habitat for basking, prey species, and nesting, and exploitation for the pet trade, which was little regulated until recently.

Status Under National Species Protection Legislation

The ringed and yellow-blotched sawbacks are listed under the U.S. Endangered Species Act (ESA). Both were among six *Graptemys* species initially investigated for listing in the late 1970s, with listings announced in 1986 and 1991, respectively (Stewart 1986a; Lohoefner 1991). Among other species, Barbour's and Cagle's map turtles were most seriously considered for protection but ultimately were not listed (see chapter 2).

In 2008, the nonprofit Center for Biological Diversity initiated status reviews of nine southern turtle species to assess whether they warrant petitions for ESA listing (J. Miller, pers. comm. 2008). Included were six species of *Graptemys*: the Alabama, Barbour's, Cagle's, Escambia, Sabine, and Texas map turtles. Conspicuous by their absence from the list were the Pascagoula and Pearl map turtles, which had been recommended for consideration for federal protection following basking surveys in the 1990s and have not improved in status since (Lindeman 1999c; Selman and Qualls 2007, 2009a). Status reviews were conducted via literature review and interviews with turtle biologists familiar with the species (J. Miller, pers. comm. 2008). A lawsuit filed against the U.S. Fish and Wildlife Service (USFWS) was settled with the agreement to review 374 southeastern aquatic species, including six *Graptemys* species: Alabama, Barbour's, Escambia, and Pascagoula (*sensu lato*) map turtles and black-knobbed sawbacks (USFWS 2011).

Under the ESA, Congress mandates that the USFWS draft and implement recovery plans for listed species. Such plans include a list of management actions that the USFWS and cooperating agencies may undertake to improve the status of the species; criteria by which to judge the effect of these actions, with objective goals that must be reached before delisting can occur; and an estimate of the time and cost required to bring about recovery. Recovery plans were completed for each sawback within a few years of its listing (Stewart 1988; Jones 1993b). Five of the six management actions listed for the ringed sawback and three of the six listed for the yellow-blotched sawback are essentially calls for basic research on the species (see chapter 6).

Delisting criteria for both species emphasize habitat protection via regulation of habitat modification, habitat acquisition, cooperative agreements with landowners, and control of water quality. The plan for the ringed sawback requires protection of 270 km of habitat in two reaches of the Pearl, with at least 48 km both above and below the Ross Barnett Reservoir near Jackson. The plan for the yellow-blotched sawback requires protection of habitat along the entire length of the Pascagoula and the lower 129 km of each of its two major tributaries, the Leaf and the Chickasawhay. Both plans require demonstration of stable or increasing populations, over ten years for the ringed sawback and fifteen years for the yellow-blotched sawback. Recovery is predicated upon basking counts averaging forty-four yellow-blotched sawbacks per river kilometer in the Pascagoula and twenty-two per river kilometer in the Leaf and Chickasawhay; there are no stipulations for minimum basking densities for the ringed sawback. The ringed sawback plan included an estimated recovery cost of $108,000 over the first three years, while the yellow-blotched sawback plan included a total estimated recovery cost of $845,000.

The USFWS has reported to Congress on its recovery efforts on a biennial schedule since 1990, except for a combined report for 1998 and 2000 (USFWS 1990, 1992; subsequent reports available at www.fws.gov/endangered/esa-library/index.html). The ringed sawback has generally been listed as stable, although it was listed as declining in 1998/2000 and 2010 and of unknown status in 2006 and 2008, and it has consistently been listed as having had 26 to 50 percent of its recovery objectives met. The yellow-blotched sawback has been listed as being of unknown population status (1992, 2002, 2004, 2006, 2008, 2010), stable (1994), or declining (1998/2000), with consistent reporting that recovery objectives are only 0–25 percent met. Both species have been consistently assigned a low priority for recovery efforts (fourteen on an eighteen-point scale).

While never considered for federal listing in the United States, the common map turtle (*G. geographica*) was designated as a species of special concern in Canada in 2002. The Canadian status report (COSEWIC 2002) lists several potential threats to populations, including shoreline development, increasing recreational use of habitat, dam impacts on habitat, increasing nest predation, and the pet trade, which when coupled with the life history of the species are the rationale for its special concern status.

Status Under State and Provincial Laws

Most U.S. states within the ranges of *Graptemys* species and Ontario and Quebec have their own lists of imperiled species, and many include map turtle and sawback species (table 5.1). The two federally listed species in the United States are also listed by their range states, and many states protect their populations of five of the nine other species characterized by restricted ranges (there is no state protection for Pascagoula, Pearl, Sabine, or Texas map turtles). Two of the three widely distributed species are listed in a few peripheral states and provinces, while the Ouachita map turtle is not listed in any range state. The primary purpose of state listings is prohibition of take, rather than protection of habitat. Notable exceptions are two state listings as species of special concern: this applied until recently to Barbour's map turtle in Florida, where there was nevertheless a possession limit of two (take was prohibited in 2009), and still applies to the false map turtle in Wisconsin, where there is nevertheless a possession limit of five (restricted to between 15 July and 30 November).

Commercial take of wild *Graptemys* is prohibited by many states, generally as part of broader prohibitions of commercial take for wild turtles or reptiles in general. One prominent exception is Louisiana, where map turtles and sawbacks are among turtles raised in farming operations for commercial export, supported by an estimated take of 1,000 animals from the wild for brood stock in one recent year (Reed and Gibbons 2004). In Arkansas, commercial harvest was long unregulated, but recent permitting processes with annual reporting have been implemented for commercial trappers. Irwin (2007b) reported a three-year average of 6,658 *Graptemys* spp. taken statewide for legal consumption, primarily in the pet trade but also for export to Asian meat markets. Two-thirds of the take was in what dealers call the "toy" class—roughly meaning juveniles—raising suspicion that some commercial collectors have used dip nets to capture turtles, possibly from brush piles at night, or have taken eggs from nests; both practices are illegal (Irwin 2007a). Similarly, a long history of commercial harvest of turtles is maintained under current state regulations in Tennessee's Reelfoot Lake, a large lake formed by earthquakes in 1811 and 1812; elsewhere in the state, commercial harvest is prohibited. The daily possession limit at Reelfoot Lake is five for all species combined.

Table 5.1. State- and province-listed species of *Graptemys*

Species	Endangered	Threatened	Special Concern[a]	Range states not listing
barbouri		GA	AL, FL	None
caglei		TX		None
ernsti			AL	FL
flavimaculata	MS			
geographica	MD	KS, QC[b]	GA, OK, ON, VT	AL, AR, IL, IN, IA, KY, LA, MI, MN, MO, NC, NJ, NY, OH, PA, TN, VA, WV, WI
gibbonsi				MS
nigrinoda	MS		AL	None
oculifera	MS	LA		None
ouachitensis				AL, AR, IL, IN, IA, KS, KY, LA, MN, MS, MO, OH, OK, TN, TX, WV, WI
pearlensis				LA, MS
pseudogeographica		SD	ND, WI	AR, IL, IN, IA, KS, KY, LA, MN, MS, MO, NE, OK, TN, TX
pulchra			AL, GA	MS
sabinensis				LA, TX
versa				TX

[a]Alternative third-tier designations used by states include Protected Nongame (AL), Rare (GA), and Level III Conservation Priority (ND).

[b]Listed as "likely to be designated" as either Threatened or Vulnerable in Quebec.

In 2007, Texas banned commercial take of nongame animals (including turtles) from public lands and waters. Commercial take of turtles was limited to sliders, softshells, and snapping turtles taken on private lands. The change in regulations stemmed primarily from concern over increased take for export to Asian meat markets. In 2008, in response to a petition from the Center for Biological Diversity, the Oklahoma Wildlife Conservation Commission placed a moratorium on commercial turtle harvest in public waters to further study the issue for three years (since extended for two additional years; Johansen 2011). For the period 1994–2010, nearly one million wild turtles were harvested, of which 1.1 percent were Ouachita map turtles and 0.2 percent were false map turtles (Johansen 2011).

Possession limits for noncommercial take have been established by many states for their unlisted map turtles. Until 2009, Florida permitted possession of two Barbour's map turtles (notwithstanding its status as a species of special concern) and two Escambia map turtles, but take of both species is now prohibited (for similarity of appearance reasons in the case of the latter species); possession of animals taken prior to the regulatory change is allowed through a special permit application. Pennsylvania recently lowered the possession limit for common map turtles from two to one. Michigan allows possession of two common map turtles. Further possession limits are four in Indiana (common, false, and Ouachita), Mississippi (Pascagoula, Pearl, Alabama, false, and Ouachita) and Ohio (common; take and possession of Ouachita map turtles is apparently not allowed); five in Virginia (common) as well as Kentucky, Missouri, and Wisconsin (common, false, and Ouachita); six in Arkansas and Oklahoma (common, false, and Ouachita) as well as in Texas (false, Ouachita, Sabine, and Texas); ten in Alabama (common and Ouachita); sixteen in Illinois (limited to a take of eight per day; common, false, and Ouachita); and one hundred in West Virginia (common and Ouachita). Conversely, take and possession of unlisted map turtle species are prohibited in Iowa, Minnesota, and New York.

Status on International Species Lists

Two international lists are the most important for protecting imperiled species: those of the Convention on International Trade in Endangered Species of Wild Fauna and Flora (CITES) and the International Union for Conservation of Nature (IUCN). Recently all species of the genus *Graptemys* were listed on CITES Appendix III, meaning that international trade (primarily in the pet trade but also in foreign meat markets) requires export permits that identify the turtles' species and ensures that the trade is not in violation of any federal or state laws. The history that led to this listing is recounted in chapter 2.

The IUCN is a global affiliation of nongovernmental organizations, government agencies, researchers, and experts with a goal of developing and promoting conservation science around the world. Their Red Lists are a set of categories used to identify species under threat. Beginning in 1994, species could be listed (in decreasing order of threat) as extinct, extinct in the wild, critically endangered, endangered, vulnerable, lower risk/near threatened, or least concern. The designations critically endangered, endangered, and vulnerable are made with reference to a series of criteria regarding population size and rate of decline, geographic range size, rate of habitat loss, and probability of extinction as assessed by quantitative methods such as a population viability analysis.

In 1996, the IUCN listed ringed and yellow-blotched sawbacks as endangered and Cagle's map turtles as vulnerable. Criteria cited for ringed and yellow-blotched sawbacks were identical, with fewer than 5,000 km^2 as the area of occurrence and continuing fragmentation and decline in habitat area and quality. Cagle's map turtle was considered to have experienced more than a 20 percent decline in total population over three decades as a result of range reduction, with less than 20,000 km^2 as the area of occurrence and severe and continuing fragmentation and decline in habitat area and quality. Six species— Barbour's, Escambia, Texas, and Pascagoula and Pearl map turtles (before the recognition of the last two as separate species) as well as black-knobbed sawbacks—were listed as near threatened, and the remaining five species were listed as least concern.

Revisions to IUCN listings in 2011 were substantial, with five species shifted to more imperiled status and four shifted to less imperiled status: Cagle's, Pascagoula, and Pearl map turtles are now listed as endangered; Barbour's map turtles and ringed and yellow-blotched sawbacks as vulnerable; Alabama and Escambia map turtles as lower risk/near threatened; and the remaining six species as least concern. The criteria for listing emphasized population declines over the last three generations, estimated to exceed 30 percent (Barbour's map turtle, yellow-blotched sawback), 50 percent (Cagle's and Pascagoula map turtles), and 70 percent (Pearl map turtle). Also cited was the continuing fragmentation of ranges: less than 2,000 km^2 for the ringed sawback and less than 500 km^2 for Cagle's map turtle.

In 2003, the Turtle Conservation Fund, a partnership that includes the IUCN's Species Survival Commission for tortoises and freshwater turtles, listed the yellow-blotched sawback on its worldwide list titled "Top 25 Turtles on Death Row." A subsequent revision to the list was made in 2007, after an extensive e-mail list-server discussion in which participating turtle researchers agreed that no U.S. turtle species was in such dire circumstances that it should take the place of the critically endangered species found elsewhere, particularly in southeastern Asia, where exploitation of turtles for meat is extraordinarily high. Instead, in addition to the top twenty-five global list, lists of the most endangered turtles were formulated for each geographic region. After several e-mail exchanges on the list server, the list for North America had the Pascagoula/Pearl map turtles combined at number three (behind the bog turtle, *Glyptemys muhlenbergii*, and the Alabama redbelly turtle, *Pseudemys alabamensis*), with the yellow-blotched sawback at number five and the ringed sawback at number six. The high ranking for the Pascagoula/Pearl map turtles recognizes their low abundance relative to the two sympatric sawbacks (Lindeman 1998, 1999c; Selman and Qualls 2009a).

Threats to *Graptemys* Populations

The most-frequently cited and best-documented threats to map turtle and sawback populations involve modifications of the rivers and lakes they inhabit and their take for the pet trade. Many other threats have been cited, although most are more suspected than confirmed as causes of population decline.

River impoundment and the loss of riparian sources of deadwood represent the most adverse forms of habitat modification, but channelization, sand and gravel mining, and contamination are also potential threats. Besides the pet trade, three other forms of direct anthropogenic mortality may threaten *Graptemys* populations: take of adults and of eggs for food, take for the trade in biological specimens, and wanton shooting. Various forms of indirect mortality, increasing populations of native predators, human disturbance, and introduced pest species may also be important threats to some *Graptemys* populations.

River Engineering

River channelization and river impoundment are expected to negatively impact map turtles and sawbacks through the loss of river bends (fig. 5.1), which concentrate their populations. In a typical bend of a river, faster current occurs on the outer bend, eroding the outer bank and depositing sediments on the less-inclined inner bank. A dramatic example of bank erosion is shown in figure 5.2. Bank erosion renews two habitat features critical for map turtles and sawbacks: deadwood from trees that fall into and become stranded in the deeper, swifter outer bends and flat sandbars that develop on the inner bends (fig. 5.3; Shealy 1976; J. B. Wallace and Benke 1984; Shankman 1993; Lindeman 1999c; Godwin 2001, 2003; Sterrett et al. 2011). The deadwood's emergent logs and branches are prime basking sites, particularly for adult females, which avoid shallower water and deadwood inclined against the bank (Waters 1974). Submerged portions of the deadwood serve as attachment points for prey communities (Waters 1974; Shively and Jackson 1985; Kofron 1991; Selman and Qualls 2008a) and may also serve as cover from potential aquatic predators (Jones 1996; Sterrett 2009). The hard surface of submerged deadwood provides a "hot spot" of invertebrate productivity and diversity (J. B. Wallace and Benke 1984; Benke and Wallace 2003; Angradi et al. 2009). Basking density generally increases with deadwood density, demonstrating the importance of deadwood as a habitat feature (Lindeman 1998, 1999c), and deadwood density may increase with increasing sinuosity of a river (McIlroy et al. 2008). The sandbars along the inner bends are the most important nesting sites (P. K. Anderson 1958; Shealy 1976; Horne et al.

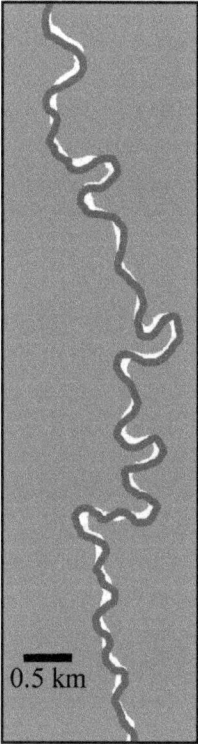

Figure 5.1. Diagram of the upper Leaf River immediately north of its confluence with the Bowie River at Hattiesburg and Petal, Mississippi, showing river bends and sandbars (white).

0.5 km

Figure 5.2. Downstream view of the Conecuh River along the former County Road 4 in Escambia County, Alabama. The river completes a sweeping leftward bend as it enters the area and has eroded the outer bend to the roadway. An aerial photograph of this reach of the river was published by Shealy (1976; his fig. 2).

Figure 5.3. Upstream view of the Conecuh River about 300 m south of the area shown in figure 5.2, showing a river bend with a flat sandbar on the inner bank and a steep, eroded outer bank with deadwood accumulation.

2003; Jones 2006). If bank erosion is blocked via channelization, riprapping, or impoundment, deadwood abundance would be expected to decline with time, sandbars will become overgrown via succession, and new oxbow lakes will not form (Shankman 1993). Removal of large pieces of deadwood from a river channel is expected to lead to decreased productivity and diversity of invertebrates and a decline in the conservation of nutrients cycled within the stream ecosystem (J. B. Wallace and Benke 1984).

Impoundment in drier climates eliminates forested river islands, thereby eliminating a major source of deadwood. In the South Dakota reach of the Missouri River, for example, a series of large dams has caused the number of cottonwood-dominated islands to decline from "dozens if not hundreds" historically to fewer than ten (Ode 2004, p. 3). Shoreline development appears to further decrease the amount of deadwood that accumulates in reservoirs, based on anecdotal comparisons of forested shorelines and those with suburban development (Lindeman 1999c; Godwin 2001, 2003). The situation is thus believed to be similar to that described for several lakes near the Wisconsin-Michigan border, where cabin density was negatively related to deadwood density in the littoral zone, presumably due to vegetational changes wrought by the development of vacation homes (Christensen et al. 1996). Habitat degradation is exacerbated by desnagging—the removal of deadwood from a river channel—a common practice in reservoirs and sections of river that receive heavy recreational use or commercial barge traffic (J. B. Wallace and Benke 1984; Babitzke 1992; Lindeman 1999c). Thus any of these common forms of river engineering is expected to lead to inexorable decline in numbers of map turtles and sawbacks.

Dams and reservoirs alter hydrology and fragment habitat, separating what might otherwise be relatively continuous populations. Of the fourteen species in the genus *Graptemys*, only the sympatric yellow-blotched sawback and Pascagoula map turtle occupy a geographic range lacking major impoundments. Smaller rivers often have dams that back water far up into tributary creeks to create branching, lake-like reservoirs. The largest rivers within the ranges of map turtles and sawbacks have been converted to series of relatively lentic pools by lock and dam operations. For example, there are twenty-nine locks and dams on the upper Mississippi River above its confluence with the Ohio River (fig. 5.4), nineteen on the Ohio River, nine on the Tennessee River, and three on the Alabama River.

Bennett (2009) recorded telemetered common map turtles that were able to pass through locks of the Trent-Severn Waterway, although shell injuries sustained by one specimen suggested that such passage may occur at serious risk to the turtles. In areas with high lock density, turtles were smaller, grew more slowly, and had smaller home ranges (Bennett 2009; Bennett et al. 2009) than those in sites with low lock density.

The case of the Alabama map turtle in the Mobile Bay drainages is an instructive example of the extent of impacts of river impoundment (fig. 5.5). The

Figure 5.4. View of Lock and Dam 11 on the Mississippi River in Dubuque County, Iowa. Twenty-nine such structures occur on the Mississippi River upstream of its confluence with the Ohio River.

three locks and dams on the Alabama River have converted about 75 percent of the river to lentic habitat in long, wide reservoirs. Impoundment occurs virtually the lengths of the Tombigbee River (twelve locks and dams, some along the artificial channels built to connect the Tombigbee to the Tennessee River via the Tenn-Tom Waterway) and the Black Warrior River (four locks and dams). Along the main stem of the Coosa River are seven major dams; these form dendritic reservoirs (used predominantly for recreation) that impound the river over approximately 85 percent of its length. The Tallapoosa River is not dammed within the range of the Alabama map turtle, but dams occur above the fall line, one of which forms Martin Lake, a large dendritic reservoir. On the upper Black Warrior River drainage, Lewis Smith Dam creates a reservoir that inundates the confluences of Sipsey Fork with Ryan, Crooked, Rock, Brushy, and Clear Creeks and dozens of smaller creeks. Even the upper part of the relatively free-flowing Cahaba River has low dams that control flooding, allowing flow over their tops; one of these, the Marvel Slab Dam, which once served as a bridge for logging and coal trucks in northern Bibb County, was recently removed. While Alabama map turtles have been recorded in the vicinity of many of the locks and dams as well as in some of the dendritic reservoirs—including Lewis Smith, Jordan, Logan-Martin, and Weiss Lakes (Mount 1975; McCoy and Vogt 1979; Godwin 2001, 2003)—the effects of these habitat alterations on their absolute and relative abundance have been little studied. Godwin (2003) did note that

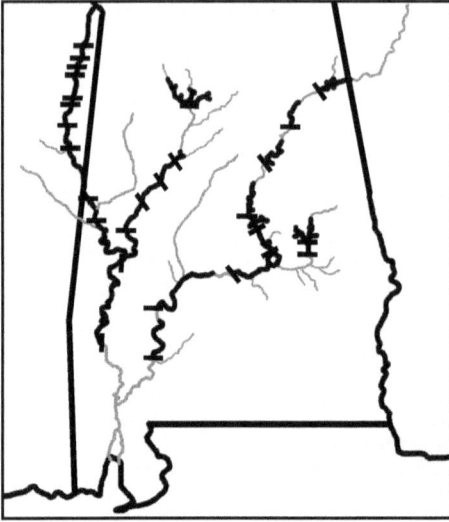

Figure 5.5. The geographic range of the Alabama map turtle (*Graptemys pulchra*), showing major dams (hash marks) and reservoirs. Black sections of streams mark the approximate portions of the drainage that are rendered lentic by inundation.

Alabama map turtles were less abundant than river cooters (*Pseudemys concinna*) in dendritic reservoirs, in contrast to rivers, where Alabama map turtles were the more abundant species.

In contrast to the persistence of the Alabama map turtle and many of its congeners in dammed rivers, Killebrew et al. (2002) found that Cagle's map turtle was absent from six impoundments within its range on the Guadalupe River, which is interrupted by nine dams. Three small impoundments in which the species persisted had little shoreline development and thus maintained streamside willows, which are important habitat for prey, particularly caddis fly larvae. Similarly, Buhlmann and Gibbons (2006) reported low incidence of Barbour's map turtles in basking surveys on an impoundment on the Chattahoochee River, with none seen in the lower, more lentic portion of the reservoir.

Much of the concern for the threatened ringed sawback lies with the proposed engineering of the middle Pearl River near Jackson that would alter its habitat. Stewart (1986a, 1986b, 1988) discussed planned flood-control modifications that would negatively impact 28 percent of the remaining habitat in the Pearl. He also felt that river channelization could extirpate populations from the Bogue Chitto River. Exceptionally severe flooding of the Pearl in Jackson in 1979 stimulated planning for flood control. In 1986, the Shoccoe Dam was authorized by Congress. It was to have been put in immediately north of Ross Barnett Reservoir and would have essentially doubled the size of impounded habitat in the middle reaches of the river (fig. 5.6), eliminating the river reach that currently contains the densest known population of ringed sawbacks (Jones and Hartfield 1995). Because of opposition from landowners and conservation interests, it is no longer considered a viable plan, although it has not been deauthorized. Likewise, a subsequent plan to build extensive levees to pro-

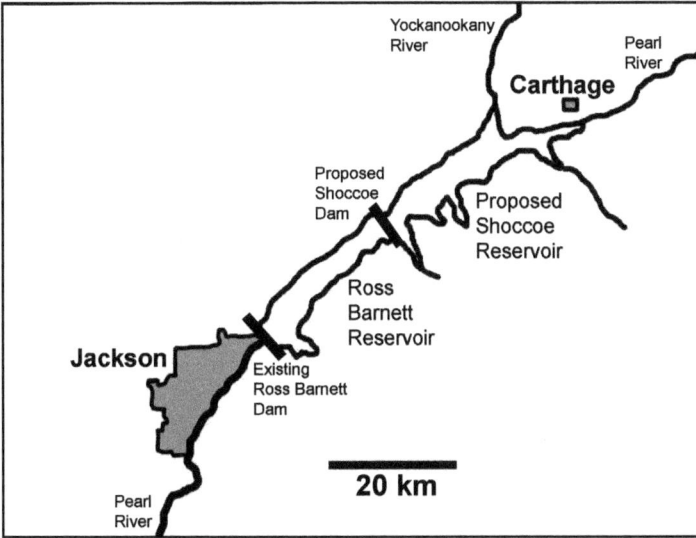

Figure 5.6. The proposed Shoccoe Dam on the upper Pearl River.

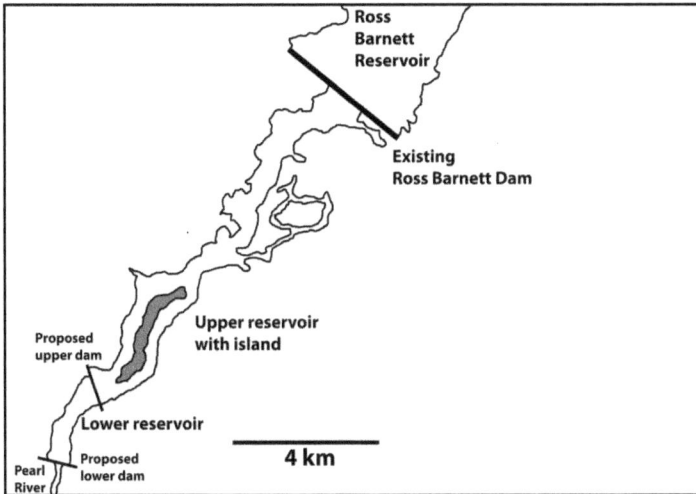

Figure 5.7. The proposed LeFleur Lakes development on the Pearl River near Jackson.

tect Jackson from flooding never drew enough support for implementation. The newest flood-control plan, the LeFleur Lakes Project, is to build two weir-like dams downstream of the Ross Barnett Reservoir. The area that would be inundated is a 1,940 ha section of the river and its largely undeveloped floodplain adjacent to Jackson. It would become the LeFleur Lakes, with a 240 ha artificial island placed in the middle of the larger, upper reservoir and slated for economic development (fig. 5.7). The plan has been opposed by the Pearl River Basin Coalition and the Mississippi Wildlife Federation, in part because of the ringed sawback, but many Mississippians appear to be enticed by the economic development aspects of the plan and support it.

Mining of Riverbeds

Sand and gravel dredging conducted in southern streams undoubtedly alters *Graptemys* habitat in a negative way. For example, a 700 ha gravel mining operation on Mississippi's lower Bowie River, just above its confluence with the Leaf River, reduced operations in 1995 (Mossa and Coley 2006); the section of the river is a chain of lakes in which slider turtles and river cooters are abundant (pers. observations 2001) but yellow-blotched sawbacks and Pascagoula map turtles are absent, as they are in artificially lentic stretches of Thompson Creek that are affected by gravel mining (Selman and Qualls 2009a). In the Bogue Chitto, Shively (1999) found low numbers of ringed sawbacks and Pearl map turtles near riverbed point bar mining operations, which tended to create shallow channels with loose substrate and little deadwood, with the deadwood that did occur being scoured clean of aquatic prey species. A survey in Louisiana and Mississippi documented 332 gravel mining sites on seventy-nine streams affecting 35,340 ha of riparian landscape (Mossa and Coley 2006). Particularly intensive mining operations occurred along the Bogue Chitto (38 sites, 3,852 ha), Pearl (27 sites, 1,425 ha), and Leaf Rivers (13 sites, 363 ha).

Contaminants

Contaminants from point and nonpoint sources may negatively impact *Graptemys* populations through either direct contamination or elimination of prey species. Pesticides entering the Pearl via drainage ditches were cited by Stewart (1988) as a possible contributor to the decline of the ringed sawback. The greatest concern over contaminants has been expressed for the yellow-blotched sawback in the Leaf and Pascagoula Rivers. Pulp mill effluent over an eleven-year period (1984–95) has been implicated in elevation of tissue contaminant levels and possibly endocrine disruption (Kannan et al. 2000; Shelby and Mendonça 2001) and may have caused depressed clutch frequency (Shelby et al. 2000; Horne et al. 2003; Shelby-Walker et al. 2009). Jones (1993b) cited dioxin contamination on the Leaf River—the subject of unsuccessful lawsuits against the pulp mill by downstream landowners in the early 1990s—as a possible contributor to the species' decline. In 1986, two years after the opening of the pulp mill, yellow-blotched sawbacks were reportedly present upstream of its discharge but absent below it (C. H. Ernst et al. 1994; C. H. Ernst and Lovich 2009), although they were present in more recent surveys (Selman and Qualls 2009a).

Levels of 2,3,7,8-tetrachlorodibenzo-*p*-dioxin (TCDD), a known endocrine disruptor, were elevated in fish from the Leaf River from 1989 to 1995 (Shelby and Mendonça 2001). In 1992, a large accidental discharge resulted in fish and turtle deaths on the lower Leaf and upper Pascagoula Rivers, with the fish showing elevated TCDD levels (Shelby and Mendonça 2001). An advisory against consumption of fish from the Leaf River was lifted in 1995, but three years later, male sawbacks on the lower Pascagoula showed low testosterone levels and, in 10 percent of samples, elevated levels of 17-β estradiol, suggesting endocrine disruption (Shelby and Mendonça 2001). Derivatives of polychlorinated biphenyls (PCBs) and the pesticide DDT, the latter believed to be from historical exposures, were present in yellow-blotched sawback tissues at low levels, compared with levels found in common snapping turtles and diamondback terrapins from other contaminated sites (Kannan et al. 2000). Several years after the discharge in the Leaf, PCB and TCDD levels in tissues were low in both the contaminated population and another population from the Chickasawhay drainage that had not been impacted, but clutch frequency appeared to be depressed in the contaminated population, suggesting a long-term negative effect of exposure (Shelby-Walker et al. 2009).

Eggs of Ouachita and false map turtles that were exposed to atrazine, an herbicide known to act as an endocrine disruptor in fish and amphibians, showed no effects with regard to embryonic survival, rate of development, or first-year growth (Neuman-Lee and Janzen 2011). Survival in a laboratory setting over the first year after hatching was significantly lower in the group exposed to the lowest concentration of atrazine, however.

The Pet Trade

Overall declared trade in exports of live turtles from the United States grew from around six hundred *Graptemys* in 1989 to more than two hundred thousand in 2000 (Telecky 2001; Maltese 2005). Quantifying the extent of trade in any one species is hampered by the tendency to lump species as "*Graptemys* spp." (including all declared exports for the genus through 1994 and from 51 to 80 percent of all exports in reported years since then), but data suggest the overall magnitude of the threat has grown rapidly (fig. 5.8; Telecky 2001; Reed and Gibbons 2004; Senneke 2006). For example, between 1996 and 2000, exports rose sharply on an annual basis for *Graptemys*, while most other turtle genera in the United States showed little in the way of a temporal trend (Reed and Gibbons 2004). Ironically, part of the increase may stem from import bans in some countries of a more heavily traded (yet more abundant and widespread) U.S. turtle species, the red-eared slider turtle, *Trachemys scripta elegans* (Maltese 2005). European nations import the largest shipments of *Graptemys*; for example, in 2005 there were two shipments to Spain numbering ten thousand individuals, a shipment to Portugal numbering seven thousand, and a shipment to Poland numbering twenty-eight hundred (Senneke 2006). Some of these were young turtles produced on farming operations, as the percentage declared to have been wild-caught ranges from less than 1 percent for *G. geographica* to 72 percent for *G. pulchra* (Senneke 2006), but the incidence of false reporting and the extent to which commercial farms continue to rely on take from the wild to replenish brood stock are not known. Most of the declared trade in specimens that were identified to species has been in the three widespread species (range 91.5 to 99.0 percent annually since 1995; fig. 5.8), these being the most-used species on

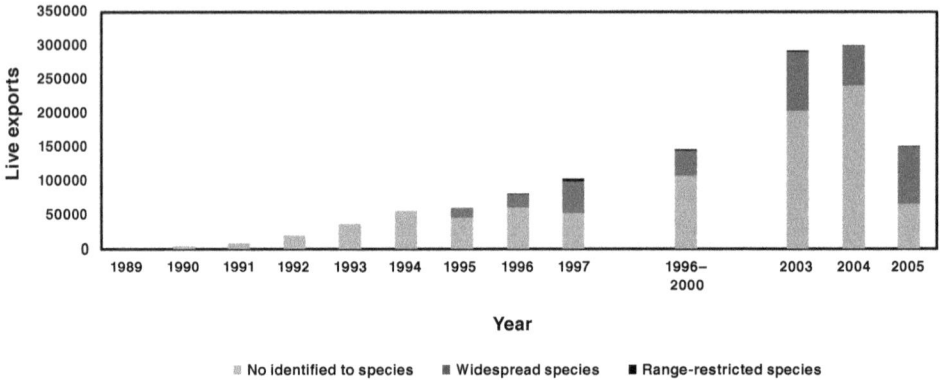

Figure 5.8. Export data reported to the U.S. Fish and Wildlife Service's Law Enforcement Management Information System, as summarized by Telecky (2001; for 1989–97), Reed and Gibbons (2004; five-year averages for 1996–2000), and Senneke (2006; for 2003–2005).

commercial turtle farms. It must be emphasized that these data do not include illegal smuggling, nor do they include commercial trade that takes place within the United States. Even recreational take of *Graptemys* may be problematic in areas with ready access and high public use (e.g., Lindeman 2004).

Muir (1984) described the adoption of Chaney and Smith's (1950) nighttime collecting technique to commercial exploitation. He witnessed a five-hour canoe expedition on the Alabama River that resulted in a take of 139 black-knobbed sawbacks (at least 20 of which were mature females) and 4 Alabama map turtles, in addition to 1–2 dozen turtles of other species. The poachers reported making numerous trips to major southeastern river drainages and regularly taking 100–200 turtles per night. At the time, dealers paid five to six dollars for black-knobbed sawbacks and up to fifteen dollars for ringed sawbacks, which had not yet received federal protection.

Reed and Gibbons (2004) ranked forty U.S. turtle species involved in legal trade according to their vulnerability to overexploitation. Their quantitative rankings were based on estimated adult survival elasticity (the contribution of adult survival rates to annual rate of population increase), geographic range size, and monetary value in retail trade. Members of the genus *Graptemys* and their overall ranking were *G. caglei* (4th), *G. gibbonsi sensu lato* (seventh), *G. nigrinoda* (8th), *G. versa* (12th), *G. ouachitensis* (including *G. sabinensis*, 15th), *G. pulchra* (21st), *G. ernsti* (25th), *G. pseudogeographica* (subspecies ranked separately, 27th and 29th), *G. barbouri* (30th), and *G. geographica* (31st). Examination of scaled values used for these rankings (fig. 5.9) reveals that rankings of high vulnerability were driven primarily by small geographic ranges, as most species had low elasticity values for adult survival and were valued below the norm. The biggest exceptions in the valuation metric were Cagle's map turtles

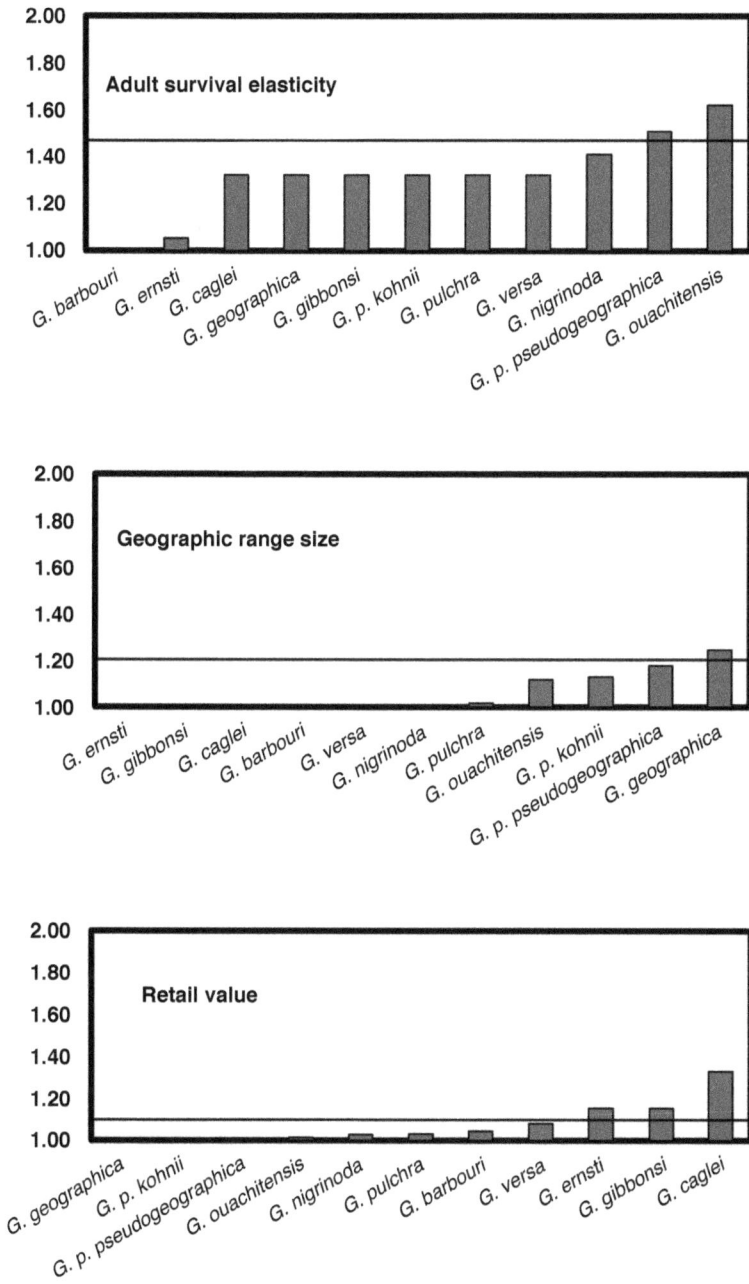

Figure 5.9. Scaled rankings of *Graptemys* species for three indices of vulnerability to trade in the analysis of Reed and Gibbons (2004). Horizontal lines are mean index values for all North American freshwater turtle species.

(median declared value $100, as high as $200) and Pascagoula/Pearl map turtles (median value $47, as high as $120; similar values were apparently substituted for the retail value of Escambia map turtles, for which the authors had no trade data).

Human Consumption

Use of map turtles and sawbacks as food has been greater than generally realized, at least historically. H. W. Clark and Southall (1920) reported substantial take for food markets in the early twentieth century. In their account of turtle consumption on the upper Mississippi and Illinois Rivers, they related how false and Ouachita map turtles were not highly esteemed as food locally, except for medicinal use by Chinese immigrants. They were, however, collected in large numbers for shipment to eastern markets such as Philadelphia, Baltimore, and Boston. Turtle buyers from these cities bought females, which they called "queen terrapins," to sell as a substitute for the declining gourmet diamondback terrapin (*Malaclemys terrapin*) of the east coast. The females were caught primarily in basking traps and sold individually for prices ranging from twenty-five cents to one dollar (or by the dozen for between seventy-five cents and three dollars), then shipped east in wooden barrels with ventilation holes. The authors recounted the case of a fisherman who caught thirty to forty turtles a day in one basking trap and described a holding pond in Grafton, Illinois, where four to five thousand sliders and map turtles were being held for eastern buyers.

Take of adults for food was reported for Barbour's map turtles in the 1960s, with one individual reportedly taking as many as fifty turtles from the Chipola River in a day (Lee 1969). McCloud (1985) listed map turtles and sawbacks as being among the "thousands" (p. 33) of live turtle shipments intercepted at the San Francisco International Airport, headed to food markets (in Asia and in California, where markets have been established by Asian Americans; Warwick and Steedman 1988). According to Warwick and Steedman, most of the turtles in McCloud's report were red-eared sliders (*Trachemys scripta elegans*). A substantial historical consumption of adults and eggs was reported for black-knobbed sawbacks on and around Gravine Island (Waters 1974; Lahanas 1982), and a 50 percent decline in common map turtle numbers in the North Fork of the White River in Missouri from 1969 to 1980 may have been due in part to targeting of large females for consumption (Nickerson and Pitt 2012). The extent to which take of adults and eggs for food remains a threat to map turtle and sawback populations is unknown, but it depends on enforcement of the generally more restrictive legislation that many states have passed since these reports were made.

Trade in Biological Specimens

As is the case for the consumption of emydid turtles in meat markets and the pet trade, the trade in biological specimens in the United States is focused on red-eared sliders, but map turtles and sawbacks are also sold. Moll and Moll (2004) reported specimens of common, false, and Ouachita map turtles among the species offered by several biological supply houses. They also related the observations of Dick Vogt concerning trade of forty to fifty thousand turtles annually in Wisconsin, with depletion reaching the point by the late 1970s that turtles began to be imported from other states to help meet demand.

Wanton Shooting

Among some outdoor enthusiasts, the shooting of basking turtles is ingrained enough as a tradition to have its own slang term, "plinking" (Shealy 1976; fig. 5.10). While this is primarily an idle form of target practice, to some degree turtles are also shot because of the misguided notion that they are "vermin," competitors for game fish (Shealy 1976). Approximately 5 percent of Cagle's map turtles captured by Flavius Killebrew in the early 1990s had shell injuries consistent with having been shot (USFWS 1993). Virtually anyone who studies freshwater turtles in the United States can recount conversations in the field with laypersons whose connection to wild freshwater turtles is plinking. (This is often admitted sheepishly as an activity they pursued when they were younger but which they have now outgrown!) The threat remains even for federally listed species. Locals on the lower Pascagoula admitted to plink-

Figure 5.10. A male Alabama map turtle killed by a gunshot wound to the middle of the right side of the carapace. Found in the Alabama River, Autauga County, Alabama.

ing during early conservation work on the yellow-blotched sawback and even shot holes in a mummified slider specimen that had been mounted on a log as a decoy over a basking trap (R. Brauman, pers. comm. 1994). Similarly, Shively (1999) reported that ringed sawbacks were sometimes shot on the Bogue Chitto River in Louisiana, and shooting at turtles was observed recently on the upper Leaf River, where most of the basking turtles are yellow-blotched sawbacks and Pascagoula map turtles (W. Selman, pers. comm. 2009).

Accidental Mortality

Indirect anthropogenic mortality is a concern with regard to commercial fishing "bycatch," strikes by boat propellers, use of all-terrain vehicles (ATVs) on sandbars during nesting season, and road mortality. H. W. Clark and Southall (1920) reported that map turtles taken as bycatch by commercial fisherman were generally killed rather than released, as they apparently were regarded as vermin. Vogt (1980a) reported drowning of turtles in commercial fishing gill nets in the Mississippi River in Wisconsin. In a study of bycatch mortality in the Mississippi River on the southern border of Missouri and Illinois, 83 percent of all turtles captured by federal fisheries biologists were an undifferentiated mix of false and Ouachita map turtles, 8 percent of which had drowned in the nets (Barko et al. 2004). Fyke nets submerged for twenty-four-hour sets had a mortality rate of 11 percent for all turtle species combined. Water temperature was the best predictor of whether captured turtles were dead or alive, cooler water being more conducive to surviving prolonged submersion. A series of vertical ropes spaced 38 cm apart in fyke net entrances was effective in lowering the capture of larger turtles, including female map turtles, without decreasing fish catch (Fratto et al. 2008).

The impact of propeller injuries (e.g., fig. 5.11) has recently become an increased focus of conservation research for *Graptemys* and other aquatic turtles in North America (Burger and Garber 1995; G. R. Smith et al. 2006; Galois and Ouellet 2007; Cecala et al. 2009; Bulté et al. 2010). In common map turtles in Ontario, a river habitat with higher boat traffic had an 8.3 percent incidence of propeller injuries, compared with a 3.8 percent incidence in a small lake with less boat traffic; in both areas, adult females had the highest incidence of injury (Bulté et al. 2010). Modeling of population viability showed that a 10 percent mortality rate for turtles struck by boats would drive the populations to extinction within five centuries, with 99 and 63 percent probabilities for the high- and low-traffic sites, respectively. Even the survivors of traumatic injuries incurred via boat strikes may contribute to decreases in long-term population viability; propeller injuries in diamondback terrapins (*Malaclemys terrapin*) were associated with relatively poor body condition in males and decreased annual survival in both sexes (Cecala et al. 2009). Exponential increase in boat and personal watercraft traffic in North America (Bulté et al. 2010) may put turtles at more risk of injury (e.g., Burger and Garber 1995). In an Indiana lake, 0–16 percent

Figure 5.11. Injury to the carapace of an adult female false map turtle (center), presumably caused by a boat propeller. Photo from LaFramboise Island in the Missouri River at Pierre, Hughes County, South Dakota.

of common map turtles captured each year over a twelve-year span showed evidence of propeller injuries; however, incidence did not increase over time, even as shoreline development and traffic on the water did (G. R. Smith et al. 2006).

Use of ATVs on sandbars of rivers with *Graptemys* populations has been reported anecdotally (Godwin 2002a; Aresco and Shealy 2006) but has not yet been identified as a factor in the mortality of nesting females, eggs, or dispersing hatchlings. Mortality due to vehicular traffic on roads occurs in some *Graptemys* populations (Ashley and Robinson 1996; Conner et al. 2005; Steen et al. 2006). Numbers of road-killed specimens are low relative to many other herpetofaunal species, undoubtedly due to the highly aquatic nature of the species, but the fact that reports involve adult females (presumably on nesting excursions; Conner et al. 2005; Steen et al. 2006) is cause for concern regarding the potential impacts on long-term population viability. Fortunately, females in most *Graptemys* habitats do not have to cross roads to reach their nesting grounds.

Predation

Some predators of map turtles and sawbacks and their eggs have experienced recent range expansions or subsidized population increases that may exacerbate population pressures (Jones 2006). Raccoons (*Procyon lotor*) thrive in landscapes disturbed by human activity (Garrott et al. 1993; Goodrich and Buskirk 1995) and prey on both turtles and their eggs (e.g., Shealy 1976; D. R. Jackson and Walker 1997; Jones 2006). The fish crow (*Corvus ossifragus*) has become a major egg predator for river-dwelling turtles in the southeastern United States (Shealy 1976; Lahanas 1982; D. R. Jackson and Walker 1997; McNair 2000; Horne et al. 2003; Jones 2006). It expanded its range inland along riparian corridors in the southeast during the late 1900s, an expansion associated from 1966 to 1979 with a range-wide 6 percent annual rate of increase in abundance, which has since slowed (McGowan 2001). Reasons for the population increase and range expansion have not been investigated but may include reduced shooting pressures under a 1972 amendment to the Migratory Bird Treaty Act, foraging at garbage dumps, and population recoveries of bird species whose nests are raided by fish crows (McGowan 2001). Nine-banded armadillos (*Dasypus novemcinctus*) used proliferating railways and roadways, in addition to transport and introduction, to spread from a confined range in southern Texas to cover much of the southeastern and south-central United States (Taulman and Robbins 1996); the species occasionally takes the eggs of map turtles and sawbacks (Jones 1993a, 2006; Moulis 1997; Godwin 2002b). These three species fit the definition of predators "subsidized" by their ability to thrive in human-dominated environments (*sensu* Boarman 1997). Their population growth and range expansion may necessitate some combination of predator population management, protection of nests, and ecosystem restoration to mitigate impact upon *Graptemys* populations (Garrott et al. 1993; Goodrich and Buskirk 1995).

Disturbance by Human Activity

Activity in map turtle and sawback habitat may depress population productivity without causing mortality. Lahanas (1982) and Jones (2006) noted that high human use of sandbars sometimes precluded their use by nesting turtles (fig. 5.12). Moore and Seigel (2006) described abandonment of 81 percent of nesting attempts by yellow-blotched sawbacks in a portion of the lower Pascagoula heavily used by recreational boaters and campers, with abandonment usually associated with human presence. They estimated that adequate time for nesting occurred on fewer than half of all regular weekdays and virtually never on weekends or weekday holidays. They also found that on average about two-thirds of basking turtles abandoned their sites when a boat or personal watercraft passed by, with about three-quarters returning to bask after an average time of twenty minutes. Females were especially prone to site abandonment and were observed at depressed densities during the middle of the day, when disturbance frequency was highest (fig. 5.13).

Selman et al. (2013; see also Selman 2010 and Selman, Qualls, and Mendonça 2008) compared disturbances at the Pascagoula site with those at an upriver site with considerably less boat traffic (six disturbances by boats per hour vs. a disturbance every two hours, respectively). They found significantly reduced bask-

Figure 5.12. Memorial Day recreation in May 2007 on a sandbar at Ratliff Ferry on the Pearl River, Madison County, Mississippi.

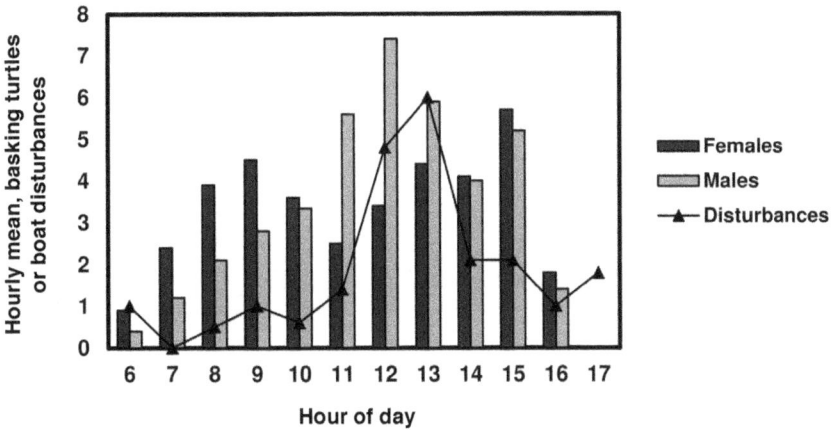

Figure 5.13. Effects of boating disturbances on basking densities of yellow-blotched sawbacks (redrawn from Moore and Seigel 2006).

ing duration at the site with more boat disturbances, particularly in spring and fall and particularly for adult females, which showed an unusual late-afternoon peak in their basking, coincident with reduced daily boat traffic. Disturbance frequency was greatest on Fridays through Mondays. Negative effects of high disturbance frequency for basking turtles may include reduced reproductive output (Moore and Seigel 2006) and increased incidence of shell disease (Selman, Qualls, and Mendonça 2008; Selman 2010; Selman et al. 2013).

Figure 5.14. Cogon grass (*Imperata cylindrica*) invading a sandbar on the upper Leaf River north of Hattiesburg, Forrest County, Mississippi.

Impacts of Non-native Species

Until recently, non-native species introductions have not been considered a major threat to map turtle and sawback populations; indeed, non-native mollusks (Asian clams, *Corbicula* spp.; zebra and quagga mussels, *Dreissena* spp.) have become important prey for females of many species (see chapter 4). At least one introduced insect and six introduced plants have negative impacts, however. Non-native fire ants (*Solenopsis invicta*) destroy eggs and hatchlings (Moulis 1997; Horne et al. 2003). Proliferation of dense stands of the submerged macrophyte hydrilla (*Hydrilla verticillata*) in the Mobile-Tensaw Delta has been associated with a steep decline over two decades in capture rate of black-knobbed sawbacks near Gravine Island (Turner 2001). Water hyacinth (*Eichhornia crassipes*) and alligator weed (*Alternanthera philoxeroides*) can impede basking behavior (Selman and Jones 2011). Cogon grass (*Imperata cylindrica*) was introduced to the southeastern United States in the early 1900s. It spreads vegetatively and via windblown seeds, forming dense stands in open, disturbed habitats—increasingly including river sandbars, rendering parts of them unsuitable for nesting (fig. 5.14; Selman, Qualls, and Mendonça 2008). Selman and Jones (2011) attributed additional negative impacts on nesting beaches to an alien cocklebur (*Xanthium stramarium*) and Chinese tallow tree (*Triadica sebifera*).

Local Extirpations of *Graptemys* Populations

In spite of the many threats faced by map turtles and sawbacks, records of local extirpations of map turtle and sawback populations are uncommon. Examples include the local loss of populations of Cagle's map turtles in reservoirs and of yellow-blotched sawbacks downstream of a pulp plant, discussed above. A population of common map turtles in a pond on South Bass Island of western Lake Erie was described as once being abundant, but the species was last observed there in 1968 (R. B. King et al. 1997). The species is also known from a single record in the Allegheny River just north of Pittsburgh, Pennsylvania, but extensive degradation of the lower river in the early 1900s apparently led to its extirpation (Atkinson 1901; Hulse et al. 2001). Similarly, populations of common map turtles in metropolitan areas along the Canadian shoreline of Lake Ontario (Toronto and Hamilton) may also be extirpated (COSEWIC 2002). Common map turtles may also be extirpated in some portions of the Verdigris River drainage in southeastern Kansas. Two specimens were taken from the drainage in 1911, but the species was not among 121 turtles trapped at eleven sites on the drainage in 1990–91 (D. Edds, pers. comm. 2008); however, recent observations of the species were made in two small tributaries, Cedar and East Buffalo Creeks (T. Taggart, pers. comm. 2009). Recent anecdotal reports suggest that harvest for the pet trade has depleted Alabama, Barbour's, Escambia, Pascagoula, and Pearl map turtles in some river reaches to the point of their possible extirpation (www.biologicaldiversity.org/programs/biodiversity/1000_species/the_southeast_freshwater_extinction_crisis/pdfs/SE_Petition.pdf), and Selman and Qualls (2009a) did not find the Pascagoula map turtle in two tributaries of the Leaf River drainage for which there were historical records. It is likely that many local extirpations that do occur are transitory events, given the continuous nature of the streams and lakeshores that *Graptemys* inhabit and their wide-ranging movements (see chapter 4). If recolonized, however, sites where populations had been extirpated may serve as sink habitats if the causes are not addressed.

6

Conservation Biology

There is little evidence of direct management of *Graptemys* populations and their habitats, even though two species are federally listed, several others have been candidates for listing (one for two decades), and several are state listed (see chapter 5). Recovery efforts have focused on further delineating the extent of and causes behind declines, via better understanding of basic ecological interactions, and on quantifying population status. The greatest threats concern habitat alteration and exploitation for the pet trade, which have been and will continue to be the primary conservation focal points. To date, there has been little emphasis on captive breeding. Most populations remain extant, and as long as this is so, *in situ* recovery will be emphasized, with *ex situ* maintenance of *Graptemys* remaining a last-gasp option on which turtle conservationists hope they never need to depend.

Basic Ecological Research

A major focus in conserving map turtles and sawbacks has been on further understanding their life history and ecological interactions. The first five of six management actions in the recovery plan for the ringed sawback (Stewart 1988) call for better understanding its limiting factors and reasons for its decline by investigating habitat, reproduction, diet, population structure, and activity patterns, including movements. The sixth item calls for protecting two stream reaches from threats. Management actions proposed for the yellow-blotched sawback (Jones 1993b) likewise constitute a call for further research and subsequent habitat protection. The first three call for range-wide assessment of populations, life-history research, and studies of water quality and habitat suitability. The other objectives are protecting habitat, education, and monitoring. Many of these objectives have begun to be met (Kofron 1991; Seigel and Brauman 1994; Jones and Hartfield 1995; Jones 1996, 2006; Lindeman 1998, 1999c; Kan-

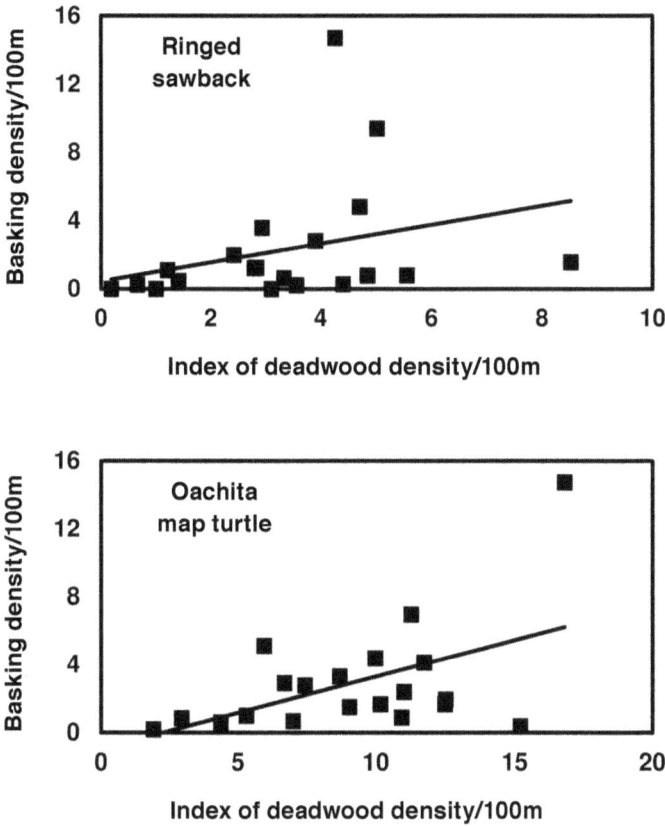

Figure 6.1. Relationship between weighted-average mean basking density and an index of deadwood abundance from replicated spotting-scope surveys. Correlation was not significant for the ringed sawback turtle in the Pearl River drainage in Mississippi and Louisiana ($p = .20$) and was significant for the Ouachita map turtle in the lower Tennessee River, Kentucky, including its impounded reservoir, Kentucky Lake ($p = .023$). Redrawn from Lindeman (1999c).

nan et al. 2000; Shelby et al. 2000; Shelby and Mendonça 2001; Horne et al. 2003; Selman and Qualls 2006; Selman, Kreiser, et al. 2007, 2013). Likewise, the COSEWIC (2002) report on the status of common map turtles in Canada recommends monitoring, including study of demographic trends and estimating population sizes.

Studies of factors that influence abundance are crucial to *Graptemys* conservation. A survey of deadwood use by five species in three river drainages was undertaken in the 1990s to determine whether basking density was correlated with deadwood density and to examine range-wide abundance of the federally listed species (Lindeman 1998, 1999c). Surveys were conducted under sunny, midday conditions in May and June, when basking frequency is high, and were replicated to account for variation in basking density. Correlation analysis examined the relationship of mean and maximum basking numbers with deadwood abundance, expressed as total potential basking substrates present or as a weighted index taking into account differential use (see chapter 4). While not all correlation coefficients were statistically significant, they showed a general pattern of positive correlation with a right-skewed triangle of data points (fig. 6.1);

this demonstrated that moderate or high turtle density was never associated with low deadwood density, although high deadwood density was not always sufficient for high turtle density (i.e., other key habitat variables may be lacking where deadwood is abundant). Sustained levels of take may also suppress abundance at some sites that provide excellent habitat.

In the mid-1990s, yellow-blotched sawbacks met federal recovery goals set for basking density (see chapter 5) at one of eight sites on the Chickasawhay, none of five on the Leaf, and all three on the Pascagoula. More recently, they met recovery goals at one of three sites on the Chickasawhay, two of four on the Leaf, and all three on the Pascagoula (Selman and Qualls 2009a).

An unexpected result of the 1994–95 surveys was the discovery that the unlisted Pascagoula and Pearl map turtles were seen only about one-fifth as frequently as their sympatric, federally listed congeners, which suggests that they too should be considered for federal protection. More recent surveys combined with population estimates suggest the status of the former has not improved (Selman and Qualls 2007, 2009a), and there are signs of increasing federal interest in its status (L. LaClaire, pers. comm. 2007).

Recent extensive basking surveys have also been conducted on Cagle's map turtle in Texas (Babitzke 1992; Killebrew et al. 2002), five species in Louisiana (Shively 1999, 2001; J. L. Carr 2001), four species in southern and central Alabama (Godwin 2000, 2001, 2003), and Barbour's map turtle in Florida (Enge and Wallace 2008). Data from these surveys should ultimately provide valuable baselines for future surveys conducted using the same methodologies. Indeed, a comparison of the Killebrew et al. (2002) and Babitzke (1992) surveys suggest a modest population increase for Cagle's map turtle over a decade, while Shively compared his 2001 surveys against data from two decades earlier (Shively 1982; Shively and Jackson 1985). Results suggested stable populations of Sabine map turtles at three sites and a possible decline (or depressed basking frequency) at a fourth site, where recreational float traffic had increased.

Habitat Preservation, Management, and Restoration

The primary way that the habitat of map turtles and sawbacks can be managed for healthy populations is through protection of riparian habitat. Preserving mature forests along unconstrained, eroding riverbanks ensures a steady supply of deadwood, a critical habitat feature. Large snags are most abundant in river reaches with unstabilized, forested banks; snags that originate from eroding riverbanks remain part of a river's physical structure for decades or centuries (Abbe et al. 2003; Benke and Wallace 2003; Angradi et al. 2004). Preservation of only a narrow riparian forest buffer can have a short-term positive impact on deadwood accumulation, as exposed trees are blown down during storms, but it will a negative impact in future decades by eliminating the source of new deadwood (Elosegi and Johnson 2003).

Figure 6.2. Satellite image of Ichauway, a private holding in Baker County, Georgia. Forested land of the reserve shows up as darker than agricultural lands, which are dominated by center-pivot irrigation; most reserve boundaries are sharply demarcated by the transition from dark to light (compare with map in L. L. Smith et al. 2006). The Flint River, the preserve's eastern boundary, is visible within the dark, diagonal riparian zone that extends from northeast to southwest. Its tributary, Ichawaynochaway Creek, flows from north to south through the reserve's western half and joins the Flint just to the right of the distance bar. Both streams are inhabited by Barbour's map turtle. Base map data © 2012 Google, Landsat.

While some species have large proportions of their ranges within federal or state habitat preserves (e.g., the Sabine map turtle and the yellow-blotched sawback), others have relatively little protected habitat (e.g., Cagle's map turtle; see species accounts in chapter 8). Private land conservation in riparian zones also plays a role in habitat preservation. For example, Ichauway, site of the Joseph W. Jones Ecological Research Center, is a reserve managed by the Robert W. Woodruff Foundation in southwestern Georgia. Its 11,700 ha are dominated by longleaf pine forests that straddle 24 km of Ichawaynochaway Creek and

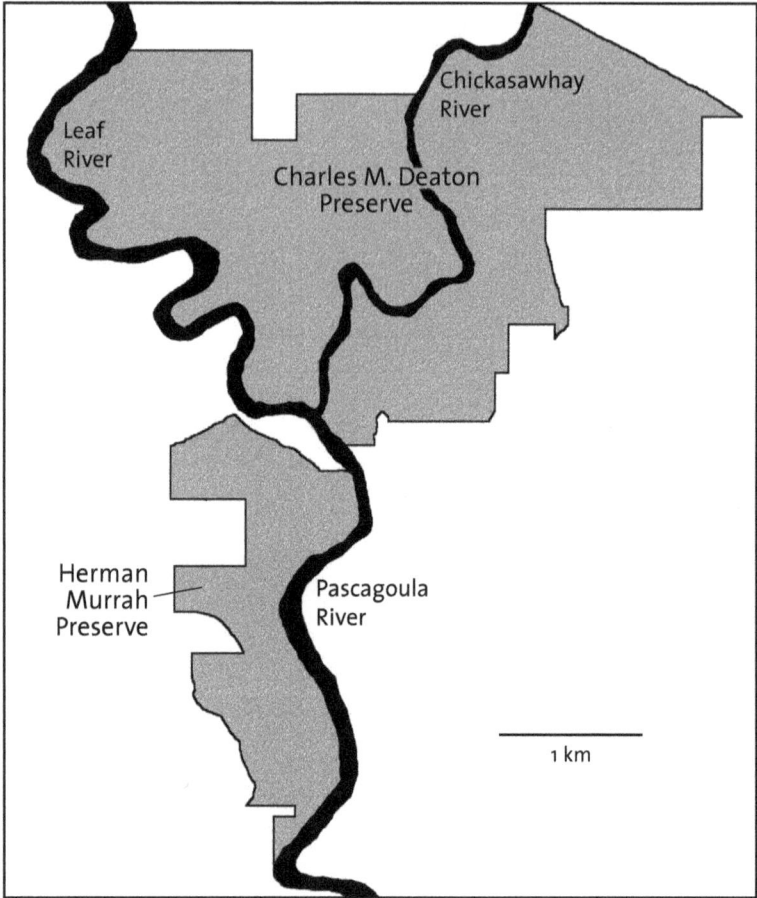

Figure 6.3. Map of two Nature Conservancy preserves at the confluence of
the Leaf and Chickasawhay Rivers, Mississippi.

border 21 km of the west side of the Flint River, habitats of Barbour's map turtle.
As a former quail-hunting reserve, much of the land has never been tilled, making it a valuable reservoir of biodiversity in an area otherwise dominated by
agriculture (fig. 6.2; L. L. Smith et al. 2006).

Another excellent example of a private initiative to preserve riparian habitat
along map turtle and sawback habitats is the Nature Conservancy's preserves at
the confluence that forms the Pascagoula River (fig. 6.3). The Charles M. Deaton
Preserve protects 1,340 ha at the confluence of the Leaf and Chickasawhay Rivers. Just downstream, the Herman Murrah Preserve protects 4.8 km along the
west side of the Pascagoula and 648 total ha. It connects the Deaton Preserve to
the northern unit of the Pascagoula River Wildlife Management Area, which is

managed by the state of Mississippi. The Conservancy removes invasive cogon grass from sandbars used by nesting turtles. Together with additional public lands farther downstream, protection is in place for a corridor of 28,000 ha of bottomland hardwood forest along the Pascagoula main stem, including riparian frontage along 80 of its total 121 river km (R. Stowe, pers. comm. 2007; W. Selman, pers. comm. 2009).

On the lower Pascagoula's state-owned Ward Bayou Wildlife Management Area, sandbars are sprayed aggressively to rid them of invasive cogon grass (L. McCoy, pers. comm. 2007). In 2003, four of the area's six sandbars were posted as closed to recreational use from 1 April to 1 October to reduce nesting disturbance, a direct result of the study by Moore and Seigel (2006; see chapter 5). The sandbars chosen were small and had little previous recreational use—hence there have been few violations and little public opposition (L. McCoy, pers. comm. 2007)—but disturbance of basking turtles by boats and degradation of sandbars by boat wakes are concerns in the area (Selman 2010; Selman, Qualls, et al. 2013).

Use of potentially harmful pesticides is regulated by the Environmental Protection Agency (EPA). To date, only a portion of the ringed sawback's range has been placed under restrictions. In two Louisiana parishes, several pesticides may not be applied within 100 yards (91 m) of habitats on the Pearl or the lower Bogue Chitto and aerial application of some pesticides is not allowed within 400 m. Instructions for reducing drift during application and runoff are provided to pesticide users in EPA brochures.

The Missouri River Recovery Program is an ambitious, joint undertaking of the U.S. Fish and Wildlife Service, the Army Corps of Engineers, state and local government agencies, and tribal nations (www.moriverrecovery.org). The river was dramatically altered from its historical flow regime by six dams along its middle reaches and extensive channelization and levee building along its lower reaches. Unprecedented flooding in 1993, together with four other major floods in the 1980s and 1990s, stimulated the transfer of floodplain habitat into public ownership and planning to restore the river's ecological integrity (Galat et al. 1998). The plan's goals for wildlife habitat restoration include three potential benefits to the false map turtles that inhabit the river. First, increased emergent sandbar area is being promoted via mechanical movement of substrate and vegetation removal, which provides additional nesting habitat (L. A. Dixon 2009). Second, restoration activities aim to raise the availability of shallow-water habitat to about 10 ha in surface area per 1.6 river km via channel widening and creation of additional side channels, which may be used by turtles (Bodie and Semlitsch 2000b). Third, restoration of erosion-dependent cottonwood forests along riverbanks and on vegetated islands will ensure a future source of deadwood for basking.

Regulation of Collection for the Pet Trade

The addition of all *Graptemys* to Appendix III of CITES makes it possible to track international trade by species; this was not possible before the listing because of trade documents that listed specimens as "*Graptemys* spp." (see chapter 5). In theory, federal officials should be able to prevent commercial trade of species where state law prohibits it, and they should also be able to determine whether any of the unprotected species are being subjected to especially heavy trade that might threaten them. Should that be the case, new state legislation, possibly combined with listing on Appendix I (no commercial trade allowed) or II (possibly with caps on the number that may be traded annually), could impose further restrictions on trade. Intercepting illegal shipments will be of vital importance, however, to reduce the threat of smuggling operations. Effective enforcement will thus require development of good identification guides for customs officials for species that are difficult to distinguish.

Control of trade requires effective monitoring of illegal activity in the field, of which there are at least two examples. On the Pascagoula River in the early 1990s, two German citizens were arrested for taking yellow-blotched sawbacks. Just a few years after the species had been federally listed—in part because of their sale for up to sixty dollars per specimen in the pet trade (Stewart 1990)—the market value overseas had soared to three thousand dollars per specimen (R. Jones, pers. comm. 1994). In Wilcox County, Alabama, an undercover operation apprehended an individual taking black-knobbed sawbacks for a market he claimed would yield five hundred dollars per specimen (J. Godwin and M. Sasser, pers. comm. 2007).

Management of Captive Populations

Captive breeding has not been a high priority in the effort to conserve *Graptemys*, perhaps because *in situ* populations, while apparently declining in many cases, only rarely disappear altogether. Populations in main-stem rivers and large lakes are in most cases rather continuously distributed, and individuals can be highly mobile, traversing distances of several kilometers (see chapter 4). The linear continuity of rivers and lakeshores and the mobility of individuals may counteract local extirpation; sites with heavy harvest for the pet trade or with high rates of anthropogenic mortality may then act as population sinks without losing populations permanently. Given a paucity of cases in which a population has disappeared from seemingly viable habitat, it is understandable that captive breeding with the goal of reintroduction has been little emphasized for these turtles.

From the standpoint of aesthetics and education, however, map turtles and sawbacks do make good animals for display in zoos and aquaria. Total holdings of 434 *Graptemys* in zoos and aquaria are summarized in table 6.1. The

Table 6.1. Summary of collections of map turtles and sawbacks in zoos and aquaria

Species	Specimens in zoos and aquaria				Total institutions	Largest collection
	Males	Females	Un-known	Total		
Common map turtle, *Graptemys geographica*	5	10	29	44	18	New York Aquarium (7)
Barbour's map turtle, *Graptemys barbouri*	6	24	9	39	10	Tennessee Aquarium (9)
Alabama map turtle, *Graptemys pulchra*	3	4	5	12	3	Henry Doorly Zoo (7)
Escambia map turtle, *Graptemys ernsti*	0	2	0	2	1	Tennessee Aquarium (2)
Pascagoula map turtle, *Graptemys gibbonsi*	0	0	0	0	0	
Pearl map turtle, *Graptemys pearlensis*	2	1	3	6	1	Tennessee Aquarium (6)
Cagle's map turtle, *Graptemys caglei*	0	1	6	7	2	Tennessee Aquarium (4)
Texas map turtle, *Graptemys versa*	1	1	2	4	2	Henry Doorly Zoo (2)
Sabine map turtle, *Graptemys sabinensis*	0	0	0	0	0	
False map turtle, *Graptemys pseudogeographica*	12	28	106	146	46	Parc Animalier de Sainte Croix (25)
Ouachita map turtle, *Graptemys ouachitensis*	10	4	9	23	10	National Mississippi River Museum and Aquarium (5)
Ringed sawback, *Graptemys oculifera*	14	11	10	35	8	Shedd Aquarium (10)
Yellow-blotched sawback, *Graptemys flavimaculata*	15	29	49	93	10	Tennessee Aquarium (45)
Black-knobbed sawback, *Graptemys nigrinoda*	7	9	7	23	7	Mississippi Museum of Natural Science (9)

Note: Data based on www.isis.org (February 2013) and correspondence with curators of three institutions not listed therein (B. Fedrick, R. McKinzie, and S. Smith, pers. comm. 2007–2008).

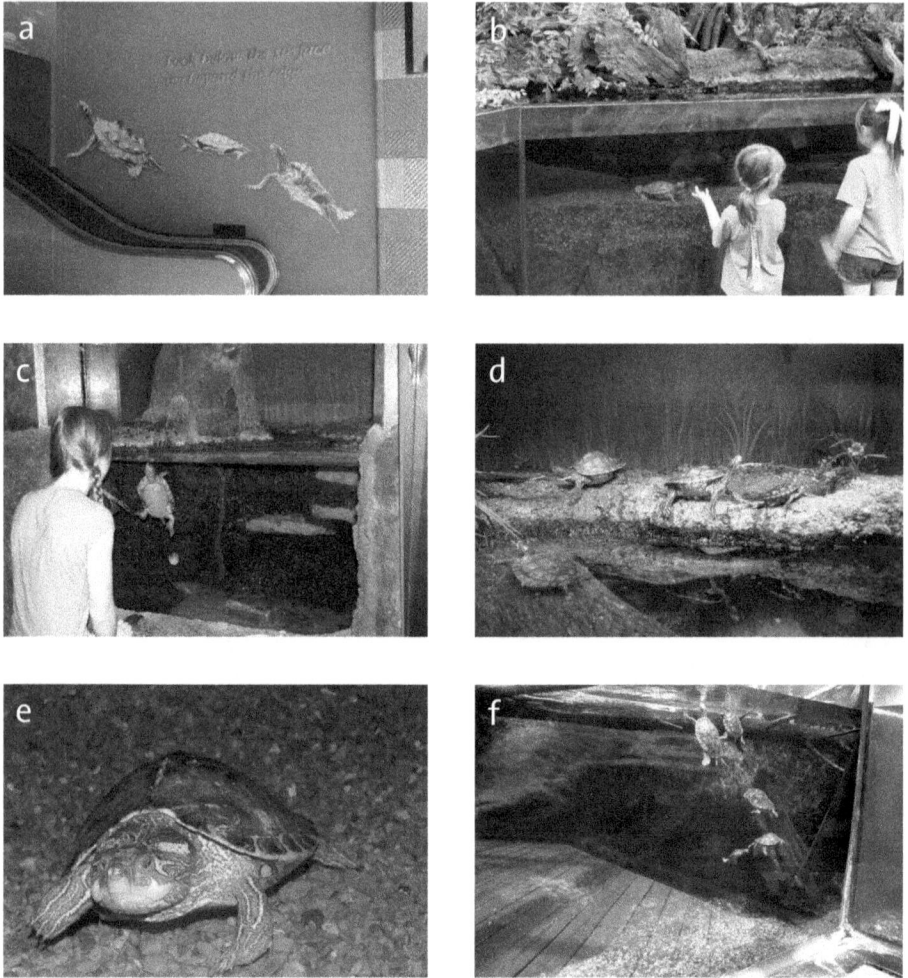

Figure 6.4. Exhibits of map turtles and sawbacks. Tennessee Aquarium: (a) entrance wall murals depicting the three sawbacks; (b) Escambia map turtle in the Delta Exhibit. Columbus Zoo: (c) Barbour's map turtle; (d) yellow-blotched sawbacks. Mississippi Museum of Natural Science: (e) Pearl map turtle in the Pearl River Aquarium; (f) three species in the sawbacks tank in the Swamp Exhibit greenhouse.

largest and longest-established collections are of common, false, Barbour's, and Ouachita map turtles and the three sawbacks. By contrast, there are much lower numbers of the other seven species in live collections. Among the best public exhibits are those at the Tennessee Aquarium in Chattanooga, the Columbus (Ohio) Zoo, and the Mississippi Museum of Natural Science (fig. 6.4). Visitors to the Tennessee Aquarium pass larger-than-life wall murals depicting swimming sawbacks. The Delta Hall, with its large, open-top exhibits, houses most of the exhibited collection of map turtles and sawbacks, primarily females of the megacephalic species. The Columbus Zoo features three tall indoor tanks with yellow-blotched sawbacks, black-knobbed sawbacks, and Barbour's map turtles. The Mississippi Museum of Natural Science has a large tank teeming with small specimens of various turtle species from Mississippi (including all three sawbacks and the Pascagoula map turtle); the ceiling-high Pearl River Aquarium exhibit, which features a large female Pearl map turtle; and a section of its greenhouse Swamp Exhibit devoted to housing adults of the three sawbacks.

The first report of successful propagation concerned a clutch of Barbour's map turtles at the Louisiana Purchase Gardens and Zoo (Stuart 1974). Goode (1997) provided a detailed summary of captive propagation at the Columbus Zoo (table 6.2). More recently, notes on hatching of Barbour's map turtles and ringed sawbacks at the Shedd Aquarium have appeared (Anonymous 2000, 2003). Clutch sizes are generally somewhat lower in captivity than in the wild (compare table 6.2 and values below with table 4.20). The winter and early spring oviposition dates at the Columbus Zoo are probably attributable to seasonal changes in lighting of exhibit areas, which have typically been changed from 12:12 light cycles to 16:8 in mid-November, but eggs have generally been fertile in spite of the unusual timing of oviposition (T. Gerold, pers. comm. 2007). Sandpits are provided for oviposition, or hormonal induction can be used for females that lay their eggs in water. At the Shedd Aquarium, reproduction is stimulated by holding adults for six to eight weeks at 10°C and then returning them to warmer water (E. Clayton, pers. comm. 2007). The Tennessee Aquarium has bred several species but places particular emphasis on yellow-blotched sawbacks and Barbour's map turtles. Both breeding programs have had some success (nine and five viable hatchlings, respectively) but have also had high rates of infertile eggs and eggs that have died as late-term embryos. Because the problem has been associated with individual clutches, nutritional deficiency is suspected; better supplementation of the commercial diet with krill, fish, and mussels, with effectiveness tracked by careful and frequent recording of female body mass, appears to improve results (D. Collins, pers. comm. 2008). Hertwig and Hohl (2001a) also stated that poor nutrition caused poor hatching rates and low viability in hatchlings that did result.

Five of the six species propagated at the Columbus Zoo have been sent to other institutions to build collections (table 6.2); the hatchling Texas map turtles and some hatchlings of the other species have also been transferred to private holdings. The Tennessee Aquarium has likewise loaned several hatchling Barbour's and Ouachita map turtles and yellow-blotched and black-knobbed sawbacks to other institutions. Captive-bred Ouachita map turtle hatchlings have been sent to other institutions from the National Mississippi River Museum and Aquarium (L. Jackson, pers. comm. 2008). Many zoos and aquaria are reluctant to breed their *Graptemys* due to concerns over finding homes for the hatchlings; they sometimes discard eggs found deposited in the exhibits.

Table 6.2. Summary of captive propagation of *Graptemys* at the Columbus Zoo

Species	Total clutches	Total females	Maxi-mum annual frequency	Years	Mean clutch size (range)	Dates of oviposi-tion	Total hatched	Institu-tions re-ceiving offspring
Barbour's map turtle, *Graptemys barbouri*	25	NR	4	8	3.4 (1–6)	13 Mar–25 Jul	9	2
Alabama map turtle, *Graptemys pulchra*	13	NR	2	5	3.5 (1–5)	21 Dec–22 Jun	8	2
Texas map turtle, *Graptemys versa*	19	NR	5	5	4.6 (1–6)	19 Feb–28 Jun	23	0
Ringed sawback, *Graptemys oculifera*	NR	NR	NR	NR	NR	NR	14	3
Yellow-blotched sawback, *Graptemys flavimaculata*	42	4	5	11	2.9 (1–5)	11 Dec–28 May	39	3
Black-knobbed sawback, *Graptemys nigrinoda*	91	55	5	6	3.2 (1–7)	2 Oct–17 Jun	78	1

Note: Data from Goode (1997), except for hatchling numbers and information on transfer to other institution provided by T. Gerold (pers. comm. 2007). NR = not reported.

European hobbyists have reported in more detail about captive propagation efforts. Often males and females are kept separated except for brief opportunities for mating in spring, ostensibly to reduce stress to females and the possibility of injury to males (e.g., Hertwig and Hohl 2000b; Hertwig 2001; Stettner 2005), although Artner (2001a, 2001b) reported success without separating the sexes. Hertwig (2001) documented the results of ten years of breeding the sawbacks and Cagle's map turtle. He achieved production of 121 black-knobbed sawback hatchlings from 177 eggs (hatching rate 68 percent) laid in forty-three clutches (mean 4.1 eggs/clutch) by two females, in addition to lesser numbers of the other three species. He kept males separate from females except for brief periods before and after a two-month forced hibernation (see below). Nijs (1999) reported a very low hatching rate for black-knobbed sawbacks (4 of 32 eggs from eight clutches, or 13 percent), which he attributed to using an overmoistened substrate for incubation. Stettner (2005) also reported successful breeding of black-knobbed sawbacks, which laid 2–3 eggs in three clutches per year. Artner (2001b) incubated 39 eggs from nine clutches (3–5 eggs/clutch) of a female Pascagoula map turtle (*sensu lato*) over two years, hatching 55 percent in the first year and 100 percent in the second. Schulz (2004) reported success with a young female Barbour's map turtle that annually laid three to four small clutches (2–4 eggs per clutch) over three years, with hatching rates from 70 to 100 percent. Artner (2001a) reported clutches averaging 3.0 eggs for three female Sabine map turtles and 3.7 eggs for two false map turtles, with one of the latter group producing four clutches in a year.

Feeding captives in zoos and aquaria is largely accomplished using a combination of commercial pellets and available vegetables and fish, but more natural foods may be preferable, particularly with regard to mollusks. At the Shedd Aquarium, for example, snails, insects, and insect larvae are given to captives to supplement a diet of Nasco Turtle Brittle, smelt, shrimp, and krill (E. Clayton, pers. comm. 2007). Nasco Turtle Brittle has been the staple of diets at the Tennessee Aquarium, but experience has shown that animal health is improved by supplementation with chopped fish, krill, and mussels (D. Collins, pers. comm. 2008). At the Columbus Zoo, diets consist of Ziegler and Masuri turtle pellets, kale, endive, collard greens, celery, sweet potato, carrot, apple, broccoli, orange, banana, pinkie mice, and smelt, but snails are also given after being frozen for seventy-two hours to kill parasites (T. Gerold, pers. comm. 2007). The Cheyenne Mountain Zoo feeds its Alabama map turtles with a combination of ReptoMin turtle pellets, fish, and crickets (R. McKinzie, pers. comm. 2007). Hobbyists commonly mix plants and insects in gelatin to feed captives (Artner 2001a, 2001b; Hertwig 2001; Schulz 2001).

Poor shell conditioning is a common problem for captive map turtles and sawbacks, which often stop shedding scutes properly and develop discolored, pitted scutes (Hernandez-Divers et al. 2009). At the Tennessee Aquarium, chronic graying of the shell associated with incomplete ecdysis, possibly due to fungal growth between scute layers, occurred frequently in the early years of collection building but was effectively treated with ultraviolet (UV) lamps that warmed basking sites to 38°C, promoting rapid and complete scute shedding (D. Collins, pers. comm. 2008). Individuals housed in a greenhouse on the roof of the building have had fewer problems with shell conditioning. The Shedd Aquarium also uses UV lights over basking sites to promote good shell conditioning (E. Clayton, pers. comm. 2007). At the National Mississippi River Museum and Aquarium, poor shell conditioning was effectively treated by installation of running-water filtration systems with either ozone or UV sterilization (L. Jackson, pers. comm. 2008). At the Mississippi Museum of Natural Science, mineral deposition in shells detracts aesthetically but does not seem to be harmful (B. Fedrick, pers. comm. 2008).

Natural cycles of light and temperature can be mimicked to keep captives on a regular reproductive cycle. Schulz (2001) kept captives at 10°C during December and January without food or basking opportunity; raised the water temperature gradually to 25°C by April, reaching a high of 27°C–29°C from May to September; and decreased temperatures during October and November. Water temperature was also lowered 3°C overnight during the summer. Daily illumination of basking platforms varied from ten hours in February and November to thirteen hours in midsummer. Egg laying occurred between April and July, similar to nesting dates in the wild. Hertwig and Hohl (2000b), Artner (2001b), and Hertwig (2001) similarly used annual cycles of light and temperature with prolonged forced hibernation periods in darkness, while Stettner (2005) simply lowered the water temperature to 15°C–18°C during winter and fed less frequently.

There have been no attempts to reintroduce captive-bred *Graptemys* to the wild. In the case of the colony of yellow-blotched sawbacks raised and bred in the Tennessee Aquarium, there has not yet been an interest from the U.S. Fish and Wildlife Service or the Mississippi Department of Wildlife, Fisheries, and Parks in repatriating any (D. Collins, pers. comm. 2008). Undoubtedly the lack of release programs relates to the fact that although many wild populations have declined, few have been extirpated. For the time being, captive propagation programs remain smaller in scale than some curators might like, as space to house offspring of these long-lived animals is scarce and placement opportunities for young animals in other zoos and aquaria quickly become saturated (D. Collins, pers. comm. 2008).

In public comments concerning the proposal to add all *Graptemys* to CITES Appendix III, it was suggested that they might be restocked in the wild using a portion of the offspring from regulated turtle farming operations. In Louisiana, this is already done with farmed alligator snapping turtles (*Macrochelys temminckii*). The U.S. Fish and Wildlife Service suggested that such programs, if implemented, would be primarily state concerns (Maltese 2005), as only two species of *Graptemys* are federally listed (and thus neither is eligible for commercial farming). Any such operation would need to be strictly controlled to prevent the release of individuals whose parents originate far from the release site or who are genetic admixtures, as they might be poorly adapted to local conditions and could dilute wild gene pools, with negative consequences. In spite of a plea for coordinated breeding of *Graptemys* among European hobbyists (Schulz 2001)—as well as a report of the beginnings of such an initiative, involving two breeders' aggregate "gene pool" of seven male and four female ringed sawbacks that have produced twenty offspring in captivity (Hertwig and Hohl 2000a; Hertwig 2001)—the concerns noted should place any future operation of captive breeding and release properly in the hands of qualified zoos and aquaria.

7

Additional Biological Aspects

Most of the literature on map turtles and sawbacks concerns species description, distribution, ecology, and conservation. The topics in this chapter are a largely unrelated set of additional topics that have been studied to varying degrees. There have been recent expansions of investigations into winter physiology, trophic morphology, and population genetics, but the remaining topics have been pursued only sporadically and generally with little depth, providing little opportunity for integration of studies; hence there is much need for continued research.

Physiology of *Graptemys*

The most detailed studies of *Graptemys* physiology concern adaptations to winter temperatures in northern populations of common map turtles. Studies under field conditions and in the laboratory have concerned the physiology of aquatic hibernation in adults and terrestrial overwintering in the nest by hatchlings. Hibernating adults efficiently utilize dissolved oxygen from the water, rather than being able to function anaerobically while under the ice. Hatchlings survive the winter in their nest cavities by being able to supercool, as opposed to tolerating freezing of their tissues. Additional topics in physiology are at more preliminary stages of investigation.

Hibernation Physiology

Graham and Graham (1992) enclosed common map turtles in respiration chambers during mid-November at a communal hibernaculum in the Lamoille River, Vermont. In water that cooled to 2°C, consumption rates were 1.38 μL O_2/g/h in a male and 0.49 and 0.58 μL O_2/g/h in two females. The authors attributed the higher rate in the male to its smaller size and to greater activity rates and responsiveness they observed in males in the hibernaculum. In the laboratory, survival in anoxic water is not possible beyond fifty to seventy-five days due to severe lactate accumulation and metabolic acidosis (Ultsch and Jackson 1995; Reese et al. 2001; Maginniss et al. 2004). Compared with other North American turtles, the ability of common map turtles to survive anoxia is comparable to that of the stinkpot (*Sternotherus odoratus*) but considerably less than that of painted turtles (*Chrysemys picta*) and common snapping turtles (*Chelydra serpentina*), which is at least partially explained by the greater ability conferred by the shells of the latter two species for buffering pH changes during lactate accumulation (D. C. Jackson et al. 2007).

Following the work of Graham and Graham (1992), studies of hibernation physiology in the common map turtle have concerned blood gases, lactate accumulation, blood pH, and blood ion concentrations. Results have been qualitatively similar for animals collected from the Vermont hibernaculum (Crocker et al. 2000) or submerged in 3°C water in the laboratory (fig. 7.1; Ultsch and Jackson 1995; Reese et al. 2001). Turtles maintained relatively stable blood gases in normoxic water, with low but steady PO_2 (1–2 mm Hg, compared with 40–50 mm Hg in animals allowed access to air) and PCO_2 between 5 and 13 mm Hg throughout submergence (similar to levels in animals allowed access to air). Whether extrapulmonary gas exchange is purely cutaneous or augmented by buccopharyngeal or cloacal ventilation is not known (D. C. Jackson et al. 2001). In anoxic water, turtles maintained a somewhat higher blood PCO_2 over fifty days of submergence. The blood concentration of lactate, an indicator of anaerobic respiration in tissues, remained low (generally below 5 mmol/L) in turtles in normoxic water but spiked to an average of 115 mmol/L over fifty days of submergence in anoxic water, triggering acidosis that reduced blood pH from a typical level near 8 (which is maintained by turtles in normoxic water) down almost to 7. Blood ions showed little change in concentration over the course of hibernation in normoxic water, although Semple et al. (1969) had earlier demonstrated slight reductions in Na, K, and Cl in winter conditions as compared with summer conditions.

Maginniss et al. (2004) discussed support for three explanations of the ability of common map turtles to survive aerobically while submerged in cold water: metabolic depression, increased efficiency of extrapulmonary gas exchange mechanisms, and increased ability of the blood to transport O_2. They assumed that metabolic depression occurs during forced submergence, given the *in situ* observations of minimal activity under winter ice (Crocker et al. 2000) and their own observations that experimental animals in cold water that were allowed access to air were more active and responsive than those in forced submergence in cold normoxic or hypoxic water. They considered gain in body mass during forced submergence to be an indicator of perfusion of extrapulmonary gas-exchange surfaces. Semple et al. (1969) also showed that common map turtles gained water during winter hibernation. Finally, the experiments by Maginniss et al. (2004) demonstrated that hibernating turtles had increased

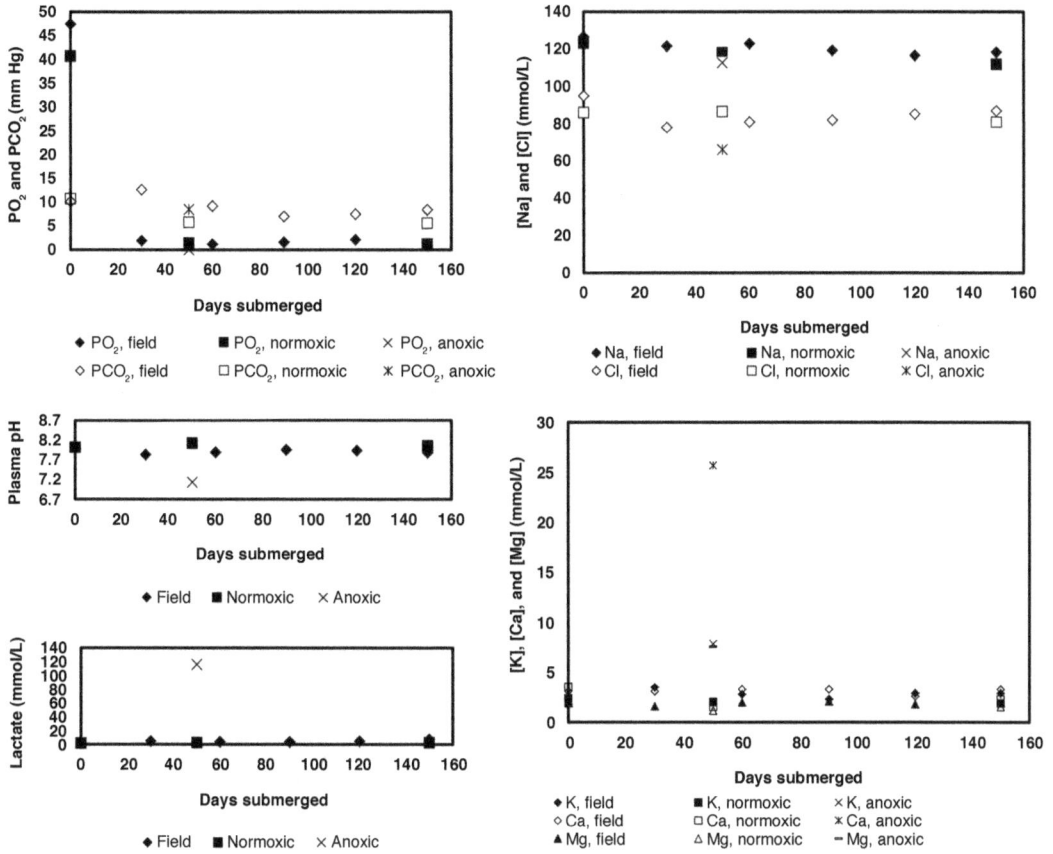

Figure 7.1. Blood parameters of hibernating common map turtles studied *in situ* in a normoxic hibernaculum in Vermont (Crocker et al. 2000) and held in the laboratory under normoxic or anoxic conditions (Ultsch and Jackson 1995; Reese et al. 2001).

concentrations of hemoglobin and increased hematocrit of the blood, with no change in red blood cell volume—that is, hemoglobin concentration and red blood cell volume rose but red blood cell size remained constant. Associated with this result was a markedly increased O_2 affinity of the blood under winter conditions (3°C; 50 percent O_2 saturation at a pressure of 6.5 torr) relative to summer conditions (22°C water; 50 percent O_2 saturation at 18.8 torr). Part of the increased O_2-binding affinity was thought to be associated with decreases in organic phosphates in the blood.

Maginniss et al. (2004) argued that increased concentration of hemoglobin was more likely accomplished by sequestration of blood plasma in peripheral organs than by production of more red blood cells, given the metabolic depression associated with entering hibernation. Turtles kept in ambient conditions simulating seasonal changes in Ontario sequestered 27 percent of their total blood plasma volume during hibernation, as demonstrated using isotopes (Semple et al. 1970). In the laboratory, sequestration of 24 percent of total blood plasma volume occurred in both gradual acclimation to cold water and sudden submergence (Stitt et al. 1970). The excess plasma was sequestered in the skin, shell, kidneys, gut, and limb muscles (Stitt and Semple 1971), and sequestration occurred in blood vessels rather than extracellular spaces (Stitt et al. 1971). Besides increasing hemoglobin concentration, closing off the circulation to peripheral and less vital inner organs may maintain adequate circulation to and perfusion of the more vital organs, as heart rate rose from four to twenty-one beats per minute when turtles were transferred from 10°C to 20°C water, with an accompanying increase in blood pressure (Semple 1964; Stitt et al. 1970; Akers and Damm [1963] also recorded a roughly linear increase in heart rate, from seven beats per minute at 10°C to ninety-two beats per minute at 40°C).

Maginniss et al. (2004) examined hemoglobin diversity and found four isohemoglobins in most turtles but as many as six. While the diversity of hemoglobins did not change with cold submergence, the authors could not assess what role this diversity might play in the increased O_2 affinity of the blood observed in winter.

The ability of hatchling common map turtles to hibernate in normoxic water appears similar to that of adults, with similar physiological responses (Reese et al. 2004). However, their tolerance of anoxic water was substantially reduced, with a survival of only fifteen days, thus possibly explaining their terrestrial overwintering strategy. Reduced buffering capacity of the less ossified shell of hatchlings was implicated in their lower survival rate in anoxic water.

Overwintering Physiology of Hatchlings in the Nest

For turtles at northern latitudes, where they overwinter in the nest and may be exposed to subzero temperatures, there are two strategies (not mutually exclusive) to avoid mortality: they may tolerate freezing of extracellular fluids or they may supercool, existing below their equilibrium freezing point but avoiding ice nucleation and subsequent freezing (Baker et al. 2003; Costanzo et al. 2008). Only some northern populations of map turtles have been reported to overwinter (Newman 1906a; Christiansen and Gallaway 1984; Pappas et al. 2000; Baker et al. 2003; McCallum 2003; Nagle et al. 2004; Parren and Rice 2004; but see Vogt 1980a). Temperatures in successful nests were recorded as low as −5.4°C in Indiana and −8.4°C in Pennsylvania (Baker et al. 2003, 2010; Nagle et al. 2004); however, overwinter mortality in the nest was very low in Indiana: just two hatchlings in twenty-three nests (Baker et al. 2010). In the laboratory, hatchlings that froze died, while substantial supercooling by other hatchlings was evident, with exposure to soil moisture and ice crystals negatively influencing supercooling capacity (fig. 7.2; Baker et al. 2003). Freeze tolerance and evaporative water loss were lower than in other turtle species, while anoxia tolerance was higher (Costanzo et al. 2001; Baker et al. 2003; Dinkelacker et al. 2005).

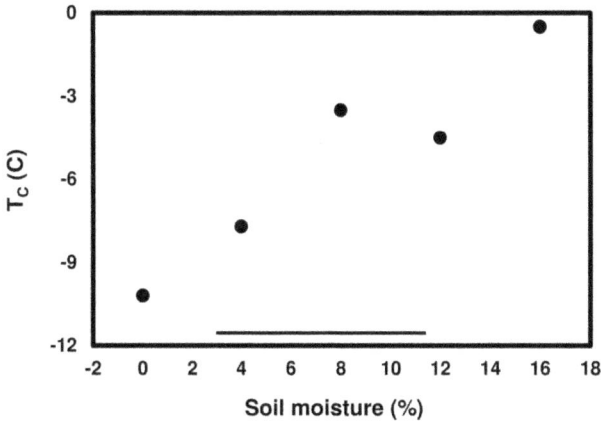

Figure 7.2. Temperature of ice crystallization in hatchling common map turtles as a function of soil moisture level in the laboratory. The black bar represents the range of soil moisture levels in field nests (Baker et al. 2003).

Thermoregulatory Physiology

Basking appears to be primarily a thermoregulatory behavior in *Graptemys*. The body temperatures of black-knobbed sawbacks captured in the water approximated the water temperature, but they were elevated above water and air temperatures for specimens caught on basking perches (fig. 7.3; Waters 1974). The mean body temperature of basking individuals was 29.1°C (range 26°C–35°C) and specimens in the laboratory achieved temperatures as high as 39°C. Barbour's map turtles captured while basking ranged from 12° to 35°C and were typically within ±3°C of air and substrate temperatures, but in most cases (97 percent of records) they were elevated over water temperatures, often by as much as 7°–11°C (34 percent of records; Sanderson 1974). In Wisconsin, six female Ouachita map turtles captured in a trammel net after evacuating from a basking site had an average cloacal temperature of 25.8°C, 3.3°C higher than the water temperature and 4.3°C higher than the air temperature, while a male's cloacal temperature matched the water temperature (Vogt 1980a). Implanted temperature recorders used on free-ranging common map turtles in Ontario demonstrated that they elevate their body temperatures several degrees above water temperatures and, in so doing, often attain preferred temperatures and maximize energy assimilation (Bulté and Blouin-Demers 2010a, 2010b).

Sexual size dimorphism has strong implications for heat gain during basking (Bulté and Blouin-Demers 2010b). Males and females of similar body sizes achieved higher temperatures than adult females and spent more time at preferred body temperatures. Minimum temperatures did not relate to body size, perhaps because turtles spend more time in water, equilibrating with the aquatic environment, than they do basking during midday, when heat gain is possible.

False map turtles basking in the laboratory had mean body temperatures of 32.7°C (range 28°C–38°C; Boyer 1965), while common map turtles allowed to move freely along an aquatic thermal gradient averaged 27.7°C (range 18°C–

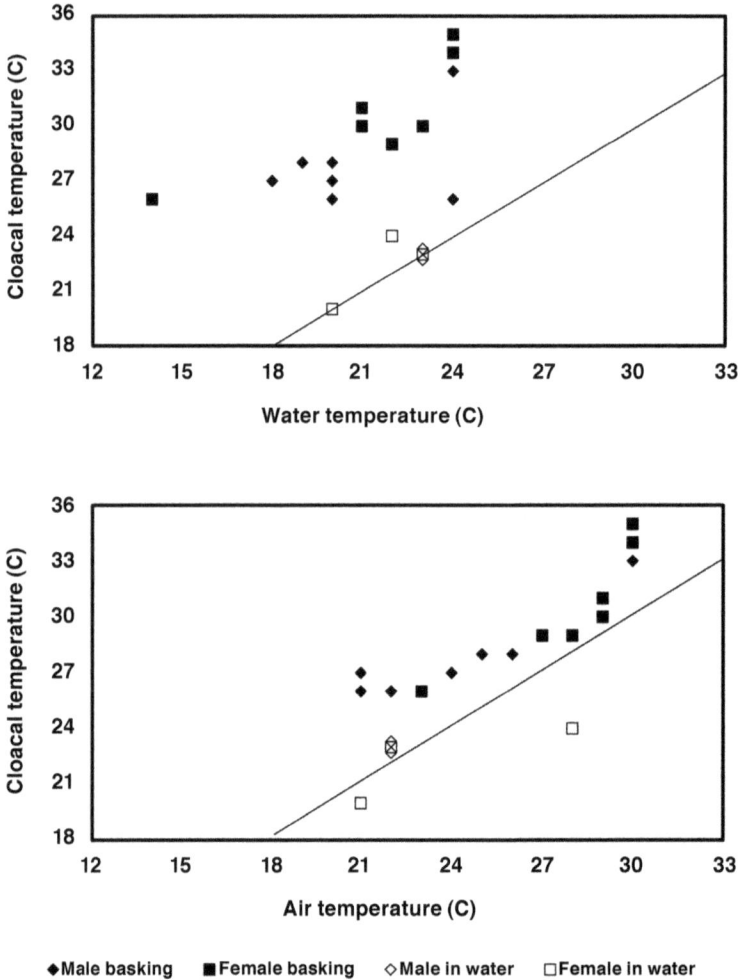

◆Male basking   ■Female basking   ◇Male in water   □Female in water

Figure 7.3. Cloacal temperatures of basking black-knobbed sawbacks as compared with water and air temperatures. Diagonal lines mark the points at which cloacal and environmental temperatures are equal. Data from Waters (1974).

33°C; Nutting and Graham 1993). The highest voluntary body temperatures approach the body temperatures at which Ouachita map turtles lost the righting response (38°C–39°C) and are a few degrees lower than their critical thermal maximum, the temperature that can cause death (40°C–42°C; Hutchison et al. 1966). In addition, common map turtles died when their core temperature was raised above 35°C (Akers and Damm 1963). Stress at high temperatures may explain the frequent submergence and re-emergence observed in *Graptemys* (Waters 1974).

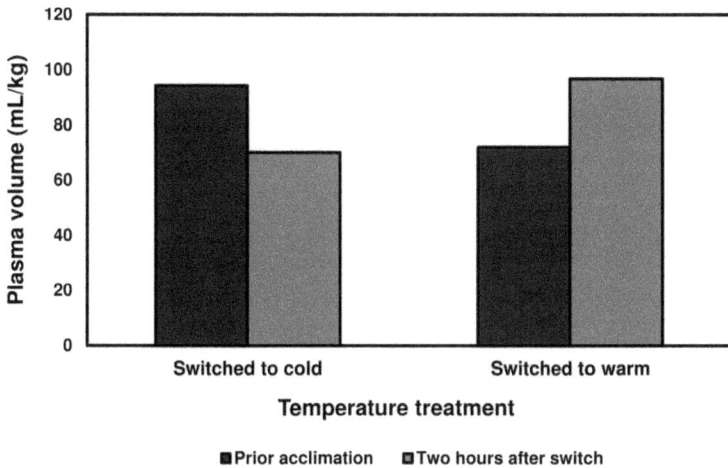

Figure 7.4. Blood plasma volume of common map turtles measured two hours after being subjected to sudden, acute changes in water temperature (20°C vs. 5°C) following at least three weeks of acclimation at the initial temperature. (Redrawn from Stitt et al. 1970.) Two hours was sufficient for turtles switched from one temperature to the other to reach plasma volumes equal to those of turtles under long-term acclimation to the other temperature.

Basking incidence of black-knobbed sawbacks was correlated highly with light intensity and, to a lesser degree, with air and water temperatures (Waters 1974). For common map turtles in a Michigan stream, the primary influence on basking incidence in autumn was the average difference between temperatures of the water and the basking substrates (i.e., the highest numbers of turtles were seen on warm days when the water was cold; Ecksdine 1985). Body size influences heat gain and loss, with larger turtles (i.e., females in *Graptemys*) heating more slowly (Boyer 1965; Waters 1974).

The plasma sequestration that appears to be an important adaptation for winter survival under ice (see above) may also play a role in thermoregulation in spring and early summer, when air temperatures are warming rapidly but water temperatures remain low. Stitt et al. (1970) showed that for common map turtles acutely cooled or warmed, plasma was fully sequestered in less than two hours and the process could be fully reversed over a similar time frame (fig. 7.4). They suggested the rapid change could be ecologically relevant in early-season basking, when a warmed basking animal might dive well below the surface into cold, deep water to avoid danger. Thus, early in the active season, turtles may frequently switch back and forth between breathing air while active at the surface or basking versus taking in oxygen via extrapulmonary means while inactive in deep water, with the latter strategy augmented by a switch to low plasma volume and increased hemoglobin concentration (Maginniss et al. 2004).

Bulté and Blouin-Demers (2008) used surgically implanted temperature re-corders to determine the field body temperatures of common map turtles in an Ontario lake. Body temperatures ranged from 7°C to 39°C, but approximately 75 percent of readings were between 20°C and 30°C. In the laboratory, both body size and water temperature were positively related to standard metabolic rate. The authors were thus able to use their body-temperature data to predict annual energy requirements of 1,490 kJ/year for an average-sized male and 22,482 kJ/year for an average-sized female. In addition, laboratory investiga-tion of standard metabolic rate in males and juvenile females that overlapped in body size showed that the females consumed 23 percent more oxygen, perhaps related to their more rapid growth (Bulté and Blouin-Demers 2009).

Ben-Ezra et al. (2008) tested the hypothesis that basking allows turtles to achieve preferred body temperatures that optimize performance, as represented by swimming speed and righting ability. Turtles were allowed to thermoregu-late by shuttling between a basking platform under a heat lamp and a cooler aquatic arena to simulate natural thermoregulation. The turtles shuttled be-tween the platform and the water frequently, with their body temperatures ris-ing and falling by several degrees with each change. Preferred temperatures ranged between 20°C and 30°C, and performance of adults and hatchlings was near optimal levels over a similar range of body temperatures. While the re-sults were thus consistent with the hypothesis, the authors cautioned that other physiological processes, such as digestion and gut passage time, may also be optimized by body temperatures achieved while basking.

Additional Topics in Physiology

Emerson (1967) recorded seasonal changes in levels of liver lipids and glycogen and blood sugar for false map turtles in South Dakota. Liver lipid content rose from June through late July and then declined to lowest levels by September. Blood sugar declined over the course of the season, while glycogen content of the liver rose, suggesting a connection between the two and that glycogen is the main storage product for hibernation. In addition, blood sugar levels were high relative to those in more southern turtles, suggesting a possible adaptation to northern climates.

Seasonal cycles in sex steroid levels have been studied in yellow-blotched sawbacks (Shelby et al. 2000). Male testosterone levels peaked in September and October, with a second, much smaller, peak in April and May. The larger fall peak was coincident with their peak testis size and peak spermiogenesis as well as congeners' mating behavior (Shealy 1976; Vogt 1980a; Lahanas 1982; Kofron 1991), but whether the smaller spring peak was indicative of a second period of spring mating could not be determined. In females, testosterone peaked in May in one year, coincident with maximum occurrence of preovulatory follicles, but in August the following year, after the nesting season. Estradiol 17-β and progesterone of females followed the typical freshwater turtle pattern: peaking coincident with follicle maturation and oviposition in May and June.

Seasonal cycles of corticosterone levels in the blood are less obvious but may be related to seasonal reproductive patterns (Selman et al. 2012). In yellow-blotched sawbacks, corticosterone levels were significantly elevated a half hour after capture in open-top basking traps, demonstrating a stress response (Selman et al. 2012).

For incubating Ouachita and false map turtle eggs, a wetter substrate (−150 kPa) was associated with weight gain during early incubation (as opposed to immediate loss of mass in a drier substrate at −950 kPa), slightly longer incubation time (averages of 49.8 vs. 48.6 days), and increased hatchling mass (by 8 percent in the Ouachita map turtle and 3 percent in the false map turtle; Janzen et al. 1995). The authors suggested selection should favor an ability of nesting females to detect substrates prone to remaining moist during incubation. Mississippi map turtles exhibit slower heart rates after hatching than as embryos at high temperatures, perhaps an adaptation to the switch from steady nourishment from yolk to uncertain nourishment from feeding (Du et al. 2010).

Bertl and Killebrew (1983) described a condition of the pleural bones they felt was evidence that the carapace acts as a calcium reservoir for egg production. In adult males and some juvenile females of four species, the pleural bones and other elements of the carapace were completely ossified, while in adult females caught after the beginning of the nesting season, the pleural bones were incompletely ossified. No further studies have assessed calcium metabolism related to egg production. In addition, while some authors have suggested that eating snails and clams is an important source of calcium for reproducing females, the importance of these prey taxa for reproduction is open to question because not all female *Graptemys* feed heavily on mollusks, many other deirochelyine turtles do not, and the amount of calcium obtained from shell fragments that pass through the digestive tract is unknown (Lindeman 2000a).

Gist and Jones (1989) found sperm-containing tubules in the oviducts of two false map turtles from Tennessee. Such tubules occurred in a diversity of turtle species and may allow for the dissociation of fall mating and fertilization in the late spring or early summer or fertilization of multiple clutches within a season following one copulation (see also Gist and Congdon 1998). Storage also likely allows for multiple paternity of clutches (Gist and Jones 1989; Pearse and Avise 2001; Pearse et al. 2001).

Morphology of *Graptemys*

Feeding Morphology

Feeding morphology was long used to partition *Graptemys* into "narrow-headed" and "broad-headed" species (e.g., Cagle 1953b, Dobie 1981; Bertl and

Killebrew 1983; C. H. Ernst et al. 1994; McCoy and Vogt 1994). More recently, allometric regressions of head width on plastron length and alveolar width on head width were used to partition females into three groups that correlate with ecology and phylogeny (fig. 7.5; Lindeman 2000a): (1) Microcephalic females (*flavimaculata*, *nigrinoda*, *oculifera*, *ouachitensis*, and *sabinensis*) are narrow headed, are always sympatric with broader-headed species, and consume few mollusks. (2) Mesocephalic females (*caglei*, *geographica*, both subspecies of *pseudogeographica*, and *versa*) have moderately broad heads, are sympatric with microcephalic species or allopatric, and tend to have a heavy intake of molluscan prey, mixed with substantial amounts of other, softer-bodied animal prey and algae. (3) Megacephalic females (*barbouri*, *ernsti*, *gibbonsi*, *pearlensis*, and *pulchra*) have exceptionally broad heads, are sympatric with microcephalic species or allopatric, and feed on mollusks to the near exclusion of other prey. In addition, microcephalic females have comparatively narrow alveolar surfaces relative to head width while megacephalic females have very broad alveolar surfaces; the alveolar surfaces of the mesocephalic females are either very broad (three populations of *geographica*) or have an intermediate width (the other three species; fig. 7.5). Males did not partition into neat, nonoverlapping groups corresponding to those of females. Instead, two less-distinct classes of relative head width and two of relative alveolar width could be recognized, without clear connection to diet or phylogenetic groups (fig. 7.5).

Evidence for the importance of trophic morphology with regard to the consumption of mollusks comes from a multivariate study of skull characters in a wide variety of emydid, geoemydid, and testudinid turtles (Claude et al. 2004). Eight emydid mollusk specialists (seven *Graptemys*, six of which were from the meso- and megacephalic categories known to have mollusk-eating females, plus the related diamondback terrapin, *Malaclemys terrapin*) showed striking similarity in skull shape with four geoemydid mollusk specialists (the Malayan snail-eating turtle, *Malayemys subtrijuga*; the spotted pond turtle, *Geoclemys hamiltonii*; and two Chinese pond turtles, *Chinemys nigricans* and *C. megalocephala*). Compared with other species classified as herbivores, carnivores, and omnivores, the mollusk specialists in the two families have large, massively developed skulls, massive upper jaws and widened palatine regions, wide alveolar surfaces, and elongated supraoccipital regions, while the olfactory region and the orbits are reduced in size. While *Graptemys* and *Malaclemys* constitute a single clade within Emydidae (Stephens and Wiens 2003), the geoemydid mollusk specialists are members of two to three separate clades (see phylogeny in Spinks et al. 2004). The fact that the same suite of skull characters has arisen in unrelated mollusk specialists thus suggests convergent evolution driven by adaptation to a diet of hard-shelled prey (Claude et al. 2004).

J. P. Ward (1980) studied the skull and cranial musculature of most deirochelyine emydid turtles, including several species of *Graptemys*, which had comparatively bulky jaw musculature. *Graptemys* also have a larger quadrate, which

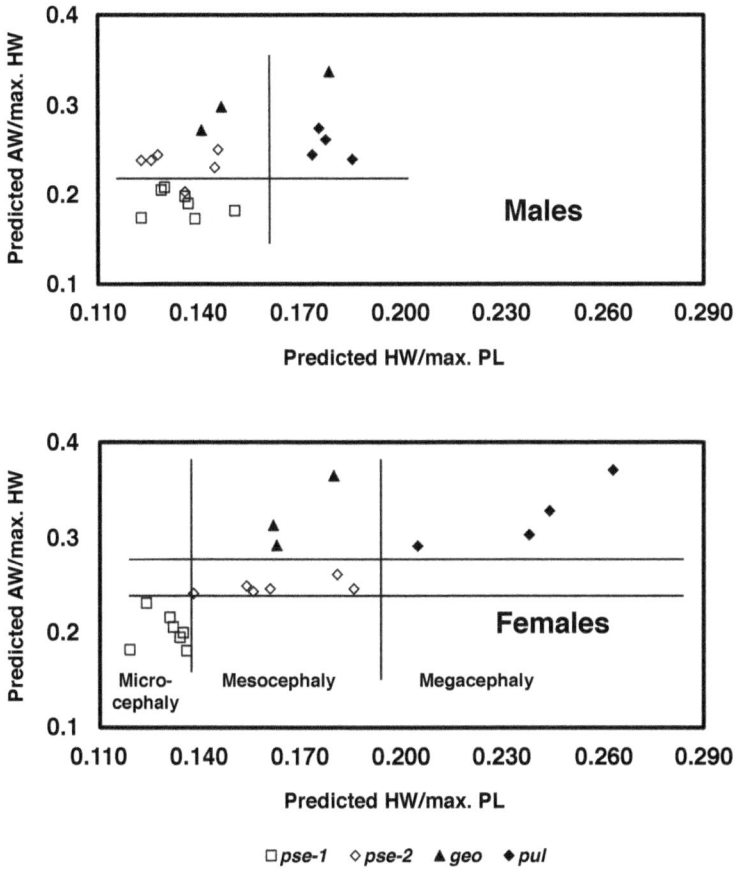

Figure 7.5. Head width (HW) as a proportion of body size (plastron length, PL) and width of the alveolar surface of the upper jaw (AW) as a proportion of HW for twenty species or populations of *Graptemys* (redrawn from Lindeman 2000a). Values depicted are predicted relative sizes of HW and AW, based on log-log regressions, at maximum observed values for PL and HW, respectively, for each of the twenty data sets. Symbols refer to four groups from three clades (*sensu* Lamb et al. 1994): the *pseudogeographica* clade, which is divided into species with microcephalic females that are sympatric with broader-headed species (*pse*-1) and species with mesocephalic females (*pse*-2); three populations of the monotypic *geographica* clade (*geo*); and the megacephalic *pulchra* clade (*pul*).

is the origin for an enlarged adductor mandibularis posterius, one of the jaw-closing muscles. The inferior temporal fossa is also enlarged and is the origin for an enlarged pseudotemporalis, another jaw-closing muscle. Both muscles are particularly enlarged in species with meso- and megacephalic females. In addition, in female Barbour's map turtles, the branchiomandibularis is mas-

Table 7.1. Results of analysis of the correlation in evolutionary changes in trophic morpholog

Variables	TIPS	Independent contrasts, gradualist		Independent contrasts, punctuational	
		LLWG	SW	LLWG	SW
Female HW and AW	.80**	.55*	.42, ns	.59**	.74**
Male HW and AW	.63**	.34, ns	.21, ns	.33, ns	.49, ns
Female and male HW	.78**	.76**	.87**	.75**	.86**
Female and male AW	.92**	.31, ns	.90**	.44, ns	.81**

Note: Analysis based on the phylogenies of Lamb et al. (1994: LLWG; results appeared originally in Lindeman and Sharkey 2001) and Stephens and Wiens (2003: SW). Comparative methods used were independent contrasts (Felsenstein 1985) and squared-change parsimony (Huey and Bennett 1987), with branch lengths in the phylogenies either used to weight estimates of ancestral character states (gradualist model) or set to one (punctuational model). TIPS = Pearson's correlation coefficient without correction for phylogenetic relatedness; HW = head width relative to plastron length; AW = alveolar width relative to head width; ns = nonsignificant.
*$p < .025$
**$p < .01$

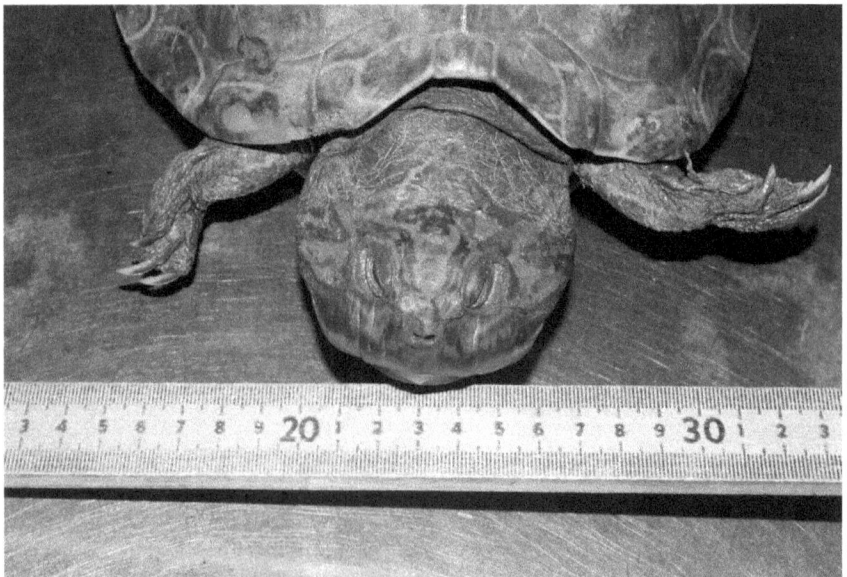

Figure 7.6. A female Barbour's map turtle (TU 13879.4) with a head width of 71.0 mm, the greatest individual head width of the 2,278 specimens of *Graptemys* spp. examined (Lindeman 2000a). Female Barbour's map turtles had the highest ratio for predicted head width as a percentage of plastron length (26.3 percent) based on species-specific allometric regressions. The ratio for this specimen is 32 percent, based on a midline plastron length of 225 mm.

Squared-change parsimony, gradualist		Squared-change parsimony, punctuational	
LLWG	SW	LLWG	SW
.62**	.75**	.59**	.72**
.38, ns	.40, ns	.31, ns	.46, ns
.43, ns	.85**	.49*	.87**
.81**	.79**	.82**	.78**

sive; it acts to open the lower jaw and pull it backward when the mouth is shut. The pterygoideus pars ventralis also becomes very enlarged in female Barbour's map turtles; it pulls the jaw forward when the mouth is closed. These females have the greatest head widths in the genus (fig. 7.6; Lindeman 2000a) and a diet dominated by medium to large snails (Cagle 1952b; Sanderson 1974), so the sliding motions of these two enlarged muscles may enhance performance in grinding and crushing prey.

Two types of comparative analyses (Felsenstein 1985, Huey and Bennett 1987) were used to analyze the intercorrelation of relative widths of the head and alveolar surfaces in map turtles and sawbacks (Lamb et al. 1994: LLWG entries, table 7.1; Lindeman and Sharkey 2001). These analyses depend on contrasts, or subtracted differences in character measurements between adjacent nodes in a phylogenetic tree, and thereby produce statistically independent data points to test a null hypothesis that evolutionary changes in two characters have not been correlated. Both methods require estimates of variables for ancestral nodes in a phylogeny, and each set of estimates may be generated by weighting averaging procedures by the phylogeny's branch lengths (gradualist model) or by setting all branch lengths to one (punctuational model; Martins and Garland 1991). The analyses revealed that relative widths of the head and alveolar surfaces have shown highly positively correlated evolutionary changes in females, but not in males, and that each character has shown positively correlated evolutionary changes between the sexes (table 7.1). Results show that the trophic variables are more important in females due to their greater molluscivory, with males tending to track conspecific females in the relative size of each variable but not showing the same tight intercorrelation typical of females (Lindeman and Sharkey 2001).

Using a phylogeny based on a larger data set (Stephens and Wiens 2003: SW entries, table 7.1) reinforces the conclusions of the earlier work based on the phylogeny of Lamb et al. (1994: LLWG entries). Relative widths of the head and alveolar surfaces showed correlated changes in females in all four comparative analyses. Males again showed no significant correlation of changes in relative widths of the head and alveolar surfaces, regardless of the method of analysis used. Correlations between the changes in males and females are significant in all methods of analysis using the Stephens and Wiens (2003) phylogeny, whereas three of eight analyses had produced nonsignificant results using the Lamb et al. (1994) phylogeny; hence the indication of genetic intercorrelation between the sexes is stronger under reanalysis.

Two recent biomechanical studies of the morphological basis for bite force in turtles, while not using *Graptemys* as subjects, nevertheless demonstrate the importance of morphology in adapting turtles to diverse diets (Herrel et al. 2002). Head dimensions of twenty-eight species from diverse lineages tended to scale isometrically to carapace length, but bite force scaled to carapace length at approximately 3:1, rather than the expected 2:1 ratio. Species that fed as carnivores or on mollusks had significantly greater bite force than did other species. The authors suggested that changes in the form of the jaw and muscle-attachment sites, rather than basic head dimensions, adapted species to their various diets. Bulté, Irschick, et al. (2008) found that bite force scaled with both plastron length and head width at allometric slopes approaching the expected 2:1 ratio in common map turtles, however, as did maximum prey hardness.

In a study emphasizing the ontogeny of bite force and head dimensions in three species (Herrel and O'Reilly 2006), hypoallometry was the rule (i.e., head

Figure 7.7. Sexual dimorphism in head width in Cagle's map turtle. Both specimens measure 63 mm in midline plastron length and were captured in their second growing seasons. The turtle on the left is a juvenile female, and the turtle on the right is an adult male.

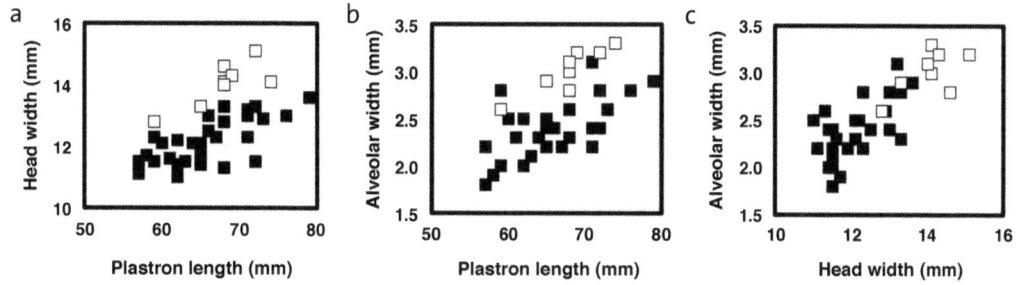

Figure. 7.8. Sexual dimorphism in head width and alveolar width in Texas map turtles, comparing adult males (filled symbols) with juvenile females (open symbols) of overlapping plastron lengths (redrawn from Lindeman 2006a). Females had significantly wider heads (a) and alveolar surfaces (b) after correction for body size in analyses of covariance, but there was no difference between the sexes in alveolar width as a function of head width (c), which is the typical pattern in the map turtles and sawbacks (Lindeman 2000a).

size grew proportionally smaller as the turtles grew larger in body size). In *Graptemys*, head width scales with strong hypoallometry to plastron length in males, with weaker hypoallometry in females of most species, but essentially with isometry in the megacephalic females (Lindeman 2000a). Alveolar width scales to head width with values suggestive of isometry or slight hyperallometry in both sexes.

Sexual dimorphism in the trophic morphology of map turtles and sawbacks is a function of head width differences that arise prior to maturation, based on statistical comparisons of juvenile females with adult males of overlapping shell lengths (fig. 7.7; Lindeman 2000a; Bulté, Irschick, et al. 2008). While there is an accompanying dimorphism in absolute width of the upper alveolar surfaces, relative to head width there is no sexual difference; thus the dimorphism is driven by divergence in relative head width alone (fig. 7.8; Lindeman 2006a). Cranial musculature is also relatively bulkier in females (J. P. Ward 1980).

Interspecific differences in relative head width that are found in sympatric species are present in hatchlings, even those that were preserved after laboratory incubation without the chance to feed (Lindeman 2000a); thus the differences clearly have a genetic basis. Whether or not diet is a secondary influence has not been studied, although Ewert et al. (2006) reported that Barbour's map turtle females raised on soft foods in a zoo grew to be just as broad headed as wild females.

The importance of relative head and alveolar width in handling mollusks was supported in an analysis of the size of Asian clam shells passed in the feces of female Texas map turtles (D. Collins and Lindeman 2006). Clam shell size was positively correlated with plastron length, width of the head, and width of the alveolar surfaces. Log-log slopes were near 1.0, suggesting isometry. When maximum shell size in each fecal sample and width of the female's alveolar

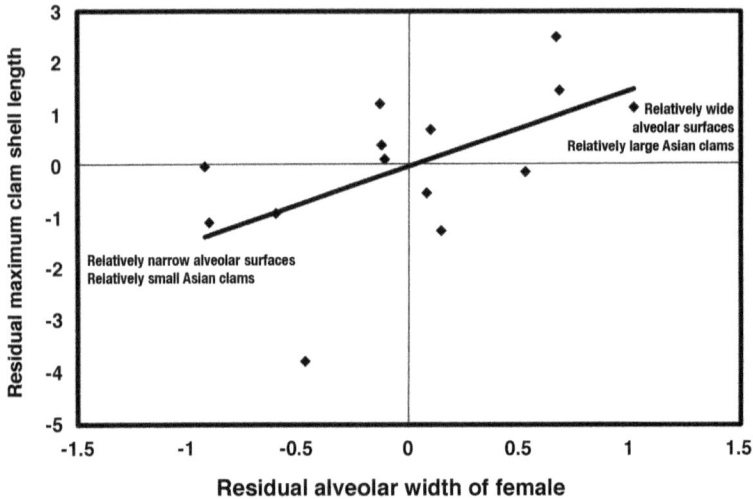

Figure 7.9. Residuals analysis of the correlation of relative size of the largest Asian clam shell in a fecal sample with relative alveolar width of female Texas map turtles, after each variable was first regressed on female head width. Note bunching of data points in or near the labeled quadrants in the upper right (both residuals positive) and lower left (both residuals negative). Redrawn from D. Collins and Lindeman (2006; from the corrected version of the figure published as an erratum in the March 2007 issue of *Herpetological Review*).

surfaces were both regressed on female head width, the resulting residuals were positively correlated and tended to lie in or near the two quadrants with like signs (fig. 7.9)—in other words, turtles with alveolar surfaces that were wider than expected based on their head width ate larger clams than expected based on head width, while turtles with narrower-than-expected alveolar surfaces ate smaller-than-expected clams.

Bulté, Irschick, et al. (2008) also demonstrated the importance of relative head width for consumption of mollusks and related their findings to a measure of fitness, body condition index (defined as residual body mass after regression on plastron length). They estimated hardness of snails taken by common map turtles in a lake in Ontario by measuring opercula passed in feces, relating operculum size to overall snail size and, in turn, relating snail size to the power required to crush the shell. They also quantified bite force with a force transducer. Females consumed snails that averaged 60 percent of the estimated maximum hardness they could crush, while males consumed snails that averaged only 28 percent; percentages declined with increased size of either sex. Turtles of either sex with relatively wide heads had higher body condition indexes (fig. 7.10), suggesting that broader heads confer higher fitness.

Three additional studies have examined the relationship of the size of mol-

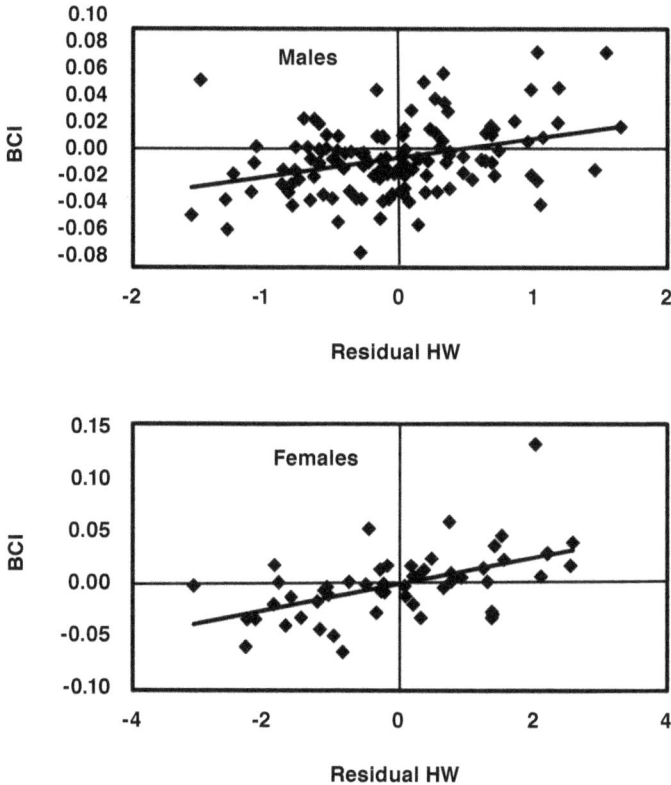

Figure 7.10. Regressions of body condition index (BCI, from residuals of log-log regression of body mass on plastron length) on relative head width (HW; residuals from log-log regression of HW on plastron length). Data occur in the lower left quadrant (showing relatively narrow heads and poor condition) and the upper right quadrant (broad heads, good condition) for 62 percent of males and 74 percent of females, and both regressions were highly significant. Redrawn from Bulté, Irschick, et al. (2008).

luscan prey with head width or body size and found significant correlation (fig. 7.11). With logarithmic transformation of variables for the data sets in figure 7.11, the log-log slope of 1.18 for Cagle's map turtle is approximately isometric, the log-log slopes of 0.66 (median clam shell lengths) and 0.68 (maximum clam shell lengths) for the Escambia map turtle are both slightly hypoallometric, and the log-log slope of 3.04 for the black-knobbed sawback is strongly hyperallometric. Hypoallometry in the Escambia map turtle may result from the distribution of available prey sizes, as the largest recorded clam size (25 mm) was taken over much of the range of female body sizes. Hyperallometry in the black-knobbed sawback may arise in part from mixing males with females for the analysis, as female black-knobbed sawbacks have significantly greater relative width of the head than males (Lindeman 2000a).

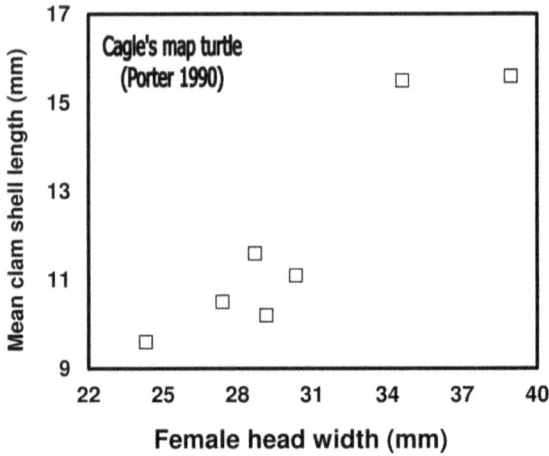

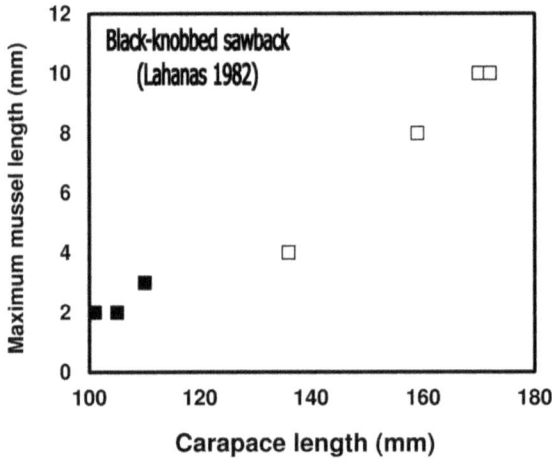

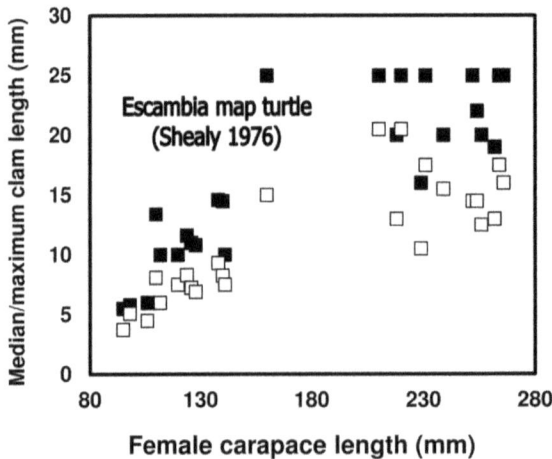

Figure 7.11. The correlation of size of molluscan prey with head width or shell length in three *Graptemys* species. For the black-knobbed sawback, open symbols represent females and filled symbols represent males. For the Escambia map turtle, open symbols represent median values and filled symbols represent maximum values.

Table 7.2. Relative tail size (mean percentage of shell length) in two species of *Graptemys*

Species	Source	Tail length as a propor- tion of	Sex	Relative length of entire tail	Relative precloa- cal tail length	Relative postcloa- cal tail length
G. barbouri	Sanderson 1974	Plastron length	Males	66	36	30
			Females	43	16	27
				53%	*125%*	*11%*
G. nigrinoda	Lahanas 1982	Carapace length	Males	46	14	22
			Females	28	8	20
				64%	*75%*	*10%*

Note: Percentages in italics below each pair of figures show how much greater relative lengths are in males than in females.

Other Sexually Dimorphic Morphological Attributes

Additional analyses of sexual dimorphism have emphasized shell shape, tail proportions, foreclaw length, and skull characters. Females have higher and wider shells as a relative proportion of shell length, probably an adaptation for greater volume for eggs (Mosimann 1958; Sanderson 1974; Lahanas 1982). Relative tail length is greater in males (table 7.2). The much longer relative precloacal tail length of males is explained by its housing of the retracted penis (Lahanas 1982). The relative postcloacal tail of males is also slightly longer, which may be an adaptation for helping to guide the tail into position during copulation (Lahanas 1982) and maintain a hold on the female (G. Bulté, pers. comm. 2009), or it may simply be a result of the overall tail enlargement necessary for housing the penis. Foreclaws are relatively longer in males in some but not all species, possibly corresponding to whether or not males use the foreclaws in courtship displays (see chapter 4); hence in these species foreclaw length may be sexually selected (Selman 2012). Four species show sharp basioccipital projections in males that are absent in females and an obtusely angled prominence along the supraoccipital crest in females that is absent in males (Bertl and Killebrew 1983), but no functional significance of the differences is known.

Coloration

Head markings and iris color may be important for species recognition during courtship. Vogt (1993) documented differences between sympatric Ouachita and false map turtles in head markings and male courtship behaviors, while shell markings showed no discernible differences. Postorbital markings are among the most frequently cited diagnostic characteristics in the genus. In the

Southeast, the sympatric species pairs differ greatly in postorbital and inter-orbital markings, these being much more extensive and often mask-like in the megacephalic clade (Myers 2008). Because courtship occurs face to face, the head markings may play a role in species recognition (Lindeman 2003a); how-ever, in trials in which specimens of the three sympatric species of the upper Mississippi drainage could see (but not smell) one another, Myers (2008) found that that neither males nor females placed with opposite-sex individuals spent significantly more time near conspecifics than they did near heterospecifics. Further, shape of the postorbital marking had only moderate heritability and was influenced by incubation temperature.

Iris color is yellow in most species, but it is white in Mississippi map turtles throughout the southern Mississippi drainage as well as in Ouachita, Sabine, Texas, and Cagle's map turtles (see color plates). Typically a black stripe runs horizontally through the iris and pupil, but in Mississippi map turtles the black stripe is usually missing (or occasionally is reduced to small black spots in front of or behind the pupil), producing a striking white ring in the eye (Lindeman 2003a). In the Calcasieu drainage, many Sabine map turtles lack the black stripe as well (S. Shively, pers. comm. 2009). In Wisconsin, Janzen et al. (1995) reported a yellow iris for Ouachita map turtles and a brown iris for false map turtles. J. F. Jackson and Shively (1983) reported a black iris for Mississippi map turtles in the Calcasieu drainage, while the same species in the Sabine drainage generally has a white iris with a black stripe (S. Shively, pers. comm. 2009). The differences may aid these sympatric species in recognizing their own species, given their overall similarity in head and shell markings. The full extent of intra- and interspecific variation in iris coloration has not been studied, however, probably largely because it cannot be recorded from preserved specimens.

Shell Morphology and Abnormalities

In many populations, hatchling and juvenile carapaces come to a very high peak with prominent midvertebral keels and particularly prominent lateral ser-rations on the marginal scutes (fig. 7.12; Folkerts and Mount 1969). The effect is especially pronounced in the sawbacks, the megacephalic clade, the Sabine map turtle, and southern populations of Ouachita and false map turtles (R. C. Wood 1977). Adult males are less domed and have moderately well-developed keels and serrations, while both keels and serrations become reduced in adult females, which retain a more domed shell, probably for increasing reproduc-tive output. The size-dependent nature of the prominence of the domed shape, keeling, and lateral serration suggest roles in defense against predators. The lati-tudinal trend suggests that it may be predators of the southeast, including the lower Mississippi River drainage, that exert the strongest selective pressures. The species with the most developed keels and serrations overlap in range with alligator snapping turtles (*Macrochelys temminckii*) and alligators (*Alligator*

Figure 7.12. (a) A male black-knobbed sawback showing height of the vertebral keels and lateral serrations of the marginal scutes of the carapace (Black Warrior River at River Road Park in Tuscaloosa, Tuscaloosa County, Alabama) and (b) a juvenile Pearl map turtle showing a high-domed carapace (Yockanookany River at Highway 16, Leake County, Mississippi).

mississippiensis), which, although not known to exert significant predation pressure on map turtles and sawbacks, do regularly take other turtles (see chapter 4). Alternatively, the keel may have a hydrodynamic advantage in current, and the spines of the keel and the serrated marginal scutes may act as holdfasts for turtles resting among submerged branches (J. P. Ward 1980).

Abnormalities and relative positioning of the scutes of the carapace and plastron in map turtles and sawbacks have been the subject of largely descriptive studies. Of 476 common map turtles, 10 percent had extra carapacial scutes, 7 percent had extra inframarginal scutes on the lower side of the bridge, and 0.4 percent had extra scutes along the midline of the plastron (Newman 1906b). Tinkle (1962) summarized species-specific patterns of seam contacts between the vertebral and marginal scutes for most North American turtle species, including ten of the fourteen species of map turtles and sawbacks. Seam contact positions varied interspecifically but seemed to offer little consistent variation useful for distinguishing among species groups. Little (1973) claimed to have used plastral seam contacts to differentiate populations of Pascagoula, Alabama, and Escambia map turtles, although he gave no details. Lovich and Ernst (1989) summarized data for the first two of these species (then considered conspecific) for plastral formulae based on the ordering of plastral scutes from longest to shortest in relative size. Most specimens' longest scutes were the abdominals, followed by the anals, and the shortest scutes were the gulars, followed by the humerals, but eighteen variations were found in *Graptemys*.

Reports of kyphosis (abnormal curvature of the vertebral column of the trunk) are rare for map turtles and sawbacks. Carpenter (1958) described an extreme case that caused distortion of the carapace and plastron in a male Ouachita map turtle. Selman and Jones (2012) reported that 0.1 percent of a large sample of female ringed sawbacks exhibited kyphosis, and they described slow growth in two specimens.

Limb Morphology

Long legs and long toes with extensive webbing are undoubtedly adaptations for swimming ability. In analyses of covariance comparing two map turtles with syntopic sliders (*Trachemys scripta*) in a Kentucky reservoir, Mississippi and Ouachita map turtles had longer hind legs, forelegs, and hind toes than the slider, and hind toes were longer in Mississippi map turtles than in both other species (fig. 7.13; Lindeman 2000b). In a long, narrow reservoir cove the two map turtles made greater use of deeper, more wave-tossed water near the main channel of the reservoir. In general, the short-limbed slider occurs in more lentic habitats, while the two map turtles use medium to large rivers and their backwaters.

In a study of evolutionary trade-offs in locomotory ability, Stephens and Wiens (2008) quantified swimming speed and endurance of males from a diverse set of sixteen emydid turtle species. The three *Graptemys* (Ouachita and

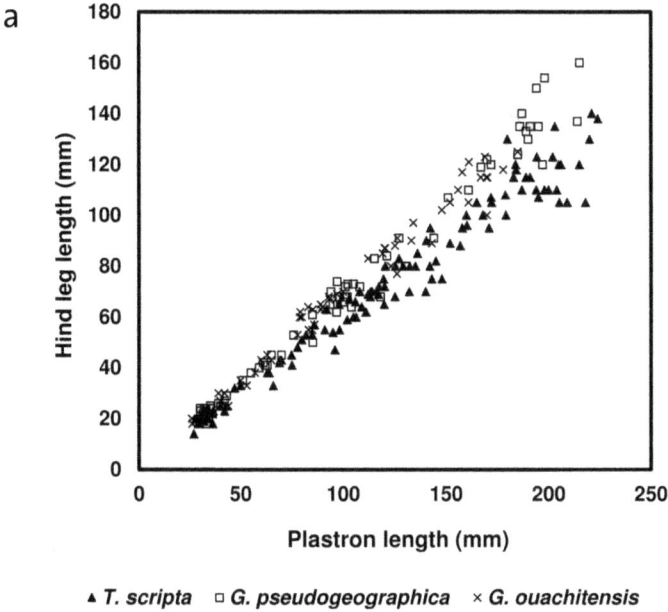

a

Hind leg length (mm) vs Plastron length (mm)

▲ T. scripta   □ G. pseudogeographica   × G. ouachitensis

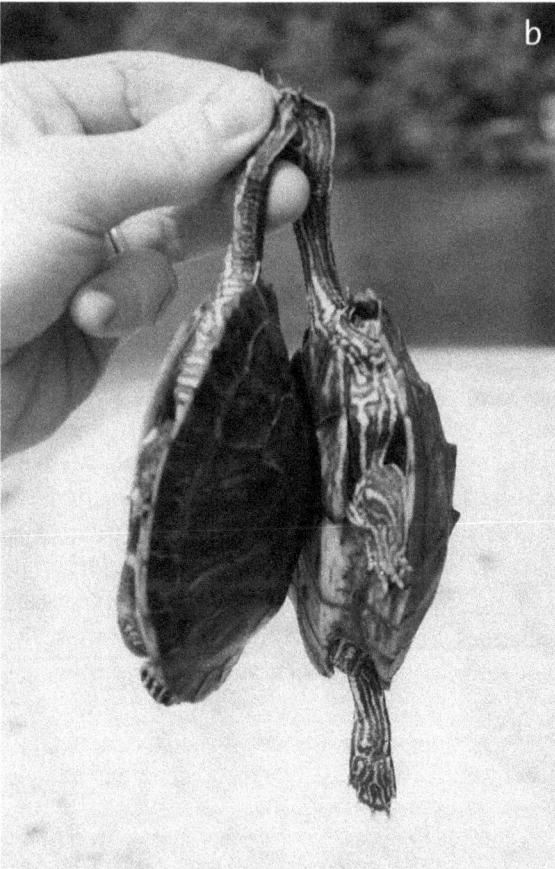

b

Figure 7.13. (a) Comparison of hind leg lengths of three syntopic turtle species in a reservoir in western Kentucky (redrawn from Lindeman 2000b). Hind legs of the two map turtles were significantly longer than hind legs of the slider, *Trachemys scripta*, in analysis of covariance. (b) A comparison of the hind legs of a slider (left) and a Ouachita map turtle (right) of similar body size (plastron lengths 171 and 169 mm, respectively), demonstrating the latter's longer hind leg (123 vs. 95 mm) and longer middle hind toe (25 vs. 23 mm).

Texas map turtles and black-knobbed sawbacks) were fastest, able to swim 4.4–5.3 times their carapace length per second. At the same time, their swimming endurance during a twenty-minute trial rated among the lowest among fourteen species, suggesting selection for rapid—but not sustained—retreat from danger, likely provided by their relatively long limbs and toes.

Population Genetics of *Graptemys*

Only recently has work begun on the population genetics of the map turtles and sawbacks. While some of the work is based on mitochondrial DNA sequencing, the application of microsatellite primers originally developed for other emydid species has been largely successful in five species of *Graptemys* and may increase the resolution of interpopulational differentiation (T. L. King and Julian 2004; Freedberg et al. 2005; Myers 2008; Selman, Ennen, et al. 2009). Most results have shown at least some degree of geographic structuring among populations.

Freedberg et al. (2005) used microsatellite data from two loci and sequenced a portion of the control region of the mitochondrial DNA as part of their analysis of nesting-area philopatry (see chapter 4). Both molecular markers showed differentiation between two major nesting areas, and at one, microsatellites showed evidence of increased differentiation with increased distance between nest sites.

Selman, Kreiser, et al. (2007, 2013; see also Selman 2010) analyzed six microsatellite loci in yellow-blotched sawbacks from the lower Pascagoula River and two major tributaries. They found a low degree of genetic subdivision, with most of the genetic variation within rather than among populations. Each population had one to four alleles not found in the other two, although all private alleles were found at low frequencies; nevertheless, they may indicate a lack of panmixis among populations. Selman, Kreiser, et al. (2007, 2013) detected no evidence of historical or recent population bottlenecks.

Myers (2008) used five microsatellite loci to examine genetic structuring of common, false, and Ouachita map turtles (sampling four, five, and five populations, respectively) in the Mississippi River drainage. She found genetic structure among populations in each species, but the structuring was particularly strong in the common map turtle, possibly due to its more specialized diet.

Bennett et al. (2010) used five microsatellite loci to study the genetic population structure of common map turtles in the Trent-Severn Waterway of Ontario. Results were consistent with panmixis in spite of recent anthropogenic fragmentation of the habitat by locks, dams, and water control structures. A. D. Brown et al. (2012) sequenced a portion of the mitochondrial DNA control region to analyze phylogeography of the Ouachita map turtle, with emphasis on the isolated population of the Scioto River in Ohio. Of three haplotype groups, two were common and broadly overlapping, both in the Scioto and in the greater Mississippi River drainage farther west. Haplotype groups may have diverged during isolation caused by glaciation during the late Pliocene or early Pleistocene, followed by wide postglacial dispersal resulting in broad overlap of groups.

Two recent genetic studies demonstrate introgressive interspecific hybridization between *Graptemys* species. In the Choctawhatchee drainage of Alabama and the Florida Panhandle, Escambia map turtles from the Escambia-Yellow drainage to the west and Barbour's map turtles from the Apalachicola drainage to the east have met in sympatry and are hybridizing introgressively. For both morphology and genetic markers, specimens from the main-stem Choctawhatchee show a predominance of Barbour's map turtle characteristics, while specimens from the major western tributary, the Pea River, show a mix of predominating characters and markers (Godwin et al. 2013). Freedberg and Myers (2012) found one common map turtle mitochondrial DNA haplotype and two nuclear gene alleles from common map turtles among an isolated lake population of false map turtles, at frequencies ranging from 3 to 10 percent. Because there are no common map turtles in their habitat, which was created by earthquakes in 1811 and 1812, these molecular markers of hybridization have persisted in the population for at least nearly two centuries. Freedberg and Myers (2012) also found evidence of introgressive hybridization between Ouachita map turtles and both common and false map turtles.

Taxonomy of *Graptemys*

The taxonomic history of the genus *Graptemys* has been characterized by considerable uncertainty regarding species limits, particularly for allopatric populations. The taxa *kohnii, oculifera, ouachitensis, sabinensis,* and *versa* have been considered conspecific with or separate from *pseudogeographica* by various authors, and the earliest specimens of *caglei, flavimaculata,* and *nigrinoda* were assigned to *pseudogeographica* (see species accounts in chapter 8). The taxa *ernsti, gibbonsi,* and *pearlensis* were considered conspecific with *pulchra* for four decades after the first extensive collections of specimens were made (Lovich and McCoy 1992). Some European authors have preferred to recognize allopatric subspecies rather than separate species in checklists, albeit without detailed analyses (e.g., *flavimaculata* and *nigrinoda* as subspecies of *oculifera* [Mertens and Wermuth 1955]; *barbouri, ernsti,* and *gibbonsi sensu lato* as subspecies of *pulchra; flavimaculata* as a subspecies of *oculifera;* and *versa* as a subspecies of *pseudogeographica* [Artner 2003]). Their taxonomies are unlikely to find acceptance, given increasing adoption of species concepts that recognize divergent, diagnosable lineages with separate binomial names (e.g., de Queiroz 2007). While most authors have favored separate-species status (see Ennen, Kreiser, et al. [2010] and Ennen, Lovich, et al. [2010] for recent arguments in favor of recognizing separate species), low levels of interspecific differentiation in mitochondrial DNA analyses led Walker and Avise (1998) to suggest that the genus had been "taxonomically oversplit" (p. 39). Analysis of multiple nuclear genes will likely eventually provide the evidence of differentiation necessary to retain separate species status for named taxa from isolated drainages.

In analysis of the *Graptemys pulchra* species group, Lovich and McCoy (1992) recognized three species distributed across four river drainages where previously only one had been recognized. They named the populations found in the Pearl and Pascagoula drainages *Graptemys gibbonsi*, declining to designate them as separate species based on their relatively high degree of overlap in discriminant function analysis. Vetter (2004) listed a "Pearl River form" and a "Pascagoula River form" (p. 5) for the species in his checklist, however, and the populations were recently split into two species with the description of *G. pearlensis* (Ennen, Lovich, et al. 2010).

There has been no critical examination of interdrainage morphological or genetic variation in five species found in multiple isolated drainages: the common map turtle (Mississippi, St. Lawrence, Susquehanna, Hudson, Delaware, and Mobile Bay drainages), false map turtle (San Bernard, Brazos, Trinity, Sabine-Neches, Calcasieu, Mermentau, and Mississippi), Sabine map turtle (Sabine-Neches, Calcasieu, and Mermentau), Barbour's map turtle (Choctawhatchee, Apalachicola, Ochlockonee, and Aucilla), and Escambia map turtle (Escambia-Yellow and Choctawhatchee). (Some of these occurrences are likely the result of anthropogenic introductions; see chapter 10.) One brief report is suggestive of possible differentiation: in addition to the unique black iris of *G. pseudogeographica* from the Calcasieu River (see above), individuals frequently had elongate chin spots that merged to form a V-shaped marking on the lower jaw (J. F. Jackson and Shively 1983), a pattern otherwise unknown for *G. pseudogeographica*.

8

Species Accounts

Introduction

I treat the Sabine map turtle as a full species, *Graptemys sabinensis*, bringing the total number of species for this account to fourteen following the recent description of *G. pearlensis* (Ennen, Lovich, et al. 2010). The Sabine map turtle was originally described as a subspecies of *pseudogeographica* (Cagle 1953b) and then retained as a subspecies of *ouachitensis* by Vogt (1978, 1993), the first author to separate *ouachitensis* from *pseudogeographica*. My rationale for species status begins with the fact that the taxon is allopatric to *ouachitensis*, and all other taxa with allopatric distributions in the genus are regarded as full species, including two taxa formerly regarded as subspecies (*versa* and *oculifera*). Cagle's decision to regard *sabinensis* as a subspecies while recognizing *oculifera*, *flavimaculata*, and *nigrinoda* as separate species (Cagle 1953b, 1954) was likely based on the latter three species' greater differentiation in carapacial markings and their more distinctive vertebral keels; however, recognizing *all* nonintergrading, allopatric, diagnosable taxa as full species is more consistent. Intergradation by species restricted to separate river drainages is unlikely.

Elevating fully allopatric subspecies to full species (e.g., J. T. Collins 1991) has been criticized (e.g., Dowling 1993) when done without detailed analysis of the true degree of morphological and genetic differentiation and the reality of geographic discontinuities; however, existing analyses support raising *sabinensis* to a full species. J. P. Ward (1980) stated that *sabinensis* is the only *Graptemys* in which the frontal bones of the skull have anterior extensions of their medial sides that fully separate the prefrontal bones (see also Cagle 1953b; Dundee 1974) and regarded it as a separate species. Stephens (1998) found that skulls of male *sabinensis* and *ouachitensis* showed no synapomorphic characteristics (female skulls were not compared) and discussed possible elevation to species. Finally, various combined analyses of morphological and molecular data placed *sabinensis* as a close relative of *caglei* and/or *versa* but not as a particularly close relative of *ouachitensis*, and none of the various analyses of partitioned data sets resolve *sabinensis* and *ouachitensis* as sister species (Stephens and Wiens 2003, 2009; Wiens et al. 2010). Two additional analyses of gene sequences further demonstrate the distinction between *ouachitensis* and *sabinensis* (Myers 2008; A. D. Brown et al. 2012), countering Vogt's (1993) lack of differentiation between the taxa in earlier allozyme studies.

I use "map turtle" as part of the common name of all but three species, for which I reinstate the vanquished term "sawback" instead. This may be controversial for those who extend to common names the same demand for uniformity that biologists generally reserve for scientific names. Within U.S. herpetology, an evolving movement seeks to standardize the common names of U.S. amphibian and reptile taxa, including species. K. P. Schmidt's (1953) checklist allowed multiple common names for species, but just one common name per species was assigned by Conant et al. (1956) and in all three editions of the Peterson field guide (Conant 1958, 1975; Conant and Collins 1991). For U.S. turtles, single common names for each species were also generally given by Pope (1939), A. Carr (1952), C. H. Ernst and Barbour (1972, 1989), Iverson (1985), C. H. Ernst et al. (1994), and C. H. Ernst and Lovich (2009). The Society for the Study of Amphibians and Reptiles (SSAR) issued a list of standardized common names in 1978 (J. T. Collins et al. 1978) and in five subsequent editions (J. T. Collins et al. 1982; J. T. Collins 1990, 1997; Crother et al. 2001, 2008, 2012); the movement later bifurcated, however, with J. T. Collins and Taggart (2002, 2009) differing in some common names in their own "fifth" and "sixth" editions, through a different publisher.

In addition to the recent schism in standardized common name lists, the effect of these lists has been a slow trickle of changes in "official" names that seem poorly grounded in any set of ground rules, such as those that govern the validity of scientific names in the *International Code of Zoological Nomenclature*. For example, the long-stable name "common map turtle" (*G. geographica*) was changed to "northern map turtle" by Crother et al. (2001, 2008, 2012), lest anyone think that the species is more abundant than it is, a change that surely inspired more eye rolling than clarification. (The change has also likely not been warmly embraced in southern range states such as Virginia, Alabama, and Arkansas.) Recently, two never-before-used common names have been introduced within the genus, as the rules for common names continue to evolve: "southern map turtle" (*G. o. ouachitensis*) and "northern false map turtle" (*G. p. pseudogeographica*; Crother et al. 2012). "Sawback" was stricken from the standardized lists and replaced with "map turtle" beginning with the first SSAR list (J. T. Collins et al. 1978). The term had previously been used for *G. oculifera*, *G. flavimaculata*, and *G. nigrinoda* by Conant et al. (1956), in the first two Peterson guides (Conant 1958, 1975), and by C. H. Ernst and Barbour (1972); indeed, it had been used by Cagle (1954) when he described the latter two species.

"Sawback" is a fitting moniker for three species that have the most pronounced vertebral keels in the genus and deeply serrated marginal scutes, either of which might evoke blade tips on a circular saw blade.

Rationales for changes in the SSAR list have only begun to be offered with the fifth edition, but the obvious interpretation for the earlier banishment of "sawback" is that the change standardizes the genus with regard to common names by making all *Graptemys* into "map turtles." Some proponents of the change might even hold that having a different common-name root for the sawback clade (Stephens and Wiens 2003; Wiens et al. 2010) would render the other eleven species in the genus, *map turtles*, paraphyletic! While few biologists would agree that we need to concern ourselves with monophyly and paraphyly issues for common names, it is nonetheless interesting to take note of the first use of "sawback." In A. Carr (1952), only *G. geographica*, the outgroup species of the genus (Stephens and Wiens 2003), was called "map turtle," while "sawback" was used in the names of all other species. This is the most *logical* arrangement for the genus: in addition to the basal phylogenetic split within the genus, only the author of *G. geographica* mentioned map-like markings on the animal's shell (LeSueur 1817), and it does not have the same degree of pointed vertebral keels that other *Graptemys* have. Nevertheless, it (unfortunately) is not the most *stable* arrangement, with few authors having followed Carr (1952); therefore I here use "sawback" for only the three species so called for much of the latter half of the twentieth century, including in two of the original descriptions. For each species, I have compiled a reasonably complete listing of common names applied by various authors.

Descriptions emphasize diagnostic features that separate the various species of map turtles and sawbacks from one another and are thus not intended to be detailed descriptions, which can be found elsewhere (e.g., C. H. Ernst and Lovich 2009; various state field guides; original descriptions). Taxonomic history and phylogeny entries include details on the collection of the first-known specimens of each species, formal description, type series and type localities, etymologies, opinions about specific or subspecific status, descriptions of subspecies, and placement in four phylogenetic trees (see figs. 3.7–3.10).

Detailed descriptions of geographic ranges are given to accompany range maps (see color plates). Emphasis is on known upstream range limits in major drainages and smaller tributaries that are inhabited by each species, with documentation from literature sources if they exist or specimen records or sightings. Downstream range limits are also given for river main stems.

Vetting of all records has not been possible due to their volume and their dispersion among many collections (see acknowledgments). The most problematic records concern specimens labeled as *pseudogeographica*, *kohnii*, *ouachitensis*, and *sabinensis*. For these I have relied heavily on the identifications of Vogt (1993) and, to a lesser degree, on whether or not data files sent by collection mangers show evidence of having been revised according to Vogt's (1993) taxonomy; I have also examined some collections on museum visits or via photographs of head patterns. A small number of records have not been mapped due to uncertainty regarding the species they represent. I have been conservative about not mapping specimens collected outside of accepted geographic ranges, as they are likely introductions.

Basic natural history information is briefly summarized in the sections on habitat, diet, reproduction and life history, and natural history; a more comparative approach to what is known about these topics appears in other chapters (especially chapter 4). The sections on populations summarize data on absolute and relative abundance collected via various modes of trapping or by observing basking turtles, as well as mark-recapture estimates of population size. These data are exceptionally difficult to compare temporally or among species, but they nevertheless represent the historical record available for conservation and ecological assessment. Sections on conservation status summarize official state, federal, and international designations for each species; population management and recovery efforts, if applicable; and federal, state, and private habitat preservation efforts. Most of the habitat preserves were not specifically intended to benefit map turtles and sawbacks but may nevertheless have strong positive effects.

Common Map Turtle, *Graptemys geographica*

Figure 8.1. Basking aggregation, Graveyard Pond in Presque Isle State Park, Erie County, Pennsylvania.

Basking seems to be their principal occupation. On warm days
they literally line the shores of the lake at certain favorable places.
Scarcely a floating board or pier lacks its quota of occupants. . . .
When basking they are decidedly gregarious, collecting in such
numbers on certain sheltered ledges that it becomes necessary for
them to pile up two or even three layers deep.

H. H. NEWMAN (1906A, P. 137)

Alternate Common Names

Lake Erie tortoise (LeSueur 1817), LeSueur's terrapin (J. E. Gray 1831), geographic tortoise (DeKay 1842), map tortoise (Hay 1892a, 1892b), geographic
terrapin (Hay 1892a), geographic turtle (Cahn 1937), hackleback (Gentry 1956),
all defunct; map turtle, an early short version of the name (e.g., Paulmier 1902;
Surface 1908; Strecker 1908, 1909; Evermann and Clark 1916; Bishop 1921), is
occasionally still used (e.g., C. H. Ernst and Barbour 1972; Kiviat and Buso
1977; J. T. Collins et al. 1978; Vogt 1981c; McCoy and Vogt 1990); northern
map turtle, dating to at least Ewert (1989) and increasingly in recent usage (e.g.,
COSEWIC 2002; McCallum 2003; Behler et al. 2004; Moriarty 2004; Browne
and Hecnar 2005) following adoption by SSAR to avoid misinterpretation of
the name "to imply abundance rather than . . . the fact that [the species] has a
broad range" (Crother et al. 2001, p. 78); *tortue géographique*, used in Quebec
(e.g., Daigle et al. 1994; D. M. Green et al. 2012). First called common map turtle
by Pope (1939).

Description and Diagnosis

Postorbital markings small, triangular with rounded angles, one pointed down-
ward, one rearward, one forward; always longer than tall. Interorbital marking
a medial longitudinal stripe, typically flanked by one to three thinner longitudinal lines on each side. Single lines on each lower side of neck curling sharply
forward and upward, their tips pointing at postorbitals. Supramandibular spots
absent. Chin marked by longitudinal lines; small submandibular spots sometimes present. Markings on head, neck, limbs, and tail light yellow to bright,
egg-yolk yellow. Females with moderately broad heads and exceptionally broad
alveolar surfaces, males with narrow to broad heads and broad alveolar surfaces. Second vertebral scute only slightly sloped, extending little higher than
the third. Carapace not highly peaked, rounded in cross section. Carapacial
keel with only low, blunted points; these often almost absent. Black pigment
along midline of carapace, particularly at keel points. Carapace marked with reticulated, black-bordered, yellowish-orange markings (brightest on pleural and
marginal scutes) with numerous points of intersection on each pleural scute,
often forming circles or ovals with concentric, alternating light and dark markings. Plastron pale yellow, with dark markings along scute seams in smaller
specimens. Male foreclaws not conspicuously elongated. Irises yellow, bisected

by black stripe. Differs from congeners by the triangular postorbitals, the low carapacial keel, and the upward-curving lower lateral neck stripes.

Taxonomic History and Phylogeny

The common map turtle was described as *Testudo geographica* by LeSueur (1817), who illustrated a specimen captured in June 1816 (fig. 8.2). No specimens are known to have been deposited in a natural history museum by LeSueur; thus the illustrated specimen stands as the lectotype (Lindeman 2009b). LeSueur gave the locality as "in a marsh, on the borders of Lake Erie" (LeSueur 1817, p. 86), apparently on Presque Isle or its enclosed bay in present-day Erie County, Pennsylvania (fig. 8.2; Lindeman 2009b). The specific epithet refers to the "stripes or markings on the disk [i.e., carapace], presenting the appearance of a geographical map" (LeSueur 1817, p. 87). No subspecies have been described. A detailed synonymy (McCoy and Vogt 1990) includes a junior synonym, *Emys megacephala* Holbrook 1836 (holotype ANSP 255), and while *Emys LeSueuri* Gray 1831 has been considered a junior synonym of the false map turtle *G. pseudogeographica* (e.g., Vogt 1995) and was the name frequently applied to that taxon in the 1800s, its description was in fact based on a common map turtle (Hay 1892b; see chapter 2); the holotype is a skeletal specimen thought to have been examined by John Gray (BMNH 1982.1297; Bour and Dubois 1983). A. E. Brown (1908, p. 114) designated the common map turtle as the type species for the genus *Graptemys* Agassiz 1857, presumably due to its status as the first-described and best-known species of the genus at that time. The species is the outgroup to all other *Graptemys* (Lamb et al. 1994; Lamb and Osentoski 1997; Stephens and Wiens 2003; Wiens et al. 2010).

Range

The common map turtle is the most widespread *Graptemys*, occurring in multiple river drainages of both the Gulf of Mexico and the Atlantic Ocean. It is widespread in the upper Mississippi River basin and some western tributaries of the lower Mississippi, although apparently absent from the lower Mississippi itself. To the south it also occurs in parts of the Mobile Bay drainage, and to the north it also occurs in the eastern Laurentian Great Lakes and St. Lawrence, Susquehanna, Delaware, and Hudson drainages.

Within the upper Mississippi drainage, the common map turtle ranges up the Mississippi to Crow Wing County, Minnesota (Oldfield and Moriarty 1994); up the Missouri River to Cole County, Missouri (Daniel and Edmond 2012); up the Ohio River into the lower Allegheny River in Armstrong County, Pennsylvania (Atkinson 1901), and the Monongahela River in Monongalia County, West Virginia (Frum 1947); and via the Tennessee River through western Tennessee and northern Alabama (Mount 1975), into several of its eastern tributaries: the Clinch, Powell, and Holston Rivers in western Virginia (Mitchell 1994); the Little Tennessee in Great Smoky Mountains National Park in eastern

Figure 8.2. (a) LeSueur's drawing of the "Lake Erie tortoise," known today as the common map turtle, published with his description of the species as *Testudo geographica* in 1817. The specimen drawn, although apparently not otherwise preserved, is the lectotype for the taxon. (b) Presque Isle on Lake Erie, where the specimen was collected in 1816. Base map data © 2012 Google, Landsat.

Tennessee (J. T. Wood 1946); the Hiwassee River to just east of the Tennessee border into North Carolina (Beane et al. 2009; Nelson et al. 2012); and Little Chickamauga Creek in Georgia (Jensen et al. 2008). While never reported from the state of Mississippi, the fact that Mississippi borders on Pickwick Lake Reservoir on the Tennessee River (home to Alabama populations of the species; Mount 1975), with embayments and the Bear Creek tributary extending into the state, makes it likely that Mississippi is also part of the range of the species (R. L. Jones, pers. comm. 1994). Other inhabited tributaries of the upper Mississippi are the St. Croix and its tributary, the Namekagon; the Minnesota; the Wisconsin; the Sugar; the Iowa and its tributary, the Cedar; and the Illinois and its tributaries, the Fox, Kankakee, Des Plaines, and Dupage Rivers (Moll 1977; Vogt 1981c; Oldfield and Moriarty 1994; Casper 1996; Vandewalle and Christiansen 1996; DonnerWright et al. 1999; C. A. Phillips et al. 1999; Brodman et al. 2002; G. Casper, pers. comm. 2008). It has been recorded in the lower Missouri and ranges up its tributaries, including the Loutres, Gasconade, and Moreau Rivers, as well as the various tributaries of the Osage, including the Marais des Cygnes, Little Osage, and Marmaton Rivers in Kansas (J. T. Collins 1993; Myers 2008; Daniel and Edmond 2012; T. Taggart, pers. comm. 2009).

In the Tennessee drainage, the species' range extends up the Duck River drainage into Bear Creek and up the Elk River drainage to Woods Reservoir (Miller et al. 2005; Haislip 2008). Upstream of the mouth of the Tennessee, additional major tributary drainages of the Ohio drainage that are inhabited include the Cumberland, including the Stones River and its east and west forks; the Tradewater; the Wabash and its tributaries, the Embarras, Vermillion, and Tippecanoe Rivers; the Green; the Kentucky; the Whitewater; the Little Miami; the Licking; the Scioto; the Great Kanawha; the Little Kanawha; and the West Fork of Little Beaver Creek (LeSueur 1827; Hay 1892a; Bailey et al. 1933; Welter and Carr 1939; Conant 1951; Conant et al. 1964; N. B. Green and Pauley 1987; C. A. Phillips et al. 1999; Poly 1999; Lindeman 2001c; Minton 2001; Rizkalla and Swihart 2006; Watson and Pauley 2006; Wynn and Moody 2006; Temple-Miller 2008; Gooley et al. 2011; Miller and Miller 2011; Niemiller et al. 2011; Weber and Layzer 2011).

Although absent from the lower Mississippi, the common map turtle inhabits several medium-sized western tributaries of the lower river, ranging via the White, Arkansas, and Ouachita drainages into Arkansas, Missouri, Kansas, and Oklahoma (Ortenburger 1929; J. T. Collins 1993; Trauth et al. 2004; Daniel and Edmond 2012; Myers 2008; T. Taggart, pers. comm. 2009; Riedle et al. 2009). Unvouchered records from an Arkansas Game and Fish Commission trapping survey in the early 1990s also place the species in the Red drainage of Arkansas (Trauth et al. 2004; S. Trauth, pers. comm. 2009), although it has not been reported from portions of the Red drainage downstream in Louisiana or upstream in Oklahoma. The species' southernmost record is from the Ouachita River in Louisiana (Douglas 1972).

Additional southerly localities occur in upstream portions of the Mobile Basin drainages of Alabama and Georgia, above the geological phenomenon known as the fall line—an arc that demarcates where upland streams drop down to the coastal plain and separates rocky, faster-flowing upstream sections from sandier, slower-flowing downstream sections. Regions of the drainage inhabited include the Cahaba River, the Black Warrior drainage's Blue and Brushy Creeks and Locust Fork, and the Coosa River upstream over the state line into its tributaries in Georgia, the Chattooga and Conasauga Rivers (Mount 1975; Jensen et al. 2008; Pauly 2010; G. Brown and Kinney 2011).

In Lake Huron, common map turtles occur primarily along the eastern shore of Georgian Bay but have also been recorded from the southernmost extent of the lake at Kettle Point as well as farther south, in Lake St. Clair and the Detroit River (Logier and Toner 1961; COSEWIC 2002). In Lake Erie, the species occurs on islands of the western basin, in the coastal marshes of Ohio, and in wetlands of the lake's major sandspit peninsulas, Point Pelee, Rondeau, Presque Isle, and Long Point (Atkinson 1901; Logier 1925; Adams and Clark 1958; Logier and Toner 1961; R. B. King et al. 1997; Gillingwater 2001; Browne and Hecnar 2005; K. M. Ryan and Lindeman 2007; Tran et al. 2007). In Lake Ontario, the species occurs in Sodus and Irondequoit Bays on the New York shore, while Ontario records are concentrated primarily around the Toronto-Hamilton metropolitan area to the west and the many long, narrow bays to the east near Belleville (Bishop 1921; Logier and Toner 1961; COSEWIC 2002).

The species is found in numerous small-to-medium streams that enter the Great Lakes and several smaller glacial lakes in Wisconsin, Michigan, Indiana, Ohio, and Ontario (Toner 1936; Lagler 1943; Mittleman 1947; Dexter 1948; Conant 1951; Wade and Gifford 1964; Iverson 1988, 1992; Casper 1996, 1999; COSEWIC 2002; G. Casper, pers. comm. 2008). Downstream of the Laurentian Great Lakes in northern New York, Vermont, and Quebec, the common map turtle occurs in Lake George, Lake Champlain, the lower Ottawa River, and lentic portions of the St. Lawrence, into which the Ottawa flows at Montreal, with recent captures extending the range to the vicinity of Quebec City (Bishop 1921; Gordon and MacCulloch 1980; Graham and Graham 1992; Daigle et al. 1994; Galois 2005).

The Susquehanna distribution includes the main-stem Susquehanna from near its mouth in Maryland into New York, as well as two tributaries, the Juniata and Conodoguinet Rivers, in Pennsylvania (Conant 1958; H. S. Harris 1969; Hulse et al. 2001; Richards-Dimitrie 2011; USNM 220872). Reports of the occurrence of common map turtles in the Delaware and Hudson Rivers are relatively recent and have thus been proposed to represent canal-mediated range expansions (Arndt and Potter 1973; Kiviat and Buso 1977); however, based on other herpetofaunal cases, the possibility that the species was merely overlooked in the Hudson for more than two centuries cannot be discounted, as the role of canals in promoting species introductions has been questioned (R. E. Schmidt et al. 2004) and common map turtle remains have been reported from three archaeological sites along the lower Hudson (R. E. Funk 1976). In the Delaware, the species occurs from Bucks County, Pennsylvania, upstream to Pike County, Pennsylvania, and Sussex County, New Jersey (Arndt and Potter 1973; Behler et al. 2004). Farther downstream, its range extends into a Delaware tributary, the Schuylkill River (Tiebout 2003). The range in the Hudson extends between Orange and Saratoga Counties, New York (Kiviat and Buso 1977; Gibbs et al. 2007).

Habitat

Detailed analyses of habitat use have been conducted on rivers in Pennsylvania, Kansas, Wisconsin, Ohio, Ontario, and Maryland and in a lentic portion of the Ottawa River in Quebec. Common map turtles were disproportionately captured (53 percent of all captures) in a slow, deep section of the Raystown Branch in Pennsylvania, which constituted only 24 percent of the study reach (Pluto and Bellis 1986, 1988). Some males moved out of the deep, slow area, which was a communal hibernaculum, into summer home ranges upstream or downstream. While some turtles traversed large areas, several individuals remained in small stream sections generally associated with favored basking sites. Compared with co-occurring turtle species, common map turtles in streams in Kansas (Fuselier and Edds 1994) and Wisconsin (DonnerWright et al. 1999) were associated with rocky- or gravel-bottomed reaches with little bank vegetation and few basking sites. In the lower 200 km of the Scioto River of Ohio, common map turtles had an overall generalized, uniform distribution, but telemetered females tended to occur near shallow sandbars (Temple-Miller 2008). In the St. Lawrence River, Ontario, turtles preferred shallow water near shore, particularly males, which also preferred areas with vegetative cover (Carrière and Blouin-Demers 2010). In the Susquehanna River, Maryland, turtles were clustered in a midstream island complex during both the active season and hibernation, although some females made extensive nesting forays away from the islands (Richards-Dimitrie 2011).

In a lentic habitat on the lower Ottawa River, six embayments inhabited by common map turtles were compared with six that were not (Flaherty 1982; Flaherty and Bider 1984). In inhabited embayments, deadwood used for basking was larger, was located farther from shore and aquatic vegetation, and in deeper water than deadwood that turtles did not use. However, the availability of potential basking sites was very similar between inhabited and uninhabited embayments. Available prey and shoreline characteristics, including nesting habitat, were also similar, suggesting the species aggregated for communal basking to facilitate predator detection (Flaherty 1982). Movement between embayments was more common in July–September (31 percent of recaptures) than in May–June (8 percent of recaptures), with much of the movement thought to represent postnesting dispersal (Flaherty 1982).

Diet

Lagler (1943) reported snails, bivalves, crayfish, and aquatic insects (primarily beetles and caddis fly larvae) as the predominant prey in Michigan. Vogt (1981a) reported that 66 percent of the diet by volume for adult females in the upper Mississippi River in Wisconsin consisted of mollusks, with fish and mayfly and damselfly nymphs of lesser importance (table 4.8); males primarily ate aquatic insects. In a small stream in Missouri, both sexes fed almost exclusively on small snails (White and Moll 1992; table 4.12). In the lower Susquehanna River in Maryland, females ate large pleurocerid snails almost exclusively, while males primarily consumed smaller physid, hydrobiid, and planorbid snails (Richards-Dimitrie et al. 2013). Early qualitative studies likewise found heavy consumption of snails, presumably also by both sexes (Garman 1890; Hay 1892b; Newman 1906a).

Anthropogenic changes to the aquatic environment can alter the diet of the common map turtle. Moll (1977) reported dramatically reduced consumption of mollusks over eight decades, coincident with habitat degradation that had reduced populations of snails and fingernail clams (tables 4.10 and 4.11). The only mollusk of any significance in more recent samples was the Asian clam (*Corbicula* sp.) in a reservoir population. Asian clams were also taken in moderate amounts by males in Maryland (Richards-Dimitrie et al. 2013). In Lake Erie, where invasions by Eurasian zebra and quagga mussels (*Dreissena polymorpha* and *D. bugensis*, respectively) have dramatically altered the biological community, females fed very heavily on the abundant new food source (Lindeman 2006b; table 4.13). Small snails were also consumed by females, while males also took some snails and mussels but fed mainly on caddis fly larvae. Similar diets were found in a small Ontario lake (Bulté, Gravel, et al. 2008). Bulté and Blouin-Demers (2008) estimated that the average adult female annually consumes 33–137 kg of zebra and quagga mussels.

Reproduction and Life History

Growth of common map turtles has been studied using plastral annuli or relative scute proportions to back-calculate sizes in Quebec and Indiana (Gordon and MacCulloch 1980; Iverson 1988) and by nonlinear modeling of growth intervals in Ontario (Bulté and Blouin-Demers 2009). Juvenile females grew more rapidly than males in all three studies and were larger by the fourth growing season in Indiana, when males had matured. Adults of both sexes grew little over one to two years in Quebec. Males mature in their second growing season in Lake Erie (Pennsylvania) but a year or two later in Wisconsin, Ontario, and Indiana; females mature as early as their eighth growing season in Lake Erie but not until about their twelfth in Ontario and perhaps not until after their thirteenth in Wisconsin (Vogt 1980a; Iverson 1988; Bulté and Blouin-Demers 2009; P. Lindeman, unpublished data). Estimated annual survival over three years was slightly higher in females (mean 87.3 percent) than in males (mean 82.6 percent; Bulté and Blouin-Demers 2009).

Average clutch size was 12.7 in Quebec (range 9–17; Gordon and MacCulloch 1980), 13.2 in Lake Erie in Ontario (7–22; Gillingwater 2001), 11.9 in Lake Erie in Pennsylvania (3–21; K. M. Ryan and Lindeman 2007), 10.6 in a stream in central Pennsylvania (6–15; Nagle et al. 2004), 15.1 in New York (10-18; Kiviat and Buso 1977), and 10.1 in Missouri (6–15; White and Moll 1991). Newman (1906a) also gave 11–14 as the range in Indiana. Two clutches a year and possibly three are laid by some females (White and Moll 1991; K. M. Ryan and Lindeman 2007). Hatchlings are apparently unique among species of *Graptemys* in that most winter in the nest and emerge in spring (Newman 1906a; Pappas et al. 2000; Baker et al. 2003; McCallum 2003; Nagle et al. 2004; Parren and Rice 2004).

In a Mississippi River population in Wisconsin, males ranged from 81 to 116 mm in midline plastron length and females reached 216 mm (Lindeman 2008a). Similar maxima were reported from an Indiana lake (112 and 211 mm, respectively; Iverson 1988). In Lake Erie, Pennsylvania, the range of adult midline plastron length was 66–117 mm in males and 169–235 mm in females (K. M. Ryan and Lindeman 2007; P. Lindeman, unpublished data). Iverson (1988) suggested there may be a latitudinal trend in body size in common map turtles, but the trend is difficult to evaluate due to a lack of data from southern populations and the diversity of lake and stream habitats studied in the north.

Natural History

Courtship has been observed in October, November, and April in Wisconsin, with males engaging in head bobbing but not foreclaw titillation (Vogt 1980a). Carrière et al. (2009) reported courtship near communal hibernacula in fall and spring, with possible copulation (G. Bulté, pers. comm. 2010). In a Lake Erie population, activity has been observed between the end of February and early November (B. S. Gray and Lethaby 2008). Parasite species of common map turtles include two turtle leeches, six coccidians, a trypanosome, one monogenean and eight trematode flukes, one enteric spiny-headed worm, and five roundworms (Van Cleave 1913; Stunkard 1915, 1919; Harrah 1922; Horsfall 1935; Hsü 1937; Hughes et al. 1942; Hopp 1946, 1954; Rausch 1946, 1947; DeGiusti and Batten 1951; Cable and Hopp 1954; Fisher 1960; Woo 1969b; Pluto and Rothenbacher 1976; Wacha and Christiansen 1976; Saumure and Livingston 1994). Only 33 percent of eggs in nests caged to protect against predation produced hatchlings, with the remainder split between dead hatchlings or embryos and eggs that showed no sign of development (Gillingwater 2001).

Populations

Relative abundance data have been reported for habitats from small streams to large rivers and lakes. In the Illinois River in Illinois, common map turtles were

23 percent of the catch in baited hoop nets and trammel nets in the river and ranked second among eight species, behind sliders (*Trachemys scripta*); in three floodplain lakes and a reservoir, however, they were considerably less abundant, at 5 percent and seventh most abundant (Moll 1977). In the upper Mississippi River, the common map turtle was 7 percent of three map turtle species captured in fyke nets, gill nets, and trammel nets in Wisconsin and 7 percent of the turtles captured in late fall at a hibernaculum site (Vogt 1980a); it was 26 percent of the catch and second in abundance among six species in Iowa (Gritters and Mauldin 1994); and it was 14 percent of the catch and ranked fifth of seven species farther south, in Illinois (R. A. Anderson et al. 2002). The latter two studies differentiated among habitats, with findings that echoed Moll's (1977) in suggesting the species is more abundant in lotic than in lentic waters. In Iowa, the species accounted for nearly two-thirds of the catch in tailwaters but only 10 percent in backwaters, while in Illinois, it was 5 percent of the catch in open river, and 21 percent of the catch in a side channel of the river, but was not represented among forty-four turtles caught in a backwater that connected to the river only during flooding.

Relative abundance of common map turtles varies considerably among smaller streams in its range. In the Black Warrior drainage above the fall line in Alabama, common map turtles were only 3 percent of the catch of eight species (Tinkle 1959). In forty-one counties in southeastern and south-central Kansas, they constituted 0.7 percent of the catch and were the ninth most frequently captured of ten species in baited hoop traps deployed in small-to-medium streams (Fuselier and Edds 1994), although much of the area was outside their Kansas range. In the upper Ottawa River, Quebec, common map turtles predominated in basking surveys at 62 percent of all turtles; relative abundance was considerably higher in rocky than in marshy habitats (Daigle et al. 1994). Their relative abundance in basking surveys declines steeply in the lower Ottawa, where painted turtles and even common snapping turtles outnumber them, the former by between 10:1 and 100:1 (Daigle and Lepage 1997; Daigle and St-Hilaire 2000). DonnerWright et al. (1999) captured turtles in the St. Croix River in Wisconsin in baited hoop nets, with 16 percent of the catch being common map turtles, third in abundance behind spiny softshells (*Apalone spinifera*) and common snapping turtles (*Chelydra serpentina*). In the upper Tradewater River in western Kentucky, the species accounted for 33 percent of turtles observed basking and 8 percent of turtles trapped in fyke nets and basking traps, ranking second and tied for fourth, respectively, in an assemblage of eight species; it was the predominant turtle, however, at the most upstream site sampled and rarer down river (Lindeman 2001c). In four small creeks in the Ouachita Mountains of Arkansas, the species accounted for 10 percent of the catch in baited hoop traps and ranked fourth among five species captured (Phelps 2004). In an urban canal connected to the White River in Indianapolis, the common map turtle predominated at 40 percent of captures in baited hoop traps and 74 percent of observations in basking surveys, although in a nearby lake its relative abundance was only 7 percent, with a ranking of third in trapping (Conner et al. 2005; Peterman and Ryan 2009). Rizkalla and Swihart (2006) captured the species in low abundance (5 percent of all turtles, fifth in abundance of eight species) in wetlands of the upper Wabash floodplain in Indiana. On the North Fork of the White River in the Ozark Mountains of southern Missouri, the common map turtle has been the strongly predominant species in a six-species assemblage over three and a half decades, at 78 percent of all captures (Pitt and Nickerson 2012).

Relative abundance data are available in several northern lakes, including small glacial lakes, the Laurentian Great Lakes, and a southern reservoir. Newman (1906a) stated that common map turtles greatly outnumbered the six other turtle species in Lake Maxinkuckee, Indiana. The species predominated in Lake Champlain, Quebec, at 77 percent of an assemblage of four turtle species (Daigle and Lepage 1997). Common map turtles were the third most abundant of seven species in Dewart Lake, Indiana, yet accounted for only 4 percent of captures in a system dominated by painted turtles (*Chrysemys picta*) and stinkpots (*Sternotherus odoratus*; Wade and Gifford 1964; L. L. Smith et al. 2006). In a bay and adjoining lagoon of Presque Isle State Park, Pennsylvania, on Lake Erie, common map turtles are the most abundant of six turtle species, accounting for 65 percent of all captures in basking traps and fyke nets from 1999 to the present (Lindeman 2006b and unpublished data). Browne and Hecnar (2007) compared data from 1972 to 1973 and 2001 to 2002 across the lake at Point Pelee National Park in Ontario, where painted turtles (*Chrysemys picta*) and common snapping turtles (*Chelydra serpentina*) predominated among seven turtle species in both time periods. Common map turtles ranked tied with stinkpots (*Sternotherus odoratus*) for fourth in relative abundance at 1 percent in the earlier sampling but surpassed stinkpots and Blanding's turtles (*Emydoidea blandingii*) to rank third in relative abundance at 11 percent in more recent sampling, a change attributed to basking traps used only in the more recent sampling. Common map turtles also ranked second behind painted turtles in basking surveys. In a fyke net survey of seventy-seven lower Great Lakes wetlands, common map turtles were fourth in abundance among five turtle species at 2 percent of the total catch and were found at 9 percent of sites trapped (DeCatanzaro and Chow-Fraser 2010). In an impoundment in Tennessee, common map turtles were the third most frequently captured of four species taken in hoop nets and basking traps (27 percent of captures) and the second most frequently seen species in basking surveys (38 percent of turtles; Weber and Layzer 2011).

Mark-recapture estimates of population size have been made for populations in lakes in Quebec and Ontario, a Tennessee reservoir, and Pennsylvania and Missouri streams. Gordon and MacCulloch (1980) used paint marks and visual sightings for recaptures one year and trap recaptures for two years. Their annual estimates of population size for an approximately 2,000 ha area in a Quebec lake ranged from 322 to 379 turtles (5–6 turtles/ha). Bulté and Blouin-

Demers (2008) estimated that 1,569 turtles occurred in a 788 ha lake in Ontario, with a biomass of 1,130 kg. Weber and Layzer (2011) estimated population at a Tennessee reservoir at 71 (2.5 turtles/ha). All three of these estimates may be biased upward, however, due to the lack of population closure that is assumed by the population estimation models that were used (see Lindeman 1990). Pluto and Bellis (1986) used two mark-recapture techniques to estimate 175 or 152 turtles in 6.6 km of a small Pennsylvania stream (26.5 or 23.0 turtles/km). Pitt and Nickerson (2012) estimated the population of common map turtles in an Ozark Mountain stream as declining from 16 turtles/ha to 8/ha over three and a half decades.

Conservation Status and Actions

Listed as a species of special concern in Canada in 2002. State-listed as endangered in Maryland, threatened in Kansas, rare in Georgia, and a species of special concern in Oklahoma and Vermont. Provincially listed as a species of special concern in Ontario and considered likely to be designated as either threatened or vulnerable in Quebec. Listed as a species of least concern with stable populations by the IUCN.

The most extensive federal preserve is the Upper Mississippi River National Wildlife and Fish Refuge along a 420 km reach of the river in Illinois, Iowa, Wisconsin, and Minnesota. Its 97,000 ha include numerous forested islands and extensive backwaters. Other prominent national wildlife refuges (NWRs) on the Mississippi are Trempealeau NWR in Wisconsin (2,521 ha), Port Louisa NWR in Iowa and Illinois (4,364 ha), Great River NWR in Missouri and Illinois (6,073 ha), and Chickasaw NWR in Tennessee (10,124 ha). In Alabama, the Tennessee River flows through Wheeler NWR (13,968 ha), and in Kansas, the Marais des Cygnes flows through Marais des Cygnes NWR (3,036 ha). In eastern Tennessee, common map turtles occur in Great Smoky Mountains National Park in tributaries of the Little Tennessee. The U.S. National Park Service also manages more extensive habitat as national rivers in five other river drainages: the St. Croix in Wisconsin and Minnesota (406 river km), Delaware Gap in Pennsylvania and New Jersey (64 km), Ozark in south-central Missouri (216 km of the Current and Jacks Fork Rivers), the Buffalo in north-central Arkansas (217 km), and the Big South Fork in Kentucky and Tennessee (50,600 ha). The species is present in Ozark streams in southern Missouri that occur within Mark Twain National Forest and in upper segments of the Ouachita drainage in Arkansas that occur within Ouachita National Forest.

Federal protection of habitat at Lake Erie occurs in the Ottawa NWR in Ohio and Point Pelee National Park in Ontario. Three prominent state and provincial parks further benefit populations in Lake Erie, at Presque Isle in Pennsylvania and Rondeau and Long Point in Ontario. In Lake Champlain, the federal government protects nearly 2,700 ha of Missiquoi Bay and the Missiquoi River delta in Missiquoi NWR; Vermont has thirteen state parks located on the lake's shoreline and on several of its major islands, while New York has three, with

Adirondack State Park adjoining much of the lake's southwestern shoreline and surrounding Lake George.

Needed Further Research

Data from southern populations are needed to examine latitudinal variation in life history. The impact of predation by common map turtles on zebra and quagga mussels has not been studied and would be an interesting research topic, given the exceptional ecological and economic costs of damage from invasive mussels. Further studies of coarse-scale habitat overlap with congeners in both the upper Mississippi and the upper Mobile Bay drainages would be a worthy addition to our understanding of how competition structures assemblages in the genus. Studies to date suggest the common map turtle may be more restricted to smaller tributaries and upstream reaches, although overlap with other species in larger downstream segments also occurs; studies of the impact of river impoundment on common map turtle distribution and abundance may also contribute greatly to this objective. Populations in the Delaware and Hudson Rivers should be examined using molecular genetic methods to determine whether they represent anthropogenic introductions.

Barbour's Map Turtle, *Graptemys barbouri*

Figure 8.3. Juvenile female, Apalachicola River at Torreya State Park, Liberty County, Florida.

The day we found *barbouri*, three or four of us were standing shivering and enviously watching Marchand as he wallowed and snorted about, searching the bottom of the Chipola River through his goggles.

ARCHIE CARR (1952, P. 153)

Alternate Common Names

Barbour's sawback turtle (Cagle 1952b), Chipola map turtle (K. P. Schmidt 1953), neither widely adopted; colloquially "big-headed shell cracker" (Lee 1969). First called Barbour's map turtle by K. P. Schmidt (1953).

Description and Diagnosis

Postorbital markings large blotches extending backward from orbits with rounded shape; lower segments curving under orbits, subsuming supramandibular spots; blotches often broken into indistinct reticulated pattern in large females. Interorbital marking an elongate, diamond-shaped blotch separate from postorbitals. Transverse light stripe on forward part of chin, subsuming submandibular spots. Markings on head, neck, limbs, and tail yellow to light orange. Females with exceptionally broad heads and exceptionally broad alveolar surfaces, males with broad heads and broad alveolar surfaces. Second vertebral scute steeply sloped, distinctly higher than the third. Carapace highly peaked at posterior of second vertebral, nearly triangular in cross section. Carapacial keels producing prominent points in males and juveniles; reduced and blunted in females. Black pigment along midline of carapace, particularly at keel points. Carapace marked with thin, black-bordered, yellowish-orange, C-shaped (left side) rings interrupted at posterior edges of pleural scutes. Marginals marked dorsally with narrow, black-bordered, yellowish-orange stripes (ca. 9 percent of fifth scute width) that curve inward and rearward, ventrally with wide dark markings (ca. 62 percent of fifth scute width). Plastron pale yellow, with black markings along scute seams in smaller specimens. Male foreclaws not conspicuously elongated. Irises yellow, bisected by black stripes. Differs from congeners by the large postorbital blotches, the transverse chin bar, and the C-shaped pleural scute markings.

Taxonomic History and Phylogeny

Two Barbour's map turtles were collected prior to the species' type series: Brimley (1910) reported a specimen collected in 1901 in Baker County, Georgia, that he identified as *G. pulchra*, and Ross Allen collected a specimen in 1937 from the Apalachicola River at Chattahoochee, Gadsden County, Florida (FLMNH 65918; A. Carr and Marchand 1942), which he labeled as *G. pseudogeographica pseudogeographica*. The species was described by A. Carr and Marchand (1942) based on a female holotype (MCZ 46251) and male allotype (MCZ 46252) collected from the "Chipola River north of Marianna, Jackson County, Florida"

(p. 98) 27–30 November 1941. Of fifty-one paratypes collected from the Chipola in Jackson and Calhoun Counties, Florida, twenty-eight were deposited in the same collection (MCZ 46253–80) and twenty-three were to be cataloged later, including eighteen sent in lots of six live specimens to each of three major eastern zoos. The six National Zoo paratypes likely were preserved, as six specimens bearing NZP labels were cataloged between 1946 and 1969 (as USNM 127235, 128176, 129395, 129541, 134437, and 220893), although none bears any labeling indicating its status as a paratype (R. Reynolds, pers. comm. 2008). Five of the Philadelphia Zoo's and four of the Bronx Zoo's paratypes are preserved (ANSP 24737–41, AMNH 64032–35). Not known is the fate of the remaining eight paratypes, including five donated to the University of Florida Biology Department (whether live or preserved is unclear). Carr and Marchand's hypodigm included twenty-three additional specimens not designated as types. They named the species in honor of Thomas Barbour (1884–1946), who served for two decades as director of the Museum of Comparative Zoology at Harvard and worked closely with Archie Carr for several years. No subspecies have been described; Artner's (2003) checklist relegated the taxon to a subspecies of *G. pulchra*, but his arrangement is unlikely to be followed by many taxonomists due to allopatry and a lack of critical analysis of variation.

Barbour's map turtle is one of five species in the megacephalic clade of *Graptemys*. Lamb et al. (1994) placed it as sister species to the clade of *G. ernsti* and *G. pulchra*, and Lamb and Osentoski (1997) placed it as sister species to *G. ernsti*. Stephens and Wiens (2003) placed it as outgroup to a clade composed of those two species plus *G. gibbonsi sensu lato* (i.e., their analyses placed it as sister species to the four megacephalic species formerly regarded as conspecific; Lovich and McCoy 1992; Ennen, Lovich, et al. 2010); however, their more recent analysis placed it as sister species to *G. pulchra* (Wiens et al. 2010).

Range

Barbour's map turtle was long regarded as endemic to the Apalachicola drainage, including its Chipola, Chattahoochee, and Flint River tributaries, in the western panhandle of Florida, southwestern Georgia, and along the Alabama-Georgia border. However, recent records from adjacent drainages bring its endemism into question. In the Apalachicola drainage, the species ranges up the Chipola in Florida nearly to the Alabama state line (A. Carr 1952); up the Chattahoochee into Russell County, Alabama, and Muscogee County, Georgia (Mount 1975; Moulis 1997; Jensen et al. 2008; pers. observation 2009); and up the Flint and its tributary, Line Creek, in Coweta, Fayette, and Spalding Counties (Wharton et al. 1973; Jensen 2011; pers. observation 2012). Although Tinkle (1959) did not find the species above the fall line in Georgia, the northernmost Flint River localities are approximately 80 km north of this zoogeographic barrier, which intersects the river in northern Taylor County (Martof 1956). Barbour's map turtle also inhabits tributaries of the Flint, including lower Ichawaynochaway, Spring, and Buck Creeks; a tributary of the middle Apalachicola, the River Styx; and a tributary of the Chipola, Spring Creek (Crenshaw and Rabb 1949; Stewart 1992; Moulis 1997; Sterrett and Grosse 2009; Sterrett et al. 2011; UF 73779). Downstream, in Franklin and Gulf Counties, Florida, the Apalachicola forms a delta region of channels that separate and rejoin. The species occurs in the river's main channel as far south as Old Womans Bluff, just below the confluence of the Jackson River, as well as in the Brothers, lower East, and upper St. Marks Rivers and the Brickyard Cut Off (Ruhl 1991).

Cagle's (1952b) reference to a Barbour's map turtle collected from the Escambia River in 1950 is apparently in error, most likely the result of transposition of collection data with a specimen of *G. ernsti* captured on the same collecting trip (Dobie 1972). More recently, however, specimens have been reported from the Ochlockonee, Choctawhatchee, and Wacissa Rivers of Florida and Alabama (Enge et al. 1996; G. E. Wallace 2000; Godwin 2002a; D. R. Jackson 2003; Enge and Wallace 2008). Whether these specimens represent previously unrecorded portions of the species' range or introductions from pet owners remains to be determined. Abundance is low in the Ochlockonee and unknown in the Wacissa, but Barbour's map turtle was abundant and widespread in surveys conducted on the Choctawhatchee and its tributary, the Pea, in Florida and Alabama (G. E. Wallace 2000; Godwin 2002a; Enge and Wallace 2008). On the Choctawhatchee, the species ranges from the border of Walton and Washington Counties, Florida, into Dale County, Alabama, while on the Pea it occurs in Geneva and Coffee Counties, Alabama. Two hatchlings (AUM 3880–81) are labeled as having been collected on the Choctawhatchee in Florida in 1965.

Two of the three Ochlockonee specimens reported by Enge et al. (1996) were juveniles, suggesting reproduction in that population. Godwin (2002a) documented reproduction in the upper Choctawhatchee drainage, and the lone specimen from the Wacissa River was a nesting female with fertile eggs (D. R. Jackson 2003).

Habitat

Qualitative descriptions of Barbour's map turtle habitat in the Chipola River agree that highest densities occur in deep, swift sections of the river with limestone substrate and abundant deadwood; turtles frequently rest in deepwater depressions in the limestone (A. Carr and Marchand 1942; Cagle 1952b; A. Carr 1952; Sanderson 1974). Lee et al. (1975) discussed prey recovered from two males captured while foraging on submerged limestone outcrops in the Chipola. Submerged limestone was the preferred habitat of aquatic caterpillars, caddis fly larvae, and a species of snail that were eaten, while a common congeneric snail species from muddier substrates was absent from the diet. Females in two Georgia creeks showed strong fidelity to deep pools with large logs and limestone shelves (Sterrett 2009). Capture success was highest in reaches with dense riparian forest cover, suggesting the importance of a source of deadwood renewal.

In the Choctawhatchee, Barbour's map turtles are considerably more abundant in upstream reaches underlain by limestone than in downstream, sandy reaches (Enge and Wallace 2008). However, appreciable populations do occur in the sandy reaches of the Apalachicola, extending into an extremely flat floodplain delta dominated by cypress-gum and mixed hardwood forests and under tidal influence (Ruhl 1991; Ewert and Jackson 1994).

Impoundments of the Apalachicola drainage that are inhabited include the former Dead Lakes on the lower Chipola River (Sanderson 1974; the dam forming this reservoir has since been removed), Lake Blackshear on the Flint (Lovich et al. 1996; fig. 11.3), Lake Eufaula on the Chattahoochee (Moulis 1997; Buhlmann and Gibbons 2006), and Seminole Lake at the confluence of the Flint and Chattahoochee Rivers and Spring Creek (Ewert et al. 2006). Godwin (2002a) saw none in an impoundment on the Pea.

Diet

Cagle (1952b) reported the feces of thirty-eight adult specimens of both sexes from the Chipola to be approximately 95 percent snail shells, with mussel shells constituting the bulk of the rest (fig. 8.4). Two females from the Flint contained

Figure 8.4. Four quart jars of snail and mussel shells from the feces of thirty-eight Barbour's map turtles collected by Cagle and his field crew and preserved in the Tulane University Museum of Natural History.

mussels up to 30 mm and snails over 10 mm in greatest shell lengths (A. Carr 1952). In the Chipola, the most important prey of unsexed juveniles, adult males, and small juvenile females was caddis fly larvae, while larger females fed almost exclusively on snails, with a much lower incidence of Asian clams (table 4.17; Sanderson 1974). Small females consumed more snails than did males, and crayfish parts were recorded in the feces of two adult females not used in the dietary study. Three juveniles from a reservoir downstream of the main study site had taken "considerable" (Sanderson 1974, p. 79) amounts of algae, in contrast to samples from the river, where algae was not as abundant. Lee et al. (1975) reported high incidence of trichopteran larvae as well as aquatic pyralid caterpillars, snails, and vegetation from two males caught foraging on underwater limestone in the Chipola. The females' strong preference for snails over Asian clams also typifies populations of the sandy lower reaches of the Apalachicola (Ewert et al. 2006) and two tributary creeks in Georgia (Sterrett 2009).

Reproduction and Life History

Growth rates from back-calculated plastron lengths based on scute proportions (Cagle 1952b) and recapture data (Sanderson 1974) show decelerating growth with age in both sexes. Growth rates peak from March to November, when the water is warmest (Sanderson 1974). The smallest mature males were in their third and fourth growing seasons (Cagle 1952b), while age at maturity in females has been loosely estimated at fifteen to twenty years (Sanderson 1974). On the lower Apalachicola, clutches average 8.8 eggs (range 4–14; Ewert et al. 2006), and clutches of 8–10 eggs have been reported on the Flint (Wahlquist and Folkerts 1973; Moulis 1997). D. R. Jackson (2003) recorded 12 eggs averaging 13.0 g in mass from the female captured in the Wacissa River. Hatchlings emerge in the fall and range from 35 to 38 mm in carapace length (Wahlquist and Folkerts 1973). Range-wide, midline plastron length ranges from 72 to 100 mm in adult males and reaches 228 mm in adult females (Lindeman 2008a).

Natural History

During flooding caused by a hurricane, Barbour's map turtles moved into slower currents of flooded river margins and were not displaced from their home ranges (Sanderson 1974). Much of the nesting recorded in the lower Apalachicola was on dredge spoil mounds, which are undergoing succession that will eventually make them unsuitable for nesting (Ewert and Jackson 1994; Ewert et al. 2006). Hybridization with the Escambia map turtle, another species only recently found in the Choctawhatchee and Pea Rivers, is indicated by intermediate markings of specimens and supported by genetic analyses (Godwin 2002a; Godwin et al. 2013). An unidentified coccidian parasite was reported by Wacha and Christiansen (1976). Scarlet snakes depredate nests, breaking open eggs and inserting their heads to gulp down the contents (Neill 1951).

Populations

A. Carr and Marchand (1942; see also A. Carr 1952) collected 76 specimens from the Chipola River over five sampling days in 1941 by swimming underwater and using a gigging pole. Marchand's field notes indicated Barbour's map turtles were two to three times more numerous than river cooters (*Pseudemys concinna*). Chaney and Smith (1950) captured an astounding 397 specimens in three nights by hand on the Chipola in 1950—a *collecting* density of approximately 62 turtles/river km for the reach they worked! On the Flint River in the mid-1950s, Barbour's map turtle made up 27 percent of the catch and was the second most abundant species, after *P. concinna* (Tinkle 1958b). On two tributary creeks in Georgia, Barbour's map turtle was the third most abundant of nine turtle species in captures made in baited traps and by snorkeling, following slider turtles (*Trachemys scripta*) and river cooters at 15 percent of the total catch (Sterrett et al. 2011).

Sanderson (1974) estimated population density at 111 turtles/km in the Chipola based on total captures and the assumption, based on declining recapture rates, that he captured 65 percent of the population. Moler (1986) conducted basking surveys on a 58.7 km stretch of the Chipola in 1985–86. Barbour's map turtle accounted for 44 percent of all identified turtles, and he saw 5.7 Barbour's map turtles/km (adjusting for unidentified turtles). Ruhl (1991) averaged seeing just 0.8 Barbour's map turtles/river km on the lower Apalachicola and its side channels, although water levels were very high and most good basking sites were submerged; spring observations during lower water at the same time of year yielded 4.5 turtles/km (Ewert et al. 2006). Moulis (1997) visited 137 sites on the Chattachoochee and Flint drainages in Georgia, finding basking Barbour's map turtles at only 10 of the 86 sites on the latter; however, lack of replication in his counts and repeated reference in his report to high water conditions that submerged most basking sites leaves open the question of how abundant the species may be in Georgia.

Enge and Wallace (2008) surveyed basking turtles on the Choctawhatchee River in 1999–2000. They did not differentiate between Barbour's and Escambia map turtles, as the latter were not yet known to occur in the river. Map turtles accounted for 35 percent of all turtles identified. They recorded 98 percent of their total sample of map turtles on an upstream stretch of the river underlain by limestone. Basking abundance in the limestone-lined reaches was 3.7 map turtles/km, and map turtles were 42 percent of all turtles seen; along sandy stretches downstream, however, basking abundance fell to just 0.1 map turtles/km and they were just 3 percent of all turtles. Godwin (2002a) recorded 0.83 basking Barbour's map turtles/km on the Choctawhatchee and Pea Rivers, with the species being second in abundance to *P. concinna* among six species, at 24 percent of all turtles. Populations in the Ochlockonee River are widely dispersed but very sparse, with only about 20 known captures or observations (Enge and Wallace 2008; D. Jackson, pers. comm. 2009).

Conservation Status and Actions

A moderately strong candidate for listing under the U.S. Endangered Species Act from the late 1970s through the early 1990s before being dropped from further consideration following basking surveys on the Apalachicola and Chipola Rivers; a new federal review was recently announced. State-listed as a protected nongame species in Alabama, a species of special concern in Florida, and threatened in Georgia. Listed as vulnerable with decreasing populations by the IUCN.

The only federal protection of the species' habitat occurs along the lower Apalachicola, a segment of which abuts the western border of Apalachicola National Forest, and in the upper end of Lake Eufaula National Wildlife Refuge on the Chattahoochee River (Buhlmann and Gibbons 2006). The lower Apalachicola and its floodplain are more extensively protected within Florida's Apalachicola River Water Management Area (31 river km; 14,700 total ha) and Apalachicola River Wildlife and Environmental Area (34,900 total ha, encompassing delta streams). Likewise, the state protects 29 river km in the Chipola River Water Management Area (2,987 total ha) and 48 river km in the Choctawhatchee River Water Management Area (23,000 total ha). Protection of shorter reaches of riparian habitat occurs in Florida Caverns State Park on the Chipola, Torreya State Park on the upper Apalachicola, Georgia Veterans State Park on Lake Blackshear, and Sprewell Bluffs State Park on the upper Flint. Ichauway is an 11,700 ha private reserve that protects 24 km of Ichawaynochaway Creek and 21 km of the west bank of the Flint (see chapter 6).

Needed Further Research

Most of the ecological information available on this species comes from the Chipola; hence there is a great need for comparative work on life history and diet in the more extensive Apalachicola, Chattahoochee, and Flint Rivers. Nesting and nest survival have not been the focus of study, having been recorded only in anecdotes. A good quantification of population status in various river drainages is also vital to proper recognition of the status of Barbour's map turtle as an imperiled species. It would also be useful to know the degree of threat still presented by illegal commerce in the pet trade or continued take of these turtles for use as food. Finally, the status of populations is unresolved for the Choctawhatchee, Ochlockonee, and Wacissa drainages as to whether these represent introductions or natural occurrences and, if the latter, whether they should be regarded as conspecific or perhaps as new species.

Alabama Map Turtle, *Graptemys pulchra*

Figure 8.5. Juvenile female, Sipsey River at County Road 12, east of Newtonville, Fayette County, Alabama.

The species comprises at least four geographical variants: one in the Pearl River drainage, a second in the Pascagoula drainage, a third in the Mobile Bay drainage, and a fourth in the Escambia and Yellow River drainages. Shell proportions separate members of the Mobile Bay drainage from all others, the adults having proportionately much flatter shells than adults from other drainages.

ROBERT SHEALY (1976, P. 59)

Alternate Common Names

Baur's terrapin (Brimley 1910), Alabama sawback turtle (A. Carr 1952), neither widely adopted. First called Alabama map turtle by K. P. Schmidt (1953).

Description and Diagnosis

Postorbital markings large blotches extending backward from orbits, with upper rear edges that are rounded or often pointed; lower segments curving under orbits to subsume supramandibular spots. Broad interorbital marking connecting by thick lines with postorbitals; hence markings mask-like. Chin with prominent midventral longitudinal stripe, to each side of which appears one small anterior spot and one elongated submandibular blotch, sometimes with transverse stripes that connect midventral stripe to submandibulars. Markings on head, neck, limbs, and tail yellow to light orange. Females with exceptionally broad heads and exceptionally broad alveolar surfaces, males with broad heads and broad alveolar surfaces. Second vertebral scute only slightly sloped, extending little higher than the third. Carapace not highly peaked, rounded to slightly triangular in cross section. Carapacial keels producing prominent points in males and juveniles; reduced and blunted in females. Black pigment along midline of carapace, particularly at keel points. Carapace marked with reticulating network of thin, black-bordered, yellowish-orange lines on pleural scutes having comparatively few intersections (one to two rings per pleural), most prominent anteriorly. Marginals marked dorsally with narrow, black-bordered, yellowish-orange lines (ca. 10 percent of fifth scute width) that curve inward and rearward and concentric yellowish-orange markings, ventrally with wide dark markings (ca. 72 percent of fifth scute width) and concentric semicircles. Plastron pale yellow, with black markings along scute seams in smaller specimens. Male foreclaws not conspicuously elongated. Irises yellow, bisected by black stripes. Differs from congeners by the mask-like head markings, the narrow light markings of upper marginals, and the comparatively low carapace.

Taxonomic History and Phylogeny

Two specimens (USNM 8808 and USNM 318254) were collected in 1876 by Tarleton H. Bean and A. L. Kumlien, with the U.S. National Museum (USNM) specimen catalog entry indicating that they were from a "lake near Montgomery, Ala[bama]," which D. M. Cochran (1961) placed in Montgomery County. Bean and Kumlien placed five fish specimens collected 14–18 July 1876 from "Round Lake" in Montgomery County in the USNM collections, so presumably the two turtle specimens were collected from this same site at this time. Unfortunately, while fish specimens from Round Lake (as well as the Alabama River) near Montgomery are also referenced in D. S. Jordan and Brayton (1878), no Round Lake occurs on present-day maps of Montgomery County; presumably the name was once used for a floodplain lake of the Alabama River. Originally the specimens were regarded as *Malacoclemmys geographicus* (i.e., common map turtles; Yarrow 1883; Lovich 1985), but Baur (1893) described the two specimens as a new species. The specific epithet is derived from the Latin term *pulcher*, meaning "beautiful."

There has been confusion regarding Baur's (1893) type series. Both of the Montgomery specimens are subadult females of similar size and were originally cataloged together as USNM 8808. The specimen retaining that catalog number today (midline plastron length 154 mm) is intact and in alcohol and was designated the lectotype by Lovich and McCoy (1992). The second (paralectotype) is a headless alcohol-preserved specimen (midline plastron length 156 mm) with a separate dry skull. The paralectotype was first recataloged as USNM OSTEO-029526 (Cagle 1952b) and then again as USNM 318254 (Lovich and McCoy 1992). Letters in the Smithsonian Archives exchanged between Baur and Leonhard Stejneger in the early 1890s indicate that skull preparation of the latter specimens had been done prior to Baur borrowing the specimens, so the association of the skull with the headless specimen is probably correct. Whereas Cagle (1952b, pp. 224–25) reported that *two* skulls were at one time cataloged together as USNM OSTEO-029526—one with features consistent with *G. pulchra*, the other being more gracile and thus believed to be more consistent with the features of *G. oculifera*—subsequent authors have referenced a single skull. Iverson (1992) gave the skull's new number as USNM 252600 and cited a personal communication from George Zug of the U.S. National Museum suggesting that the skull should not be considered part of the type series of *G. pulchra*; that skull is in fact now associated with USNM 15510, one of Baur's (1890) syntypes of *G. oculifera* (Reynolds et al. 2007). (Association of that skull with 15510 is almost certainly correct, as the remainder of 15510 consists of a dry carapace and plastron and all remaining bones of the skeleton; also, a letter in the Smithsonian Archives shows that Stejneger shipped the two specimens labeled 8808 and the skull of 15510 to Baur on 4 December 1890, hence the opportunity for later confusion noted by Cagle.) Lovich and McCoy (1992) examined a skull labeled OSTEO-029526, found its characteristics consistent with those of *G. pulchra* rather than *G. oculifera*, and concluded that it probably is the skull from USNM 318254, with which it is now associated; recataloging of the *G. oculifera* skull had been done in 1985 (Reynolds et al. 2007), possibly prior to examination of the type specimens by Lovich and McCoy (1992). Hence the first part of the type series consists of the two fluid-preserved subadult females, one intact and the other with a separate dried skull, from Montgomery County, Alabama (fig. 8.6).

Baur (1893) also stated that he examined a living specimen and a skull he received from Gustave Kohn. Kohn's letter to Baur dated 3 November 1891 (Smithsonian Archives) indicates he had purchased the specimens in a New Orleans market in May 1891. The whereabouts of these specimens have not been noted in recent taxonomic works (e.g., Lovich and McCoy 1992; Iverson 1992; Reynolds et al. 2007), yet both are extant: the live specimen is now USNM 220884, which consists of a complete plastron, disarticulated carapace and skeleton, and skull of a large adult female, while the skull is FMNH 22171, also from a large adult female (fig. 8.6). USNM 220884 was cataloged on 22 June 1891. It had apparently been shipped alive to Baur the month before, based on a letter to Baur from Kohn dated 14 October 1891 inquiring about the specimen he had sent and Baur's list of turtles he received from Kohn between 1890 and 1892. Kohn sent the skull to Baur on 3 November 1891, and it was later cataloged as FMNH 22171 with a collecting date of 1891. Hence the two specimens should be treated as paralectotypes; indeed, Marx (1958) listed FMNH 22171 as a "cotype" (p. 449) and the Field Museum treats it as type material. It should be further noted that both of the skeletal specimens likely represent *G. pearlensis*,

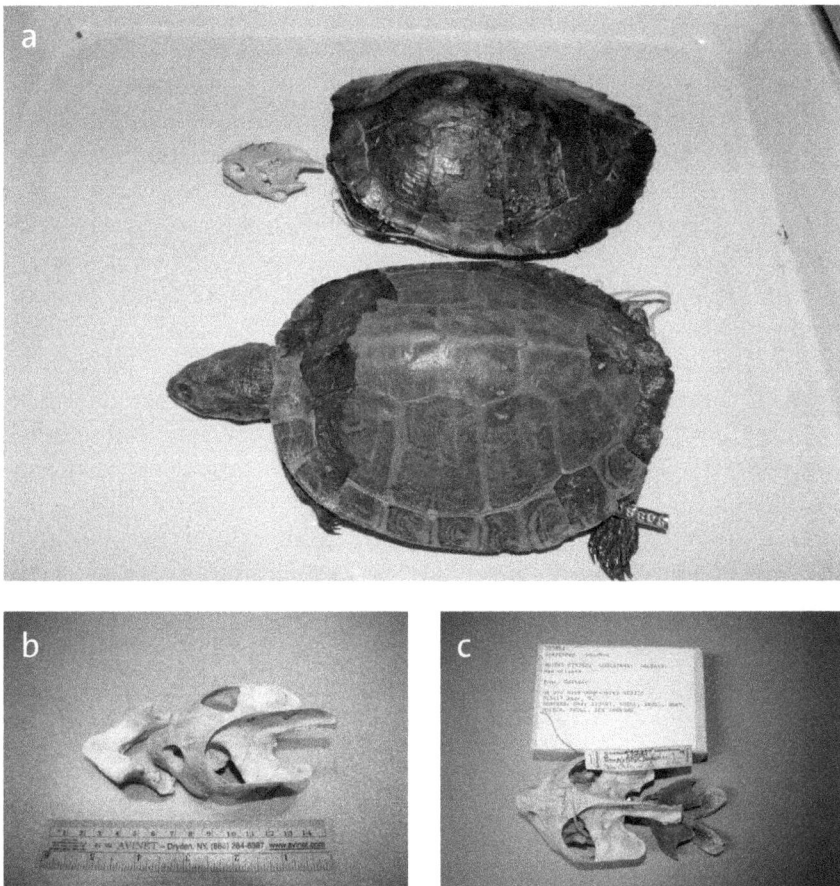

Figure 8.6. Type series of the Alabama map turtle, *Graptemys pulchra* Baur 1893. (a) USNM 8808 (lectotype, foreground) and USNM 318254, from Montgomery County, Alabama. (b) FMNH 22171, originally from the Gustave Kohn collection. (c) Skull of USNM 220884, also originally from the Kohn collection. FMNH 22171 and USNM 220884 are likely *G. pearlensis* (see text).

given their procurement by Kohn in New Orleans (see species account for *G. pearlensis* and chapter 2).

   Following Baur's (1893) description, no further information on the species was available until 1950, when Cagle (1952b) and his students trapped in the Pearl and Escambia Rivers, collecting specimens they regarded as conspecific with Baur's two specimens but which are now treated as separate species (Lovich and McCoy 1992; Ennen, Lovich, et al. 2010). Additional specimens of Alabama map turtles (*sensu stricto*) from the Mobile Bay drainage were first collected in 1952 (Tinkle 1958b, 1959; unpublished notes of F. Cagle). Over the next four decades, several authors commented on interdrainage distinctions of specimens from the Mobile Bay, Escambia Bay, Pearl, and Pascagoula drainages. Cagle

(1952b) commented on differences in lower jaw markings between specimens from the Escambia and Pearl Rivers; his later field notes contain references to morphological differences between the Pascagoula and Mobile Bay drainages as well. Tinkle (1962) used carapacial seam contacts to compare specimens from the Escambia, Pearl, and Pascagoula Rivers. He found interdrainage differences and concluded that specimens from the first two rivers were most similar. Little (1973) claimed that specimens from the four drainages had consistent differences in plastral scute contact points, although his conference abstract does not give details. Mount (1975) commented on interdrainage differences in markings on the head and shell. Shealy (1976) noted variation in coloration among the four drainages as well as the flatter shell of specimens from the Mobile Bay drainage. Lovich and McCoy (1992) analyzed differences in markings of the head and marginal scutes and the relative sizes of plastral scutes and described two new species based on specimens from the Escambia Bay drainages (*G. ernsti*) and the Pearl and Pascagoula drainages (*G. gibbonsi*), restricting *pulchra* to specimens from the Mobile Bay drainages. Phylogenetic analyses placed the Alabama map turtle within the megacephalic clade as sister species to *G. ernsti* (Lamb et al. 1994), to a clade composed of the other four species (Lamb and Osentoski 1997), to a clade composed of *G. ernsti* and *G. gibbonsi sensu lato* (Stephens and Wiens 2003), and to *G. barbouri* (Wiens et al. 2010).

Range

The Alabama map turtle is restricted to the drainages that converge at the Mobile-Tensaw Delta. Its range encompasses most of southern and central Alabama and portions of northeastern Mississippi and northwestern Georgia. It occurs downstream into the Mobile-Tensaw Delta in Baldwin County, Alabama (Mount 1975; Godwin 2001, 2003). Its range extends upstream in the Tombigbee drainage and its tributaries, the Noxubee and Tibbee Creek in Mississippi and the Sipsey in Alabama (McCoy and Vogt 1979; Jones et al. 1996; Thomas and Bradford 1997). It ranges the length of the Black Warrior in Alabama and upstream into Locust Fork (USNM 247947), Mulberry Fork (Mount 1975), Sipsey Fork (Dodd 1988a), and Brushy Creek (FLMNH 63021). It ranges up the Cahaba into suburban southeastern Birmingham and up the Coosa over the border into Georgia and up a tributary, the Oostanaula, and its northern tributary, the Conasauga, to within 7.5 km of the Tennessee border (Mount 1975; J. L. Harris et al. 1982; Santhuff and Wilson 1990; Lovich and McCoy 1992; Jensen et al. 2008; Lindeman 2008b; G. Brown et al. 2011). It also occurs in the lower Tallapoosa River and several of its tributaries, including Line Creek, Cubahatchee Creek, and Uphapee Creek and its two tributaries, Opintlocco and Chewacla Creeks (Mount 1975; Shealy 1976; Lovich and McCoy 1992; AUM 12592). Additional smaller tributaries in Alabama that house the species in their lower reaches include Big Flat, Bogue Chitto, and Pintlala Creeks and the Little River (AUM 5607, ANSP 31275, AUM 11561, AUM 11095). In the Black

Warrior, Cahaba, and Coosa Rivers, the Alabama map turtle ranges upstream of the fall line (Tinkle 1959), but it apparently does not occur above the fall line in the Tallapoosa (Mount 1975).

Habitat

Alabama map turtles occur in a remarkable diversity of stream habitats, from large, deep, wide rivers in southern portions of the range to considerably smaller streams to the north that become shallow and narrow. Streams below the fall line are dominated by sand and gravel substrates, while the generally smaller streams above the fall line have rockier substrates and cut through large rock formations. In small, rocky streams males and juveniles may occur in very shallow water, while females are more confined to deep pools (C. H. Ernst and Lovich 2009). Deadwood is likely an important habitat feature.

Alabama map turtles are somewhat abundant in impoundments of the Coosa (Shealy 1976; Godwin 2001, 2003) and also occur in relatively lentic pools behind dams on the Black Warrior, the Tombigbee, and the Alabama. Besides the type specimens from "Round Lake" (see above), there is just one other record from a floodplain habitat: four specimens from a pond near Bogue Chitto Creek (ANSP 31278–81). A more focused search of such habitats may produce more records in naturally lentic floodplain habitats.

Diet

The only record of diet for the Alabama map turtle is an anecdotal account by Shealy (1976, p. 95), who reported shells of native mussels measuring 30–50 mm in the feces of females, with one passing fish vertebrae. There are no reports of the diet of males or juveniles.

Reproduction and Life History

Clutch size ranges from four to seven eggs (P. Lindeman, unpublished data). Range-wide, midline plastron length ranges from 74 to 100 mm in adult males and reaches 222 mm in adult females (Lindeman 2008a).

Natural History

Nest predation rates were approximately 50–75 percent in two small streams of the Tallapoosa drainage, and incidence of infertile eggs or undeveloped embryos was high (Shealy 1976). Turtle leeches and a parasitic cnidarian have been reported from Alabama map turtles (C. H. Ernst and Barbour 1972; Dodd 1988b).

Populations

Alabama map turtles were the predominant turtle captured on the Tombigbee River (50 percent of all turtles) but had much lower prevalence in the Black Warrior (6 percent) and Coosa Rivers (4 percent; Tinkle 1958b). Abundance relative to other emydids and kinosternids in each of the Tombigbee, Black Warrior, and Alabama/Coosa drainages in 1953–56 was approximately equal below and above the fall line (all values within the range 4-9 percent of total catch), although capture rates using traps and nighttime searches were substantially greater below the fall line (an Alabama map turtle every 15, 11, or 20 hours, respectively, vs. one every 104, 89, or 32 hours above the fall line in these drainages; Tinkle 1959).

Godwin (2001, 2003) recorded Alabama map turtle basking densities and relative abundance in the seven major inhabited drainages in Alabama. Highest densities were observed in the Coosa River (up to 9.5/km, weighted mean 4.8/km), followed by the Alabama, Tallapoosa, and Cahaba Rivers (highest densities 2.5–7.7/km, weighted mean densities 1.8–2.7/km), with low densities in the Tombigbee, Tensaw, and Black Warrior Rivers (all surveys <1/km). The Alabama map turtle was the predominant basking turtle in the Coosa River (49 percent of all turtles seen) but a distant second to the black-knobbed sawback (G. nigrinoda) in the other drainages (range 3–24 percent).

Conservation Status and Actions

Currently under review for federal listing. State-listed as a protected nongame species in Alabama and as rare in Georgia. Listed as near threatened with unknown population trends by the IUCN.

Choctaw National Wildlife Refuge protects riparian habitat, including numerous backwaters and oxbows, on a portion of the west side of the lower Tombigbee. A stretch of the upper Noxubee is protected in Noxubee National Wildlife Refuge, a section of the middle Cahaba is in or on the border of Talladega National Forest, the upper tributaries of the Black Warrior are in William B. Bankhead National Forest, a short section of Uphapee Creek is in Tuskegee National Forest, and a short section of the upper Conasauga is bordered by the Chattanooga National Forest. The Cahaba River National Wildlife Refuge protects 5.6 km of the middle part of the river. Alabama protects much of the Mobile-Tensaw Delta (37,850 ha) in two large wildlife management areas and 1,214 ha of bottomland swamp forest along the east bank of the middle Sipsey River in a preserve and recreation area. In addition, Roland Cooper State Park is on the southern bank of the lower Alabama, and Fort Toulouse-Jackson State Historical Park is located at the confluence of the Coosa and the Tallapoosa. The Nature Conservancy owns two small preserves along the lower Cahaba, one of which includes one of the river's largest nesting beaches, Barton Beach.

Needed Further Research

With recent publication of studies on the life history and ecology of the Texas map turtle (G. versa; see species account), which had held the distinction of having the shortest species account of any North American turtle species in the compilations of C. H. Ernst et al. (1994; see fig. 2.21), the Alabama map turtle assumes the status of the least-known species of the genus Graptemys. Studies of any and every aspect of its biology are much needed. In particular, conservation status should be assessed, as the relative lack of attention paid by listing entities may be more a case of benign neglect than of confidence in the health of populations. Variation in life history and diet related to the diversity of streams inhabited by the species should also be an interesting topic of research.

Escambia Map Turtle, *Graptemys ernsti*

Figure 8.7. Adult female, Conecuh River at old County Road 4 bridge, Escambia County, Alabama.

> The greatest density of turtles observed . . . was associated with an extreme abundance of molluscs, including several species of pelecypods and at least one species of gastropod. . . .
>   Few G. [*ernsti*] were found in those streams supporting scant mollusc populations, and none at all were found in streams in which molluscs were entirely absent. Conversely, no stream within the range contains large mollusc populations and no G. [*ernsti*].
> ROBERT SHEALY (1976, P. 110)

Alternate Common Names

Ernst's map turtle (Buhlmann et al. 2008). First called Escambia map turtle by Lovich and McCoy (1994a).

Description and Diagnosis

Postorbital markings large blotches extending backward from orbits, with upper rear edges that are rounded, lower segments sometimes curving under orbits to subsume supramandibular spots, or submandibulars separate. Interorbital marking trident-like with anterior points, lacking connection to postorbitals. Two dorsal neck stripes ending anteriorly on back of head in expanded, oval tips, these sometimes separate as supraoccipital spots. Chin often with prominent midventral longitudinal stripe, to each side of which appears one small anterior spot and one elongated submandibular blotch, sometimes with transverse stripes that connect midventral stripe to submandibulars; occasionally transverse bars or spots replace midventral stripe. Markings on head, neck, limbs, and tail light yellow or, rarely, yellow orange. Females with exceptionally broad heads and exceptionally broad alveolar surfaces, males with broad heads and broad alveolar surfaces. Second vertebral scute steeply sloped, distinctly higher than the third. Carapace highly peaked at posterior of second vertebral, nearly triangular in cross section. Carapacial keel producing prominent points in males and juveniles, reduced and blunted in females. Black pigment along midline of carapace, particularly at keel points. Carapace marked with reticulating network of thin, black-bordered, yellowish-orange lines on pleural scutes having comparatively few intersections (one to two rings per pleural), most prominent anteriorly. Marginals marked dorsally with wide, black-bordered, yellowish-orange stripes (ca. 17 percent of fifth scute width) that curve inward and rearward, ventrally with wide dark markings (ca. 59 percent of fifth scute width). Plastron pale yellow, with black markings along scute seams in smaller specimens. Male foreclaws not conspicuously elongated. Irises yellow, bisected by black stripes. Differs from congeners by the interorbital trident, the lack of connection of the interorbital to the postorbitals, and the supraoccipital spots.

Taxonomic History and Phylogeny

The first specimens recorded for this species were a dozen animals collected in 1950 by Cagle (1952b) and his students and regarded as *G. pulchra*, a species complex that until then had a poorly known range. Over four decades, several authors commented on the distinctiveness of specimens from the Escambia Bay drainages relative to those from the Pearl, Pascagoula, and Mobile Bay drainages (see species account for *G. pulchra*). Lovich and McCoy (1992) examined morphometrics and markings of specimens from all four drainages and described the Escambia drainage populations as *Graptemys ernsti*, honoring Lovich's M.S. thesis advisor and noted expert on North American freshwater turtles, Carl H. Ernst. The type series includes a holotype (CM 122408) collected from the "Conecuh River, 1 mile upstream from County Road 4 Bridge, 14 km east of East Brewton, Escambia County, Alabama, USA" (Lovich and McCoy 1992, p. 300) on 30 September 1988 by Jeffrey E. Lovich, Anthony M. Mills, and Joshua Schachter, as well as ten paratypes from the type locality (CM 122403–407, 122409–11, USNM 300604–605). No subspecies have been described; Artner's (2003) checklist relegated the taxon to a subspecies of *G. pulchra*, but his arrangement is unlikely to be followed by many taxonomists due to allopatry and a lack of critical analysis of variation. Phylogenetic analyses show the Escambia map turtle to be nested within the clade of five megacephalic species but differ as to whether it is sister species to *G. pulchra* (Lamb et al. 1994), *G. barbouri* (Lamb and Osentoski 1997), or *G. gibbonsi sensu lato* (Stephens and Wiens 2003).

Range

The Escambia map turtle occurs in the drainages of Escambia Bay in western Florida and southeastern Alabama. It occurs in the Escambia and its major tributary, the Conecuh, ranging well upstream of Gantt Reservoir into Pike County, Alabama (Godwin 2000; pers. observations 2012). Smaller tributaries inhabited include the Sepulga River and its tributary, Persimmon Creek, as well as Patsaliga Creek and Murder Creek (Lovich and McCoy 1992; Godwin 2000; Aresco and Shealy 2006). The downstream range limit in the Escambia is several kilometers south of Highway 184 between Escambia and Santa Rosa Counties, Florida (Aresco and Shealy 2006). The species also occurs in the Yellow to the east, between Covington County, Alabama, and Santa Rosa County, Florida, and has been sighted in the lower Shoal, a tributary of the Yellow (Iverson and Etchberger 1989; Lovich and McCoy 1992; Godwin 2000; Aresco and Shealy 2006; J. Lovich, pers. comm. 2008).

Recently Escambia map turtles were recorded in the Pea and in the Choctawhatchee downstream of its confluence with the Pea, in Coffee and Geneva Counties, Alabama, including a short section of the Pea that winds over the Florida state line and back again (Godwin 2002a; Lechowitz and Archer 2007). It is not clear whether this drainage is a natural part of the species' range or whether the species has been introduced via the pet trade. Apparently the species does not occur downstream in Florida reaches of the Choctawhatchee (Ewert et al. 2006).

Habitat

Shealy (1976) and Godwin (2000) described two types of reaches inhabited by Escambia map turtles (fig. 8.8). The first is characterized by relatively straight river channels with steep banks of exposed limestone, bedrock or gravel substrate, and few, mostly small, sandbars. Slow bank erosion contributes little deadwood usable for basking. The second is characterized by a meandering river channel, lower sandy banks and larger sandbars, and more rapid bank erosion with more deadwood. Two impoundments, the Point A and Gantt Reservoirs on the Conecuh, are inhabited (Shealy 1976; Godwin 2000), as is a small impoundment of the Pea (Godwin 2002a).

Diet

Cagle (1952b) reported mussels and snails from two adult females from the Escambia. In the Conecuh, females fed almost exclusively on exotic Asian clams

Figure 8.8. Contrasting habitats of the Escambia map turtle. (a) The Sepulga River (Highway 42 near Brooklyn, Conecuh County, Alabama), with prominent limestone formations and steep banks, typical of upstream reaches. (b) The Escambia River (County Road 4 west of Jay, Escambia and Santa Rosa Counties, Florida), with large sandbars and a wider, meandering channel, more bank erosion, and greater deadwood density, typical of downstream reaches.

(*Corbicula* spp.) and took small amounts of native mussels, snails, and crayfish (Shealy 1976). Males and small juvenile females fed on caddis fly larvae, terrestrial beetles, dragonfly nymphs and adults, snails, small mussels and clams, hymenopterans, and millipedes. Lovich et al. (2011) also reported native mussel shells passed by a female from the Pea River.

Reproduction and Life History

Male Escambia map turtles exhibit the typical von Bertalanffy pattern of decelerating growth rate with increasing age and body size (Shealy 1976; Lindeman 1999b). Females have a different pattern of growth, however, best described by a Gompertz curve, with slight acceleration occurring in large juvenile females and deceleration in adults. Maturity is reached in the third season in males but not until around age thirteen to eighteen in females (Shealy 1976; Lindeman 1999b).

Average clutch size is 7.2 eggs (range 3–12) and increases with female body size (Shealy 1976). Based on dissections, it appears most females lay four clutches (Shealy 1976). One exceptional female (the largest by 15 mm that Shealy examined) had thirty-three corpora lutea, twelve oviductal eggs, and twenty-six enlarged follicles on 4 July, indicating six or seven potential clutches. Range-wide, midline plastron length ranges from 68 to 100 mm in adult males and reaches 231 mm in adult females (Lindeman 2008a).

Natural History

Courtship has been observed between September and November, with court-ing males engaging in head bobbing but not foreclaw titillation (Shealy 1976). Hybridization with Barbour's map turtle is indicated by intermediate markings possessed in specimens from the Pea and Choctawhatchee Rivers and is sup-ported by genetic analyses (Godwin 2002a; Godwin et al. 2013). Shealy (1976) gave anecdotal accounts of parasites and predators. Turtle leeches were abun-dant in the spring, while endoparasitic spiny-headed worms and trematode flukes were found only at low frequencies. The major nest predator was the fish crow. Several adults had smooth, wedge-shaped wounds to the legs, tail, or cara-pace margin, possible results of encounters with alligator snapping turtles, and one nesting female was found that had been eviscerated by a raccoon.

Populations

The Escambia map turtle constituted 61 percent of the catch of five turtle spe-cies on the Escambia River in the mid-1950s (Tinkle 1958b). Shealy (1976) con-sidered it to be the most abundant turtle species in the lower Conecuh River. He estimated densities of 260–300 individuals/river km based on the assump-tion that he saw 25 percent of the population during basking surveys and 50 percent during dives, but he also suggested these were underestimates of total population densities. In basking surveys on the Conecuh, Escambia map turtles greatly outnumbered all other turtle species combined, but relative abundance was much lower in smaller tributaries and in the Yellow River (Godwin 2000). Godwin (2000) tallied up to 63 Escambia map turtles/km on the Conecuh, while basking densities on the Yellow were no higher than 7 turtles/km. River segments with narrow channels, steep rocky banks, low deadwood densities, and bedrock or gravel substrates averaged 26 turtles/km, while segments with broad channels, frequent sandbars, more deadwood, and sandy substrates aver-aged 14 turtles/km. Godwin also saw low numbers (<9 turtles/km) in the Point A Reservoir, Patsaliga Creek, and the Sepulga River. In basking surveys con-ducted in the Choctawhatchee drainage, Escambia map turtles were the fifth most abundant of six species observed and were outnumbered by Barbour's map turtles by a ratio of 8.5:1, although 16 percent of all map turtles could not be confidently identified (Godwin 2002a). In Florida, Aresco and Shealy (2006) recorded 5 basking Escambia map turtles/km in a late-summer survey on the lower Escambia.

Conservation Status and Actions

Currently under review for federal listing. State-listed as a protected nongame species in Alabama; although not state-listed in Florida, protected from take due to its similarity to Barbour's map turtle. Listed as near threatened with decreasing populations by the IUCN.

Protection of riparian habitat for this species occurs in short segments of the Conecuh and the Yellow that serve as the borders of the Conecuh National Forest in Alabama. The riparian zone on the east side of the lower Yellow is also largely spared from development by Eglin Air Force Base. Florida protects 48 river km (14,200 ha) in the Escambia River Water Management Area and 31 river km of floodplain habitat (7,170 ha) in the Yellow River Water Management Area.

Needed Further Research

Further studies of the species in Florida are necessary to determine its conservation status there; densities observed by Aresco and Shealy (2006) are quite low relative to Alabama survey data, although low densities may have resulted from the late-summer timing. A comparative study of life history and diet in limestone-rich and sandy habitats would yield interesting insights into the biology of the species. Status of the population in the Choctawhatchee River drainage requires further study (see Barbour's map turtle account).

Pascagoula Map Turtle, *Graptemys gibbonsi*

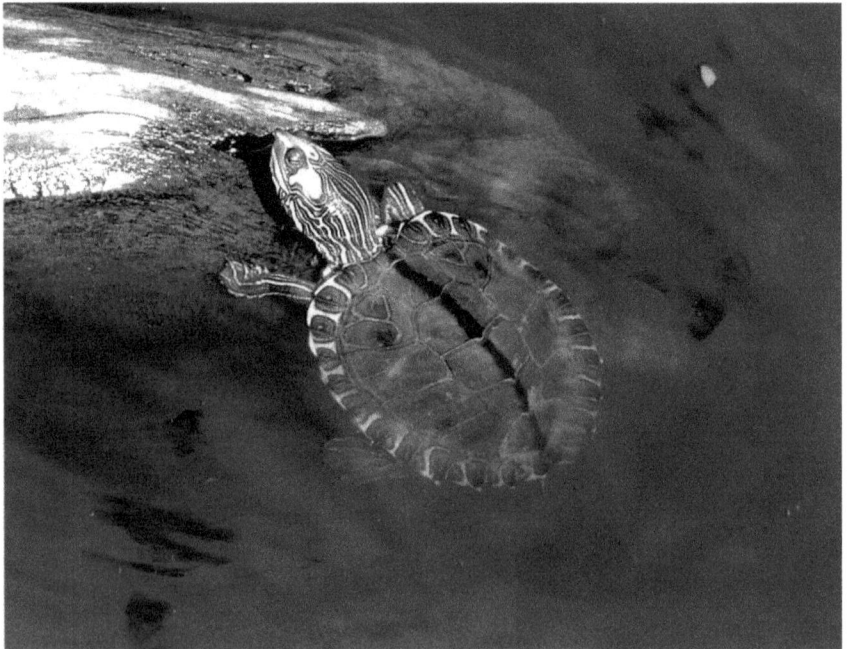

Figure 8.9. Adult female, Pascagoula River at Highway 596 in Merrill, George County, Mississippi.

The conservation status of *G. gibbonsi* [*sensu lato*] needs attention
... two [federally] listed species outnumbered it in total basking
counts by 5.1:1.

PETER LINDEMAN (1999C, PP. 39–40)

Alternate Common Names

Gibbons' map turtle (Buhlmann 2006). First called Pascagoula map turtle by
Lovich and McCoy (1994b).

Description and Diagnosis

Postorbital markings large blotches, with upper rear edges that are rounded (often deeply dissected from rear margin, especially in females, forming large upper and small lower lobes), curving below orbits to subsume supramandibular
spots. Interorbital marking a blotch, which may or may not be trident shaped,
connecting by thick lines with postorbitals. Chin with prominent midventral
longitudinal stripe, to each side of which appears one small anterior spot and
one elongated submandibular blotch, sometimes with transverse stripes that
connect midventral stripe to submandibulars. Markings on head, neck, limbs,
and tail yellow to light orange. Females with exceptionally broad heads and exceptionally broad alveolar surfaces, males with broad heads and broad alveolar
surfaces. Second vertebral scute steeply sloped, distinctly higher than the third.
Carapace highly peaked at posterior of second vertebral, nearly triangular in
cross section. Carapacial keels produce prominent points in males and juveniles; reduced and blunted in females. Black pigment along midline of carapace
generally broken, most prominent at keel points. Carapace marked with reticulating network of thin, black-bordered, yellowish-orange lines on pleural scutes
having comparatively few intersections (one to two rings per pleural), most
prominent anteriorly. Marginals marked dorsally with wide yellowish-orange
stripes (ca. 20 percent of fifth scute width) that curve inward and rearward, ventrally with alternating narrow dark and light markings (ca. 40 percent of fifth
scute width); dorsal surface of twelfth marginal with light edge and light medial
stripe extending forward over 75 percent of scute length. Plastron pale yellow,
with black markings along scute seams in smaller specimens. Male foreclaws
not conspicuously elongated. Irises yellow, bisected by black stripes. Differs
from congeners by the wide light markings of upper marginals; the connection
of the interorbital, postorbital, and supramandibular markings; the interrupted
dark pigmentation along the carapacial keel; and the shape of the light pigmentation of the marginals.

Taxonomic History and Phylogeny

The oldest preserved specimens of this species are sixteen specimens collected
by Cagle's Tulane field crew in August 1952 and regarded as *G. pulchra*, a spe-

Figure 8.10. CM 94979, the holotype of *Graptemys gibbonsi*.

cies complex that up until then had a poorly known range. Over four decades, several authors commented on the distinctiveness of specimens from the Pascagoula drainage relative to those from the Escambia, Mobile Bay, and Pearl drainages (see species account for *G. pulchra*). Lovich and McCoy (1992) examined morphometrics and markings of specimens from all four drainages and described the Pascagoula and Pearl drainage populations as *Graptemys gibbonsi*, honoring Lovich's doctoral advisor, noted expert on North American freshwater turtles, and teenage member of Cagle's field crews, J. Whitfield Gibbons. The type series includes a holotype (CM 94979; fig. 8.10) collected from the "Chickasawhay River, Leakesville, Greene County, Mississippi, USA" (Lovich and McCoy 1992, p. 302) on 21 July 1978 by Richard C. Vogt, Michael Pappas, and Paul S. Freed, as well as seventeen paratypes from the type locality (CM 94966–67, 94970–73, 94976–78, 94980–81, 94983, 95361–62, 95559, 95561, 95577). No subspecies have been described, but Vetter (2004) informally noted a "Pearl River form" and a "Pascagoula River form" and Ennen, Lovich, et al. (2010) described the Pearl drainage form as *G. pearlensis*. Artner's (2003) checklist relegated the Pascagoula map turtle *sensu lato* to a subspecies of *G. pulchra*, but his arrangement is unlikely to be followed by many taxonomists due to allopatry and a lack of critical analysis of variation. Phylogenetic analyses placed the Pascagoula map turtle (*sensu lato*) within the megacephalic clade, as sister species to a clade composed of the other three species (Lamb et al. 1994), to a clade composed of *G. barbouri* and *G. ernsti* (Lamb and Osentoski 1997), to

G. ernsti (Stephens and Wiens 2003), and to a clade composed of *G. barbouri* and *G. pulchra* (Wiens et al. 2010).

Range

The Pascagoula map turtle is restricted to the Pascagoula drainage in southern Mississippi. It extends up the Leaf River into Smith County and up the Chickasawhay River into Okatibbee Creek in Lauderdale County and the Chunky River in Newton County (Selman and Qualls 2009a), as well as a tributary of the latter, Okahatta Creek (MMNH 4041). Several other tributaries are inhabited, including Black and Red Creeks for the Pascagoula main stem; the Bowie River and the Bogue Homa, and Okatoma, Oakohay, Tallahala, Tallahoma, Thompson, and Gaines Creeks for the Leaf drainage; Buckatunna and Souinlovey Creeks for the Chickasawhay drainage; and the lower Escatawpa River on the lower Pascagoula (Cliburn 1971; McCoy and Vogt 1979; Lovich and McCoy 1994b; Lindeman 1998, 1999c; Selman and Holbrook 2010; Selman and Qualls 2008c, 2009a). The downstream range limit on the Pascagoula is in the vicinity of Vancleave in Jackson County (Selman and Qualls 2009a).

Habitat

Rivers inhabited by the Pascagoula map turtle primarily have sand and gravel substrates; the upstream transition to rocky substrates in rivers cutting through limestone formations that is typical of other members of the megacephalic clade to the east is much less evident for this species, as its range lies just west of the area of the Gulf Coast affected by the fall line. Greatest densities occur near the same dense accumulations of deadwood and large sandbars that are prime habitat for the yellow-blotched sawback, with which Pascagoula map turtles are sympatric, but it ranges farther upstream into most of the smaller streams in its range than yellow-blotched sawbacks do. In replicated spotting scope counts at twenty-one sites, the correlation of basking density to deadwood density was nonsignificant but trended positive for three of four analyses (Lindeman 1999c). No reports exist of its occurrence in oxbow lakes or other floodplain habitats.

Diet

Cagle (1952b) reported insects (identified as caddis fly larvae in his field notes; C. H. Ernst and Lovich 2009) in two males and mussels and snails in a juvenile female. Adult females pass shell fragments of Asian clams (Ennen et al. 2007) and moderately large native mussels in their feces (R. Jones, pers. comm. 1994), consistent with their megacephaly. For combined samples from both sexes, mollusks were 82 percent of the volume of prey on the lower Chickasawhay but only 25 percent on the middle Pearl, where fish constituted 44 percent (Lovich et al. 2009).

Reproduction and Life History

Mean clutch size was 7.5 on the lower Chickasawhay (Lovich et al. 2009). Range-wide, midline plastron length ranges from 72 to 101 mm in adult males and reaches 218 mm in adult females (Lindeman 2008a).

Natural History

Selman, Strong, et al. (2008) described voluntary release by a turtle leech they observed on a female basking on a hot, sunny day.

Populations

Relative abundance in both trapping and basking surveys has been reported over five decades. Data indicate a probable decline in abundance of Pascagoula map turtles relative to the sympatric, microcephalic yellow-blotched sawback *G. flavimaculata*, which is federally listed as threatened. The species generally predominated in older trapping studies but has lagged behind its congener in more recent basking surveys and trapping (Lindeman 1999c; Selman and Qualls 2009a).

Yellow-blotched sawbacks predominated at 1.4:1, but Pascagoula map turtles were the second most abundant of four species, at 33 percent of the catch, in trapping studies summarized by Tinkle (1958b). Cliburn (1971) reported *Graptemys* ratios in captures from widely dispersed sites on the Pascagoula drainage, with Pascagoula map turtles outnumbering yellow-blotched sawbacks 1.03:1.

In range-wide replicated basking surveys in 1994–95, Pascagoula map turtles were 22 percent of the turtles seen, at 6.4/river km (Lindeman 1998, 1999c). Their relative abundance nearly matched that of the river cooter (*Pseudemys concinna*), but these two species lagged far behind the yellow-blotched sawback (53 percent). Highest densities of Pascagoula map turtles were observed on the main-stem Pascagoula (9.0/km), the Chickasawhay River (8.0/km), the Chunky River (7.8/km), and Black Creek (13.4/km). In 2006–2008, Selman and Qualls (2009a) recorded basking densities averaging 11.7 Pascagoula map turtles/ river km on the Pascagoula, 15.3/river km on the Leaf, and 17.6/river km on the Chickasawhay and ranging between 0 and 7.5/river km on various smaller tributaries. Overall they saw 2.6 yellow-blotched sawbacks for every Pascagoula map turtle. Annual ratios of *Graptemys* in basking trap captures ranged between 1.6 and 2.0 to 1 on the lower Chickasawhay, between 2.1 and 3.7 to 1 on the upper Leaf, and between 15 and 102 to 1 on the lower Pascagoula (Selman and Qualls 2009a).

Selman and Qualls (2009a) also estimated population size at the Leaf River site using a capture-mark-resight technique. Their estimates were 34–44 Pascagoula map turtles/river km and 80–120 yellow-blotched sawbacks/river km, with ratios of estimated population size (2.8:1 and 2.0:1) approximating the ratios in their basking surveys.

Conservation Status and Actions

Currently under review for federal listing. Recent survey work resulted in a recommendation of federal and state listing (Selman and Qualls 2009a). Listed as endangered with decreasing populations by the IUCN.

Portions of some of the minor tributaries of the Pascagoula occur within DeSoto National Forest, including 34 km of Black Creek protected as a National Wild and Scenic River, although populations are small in these tributaries. Extensive state protection of riparian habitat occurs within the Pascagoula River and Ward Bayou game management areas (26,000 ha). The Nature Conservancy protects 1,988 ha near the confluence of the Leaf and the Chickasawhay in the Charles M. Deaton and Herman Murrah Preserves (see chapter 6).

Needed Further Research

There is a critical need to further assess the population status and trends for Pascagoula map turtles, given their low relative abundance (and apparently increasing disparity in numbers) relative to the sympatric yellow-blotched sawback, which is federally listed as threatened. Detailed studies of diet and reproduction are also lacking.

Pearl Map Turtle, *Graptemys pearlensis*

Figure 8.11. Adult female, Pearl River at Highway 19 near Philadelphia, Neshoba County, Mississippi.

There is very little concrete information on the exploitation of
G. gibbonsi [including *G. pearlensis*], although it is reported that
hundreds were collected in the Pearl River Basin in 2006.

JEFFREY LOVICH ET AL. (2009, P. 6)

Alternate Common Names

Big-headed map turtle (Cagle 1952a), never adopted; Pearl River map turtle
(Lovich et al. 2009). First called Pearl map turtle by Vetter (2004).

Description and Diagnosis

Postorbital markings large blotches, with upper rear edges that are rounded (often deeply dissected from rear margin, especially in females, forming large upper and small lower lobes), curving below orbits to subsume supramandibular spots. Interorbital marking a blotch, which may or may not be trident shaped, connecting by thick lines with postorbitals. Chin with prominent midventral longitudinal stripe, to each side of which appears one small anterior spot and one elongated submandibular blotch, sometimes with transverse stripes that connect midventral stripe to submandibulars. Markings on head, neck, limbs, and tail yellow to light orange. Females with exceptionally broad heads and exceptionally broad alveolar surfaces, males with broad heads and broad alveolar surfaces. Second vertebral scute steeply sloped, distinctly higher than the third. Carapace highly peaked at posterior of second vertebral, nearly triangular in cross section. Carapacial keels producing prominent points in males and juveniles, reduced and blunted in females. Black pigment generally continuous along midline of carapace, including keel points. Carapace marked with reticulating network of thin, black-bordered, yellowish-orange lines on pleural scutes having comparatively few intersections (one to two rings per pleural), most prominent anteriorly. Marginals marked dorsally with wide, yellowish-orange stripes (ca. 20 percent of fifth scute width) that curve inward and rearward, ventrally with solid dark markings (ca. 40 percent of fifth scute width); dorsal surface of twelfth marginal with light edge having little forward extension or with stripe extending forward over half of scute length on outer half of scute. Plastron pale yellow, with black markings along scute seams in smaller specimens. Male foreclaws not conspicuously elongated. Irises yellow, bisected by black stripes. Differs from congeners by the wide light markings of upper marginals; the connection of the interorbital, postorbital, and supramandibular markings; the continuous dark pigmentation along the carapacial keel; and the shape of the light pigmentation of the marginals.

Taxonomic History and Phylogeny

The oldest preserved specimens of this species are six adult females collected by Kohn in the 1890s (two skeletal syntypes noted in the *G. pulchra* account and four taxidermic mounts at Tulane; fig. 8.12). The next specimens were a large

Figure 8.12. TU 7680, collected in 1891; one of Gustave Kohn's four taxidermic mounts of female Pearl map turtles in the Tulane University Museum of Natural History.

series taken in 1950 by Cagle (1952b) and his students and regarded as *G. pulchra*, a species complex that until then had a poorly known range. Over four decades, several authors commented on the distinctiveness of specimens from the Pearl drainage relative to those from the Escambia, Mobile Bay, and Pascagoula drainages (see species account for *G. pulchra*). Lovich and McCoy (1992) examined morphometrics and markings of specimens from all four drainages and described the Pascagoula and Pearl drainage populations as *Graptemys gibbonsi*. Subsequently, Ennen, Lovich, et al. (2010) described Pearl drainage specimens as *Graptemys pearlensis*. The type series includes a holotype (CM 62162) collected from "Mississippi, Copiah County, Pearl River at State Highway 28, near Georgetown" (Ennen, Lovich, et al. 2010, p. 104) on 23 September 1967 by T. E. Majers, as well as fourteen paratypes from various Pearl River localities (AUM 21975, 32438; CM 67474, 67480, 94904, 94909, 94916, 94940, 95050, 95055, 95059, 95663, 95632, 95674). No subspecies have been described. Artner's (2003) checklist relegated *G. gibbonsi sensu lato* to a subspecies of *G. pulchra*, but his arrangement is unlikely to be followed by many taxonomists due to allopatry and a lack of critical analysis of variation. Phylogenetic analyses have not differentiated between the Pearl and Pascagoula map turtles and have placed them within the megacephalic clade, as sister species to a clade composed of the other three species (Lamb et al. 1994), to a clade composed of *G. barbouri* and *G. ernsti* (Lamb and Osentoski 1997), to *G. ernsti* (Stephens and Wiens 2003), and to a clade composed of *G. barbouri* and *G. pulchra* (Wiens et al. 2010).

Range

The Pearl map turtle is restricted to the Pearl drainage in southern Mississippi and two parishes of southeastern Louisiana. Dundee and Rossman's (1989) record from the Tickfaw River in Livingston Parish, Louisiana, is based on NLU 5596, but it was collected on the same dates (5–7 January 1965) as two specimens taken from the Bogue Chitto River near Franklinton (NLU 5594–95) and likely comes from that same locality.

The species occurs upstream in the Pearl into the Nanih Waiya Wildlife Management Area in Neshoba County, Mississippi (Cliburn 1971; Lindeman 1998, 1999c; Keiser 2000). Its most extensive occurrences in tributaries are in the Bogue Chitto River, in which it extends upstream north of the Louisiana-Mississippi state line into Walthall and Pike Counties (Cliburn 1971; Shively 1999; Lindeman 2010; pers. observations 2010), and the Strong River in Simpson and Rankin Counties (Lindeman 1998, 1999c; pers. observations 2008). It also occurs in the lower reaches of four smaller tributaries: the Yockanookany River and Lobutcha Creek on the upper Pearl (Lindeman 1998, 1999c; MMNH 15516; pers. observations 2010), Pushepatapa Creek on the lower Pearl (J. L. Carr and Messinger 2002), and Topisaw Creek on the upper Bogue Chitto (pers. observations 2010). Downstream, the species occurs in the braided channels of the East and West Pearl Rivers, the Pearl River Canal, and the Porter River to the vicinity of the towns of Pearl River in St. Tammany Parish and Napoleon in Hancock County (Cagle 1952b; Lindeman 1998, 1999c; W. Selman, pers. comm. 2008).

Habitat

Rivers inhabited by the Pearl map turtle primarily have sand and gravel substrates; the upstream transition to rocky substrates in rivers cutting through limestone formations that is typical of other members of the megacephalic clade to the east is much less evident for this species, as its range lies just west of the area of the Gulf Coast affected by the fall line. Greatest densities occur near the same dense accumulations of deadwood and large sandbars that are prime habitat for the ringed sawback with which the Pearl map turtle is sympatric, but it ranges farther upstream into most of the smaller streams in its range than the ringed sawback does. In replicated spotting scope counts at twenty sites, the correlation of basking density to deadwood density trended positive and was statistically significant for two of four analyses (Lindeman 1999c).

No reports exist of their occurrence in oxbow lakes or other floodplain habitats. The species was notably absent from Mayes Lake, where ringed sawbacks are common (Lindeman 1998, 1999c; pers. observations 2007, 2010). It occurs at the upstream end of the single large impoundment within its range, Ross Barnett Reservoir (Boyd and Vickers 1963; Lindeman 1998, 1999c).

Diet

Cagle (1952b) reported insects (identified as caddis fly larvae in his field notes; C. H. Ernst and Lovich 2009) in two males and mussels and snails in a juvenile female. Adult females pass moderately large native mussels in their feces (R. Jones, pers. comm. 1994), consistent with their megacephaly. For combined samples from both sexes, mollusks were 25 percent of prey volume, while fish constituted 44 percent (Lovich et al. 2009).

Reproduction and Life History

Cagle (1952b) noted a clutch of three eggs in a small dissected female with enlarged follicles that may have represented two additional three-egg clutches for the same year. The three oviductal eggs average 45.3 × 25.9 mm. Mean clutch size is 6.4 (Lovich et al. 2009). Range-wide, midline plastron length ranges from 58 to 98 mm in adult males and reaches 215 mm in adult females (Lindeman 2008a).

Natural History

Nests are concentrated near sandbars' elevated vegetation lines on their inner edges (P. K. Anderson 1958). J. L. Carr and Messinger (2002) reported a hatchling in the stomach of a spotted bass.

Populations

Relative abundance in both trapping and basking surveys has been reported over five decades. These data indicate a probable decline in abundance of Pearl map turtles relative to the sympatric, microcephalic ringed sawback G. oculifera, which is federally listed as threatened. The species generally predominated in older trapping studies but has lagged behind its congener in more recent basking surveys and trapping (Lindeman 1999c).

Pearl map turtles predominated in nighttime hand captures in 1950–51, at 53 percent of all turtles and 67 percent of all map turtles and sawbacks (Cagle 1953a). Tinkle (1958b) captured more Pearl map turtles than five other species combined (51 percent relative abundance), and his ratio of Pearl map turtles to ringed sawbacks was 1.9:1. Cliburn (1971) reported *Graptemys* ratios for captures on widely dispersed sites in the Pearl, with Pearl map turtles outnumbering ringed sawbacks 2.1:1. The species constituted 15 percent of the catch in fyke nets employed for three hundred trap nights on the lower Pearl River, fourth in abundance behind sliders (*Trachemys scripta*), ringed sawbacks, and razorback musk turtles (*Sternotherus carinatus*), with the *Graptemys* ratio being 1.3:1 (Vogt 1980b).

In range-wide, replicated basking surveys in 1994–95, Pearl map turtles were 8 percent of the turtles seen, at 4.8/river km (Lindeman 1998, 1999c). Their relative abundance nearly matched that of the river cooter (*Pseudemys concinna*), but these two species lagged far behind the ringed sawback (75 percent). Highest densities were observed on the Pearl south of Jackson (10.2/km). At three of four sites on the Pearl, trapping ratios of the two *Graptemys* species in 1988–90 (Jones and Hartfield 1995; R. Jones, pers. comm. 1997) were imbalanced toward ringed sawbacks to a lesser extent than basking surveys were in 1994–95 (Lindeman 1999c). In basking surveys on the Bogue Chitto River in 1999, Pearl map turtles were outnumbered by ringed sawbacks 1.4:1 and constituted 24 percent of all turtles identified, being the third most abundant species after river cooters and ringed sawbacks (Shively 1999).

Conservation Status and Actions

No federal or state listings. Listed as endangered with decreasing populations by the IUCN.

Federally protected habitat occurs chiefly in 14,500 ha of the Bogue Chitto National Wildlife Refuge. A portion of the Natchez Trace Parkway (managed by the National Park Service) provides a relatively undeveloped western shoreline along the upper Pearl, lower Yockanookany, and Ross Barnett Reservoir. Mississippi maintains the Ringed Sawback Turtle Refuge on the Pearl above the reservoir.

Needed Further Research

There is a critical need to further assess the population status and trends for Pearl map turtles, given their low relative abundance (and apparently increasing disparity in numbers) relative to the sympatric ringed sawback, which is federally listed as threatened. In particular, occurrence in undocumented tributaries and floodplain habitats should be investigated. Detailed studies of diet and reproduction are also lacking.

Cagle's Map Turtle, *Graptemys caglei*

Figure 8.13. Adult male, unnamed oxbow lake of the San Marcos River, Palmetto State Park, Gonzales County, Texas.

We name it in memory of Fred R. Cagle whose research greatly increased our knowledge of *Graptemys* and turtles of the southern states in general.

DAVID HAYNES AND RONALD MCKOWN (1974, P. 143)

Alternate Common Names

None; common name first used by Haynes (1976).

Description and Diagnosis

Postorbital markings typically narrow crescents curving forward under orbits to subsume supramandibular spots, less often vertical bars separate from supramandibulars, allowing one or more lines to extend to orbit; continuous with lines that converge posteriorly to form V shape on top of neck. Behind each bar or crescent, typically two additional crescent- or bar-shaped markings that may be continuous with neck stripes posteriorly. Interorbital marking a medial longitudinal stripe widening slightly posteriorly, typically flanked by one to three thinner longitudinal lines on each side. Chin with transverse light stripe along forward edge, subsuming submandibular spots. Markings on head, neck, limbs, and tail light yellow. Females with moderately broad heads and moderately broad alveolar surfaces, males with narrow heads and broad alveolar surfaces. Second vertebral scute only slightly sloped, extending little higher than the third. Carapace not highly peaked, rounded in cross section. Carapacial keels producing prominent points in males and juveniles, reduced and blunted in females. Black pigment along midline of carapace, particularly at keel points. Carapace marked with series of concentric, alternating light and dark lines arranged in reticulating network of circles or ovals and continuing onto marginals. Plastron pale yellow, with dark markings along scute seams in smaller specimens. Male foreclaws not conspicuously elongated. Irises white, bisected by black stripes. Differs from congeners by the V-shaped head marking, the triple crescent or bar on each side of the head, the transverse chin bar, the concentric markings of the carapace, and the bisected white irises.

Taxonomic History and Phylogeny

Three specimens of Cagle's map turtle were collected in 1957–58 from an unnamed oxbow lake in Palmetto State Park in Gonzales County, Texas (fig. 8.14), and considered by Raun (1959) to represent a disjunct population of the false map turtle, *G. pseudogeographica*. They were later referred to the Texas map turtle, *G. versa* (Webb 1962), and then called an undescribed species (Raun and Gelbach 1972). The species was described by Haynes and McKown (1974) based on a holotype (TNHC 36061) they collected "from the Guadalupe River, 8 km NW of Cuero, DeWitt Co., Texas" (p. 143) on 29 June 1967. They designated thirty-eight paratypes from the type locality (TNHC 36056, 36068–70, 36072, 36074–91, 36093–98, 36100–105, 36107–109) and two from a locality in Kerr

Figure 8.14. An unnamed oxbow lake of the San Marcos River in Palmetto State Park, Gonzales County, Texas, where the first specimens of Cagle's map turtle were collected in the late 1950s and where the species persists today. The bridge in the background is Highway 2091.

County, Texas (TNHC 36106, 41223). The specific epithet honors Fred R. Cagle (1915–68), whose many contributions to map turtle research as a professor at Tulane University in the 1940s and 1950s are chronicled throughout this book. No subspecies have been described, nor has the species ever been suggested to be conspecific with any other *Graptemys* since its description. Phylogenetic analyses placed the species in an outgroup position relative to a large clade of species with micro- and mesocephalic females (Lamb et al. 1994) and relative to the clade of species with megacephalic females (Stephens and Wiens 2003) and as sister species to *G. ouachitensis* (Lamb and Osentoski 1997) and *G. versa* (Wiens et al. 2010).

Range

Cagle's map turtle is apparently restricted to the Guadalupe River drainage of south-central Texas. In the Guadalupe, it occurs from Victoria County (Killebrew and Porter 1991) upstream into the South Fork of the Guadalupe in Kerr County (Haynes and McKown 1974). It has been recorded from deep pools and adjoining springs of the Blanco River as well as in lower portions of the river it joins, the San Marcos, which subsequently flows into the Guadalupe (Raun 1959; Haynes and McKown 1974; Babitzke 1992; T. R. Simpson and Rose 2007).

Unquestioned specimen records do not exist from the San Antonio River, which joins the Guadalupe from the west, about 35 km downstream of localities in Victoria County, or from the Medina River, a western tributary of the San Antonio. Haynes and McKown (1974, p. 150) claimed sight records for both rivers. Haynes (1976) mapped the San Antonio River site just south of the city of San Antonio in Bexar County (Mission Road Bridge between Highway 410 and the confluence of the Medina River, where a basking male was seen with binoculars; Vermersch 1992; D. Haynes, pers. comm. 2006). A specimen listed as having been collected by G. W. Marnock in Bexar County (SM 7359, a dry carapace and plastron) is not a valid county record of Cagle's map turtle for two reasons. First, it resembles G. *versa* more than it does G. *caglei* in having a low carapacial keel. Second, Marnock, an amateur collector, lived for about four decades on Helotes Creek, northwest of San Antonio in Bexar County; upon his death in 1920, his specimens were bought by John K. Strecker (Strecker 1922), and it appears some were falsely attributed to the Bexar County homesite (Raun and Gelbach 1972). Olson (1959) reported collecting a juvenile specimen (identified as G. *pseudogeographica* but clearly outside the range of that species) from the southern outskirts of San Antonio in Bexar County, where he stated the species has "a well-established colony" (p. 48). The specimen has apparently been lost, as it was not among specimens transferred from the Witte Museum to the Texas Cooperative Wildlife Collection (K. Vaughan, pers. comm. 2006). Vermersch (1992) stated that in the 1980s a specimen "was possibly collected on the Medina River in southern Bexar County" (p. 108) but gave no details or source of his information. Survey work on the San Antonio in Karnes County was negative for Cagle's map turtle (Babitzke 1992). Whether or not it once occurred in any abundance in the San Antonio drainage and was eliminated by urbanization and degradation of water quality is a matter of conjecture, given the uncertainty of records and the recent description of the species.

Habitat

Killebrew et al. (2002) completed a detailed assessment of the habitat requirements of Cagle's map turtle. They measured physical and biological parameters along river stretches with widely varying density estimates to examine the correlation of abundance with these parameters. Moderately high water velocities (0.5–0.8 m/s) and presence of leptocerid caddis fly larvae were associated with the highest population densities. Significant positive correlations related population density to percentage of the substrate composed of cobble and densities of basking sites and shoreline willow trees (*Salix* sp.). Water velocity, cobble substrate, and the underwater root systems of the willows may all develop dense concentrations of the insect prey of small individuals. In particular, caddis fly larvae were more strongly associated with willow roots than with the roots of other shoreline trees.

Impoundment generally eliminates Cagle's map turtles. None are known to inhabit the 30 km Canyon Lake reservoir (Babitzke 1992; Connor 1993; Killebrew et al. 2002); nor are they found in five smaller downstream reservoirs, which range from 4 to 14 km in length. Populations occur in the upstream, more river-like sections of three small impoundments (Babitzke 1992; Killebrew et al. 2002) and in a disjunct oxbow of the San Marcos (fig. 8.14; Raun 1959).

Diet

Haynes and McKown (1974) reported on the diet of unsexed juveniles, adult males, and juvenile females anecdotally, while W. Lehmann (1979, cited in Bertl and Killebrew 1983) reported on the diets of young specimens she identified as ranging in age from one to five years. Both sources reported a preponderance of aquatic insects, particularly caddis fly larvae, flies, hemipterans, and drag-onfly nymphs; snails also occurred in the juvenile females of the former study. Porter (1990) conducted a more detailed analysis that included a breakdown of samples from all four seasons for juveniles, males, and females (table 4.9). The male diet was dominated by caddis fly larvae, with considerably lower reliance on snails, damselflies and dragonflies, and mayfly nymphs, while the female diet was dominated by Asian clams. Four juvenile females were similar to adult females in diet, although their use of snails was somewhat greater and their use of bivalves somewhat less. Six juvenile males likewise were most similar in diet to adult males, although they fed much more heavily on snails and less heavily on insects. Seasonal dietary diversity for each sex changed chiefly as a function of the abundance of their primary prey: males had highest dietary diversity in the summer, when caddis fly larvae were least prevalent, and lowest diversity in the winter, when they were most prevalent, while female reliance on Asian clams rose steadily from spring through winter, with an inverse relationship to overall dietary diversity.

Reproduction and Life History

In males, growth follows the von Bertalanffy pattern of rapid juvenile growth that slows following maturation in the second growing season; growth and maturation have not been studied for females (Lindeman 1999b). Killebrew and Babitzke (1996) reported an average clutch size of 3.8 (range 1–8) with a nesting season ranging from late March to late August, based on gravid female captures. Hatchlings emerge in late summer and fall (Haynes and McKown 1974). Range-wide, midline plastron length ranges from 63 to 94 mm in adult males and reaches 172 mm in adult females (Lindeman 2008a).

Natural History

Courting males engage in head bobbing but not foreclaw titillation (Hertwig 2001). Craig (1992) reported activity in all months of the year, with peak basking counts in February–May and October–November. Five species of coccidian parasites have been reported in Cagle's map turtles (McAllister et al. 1991).

Populations

Greatest population densities occur in the Guadalupe River in a 233 km section between Seguin (Guadalupe County) and Cuero (Dewitt County), where it was estimated that about 70 percent of all Cagle's map turtles occurred in 1991 (Babitzke 1992; Connor 1993) and 75 percent in 2000–2001 (Killebrew et al. 2002). Densities are considerably lower downstream of Cuero to Victoria; in the lower San Marcos River; and upstream of Seguin to New Braunfels (Comal County), a section of the river with a series of impoundments. Small, highly discontinuous populations occur farther upstream into Kerr County.

Babitzke (1992) conducted population estimates using mark-recapture estimates from a population near Cuero and extrapolating, via basking surveys, to stream sections totaling 89 percent of the species' range in the Guadalupe (543.0 of 608.5 km) and 9.9 km of the lower San Marcos. He averaged seeing 2.8 percent of the estimated population of his main study site, and from this figure he extrapolated that the total population in 1991 was 11,717 for the surveyed sections of the Guadalupe River and 284 in the lower San Marcos River, averages of 21.6 and 28.7 turtles/river km, respectively. New estimates for 2000–2001 were based on the entire range within the Guadalupe and average sighting rates of 3.0 percent at the main study site, with a total estimated population of 13,468, or 22.1 turtles/river km (Killebrew et al. 2002). Along one stretch in Comal County, a decline from 36 turtles/km to none observed followed intensive residential development. A dam brought back online in 1993 in DeWitt County may have been associated with population movements that caused changes in abundance in two river stretches upstream of the reflooded reservoir, as the stretch closer to the reservoir and dam had its estimate approximately halved while a stretch farther upstream had its estimate approximately doubled.

The main other species with a proclivity for basking in the Guadalupe is the more numerous Texas river cooter (*Pseudemys texana*). Babitzke's (1992) basking surveys yielded a ratio of the two species of 3.0:1. Using the same mark-recapture and extrapolation method, he estimated a total population of 41,349 Texas river cooters in the Guadalupe and lower San Marcos Rivers, for a range-wide ratio of the two species of 3.5:1. In the Blanco River, T. R. Simpson and Rose (2007) reported approximately equal numbers of Texas river cooters and sliders (*Trachemys scripta*), with both greatly outnumbering Cagle's map turtle, of which they found only 13 in 453 search hours.

Conservation Status and Actions

Listed as a candidate species under the U.S. Endangered Species Act beginning in 1982 but not listed for many years due to assignment of low listing priority, then dropped from candidate status in 2006. State-listed by Texas as threatened. Listed as endangered with decreasing populations by the IUCN.

There is no federal habitat protection. Short, protected riparian reaches oc-cur in Kerrville-Schreiner City Park and Guadalupe State Park on the Guada-lupe and in Palmetto State Park on the lower San Marcos.

Needed Further Research

Additional data on the life history of Cagle's map turtle are necessary for fur-ther evaluation of its conservation status. Given its restricted range and the very small population sizes over much of its range, it is important to better under-stand the relationship between habitat conditions and density and to continue to monitor population persistence. The restricted range also makes this species particularly vulnerable to exploitation; thus data from the illegal animal trade would be of particular interest and importance for Cagle's map turtle.

Texas Map Turtle, *Graptemys versa*

Figure 8.15. Adult male, South Llano River in Junction, Kimble County, Texas.

In an exhaustive compendium of literature citations on the 56 species of turtles native to North America . . . , *G. versa* has the distinction of having the shortest species account with the fewest citations of ecological studies.

PETER LINDEMAN (2005, P. 378)

Alternate Common Names

Texan map turtle (B. C. Brown 1950), Texas sawback turtle (A. Carr 1952), neither widely adopted. First called Texas map turtle by Pope (1939).

Description and Diagnosis

Postorbital markings thin, short bars with long extensions from lower ends posteriorly onto neck; vertical sections tend to tilt slightly forward and curve around tops of orbits, producing overall stylized S shape (right side). Interorbital marking a medial longitudinal stripe widening slightly posteriorly, typically flanked by one to three thinner longitudinal lines and surrounded posteriorly by U-shaped line. Supramandibular spots very small. Chin with one small forward medial spot, usually one slightly larger medial spot behind it, and two small submandibular spots, each surrounded by concentric rings. Markings on head, neck, limbs, and tail yellow to light orange. Females with moderately broad heads and moderately broad alveolar surfaces, males with narrow heads and narrow alveolar surfaces. Second vertebral scute only slightly sloped, extending little higher than the third. Carapace not highly peaked, rounded in cross section. Carapacial keels not producing prominent points. Carapacial keels distinctly lighter than rest of carapace and horn colored. Carapace marked with series of concentric, alternating light and dark lines arranged in reticulating network of circles or ovals and continuing onto marginals. Pleural and marginal scutes with prominently raised central bosses and sunken sutures, giving carapace a quilted appearance, particularly in juveniles and males. Plastron pale yellow, with black markings along scute seams in smaller individuals. Male foreclaws not conspicuously elongated. Irises white, bisected by black stripes. Differs from congeners by the lower rearward projections of the postorbital markings; the low, horn-colored carapacial keels lacking dark pigmentation; the concentric markings of the carapace; and the bisected white irises.

Taxonomic History and Phylogeny

A turtle cataloged as *"Malacoclemys LeSueuri"* and described as a live specimen collected by G. Stolley on 23 July 1883 in Austin, Texas (USNM 13339) may have been the first museum specimen of a Texas map turtle, although it has been lost. Strecker (1909, 1930) and Strecker and Williams (1927) discussed specimens from the early 1900s that Strecker and a correspondent, C. S. Brimley, identified as *G. geographica* and *G. oculifera*, from the Colorado drainage in Burnet and Travis Counties, Texas. These specimens may in fact have been Texas map turtles (H. M. Smith 1948; H. M. Smith and Sanders 1952), although records in the USNM suggest that the former was applied to Brimley's specimens of *Pseudemys texana* and the latter to his specimens of *G. versa*. Stejneger (1925) described the species on the basis of a holotype (USNM 27473) and seven paratypes (USNM 27474–79, MCZ 42346) from "Austin [Travis County], Texas"

(p. 463), a city located on the Colorado River, in which the species occurs. The type series had been collected in July 1900 by an unknown collector and donated to the museum by Herbert H. Brimley and Clement S. Brimley (D. M. Cochran 1961; Reynolds et al. 2007). All eight types are adult males, ranging in midline plastron length from 60 to 80 mm. Stejneger named the taxon *Graptemys pseudogeographica versa*, with the subspecific epithet from the Latin *vers* (to change), thought to refer to the difference in head pattern between this taxon and *pseudogeographica* (Vogt 1981b), the sole diagnostic characteristic that Stejneger noted. H. M. Smith (1948) found *versa* "shows no kinship with *G. pseudogeographica* sufficient to warrant subspecific nomenclature" (p. 60) and raised *versa* to full species status, later giving a more detailed account of his rationale and redescribing the species (H. M. Smith and Sanders 1952). While some authors continued to consign *versa* to subspecific status (A. Carr 1949, 1952; K. P. Schmidt 1953), its status as a separate species was recognized in the first edition of the Peterson field guide (Conant 1958) and verified in taxonomic studies by Vogt (1978, 1981b, 1993), although Artner (2003) persisted in listing it as a subspecies. No subspecies have been described. Phylogenetically, the species was placed as sister species to a clade consisting of *G. ouachitensis*, *G. sabinensis*, *G. flavimaculata*, and *G. oculifera* (Lamb et al. 1994); in an unresolved polytomy involving eight species (Lamb and Osentoski 1997); as sister species to *G. sabinensis* in a clade that is sister to a clade composed of *G. flavimaculata*, *G. oculifera*, *G. nigrinoda*, *G. ouachitensis*, and *G. pseudogeographica* (Stephens and Wiens 2003); and as sister species to *G. caglei* (Wiens et al. 2010).

Range

The Texas map turtle is restricted to the Colorado River drainage of Texas, largely in a region known as the Edwards Plateau that is defined to the east by the Balcones Escarpment (H. M. Smith and Buechner 1947). On the main-stem Colorado, the species occurs from Colorado County (UTA 30146–59)—that is, east and downstream of the escarpment—upstream onto the plateau as far as Coke County (J. R. Dixon 2000). Major western tributaries of the system that are inhabited, from north to south, are the Concho River, as far upstream as Tom Green and Irion Counties in its tributaries (Marr 1944; J. R. Dixon 2000); the San Saba River to the border of Schleicher and Menard Counties (Kizirian et al. 1990); the Llano River and its various tributaries, as far upstream in the South Llano River as northeastern Edwards County (Daugherty 1942; Lindeman 2004, 2005); and the Pedernales River into Gillespie County (Olson 1967; J. R. Dixon 2000). A single large eastern tributary, Pecan Bayou, is inhabited into Brown County (UNM 42642, 42683). A multitude of small tributary creeks throughout the system also house populations (see below).

Habitat

Streams that join the Colorado River from the west are fed by springs connected to the massive Edwards Aquifer. Their water is very clear, cool, and shallow, and riffles occur frequently. Occurrence in shallow water (1–2 m) is much more common than for species in drainages farther east. The Colorado itself and its major eastern tributary, Pecan Bayou, are murkier streams that cut deeper and narrower gorges.

The substrate in the South Llano is a mix of bedrock and loose, cobble-sized rocks in swift-flowing riffle areas, with muddy bottoms in the longer, deeper pools. Texas map turtles occur in both pools and riffles (Lindeman 2003b). Basking sites in the South Llano are relatively few due to sparse riparian tree cover in the dry Texas Hill Country; the species occurs in the most arid region of any *Graptemys*. Lack of tree cover and the resulting penetration of sunlight may relate to the tendency of the species to occur in smaller tributaries of the drainage not named above, as literature accounts (Smith and Sanders 1952; Rakowitz et al. 1983; Lindeman 2004, 2005) and numerous unpublished specimen records place the species well upstream of the mouths of several small tributary creeks (fig. 8.16). The relatively steady flow of water from the Edwards Aquifer probably also contributes to habitat suitability in smaller creeks.

Reservoirs on the upper Colorado drainage (E.V. Spence Reservoir, O.C. Fisher Lake, Lake Buchanan, and Inks Lake) are inhabited by Texas map turtles, although there are no reports of their abundance in reservoirs relative to rivers. The species occurs in Buck Lake, a disjunct oxbow of the South Llano (pers. observation 2006).

Figure 8.16. Four small tributary creeks that are inhabited by Texas map turtles: (a) White Oak Creek, Tierra Linda Ranch, Gillespie County; (b) Live Oak Creek, Ladybird Johnson Municipal Park, Gillespie County; (c) Rough Creek, Highway 580, San Saba County; and (d) Elm Creek, Ballinger City Park, Runnels County.

Diet

W. Lehmann (1979, cited in Bertl and Killebrew 1983) reported on the diets of young specimens she identified as ranging in age from hatchling to four years, finding primarily aquatic insects, with mayfly nymphs and caddis fly larvae being most important. Kizirian et al. (1990) reported large quantities of gastropod shells in a female specimen of record size from the San Saba. On the South Llano, substantial change occurred in female diets, but not in male diets, over a fifty-year period associated with the invasion of the habitat by Asian clams (*Corbicula* spp.). Whereas adult females had once consumed a diverse diet of caddis fly larvae, native sphaeriid clams, bryozoan colonies, sponges, and filamentous algae, their diet now consists almost entirely of exotic clams (table 4.16; Lindeman 2006a). Males feed most heavily on caddis fly larvae, mayfly nymphs, and snails.

Reproduction and Life History

The only published life-history data for Texas map turtles are from the South Llano River, from a comparison of museum specimens collected in 1949 with specimens captured in 1998–2000 (Lindeman 2005). In females, growth fit a von Bertalanffy curve with comparatively rapid approach to asymptotic size, consistent with their rapid maturation (seventh growing season) and small size at maturation (115 mm). Males matured in their second or third full growing seasons at 57 mm midline plastron length. Females averaged 5.6 eggs/clutch (range 4–9) and may have produced as many as four annual clutches. Eggs averaged 35.1 × 20.7 mm. Neither body size nor clutch size showed a change between time periods, in spite of a change in female diet. Range-wide, midline plastron length ranges from 55 to 95 mm in adult males and reaches 180 mm in adult females (Lindeman 2008a), although maximum sizes in the South Llano were considerably smaller, at 79 mm for males and 163 mm for females (Lindeman 2005). Sizes for both sexes in White Oak Creek, a tributary of the Pedernales, were similar to those of the South Llano population (to 79 mm in males and to 166 mm in females; J. Dobie, pers. comm. 2008).

Natural History

In White Oak Creek, males court females primarily in October and early November (four observations), but pairs have also been observed in April (four observations) and even once in July (J. Dobie, pers. comm. 2008). Behavior resembling courtship exhibited among captive males involves head bobbing but not foreclaw titillation (C. H. Ernst and Barbour 1972; Fritz 1991). Female Texas map turtles had considerable parasite loads, mainly of an enteric spiny-headed worm, while male parasite loads were far lower (Lindeman and Barger 2005). Two species of coccidian parasites have been reported (McAllister et al. 1991).

Populations

Little has been published on the abundance of Texas map turtles. McKinney (1987) captured 42 in a total sample of 170 turtles (25 percent) in the lower South Llano River in Kimble County in 1986. Farther upstream, in 1998–2000, they constituted 26 percent of 424 turtle captures made using fyke nets, basking traps, and hand captures, with the Texas river cooter (*Pseudemys texana*) predominating (56 percent of all captures; Lindeman 2005, 2007, and unpublished data).

A historical decline in abundance on the South Llano was suggested by a comparison of specimen collecting one morning in 1949, when forty-two specimens were captured by hand by three wading searchers, with hand capture rates in 1998–2000, when fewer than one turtle was captured per searcher per day (Lindeman 2004). Hand capture rates in 1998–2000 were higher on stream reaches flanked by private ranches than on those adjacent to public access provided by highway crossings, suggesting that collecting by hobbyists has played a role in population declines.

Conservation Status and Actions

Briefly considered for listing under the Endangered Species Act in the late 1970s and early 1980s but dropped from further consideration. Listed as a species of least concern with stable populations by the IUCN.

Federal habitat protection occurs at the Lyndon B. Johnson National Historical Site, managed by the National Park Service on the Pedernales across from Lyndon B. Johnson State Park. Additional short stream reaches are protected at Colorado Bend State Park, Fort McKavett State Historical Park on the San Saba, Pedernales Falls State Park, San Angelo State Park on the North Concho River and O. C. Fisher Reservoir, and South Llano River State Park.

Needed Further Research

The Texas map turtle has several intriguing biological features that are worthy of investigation: the factors that allow it to range far up into small creeks, nesting biology, interpopulational variation in life history, and the role of private lands in conserving the species. A study of the factors regulating abundance in a network of creeks joining one of the main-stem tributaries of the Colorado drainage would be of great interest; it could possibly be combined with testing the hypothesis that populations are relatively more depleted on stream stretches with public access. Given the absence of sandbars in inhabited streams, a determination of nesting habitat combined with studies of nest survival rates and temperature-dependent sex determination in the wild promise an interesting contrast to the nesting biology of congeners. The South Llano River population should be compared with other populations to determine the full extent of variation in body size and its ecological and reproductive correlates.

Sabine Map Turtle, *Graptemys sabinensis*

Figure 8.17. Adult male, Sabine River at Palmer Lake Road, Beauregard Parish, Louisiana.

> *Graptemys ouachitensis* may . . . consist of multiple species. The two subspecies of *G. ouachitensis* were not closely related in our preferred tree . . . and did not appear as sister-taxa in any analyses. . . . Furthermore, no hybrids or areas of natural sympatry between *G. o. ouachitensis* and *G. o. sabinensis* have been reported.
> PATRICK STEPHENS AND JOHN WIENS (2003, PP. 596–97)

Alternate Common Names

None; common name first used by Cagle (1954).

Description and Diagnosis

Postorbital markings small ovals, rectangles, or chevrons, under which several thin neck stripes extend toward orbits. Interorbital mark a medial longitudinal stripe widening slightly posteriorly, typically flanked by one to three thinner longitudinal lines. Supramandibular spots subsumed by often prominent ventrolateral neck stripes. Chin with transverse bar anteriorly, subsuming submandibular spots. Markings on head, neck, limbs, and tail yellow. Females and males with narrow heads and narrow alveolar surfaces. Second vertebral scute steeply sloped, distinctly higher than the third. Carapace highly peaked at posterior of second vertebral, nearly triangular in cross section. Carapacial keels producing prominent points in males and juveniles, reduced and blunted in females. Black pigment along midline of carapace, particularly at keel points. Carapace marked with reticulated, black-bordered, yellowish-orange markings with numerous points of intersection on each pleural scute, often forming circles or ovals with concentric, alternating light and dark markings. Plastron pale yellow, with large oblong central figure of undulating black and yellow lines arranged alternately in symmetrical pattern. Male foreclaws conspicuously elongated. Irises white or occasionally yellow, often but not always bisected by black stripes. Differs from congeners by the transverse chin bar, the small postorbitals allowing several neck stripes to reach the orbits, the neck stripes that subsume the supramandibular spots, and the high, peaked carapace.

Taxonomic History and Phylogeny

Specimens of this species were first collected in Louisiana from the Mermentau River and from near the town of Vinton (probably the Sabine River) in 1893–94 but were never formally described (fig. 8.18; see chapter 2). A single specimen was collected from the Calcasieu River by a Tulane field crew in 1948 (TU 3473). Specimens in the USNM that are listed as having been collected in 1935 from the False and Red Rivers, Louisiana, are definitely *sabinensis*, but Cagle (1953b, p. 9) questioned the distribution records, noting that the specimens may have been purchased at these sites rather than collected there. The species was described by Cagle (1953b), who designated as holotype an adult female (UMMZ 104351) collected "from the Sabine River, 8 miles southwest of Negreet [Sabine Parish], Louisiana" (p. 2) on 5 July 1950. The specific epithet means "from the Sabine." Cagle also designated 158 paratypes from the type locality: FMNH 67105–16; TU 13110–11, 13115, 13119–21, 13127–28, 13131, 13139, 13141–42, 13148–49, 13152, 13160, 13166, 13172, 13175, 13177–79, 13181, 13185–86, 13190, 13194–95, 13197, 13200, 13202–204, 13206–209, 13253, 13258, 13261–62, 13510 (2 specimens), 13564 (3), 13740 (13), 13741 (7), 13743 (8), 13744 (7), 13745 (4), 13746 (5), 13747 (2), 13748 (17), 13760 (15); UIMNH 26718–22; and UMMZ 104352–69. The primary collecting technique used was nighttime hand capture (Chaney and Smith 1950).

While *Graptemys sabinensis* Cagle 1953 could properly be regarded as a junior synonym of *Graptemys intermedia* (Beyer 1900), as described in chapter 2, a more sensible course of action would be a petition to suppress the never-again-used *intermedia* in favor of *sabinensis*, which has been in common usage for over half a century. Action on this point should be withheld, however, until such time as an analysis of morphological and genetic variation can be conducted on specimens from the Sabine-Neches, Calcasieu, and Mermentau drainages. Should the populations prove to be distinct, the epithet *sabinensis* would be retained for the Sabine-Neches drainage and *intermedia* would be the proper epithet applied to populations from the Mermentau.

Figure 8.18. (a) Kohn's specimens of the Sabine map turtle from the Mermentau River (TU 7632, 7634, 7640, and 7690). (b) View looking downstream at the Mermentau River from Highway 3166 near Silverwood in Jefferson Davis Parish, Louisiana.

Cagle described the Sabine map turtle as a subspecies of the false map turtle, *G. pseudogeographica sabinensis*, but discussed the possibility that it might instead be a species separate from *pseudogeographica*, in a "species group" with *versa* and *ouachitensis*, taxa he also considered subspecies of *pseudogeographica*. Vogt (1978, 1993) recognized *ouachitensis* and *sabinensis* as conspecific but separate from *pseudogeographica*. He noted distinct differences in head and shell patterns overall but found some Sabine River specimens had head patterns indistinguishable from those of *ouachitensis* and considered them intergrades. Also, he found no consistent allozyme differences between the taxa. Because both *ouachitensis* and *sabinensis* had been described simultaneously (Cagle 1953b), Vogt, as first reviewer, was free to choose either name as the specific epithet; he chose the former, designating a widespread nominate subspecies and another subspecies, *sabinensis*, with a more restricted distribution.

J. P. Ward (1980) examined skull anatomy and found *sabinensis* to be "so unlike any other [*Graptemys* species] that it represents a distinct entity" (p. 302) and regarded it as a separate species, based primarily on the fact that it was the only species of *Graptemys* that has prefrontal bones prevented from medial contact by frontal bones that extend along their entire lengths. Dundee and Rossman (1989) criticized Ward's conclusion due to the small sample size (two or three skulls) and consequent lack of consideration of sexual dimorphism. More recently, Vetter (2004) and Buhlmann et al. (2008) recognized *sabinensis* as a separate species without further comment. While analysis of mitochondrial DNA sequences initially placed the Sabine map turtle as sister species to *G. ouachitensis* (Lamb et al. 1994), reanalysis with more outgroup taxa resulted in a polytomy that also included *G. oculifera* and *G. flavimaculata* (Lamb and Osentoski 1997). Analysis of larger sets of molecular and morphological data placed the species as sister species to *G. versa* (Stephens and Wiens 2003) and as sister species to a clade composed of *G. caglei* and *G. versa* (Wiens et al. 2010), with a more distant relationship to *G. ouachitensis* in both cases.

Range

The Sabine map turtle inhabits the Sabine-Neches, Calcasieu, and Mermentau drainages in eastern Texas and southwestern Louisiana. (It is often erroneously labeled a Sabine endemic, in some cases even as authors discuss its occurrence in other drainages; e.g., Vogt 1993, C. H. Ernst et al. 1994; Lamb and Osentoski 1997; Moll and Moll 2004; Vetter 2004; C. H. Ernst and Lovich 2009; A. D. Brown et al. 2012.) It occurs in the lower Neches River and a tributary, Village Creek (Rudolph 1983; J. R. Dixon 2000). In the Sabine River the species occurs from Calcasieu Parish, Louisiana (Dundee and Rossman 1989), upstream to the vicinity of Silver Lake, a reservoir in Van Zandt County, Texas, where recent upstream range expansion is suspected to have occurred due to altered flow regime (Harvey 1992). It occurs in two Sabine tributaries, Anacoco Bayou and

Bayou Toro (Shively 2001). Harvey (1992) also noted occurrence in Caddo Lake, a natural water body on the Texas-Louisiana border that lies near the Sabine but within the Red River drainage; however, collections in the lake by Tulane field crews yielded only *G. ouachitensis* and *G. pseudogeographica kohnii* (see Vogt 1993).

The Sabine map turtle occurs in the Calcasieu River and seven of its tributaries: the West Fork Calcasieu, Houston, Little, and Whiskey (= Ouiska) Chitto Rivers; English and Indian Bayous; and Bundick Creek (Cagle 1953b; Shively and Jackson 1985; Dundee and Rossman 1989; Shively 2001; pers. observations 2008, 2011–12). It occurs in the Mermentau drainage from below Lake Arthur upstream into five major tributaries: the bayous Lacassine, Queue de Tortue, Nezpique, des Cannes, and Plaquemine Brule (Shively 2001; pers. observations 2010; fig. 8.18).

Habitat

Shively and Jackson (1985) found that Sabine map turtle density was positively related to stream width, flow rate, light levels, chlorophyll levels, and insect biomass on submerged limbs over six sites on the Whiskey Chitto River. They assessed the influence of abiotic and biotic environmental parameters on the species' declining abundance and eventual disappearance at upstream localities. Their model suggested that progressive narrowing of the creek at upstream sites blocked sunlight that reduced the density of algae, moss, and insects that the turtles were observed grazing from submerged deadwood. They also classified short (23 m) segments of two stream reaches with intermediate turtle densities as favorable (occupied) or unfavorable (unoccupied) habitat and plotted the distribution of favorable habitat over all sites. Favorable reaches were shorter and more widely spaced upstream, suggesting that increasing dispersal distances among patches of decreasing size may set the upstream range limit. The Mermentau drainage and the lower Calcasieu drainage are sluggish stream systems with relatively abrupt banks that lack sandbars.

Sabine map turtles are abundant in Lake Arthur, a natural widening of the lower Mermentau (pers. observations 2010). One specimen (TCWC 48495) was taken from the south end of the large Toledo Bend Reservoir on the Sabine River. There are no records from oxbow lakes or backwaters.

Diet

In the Whiskey Chitto River, filamentous algae and caddis fly larvae predominated in the diet, the former being about two-thirds of the dry mass recovered from adult females and the latter being more than two-thirds of that recovered from adult males and juveniles (table 4.7; Shively 1982; Shively and Jackson 1985). Additional important prey were Asian clams (3–13 percent of dry mass of the three classes) and mosses (6 percent of females' dry mass); snails were absent from the diet. Based on relative abundance in dietary samples versus

scrapings from submerged deadwood, *Hydropsyche* caddis fly larvae were selectively grazed.

Reproduction and Life History

Reproductive data have been reported from the Sabine and Calcasieu Rivers
(Ewert, Doody, et al. 2004). Clutch size ranged from 1 to 4 with a mode of 2 eggs
and a mean of 2.3. Eggs averaged 9.9 g, and hatchlings averaged 7.4 g. Nesting
was observed between 13 June and 16 July over six years. Most nests were in
open locations on sandbars and sandy riverbanks, although some were partially
shaded by understory (S. Doody, pers. comm. 2010). Females from the Mermentau and lower Calcasieu drainages have larger clutch sizes, ranging up to 7
eggs (P. Lindeman, unpublished data). Range-wide, reported midline plastron
length ranges from 64 to 92 mm in adult males and reaches 176 mm in adult
females (Lindeman 2008a), although a specimen collected in 1893 near Vinton,
Louisiana, measures 202 mm and females as large as 189 mm have been captured recently (P. Lindeman, unpublished data).

Natural History

Courting males drum the backs of their forefeet claws against the ocular regions of females (Artner 2001a). Coleman and Gutberlet (2008) reported activity in all months of the year, with peak basking counts in March–May and
September–November.

Populations

Sabine map turtles dominated in the trapping data of Tinkle (1958b) on the Sabine River, constituting 78 percent of his catch of five species. In basking counts
on the upper Sabine, Sabine map turtles outnumbered Mississippi map turtles
by a ratio of 8.4:1 (Coleman and Gutberlet 2008). They constituted 13 percent of
turtle captures in baited traps, ranking third of nine species, in the same region
of the river (Hively 2009).

On the Whiskey Chitto River, Shively and Jackson (1985) marked individuals with enamel paint over three days and calculated averages of 10 counts of
marked and unmarked basking turtles on the fourth day at each site. Total estimated population for two lower reaches averaged 150 and 65 turtles/ha of river
surface, while two upper reaches averaged 40 and 28 turtles/ha. Basking abundance averaged 27.2 turtles/river km in the upper Sabine River (Coleman and
Gutberlet 2008).

Conservation Status and Actions

Briefly considered for federal listing under the Endangered Species Act in the
late 1970s and early 1980s but dropped from further consideration. Listed as a
species of least concern with stable populations (albeit as part of *G. ouachitensis*) by the IUCN.

There is extensive federal protection of habitat in Texas. Almost the entire Neches River riparian zone within the species' range is protected as part of the Big Thicket National Preserve, managed by the National Park Service. The Sabine National Forest protects most of the west shoreline of the large Toledo Bend Reservoir, the largest impoundment of the Sabine. Parts of Village Creek, a Neches tributary, are protected within Village Creek State Park and the Nature Conservancy's Roy E. Larsen Sandyland Sanctuary. Habitat protection in Louisiana is limited to Sam Houston Jones State Park on the West Fork Calcasieu.

Needed Further Research

Life history has received only a cursory examination and warrants further study, as some populations of the species are at an extreme for small body size and small clutch size within the genus. There has been no examination of interdrainage variation in morphology or genetic markers, which may be of taxonomic interest. Additionally, the Mermentau and lower Calcasieu habitat appears different enough to possibly relate to differences in diet or life history. While it is clear that Sabine map turtles can be abundant at some locations, it is not clear whether the best population strongholds are in the Neches, Sabine, Calcasieu, or Mermentau drainages, nor is it known where the species is most abundant within each; thus identification of the best population strongholds is a priority. At 105 km in length with nearly 2,000 km of shoreline along its embayments, the Toledo Bend Reservoir has greatly altered the middle Sabine. Since its construction in the 1960s, the reservoir has been the source of just one specimen, but whether extensive searches have been made on the reservoir is unclear. An investigation of the reservoir ecology of the species would be a valuable addition to the conservation biology of the genus.

False Map Turtle, *Graptemys pseudogeographica*

Figure 8.19. Adult female, Missouri River at LaFramboise Island, Hughes County, South Dakota.

> The false map turtle complex . . . has perplexed taxonomists and confused ecologists since the time of the original descriptions.
>
> RICHARD VOGT (1980A, P. 17)

Alternate Common Names

Pseudo-geographic tortoise (DeKay 1842), LeSueur's map tortoise (Hay 1892a), LeSueur's terrapin (Ditmars 1907), midland sawback turtle (A. Carr 1952), none widely adopted; Mississippi map turtle (K. P. Schmidt 1953) remains in usage for the subspecies *kohnii* (although originally assigned to the nominate subspecies by Pope 1939); Kohn's terrapin (Beyer 1900), Kohn's false map turtle (Peterson 1950), Gulf Coast map turtle (B. C. Brown 1950), Kohn's map turtle (Cagle 1952a), Mississippi sawback turtle (A. Carr 1952), bayou map turtle (K. P. Schmidt 1953), all likewise for the subspecies *kohnii* but none widely adopted; northern false map turtle (Crother et al. 2012) for the nominate subspecies. First called false map turtle by Cagle (1952a).

Description and Diagnosis

Postorbital markings typically narrow vertical bars not contacting small supramandibular spots (northern populations), or narrow crescents curving forward under eyes and subsuming supramandibular spots (southern populations); may or may not be continuous with neck stripes. Bars allow neck stripes to extend to orbits; crescents do not. Interorbital mark a medial longitudinal stripe, typically flanked by one to three thinner longitudinal lines. Chin marked with small anterior spot and small submandibular spots. Markings on head, neck, limbs, and tail yellow to light orange. Females with moderately broad heads and moderately broad alveolar surfaces, males with narrow heads and broad alveolar surfaces. Second vertebral scute only slightly sloped, extending little higher than the third. Carapace not highly peaked, rounded in cross section in northern populations; somewhat highly peaked in southern populations. Carapacial keels producing prominent points in males and juveniles, reduced and blunted in females. Black pigment along midline of carapace, particularly at keel points. Carapace marked with reticulating network of thin, black-bordered, yellowish-orange lines on pleural scutes having comparatively few intersections (one to two rings per pleural), continuing onto marginals. Large specimens of both sexes often with large, melanistic blotches on carapacial scutes. Plastron pale yellow, with large oblong central figure of undulating black and yellow lines arranged alternately in a symmetrical pattern. Male foreclaws conspicuously elongated. Irises yellow, bisected by black stripes in northern populations; white and usually without black stripes in most southern populations; nearly black in the Calcasieu drainage. Differs from congeners by the narrowness of the postorbital markings and the small size of the supra- and submandibular spots and chin spots.

The southern subspecies *kohnii* differs from the nominate, northern subspecies with regard to the crescents (fused postorbital markings and supramandibular spots) and the white, unstriped iris color. Purported intergrades have interrupted crescents that allow few neck stripes to extend to the orbits and may have dark spots, or rarely stripes, on white irises (Vogt 1993; Lindeman 2003a).

Taxonomic History and Phylogeny

The first specimens of this species were from the Wabash River in Indiana. They were described by LeSueur (1827), who found them to be distinct yet did not propose a name. The species was inadvertently authored as *Emys pseudogeographica* by J. E. Gray (1831) in his description of a junior synonym of *Graptemys geographica*, *G. lesueurii* (see Hay 1892b), in which he mentions "LeSueur Mss. (Mus. Paris)" containing the name *pseudogeographica*. This is probably a reference to the handwritten specific epithet of LeSueur on some of his Wabash River specimens' plastra (Bour and Dubois 1983; see chapter 2). The name alludes to a resemblance of the species to *geographica* (at the time, no other *Graptemys* had been described). The type series of specimens was not noted by Gray, but Bour and Dubois examined nineteen specimens collected by LeSueur in the Muséum National d'Histoire Naturelle in Paris. Only four could be positively identified as likely candidates for the syntypes Gray had examined during his visit to Paris in 1830; most of the others are of uncertain collecting dates (Bonnemains and Bour 1996). Of the four specimens collected in 1827 or 1828, one taxidermic mount (MNHN 9146) was identified by Bour and Dubois as a *G. ouachitensis*

Figure 8.20. TU 16409, apparently the sole surviving syntype of *Malacoclemmys kohnii* Baur 1893 (now *Graptemys pseudogeographica kohnii*). The specimen is herein designated as lectotype for the taxon.

and two carapaces (MNHN 9136 and 9137) were considered of uncertain species identity, leaving only MNHN 9147, a taxidermic mount they designated the species' lectotype. Bonnemains and Bour (1996) also listed MNHN 9146 as *G. ouachitensis* but concluded (without elaboration) that MNHN 9136 and 9137 were *G. pseudogeographica*; they further suggested the existence of one additional syntype, MNHN 9139, another carapace they identified as *G. pseudogeographica*. The four syntypes other than MNHN 9147 should be regarded as paralectotypes.

A taxon now regarded as conspecific with *pseudogeographica* was described by Baur (1890) as *Malacoclemmys kohnii*. He named as types a pair of specimens from "Bayou Lafourche [LaFourche Parish], L[ouisian]a; Bayou Teche, St. Martinsville [St. Martin Parish], L[ouisian]a" (p. 263) and also examined a specimen labeled as being from Pensacola, Florida. All three were in the private collection of Joseph Gustave Kohn (1837–1906), whom the patronym honors. Kohn authorized Baur to deposit types in the USNM, and Baur promised USNM curator Leonhard Stejneger one stuffed specimen, with a second stuffed specimen to be returned to Kohn, but no specimen sent to the USNM can be traced. Apparently the only remaining syntype is the Bayou Teche specimen (TU 16409, a male collected in July 1874; fig. 8.20), which I hereby designate the lectotype of *M. kohnii* Baur in order to ensure stability in name application in the taxonomically volatile *pseudogeographica-kohnii-ouachitensis-sabinensis* complex.

Stejneger and Barbour (1917, 1923, 1933, 1939, 1943) regarded *kohnii* and *pseudogeographica* as conspecific, as did A. Carr (1949, 1952). Cagle (1953b) acknowledged possible conspecific status but chose to revert to recognition of *kohnii* as a distinct species, an opinion followed in the three editions of the Peterson field guide (Conant 1958, 1975; Conant and Collins 1991). Vogt (1978, 1993) regarded the two taxa as conspecific subspecies and published a dot distribution range map showing putative areas of intergradation in northeastern Kansas, Missouri, southwestern Illinois, western Kentucky, and western Tennessee. Specimens from the lower Tennessee River in western Kentucky showed primarily *kohnii* characteristics, with a few specimens showing *pseudogeographica* characteristics (Lindeman 2003a). Close affinity of the taxa has been borne out in phylogenetic studies (Lamb et al. 1994; Stephens and Wiens 2003). Three additional taxa (*versa*, *sabinensis*, and *ouachitensis*) were originally described as subspecies of *pseudogeographica* (Stejneger 1925; Cagle 1953b), and *oculifera* was relegated to the same status by Stejneger and Barbour (1917, 1923, 1933) in the first three editions of their checklist; all four taxa are now considered separate species. Phylogenetic analyses placed the false map turtle in poorly resolved basal positions relative to *G. oculifera*, *G. flavimaculata*, *G. ouachitensis*, *G. sabinensis*, *G. nigrinoda*, *G. versa*, and in one case *G. caglei* (Lamb et al. 1994; Lamb and Osentoski 1997) and as sister species to a clade composed of *G. flavimaculata*, *G. oculifera*, *G. nigrinoda*, and *G. ouachitensis* (Stephens and Wiens 2003; Wiens et al. 2010).

Range

The false map turtle occurs widely in the Mississippi drainage and several smaller drainages to the west. Records mapped by Vogt (1993) for the St. Joseph River, which drains into Lake Michigan in northern Indiana, and the Maumee River, which drains into Lake Erie in northern Ohio, are problematic. A specimen from the St. Joseph (USNM 194611) was identified as *G. ouachitensis* by Burling (1972) but regarded as *G. pseudogeographica pseudogeographica* by Vogt (1993), who also entered a note in the USNM catalog that the specimen showed evidence of having been a released captive. Dolley (1933) also noted false map turtles were one of four "abundant" turtles in the St. Joseph but did not preserve specimens. While several specimens of common map turtles have been collected from the Maumee, MCZ 1727 is the only record of a false map turtle in the drainage; its exact collecting date is not known, but it is of sufficient antiquity to regard it dubiously. With no further specimens, I chose to not map these localities.

The species' upstream range in the Mississippi reaches Hennepin County, Minnesota (Oldfield and Moriarty 1994). Tributaries of the upper Mississippi that are inhabited include the lower St. Croix River along the Minnesota-Wisconsin border (C. H. Ernst 1973; DonnerWright et al. 1999); the Chippewa River and its tributary the Red Cedar River in Wisconsin (Casper 1996; G. Casper, pers.

comm. 2008); the Minnesota River into southwestern Minnesota (C. H. Ernst 1973; Gamble and Moriarty 2006); the Iowa River and its tributary, the Cedar River, in Iowa (Vogt 1993; Vandewalle and Christiansen 1996); the Illinois River (Lamer et al. 2008), with an old record (ca. 1905) far up a tributary, the Fox River (Vogt 1993); and the Big Muddy River into a tributary that is dammed as Crab Orchard Lake in Williamson County, Illinois (Vogt 1993).

False map turtles range far up the Missouri drainage, into south-central North Dakota, and have been reported from two major tributaries, the Vermillion and James Rivers in South Dakota (Timken 1968a, 1968b; Lynch 1985; Jundt 2000; A. Gregor, pers. comm. 2008; LeClere et al. 2009). In lower Missouri tributaries, the species has been reported from the Platte River in Nebraska (Hudson 1942); Branched Oak Reservoir in the Salt Creek drainage and the upper Logan Creek drainage, both part of the Platte drainage (Lynch 1985; Ballinger et al. 2010); the Kansas River drainage into its tributaries, the Saline, Solomon, and Big Blue Rivers, in north-central Kansas (Taggart 1992; KU 224654, a specimen which was misidentified in Sievert and Collins 1998; KU 218789); and the Osage drainage in central Missouri (Daniel and Edmond 2012) upstream via its tributary, the Marais des Cygnes River, into Long Creek in Kansas (Clarke 1953; Vogt 1993).

In the Ohio River the false map turtle occurs to about its confluence with the Wabash River (C. A. Phillips et al. 1999). It ranges into three tributaries of the lower Ohio: the Tennessee River upstream into Bear Creek in northeastern Mississippi (AUM 7431), the Cumberland River upstream into western Tennessee (Vogt 1993), and the Wabash River along the Illinois-Indiana border (Ewert 1979, 1989) and three of its tributaries, the Embarras River in Illinois (Peters 1942; Vogt 1993) and the Tippecanoe and Mississinewa Rivers in Indiana (Rizkalla and Swihart 2006; C. Rizkalla, pers. comm. 2007). Vogt's (1993) record of the species in the upper Scioto River in Franklin County, Ohio (USNM 131834, collected in the mid-1800s), is disregarded, as notes in the USNM catalog suggest uncertainty regarding whether the locality is really in Ohio or Mississippi. A recent photograph of a basking individual in the Great Miami River (Krusling et al. 2010b), if confirmed to represent a native population, would extend the range into Ohio, however.

The species has been found in the Mississippi as far downstream as northern Plaquemines Parish, Louisiana, 126 river km from the mouth of the river (TU 21475). Five eastern tributaries of the river in the state of Mississippi are inhabited: the Buffalo River (TU 30312); the Homochitto River (MMNH 3891); Bayou Pierre (MMNH 4020); the Big Black River upstream to Choctaw and Webster Counties (USNM 318527), with occurrences in two tributaries, Fourteenmile Creek (TU 21744) and Doaks Creek (TU 21479); and the Yazoo River and its upstream tributaries, the Big Sunflower River (MMNH 7655), the Coldwater River (APSU 17549), and the Yalobusha River (TU 21480). The species also occurs in the lower Comite River in Louisiana (Myers 2008).

On the west side of the lower Mississippi, the false map turtle occurs extensively in the St. Francis and White drainages of Arkansas and Missouri (Shipman and Riedle 1994; Trauth et al. 2004; Daniel and Edmond 2012). The biggest western tributary that is inhabited is the Arkansas River, in which the false map turtle ranges into several tributaries. Inhabited northern tributaries entering Kansas and Missouri include the Neosho River and three of its tributaries, the Cottonwood and Spring Rivers and Labette Creek; the Verdigris River; and the Walnut River and its major tributary, Grouse Creek (Clarke 1956; J. T. Collins 1993; Vogt 1993; Fuselier and Edds 1994; D. Edds, pers. comm. 2007; Young and Thompson 1995; Riedle and Hynek 2002). Southern tributaries of the Arkansas that are inhabited include the Fourche La Fave, Petit Jean, Poteau, and Canadian drainages in Arkansas and Oklahoma (Trauth et al. 2004; Riedle et al. 2009).

Farther south along the western side of the Mississippi, the species inhabits the Black-Ouachita-Tensas river drainage and several of its tributaries, including bayous Bartholomew and D'Arbonne and the Boeuf, Little, Saline, and Little Missouri Rivers, in Louisiana and Arkansas (Haynes and McKown 1974; Dundee and Rossman 1989; Vogt 1993; J. L. Carr 2001; Trauth et al. 2004). In the main-stem Red River, the species occurs upstream to Grayson County, Texas (J. R. Dixon 2000), although specimens are also known from farther upstream, on Medicine Creek, a tributary of East Cache Creek that joins the Red River in Comanche County, Oklahoma (OKMNH 35453), and from a farm in Hardeman County, Texas (Killebrew et al. 1996). Additional Red tributaries with records include Bois d'Arc and Coffee Mill Creeks and the Sulphur River in northeastern Texas, Caddo Lake and Ferndale Lake on the Cypress Creek drainage in Texas and Louisiana, and the Little River and its tributaries in Oklahoma (Cagle and Chaney 1950; McAllister et al. 1983; J. S. McCord and Dorcas 1989; Vogt 1993; TCWC 85860; UTA 38634). The species is widespread in the drainages of the Atchafalaya River, a westward-diverging offshoot of the lower Mississippi in Louisiana, with records from Bayou Teche, Bayou Courtableau, Little Alabama Bayou, Henderson Swamp, and the Butte La Rose Canal (Vogt 1993).

Six drainages are inhabited west of the Mississippi in Louisiana and Texas. A specimen was collected from the Mermentau River in the 1890s (TU 7631), and two specimens were collected in or near Rayne in 1960, presumably from a tributary, Bayou Wikoff (Dundee and Rossman 1989); the species also occurs in bayous Nezpique and Plaquemine Brule (pers. observations 2010). The Calcasieu River and its five tributaries—the Houston, West Fork Calcasieu, and Whiskey Chitto Rivers; Indian Bayou; and Bundick Creek—are inhabited (Shively and Jackson 1985; Shively 2001; pers. observations 2008, 2012). In the Sabine portion of the Sabine-Neches drainage, the species occurs from Orange County upstream to Rains and Van Zandt Counties and also ranges into two tributaries in Louisiana, Bayou Anacoco and Bayou Toro (J. R. Dixon 2000;

Shively 2001). In the Neches River, it ranges from the Jasper-Hardin County line upstream to Henderson County, with occurrences in two tributaries, Pine Island Bayou and Village Creek (Haynes and McKown 1974; Rudolph 1983; Vogt 1993; J. R. Dixon 2000; TU 21477). In the Trinity River, it ranges from Liberty County upstream to Dallas County (J. R. Dixon 2000; S. Shively, pers. comm. 2009). In the Brazos drainage, it occurs from Washington County upstream to Young County in the main-stem river and ranges into several Brazos tributaries, including the Navasota, Nolan, and Little Rivers; the Clear Fork of the Brazos; Pin Oak Creek; and two impoundments of small creeks, Wickson Lake on Allcorn Creek and Lake Somerville on Yegua Creek (Tinkle and Knopf 1964; Haynes and McKown 1974; Vogt 1993; J. R. Dixon 2000; S. J. Taylor et al. 2003; D. J. Brown et al. 2008; TCWC 82374). The species was recently reported for the first time in the small San Bernard drainage, just west of the lower Brazos (Rodriguez et al. 2006).

Habitat

Habitat studies of the false map turtle show it to be largely tied to medium-to-large rivers, but it also uses floodplain habitats and persists in impoundments. Compared with a sympatric congener, it ranged considerably farther upstream, into narrow, shaded stretches of a small stream in Louisiana (Shively and Jackson 1985). Fuselier and Edds (1994), trapping in several streams in eastern Kansas, and DonnerWright et al. (1999), trapping along the St. Croix in Wisconsin, found the species associated with high abundance of emergent deadwood and muddy substrates, with river width also directly related to abundance in the latter study. Correlation of basking abundance to deadwood density was not significant across twenty sites on the lower Tennessee River and its impoundment, Kentucky Lake, however (Lindeman, 1999c). High abundance in a Louisiana bayou occurred in an area with abundant molluscan prey (Shively 2001).

In a complex mosaic of habitats along the lower Missouri River, telemetry locations and recaptures indicated that males and females moved freely between the river and floodplain habitats (Bodie and Semlitsch 2000b). They made greatest use of the river channel, backwater sloughs, and scours created by levee breaches during major flooding but also occupied flooded forests and agricultural lands.

False map turtles occur in the coves of dendritic reservoirs (e.g., Lindeman 1999c). A large population occurs in Reelfoot Lake in Tennessee (Vogt 1993), which was created during a series of major earthquakes in 1811 and 1812. Many specimen records come from oxbow lakes along the lower Mississippi River, and the species is commonly reported from sloughs, bayous, and lentic habitats of floodplains such as oxbow lakes (e.g., Moll 1977; Vogt 1980a; Bodie and Semlitsch 2000b; J. L. Carr 2001; R. A. Anderson et al. 2002; Dreslik et al. 2005; Dreslik and Phillips 2005; Rizkalla and Swihart 2006).

Diet

Moll (1976b) reported the diet of false map turtles from a floodplain lake in Illinois and the Mississippi River in Tennessee (specimens from both sites were misidentified as Ouachita map turtles; D. Moll, pers. comm. 2010). He later gave a more detailed breakdown of the diets of the lake population and a population in the Illinois River's main channel (table 4.14; Moll 1977). In the lake, heavy siltation caused a historical change from a diverse diet of insects and aquatic vegetation to one dominated by benthic midge larvae. The turtles in the Mississippi River fed primarily on terrestrial vegetation submerged by high water, while those in the Illinois River fed exclusively on caddis fly larvae.

Vogt (1981a) reported the diet of specimens from the Mississippi and Wisconsin Rivers, Wisconsin, the White River in Arkansas, and unnamed sites in Louisiana, concentrating on adult females. Vegetation—principally pondweed, duckweed, and algae—was the primary food category for females in Wisconsin, followed by mollusks and mayfly nymphs (table 4.8). In Arkansas and Louisiana, females had broader heads and fed more heavily on mollusks. Males in both regions fed primarily on insects and did not eat vegetation. Shively and Vidrine (1984) reported nearly exclusive consumption of Asian clams by females in Louisiana, although in a study of diet in a Kentucky reservoir, the index of relative importance value for Asian clams was just 44 of a possible 100 for adult females and 33 for smaller females (table 4.15; Lindeman 1997c, 2000b). Additional major prey items of adult females in Kentucky were fish carrion, mayfly nymphs, and algae. Smaller females also took snails, algae, and bryozoan colonies as their major prey. Males took chironomid egg cases, algae, Asian clams, and sponges as their major prey. Small snails were the major prey of unsexed juveniles, which also took much smaller quantities of insects (primarily adult flies and caddis fly larvae). J. L. Carr (2008) observed repeated brief bouts of terrestrial foraging, apparently on herbaceous plants, along a riverbank in Louisiana.

Reproduction and Life History

Growth in Wisconsin and Kentucky follows a von Bertalanffy pattern of steadily declining growth rate with age (Vogt 1980a; Lindeman 1999b). Males matured in their third full growing season in Kentucky and fourth in Wisconsin. The youngest mature females in Wisconsin were in their eighth growing season, while growth curves suggested maturation occurred during the eleventh growing season in Kentucky.

In the Wabash, LeSueur (1827) anecdotally reported clutch sizes (presumably of mixed false and Ouachita map turtles; see Bour and Dubois 1983) that ranged as high as 20–24 eggs, but more recent accounts give smaller average clutch sizes: 12.3, 10.8, and 10.9 eggs in South Dakota (overall range 6–18; Timken 1968a; L. A. Dixon 2009; A. Gregor, pers. comm. 2008); 14.1 and 12.8

eggs in Wisconsin (range 8–19; Vogt 1980a; Janzen et al. 1995); and 6.6 eggs in Louisiana (range 4–11; J. L. Carr 2001). In Wisconsin and Louisiana, clutch frequency based on dissections was generally two but occasionally three. In South Dakota, small adult females tended to show evidence of a single annual clutch while large adult females often showed evidence of two (Timken 1968a). Freedberg et al. (2005) stated that nesting females produced two to four clutches annually in western Tennessee.

There may be a slight latitudinal trend in adult body size. In Wisconsin, midline plastron length ranged from 109 to 121 mm in males and reached 239 mm in females; in Kentucky, males ranged from 77 to 121 mm and females reached 215 mm; and in Louisiana, males ranged from 80 to 110 mm and females reached 228 mm (J. L. Carr 2001; Lindeman 2008a).

Natural History

Courtship has been observed in October, November, and April, with courting males drumming the backs of their forefeet claws against the ocular regions of females in repeated half-second bouts, averaging about ten strokes per bout (Vogt 1980a, 1993). In the upper Mississippi River, nesting occurred primarily on open sandy beaches of islands (Vogt and Bull 1984). Bodie and Semlitsch (2000a) reported a mass midwinter die-off of juvenile false map turtles in an ephemeral wetland in the Missouri River floodplain. Coleman and Gutberlet (2008) reported activity in all months of the year, with peak basking counts in March–April, June, and September–November. Numerous internal parasites have been reported, although many older records are unreliable, as they may concern Ouachita or Sabine map turtles (see chapter 4). The more reliable records concern monogenean and trematode flukes (Acholonu 1969b; Brooks 1975; Brooks and Mayes 1975, 1976), spiny-headed worms (Acholonu 1969a), and roundworms (Ash 1962; Acholonu and Arny 1970; Dyer and Wilson 1997). Emerging hatchlings in Wisconsin were preyed upon heavily by ring-billed gulls, American crows, grackles, and red-winged blackbirds (Vogt 1980a). Vogt (1979) described a symbiotic relationship in which basking turtles allowed grackles to approach and pull leeches from them.

Populations

Relative abundance data vary widely in the northern range of the species, but the false map turtle is generally among the predominant turtle species. Using primarily baited wire funnel traps in Missouri River impoundments and tributaries, Timken (1968a) captured 66 percent false map turtles in an assemblage of six species. In the Mississippi River in Wisconsin, Vogt's (1980a) sample of three map turtle species in fyke, gill, and trammel nets was approximately 45 percent false map turtles, and 20 percent of captures at a hibernaculum site in the fall were false map turtles. In the Mississippi in Illinois, the species accounted for only 5 percent of all map turtles and less than 1 percent of all turtles captured in baited traps (R. A. Anderson et al. 2002). Undifferentiated mixes of false and Ouachita map turtles (*G. ouachitensis*) constituted just 9 percent of a sample of seven species taken in fyke nets in the Mississippi in Iowa (Gritters and Mauldin 1994) but 82 percent of a sample of six species taken in hoop, fyke, and gill nets from the Mississippi above its confluence with the Ohio River (Barko et al. 2004). Bodie et al. (2000) used baited traps and fyke nets in the lower Missouri in 1996–98 to capture six turtle species, with false map turtles being the most abundant species, at 46 percent of the sample; at three floodplain hibernacula that dried, causing winterkills, false map turtles were the most abundant of four species trapped live the following spring (60 percent) and the second most abundant species found as carcasses (44 percent; Bodie and Semlitsch 2000a). In the lower St. Croix River in Wisconsin, 6 percent of 663 turtles captured in baited traps were false map turtles, making them the fifth most commonly captured of seven species (DonnerWright et al. 1999). Rizkalla and Swihart (2006) captured the species in low abundance (1 percent of all turtles, sixth in abundance of eight species) in wetlands of the upper Wabash floodplain in Indiana. In floodplain habitats of the Mississippi in southern Missouri, false map turtles were the second most abundant of eight species (behind slider turtles, *Trachemys scripta*), constituting 28 percent of all turtles taken in unbaited hoop nets, fyke nets, and basking traps (J. E. Wallace et al. 2007).

Relative abundance in the southern part of the range is likewise variable. In the mid-1950s in Louisiana, Mississippi map turtles constituted 7 percent of the catch of five species (tied for second most abundant) on the Sabine River and 37 percent of the catch of six species (most abundant) on the Tensas River (Tinkle 1958b). They were 2 percent of the catch and ranked seventh of ten species in baited hoop traps deployed in forty-one counties in Kansas (Fuselier and Edds 1994). In Kentucky Lake and the Tennessee River below its dam, false map turtles were 17 percent of the turtles seen in spotting scope surveys and 15 percent of the turtles captured, primarily in basking traps and fyke nets; in both cases they ranked third among eleven species (Lindeman 1999c and unpublished data). In northern Louisiana, Mississippi map turtles were seen basking at sixteen of seventeen aquatic habitats surveyed by J. L. Carr (2001) in 1998–2000. They accounted for 34 percent of 929 emydids seen, with approximately equal relative abundance on five rivers (33 percent of emydids) and twelve lakes and bayous (35 percent). In basking counts on the upper Sabine River in east Texas, Mississippi map turtles were outnumbered by Sabine map turtles (*G. sabinensis*) by a ratio of 8.4:1 (Coleman and Gutberlet 2008). In the tributary creeks of the Arkansas River in Oklahoma's Sequoyah National Wildlife Refuge, Mississippi map turtles were the eighth most abundant of nine species in baited hoop nets, at 0.6 percent of captures (Riedle et al. 2008). In an assemblage of six basking species in the Brazos River in Texas, the Mississippi map turtle was the third most frequent species in both visual surveys (5 percent) and captures in basking traps (3 percent; Hill 2008). In the Sabine River and its former river channel in Texas, Mississippi map turtles were fifth of nine species in abundance, at 6 percent of captures in baited traps (Hively 2009). In baited hoop net surveys in eastern Oklahoma, it was the ninth most abundant of thirteen species, at 0.9 percent of all captures (Riedle et al. 2009).

The only mark-recapture estimate of population size for this species is an estimated population size of 71 at a nesting island in the Red River of Louisiana (J. L. Carr 2001). Mean basking densities at twenty sites in Kentucky Lake and the Tennessee River ranged as high as 61/km of shoreline with an overall mean of 14.3/km (Lindeman 1998, 1999c). Basking density ranged from 2 to 43 turtles/river km at sites in northern Louisiana (J. L. Carr 2001) and averaged 3.2/km in the Sabine (Coleman and Gutberlet 2008) and 3.6/km in the Brazos (Hill 2008).

Conservation Status and Actions

Populations in five northern states were listed as candidates for federal listing in 1994 but subsequently dropped from further consideration. State-listed as conservation priority level III in North Dakota (for its peripheral status), threatened in South Dakota, and a species of special concern in Wisconsin. Listed as a species of least concern with unknown population trends by the IUCN.

The most extensive federal habitat preserve is the Upper Mississippi River National Wildlife and Fish Refuge, along a 420 km reach of the river in Illinois, Iowa, Wisconsin, and Minnesota. The refuge's 97,000 ha include numerous forested islands and extensive backwaters. Other prominent national wildlife refuges (NWRs) on the Mississippi are Trempealeau NWR in Wisconsin (2,521 ha), Port Louisa NWR in Iowa and Illinois (4,364 ha), Great River NWR in Missouri and Illinois (6,073 ha), and Chickasaw NWR in Tennessee (10,124 ha). Reelfoot NWR (4,222 ha) contains a large population in Tennessee, and White River NWR (64,775 ha) protects 145 km of the lower White in Arkansas near its confluence with the Mississippi. Additional federal refuges of importance include Black Bayou Lake NWR (1,862 ha), D'Arbonne NWR (7,052 ha), Tensas NWR (28,300 ha), and Upper Ouachita NWR (17,611 ha) in northern Louisiana; Little River NWR (6,073 ha) in southeastern Oklahoma; Felsenthal NWR (26,316 ha) on the Ouachita and Saline Rivers in southern Arkansas; Sequoyah NWR (8,421 ha) on the Arkansas River in eastern Oklahoma; Meredosia NWR (1,560 ha) on the Illinois River in Illinois; and Boyer Chute NWR (1,356 ha) and DeSoto NWR (3,384 ha) on the Missouri River in Iowa. In Wisconsin and Minnesota, habitat is protected by the National Park Service in the St. Croix National Scenic Riverway (406 river km). The Park Service also protects 39,355 ha along most of the inhabited reaches of the Neches in the Big Thicket National Preserve. Almost the entire western shore of Toledo Bend Reservoir on the Sabine River occurs within Sabine National Forest in Texas. Land Between the Lakes National Recreation Area protects 480 km of forested reservoir shoreline along impoundments of the lower Tennessee and Cumberland Rivers in Kentucky and Tennessee. In Arkansas, a 1,175 ha bottomland forest along a 10 km reach in the Ouachita River Nature Preserve is managed under a conservation easement by the Nature Conservancy.

Needed Further Research

The most inviting target of future research is further determining the relationship of the two recognized subspecies. A phylogeographic study with large genetic samples from the range of the species should reveal much about its history, including past episodes of divergence and coalescence, the possibility of hybridization in areas such as Reelfoot Lake (chapter 10), and the relationship of isolated populations in the smaller western drainages of the Gulf Coast to those in the Mississippi drainage. Latitudinal trends in life-history parameters should be examined with an eye toward comparisons with other widespread North American turtles that have been more extensively studied.

Ouachita Map Turtle, *Graptemys ouachitensis*

Figure 8.21. Adult male, Caney River at W3200 Road, Washington County, Oklahoma.

> The side of the head is conspicuously three-spotted, besides being longitudinally striped. The uppermost spot is the postocular spot of typical *pseudogeographica* but here it is usually detached from all neck and head stripes. The second spot is beneath the eye and the third on the mandible beneath the second.
>
> ARCHIE CARR (1952, P. 211)

Figure 8.22. Adult male Ouachita map turtle exhibiting fusion of its postorbital and supramandibular spots to form a crescent, as commonly seen in northern populations. A postorbital crescent in a Ouachita map turtle is wider than the crescent typical of Mississippi map turtles (*Graptemys pseudogeographica kohnii*). Photo from backwater of the Mississippi River in Pettibone Park on Barron Island, La Crosse County, Wisconsin.

Alternate Common Names

Southern map turtle (Crother et al. 2012); common name first used by Cagle (1954).

Description and Diagnosis

Postorbital markings typically wide vertical bars which may or may not be continuous with upper neck stripes. Interorbital mark a medial longitudinal stripe, typically flanked by one to three thinner longitudinal lines. Large supramandibular spots; in some specimens, particularly in the north, these join postorbitals as wide crescent-like markings (e.g., fig. 8.22). Chin marked with large anterior spot and two large submandibular spots. Markings on head, neck, limbs, and tail yellow to light orange. Females and males with narrow heads and narrow alveolar surfaces. Second vertebral scute only slightly sloped, extending little higher than the third. Carapace not highly peaked, rounded in cross section in northern populations; somewhat highly peaked in southern populations. Carapacial keels producing prominent points in males and juveniles, reduced and blunted in females. Black pigment along midline of carapace, particularly at keel points. Carapace marked with reticulating network of thin, black-bordered, yellowish-orange lines on pleural scutes having comparatively few intersections (one to two rings per pleural), continuing onto marginals. Large specimens of both sexes often with large, melanistic blotches on carapacial scutes. Plastron pale yellow, with large oblong central figure of undulating black and yellow lines arranged alternately in symmetrical pattern. Male foreclaws conspicuously elongated. Irises white, bisected by black stripes. Differs from congeners by the wide postorbital bars and the large size of the supramandibular, submandibular, and chin spots.

Taxonomic History and Phylogeny

The Ouachita map turtle was long part of a cryptic species complex. Specimens had been collected from several localities before the type series was collected in 1950. At least one of the syntype series of *G. pseudogeographica* from the Wabash River (LeSueur 1827; J. E. Gray 1831) was later recognized to be a Ouachita map turtle, as were five additional specimens LeSueur collected from the Wabash (Bour and Dubois 1983; Bonnemains and Bour 1996). Baur examined specimens thought to be from the Saline River in Arkansas in the early 1890s and appeared to recognize them as a distinct species but did not publish a name and description (see chapter 2). A. Carr (1949, 1952) commented on the distinctiveness of six specimens collected in 1944 in Comanche County, Oklahoma. The species was represented among specimens from Caddo Lake identified as *G. pseudogeographica pseudogeographica* by the first Tulane field crew in 1947 (Cagle and Chaney 1950; most of their series was *G. pseudogeographica kohnii*). Cagle (1953b) described the taxon as *Graptemys pseudogeographica ouachitensis* but suggested it might warrant separate species status and gave several additional localities. The holotype (UMMZ 104345) was collected "from the Ouachita River, four miles northeast of Harrisonburg, Louisiana" (Cagle 1953b, p. 9) on 10 June 1950 by Allan H. Chaney and Clarence L. Smith; thus the specific epithet, which means "from the Ouachita," refers to just one of the many tributaries of the Mississippi drainage inhabited by the species. An additional thirty-two paratypes were taken from the type locality (FMNH 67101–104; TU 12536, 12545, 12631, 12643, 12655, 12658, 12664–67, 12670–71, 12686, 12695, 12701, 12705, 12710, 12783, 12975 [3 specimens]; UIMNH 26716–17, UMMZ 104346–50).

The taxon was first raised to full species status by Vogt (1978, 1993), who also included *sabinensis* as a subspecies of *ouachitensis*. The change was not universally embraced in the decade and a half between Vogt's dissertation and his publication, but it has been more accepted since the latter date (see chapter 2). Nevertheless, even now some authors cannot or choose to not make the distinction between *pseudogeographica* and *ouachitensis* (e.g., Minton 2001; Barko et al. 2004), in spite of molecular genetic phylogenetic studies supporting Vogt's separation of the taxa (Lamb et al. 1994; Lamb and Osentoski 1997; Stephens and Wiens 2003). Phylogenetic analyses placed the Ouachita map turtle as sister species to *G. sabinensis* in a clade that is sister to a clade composed of *G. flavimaculata* and *G. oculifera* (Lamb et al. 1994), as sister species to *G. caglei* (Lamb and Osentoski 1997), and as sister species to the clade composed of the three sawbacks (*G. nigrinoda*, *G. oculifera*, and *G. flavimaculata*), nested within a larger clade that also contains *G. pseudogeographica* (Stephens and Wiens 2003; Wiens et al. 2010).

Range

The Ouachita map turtle is restricted to the Mississippi drainage. Its upstream range limit is in Dakota County, Minnesota (Oldfield and Moriarty 1994; McCarthy 2010). It ranges into five tributaries of the upper Mississippi: the Wisconsin River; the Cedar and Skunk Rivers in Iowa; the Illinois River and its tributaries, the Somonauk and Fox Rivers; and the Kaskaskia River in Illinois (Vogt 1981c, 1993; C. A. Phillips et al. 1999; Myers 2008; Dolan et al. 2011). Vogt's (1993, 1995a) record of an occurrence in the St. Croix River is apparently in error, as no St. Croix specimen is cited in the appendix of the first publication, in his account of the species' distribution in Wisconsin (Vogt 1981c), or in an account of the species' Minnesota distribution (Oldfield and Moriarty 1994). The species was also not reported in extensive sampling of turtles of the St. Croix by DonnerWright et al. (1999).

In the Missouri drainage, the species has been recorded in the main-stem river only in Calloway County, Missouri (TU 21908, 30123). Records exist for three tributary drainages, however: the Gasconade River in Missouri (Fratto and Swallow 2007) and, in Kansas, the Marais des Cygnes River (Fuselier and Edds 1994; D. Edds, pers. comm. 2007; Irwin and Collins 2005) and the Kansas River plus its tributaries, the Republican River and Stranger Creek (Vogt 1993; Busby and Parmelee 1996; Tollefson 2004). A record by Sievert and Collins (1998) from a reservoir on another Kansas River tributary, the Saline River, is a misidentified *G. pseudogeographica*.

In the Mississippi River, the Ouachita map turtle has been recorded as far downstream as northern Plaquemines Parish, Louisiana, 126 river km from the mouth of the river (TU 21912). It ranges extensively in western tributaries of the lower Mississippi. It occurs in the Meramec, L'Anguille, St. Francis, and White drainages in Missouri and northeastern Arkansas (Shipman and Riedle 1994; Trauth et al. 2004; Daniel and Edmond 2012). Its range extends up numerous tributaries of the Arkansas, including the Chikaskia, Pawnee, Little Arkansas, South Fork of the Ninnescah, Walnut, Big Caney, Little Caney, Fall, Verdigris, Neosho, and Cottonwood Rivers and Labette Creek into Kansas, as well as Skeleton Creek, the Canadian River drainage, the Illinois River, and the Poteau River in Oklahoma (Ortenburger 1929; Black et al. 1987, 1993; Edds et al. 1991; J. T. Collins 1993; Vogt 1993; Fuselier and Edds 1994; D. Edds, pers. comm.

2007; Lardie 1999; Riedle and Hynek 2002; Casley and Sievert 2006; Riedle et al. 2009; A. D. Brown et al. 2012). The species ranges up the Black-Ouachita-Saline River drainage into Arkansas (Vogt 1993; Dundee and Rossman 1989; J. L. Carr 2001; Trauth et al. 2004). In the Red drainage it ranges as far upstream as the Fort Sill area in Comanche County, Oklahoma (A. Carr 1952; Vogt 1993; farther upstream, J. R. Dixon's [2000] record for Baylor County, Texas, appears to be in error, based on a specimen number provided by J. R. Dixon, pers. comm. 2007). The species also occurs in several Red tributaries in Oklahoma and Arkansas: the Washita, Muddy Boggy, Kiamichi, and Little Rivers; Hickory Creek; and two tributaries of the Little, the Mountain Fork and Glover Rivers (Vance 1986; Black et al. 1987, 1993; Tyler 2000; Trauth et al. 2004; Riedle et al. 2009). It occurs on the Louisiana-Texas border in Caddo Lake, which joins the Red River via Cypress Creek (Cagle and Chaney 1950; Vogt 1993). It ranges into the Atchafalaya River, which diverges from the Mississippi westward in Louisiana, and two tributaries, Bayou Pigeon and Bayou Teche (Vogt 1993).

The Ouachita map turtle's distribution in eastern tributaries of the lower Mississippi has either been poorly described or is almost entirely lacking, as it has been reported from only the Hatchie River in western Tennessee, near the mouth of the Quiver River of the Yazoo-Big Sunflower drainage in Mississippi, and the lower Comite River in Louisiana (Norton and Harvey 1975; George et al. 1995; Myers 2008). It occurs in the Tennessee drainage in Alabama (Mount 1975) and farther upstream into the Little River in Tennessee (Daniels et al. 2012; TU 21909). In the Cumberland River it ranges upstream almost to the Kentucky-Tennessee border (APSU 833).

In the Ohio River basin just above the confluences of the Tennessee and Cumberland Rivers, the Ouachita map turtle has been recorded in the Tradewater River in Kentucky, the Wabash River in Illinois and Indiana, and in the two forks of the White River, a tributary of the Wabash in southern Indiana (Ewert 1979, 1989; Vogt 1993; Lindeman 2001c; Minton 2001). There are no records of the species from a long stretch of the Ohio between sight records at Louisville, Kentucky, and photographed specimens in West Virginia (Ewert 1979; Ewert 1989; Vogt 1993; Minton 2001; pers. observations 2008; Gooley et al. 2011). Upper Ohio basin populations may therefore be disjunct: the Whitewater and Great Miami River drainage in Indiana and Ohio (Hay 1892a; Krusling et al. 2010a), the Scioto River in Ohio (Conant et al. 1964; Temple-Miller 2008; A. D. Brown et al. 2012), the Great Kanawha and Little Kanawha Rivers in West Virginia (Richmond 1953; Watson and Pauley 2006; Gooley et al. 2011), and the Muskingum River in Ohio (Davis and Krusling 2010). A record farther upstream in the Scioto from Franklin County (USNM 7751; Conant et al. 1964) is disregarded, as notes in the USNM catalog suggest uncertainty regarding whether the locality is really in Ohio or Mississippi.

Habitat

As with the sympatric (and generally syntopic) G. *pseudogeographica*, the Ouachita map turtle is largely tied to medium-to-large rivers, but it also occurs in floodplain lentic habitats and can persist in river impoundments. Fuselier and Edds (1994) trapped Ouachita map turtles in several streams of eastern Kansas and found them associated with wide, unshaded stretches of river with abundant basking sites, with substrates ranging from sand to mud to rock. Correlation of basking abundance to deadwood abundance was positive and significant in the lower Tennessee and its impoundment, Kentucky Lake (Lindeman 1999c), and the species was strongly associated with deadwood in a variety of habitats of the lower Kansas drainage (Tollefson 2004). In the lower 200 km of the Scioto in Ohio, Ouachita map turtles had a clumped distribution and telemetered females tended to occupy deep, wide sections of the river (Temple-Miller 2008).

The Ouachita map turtle can be abundant in coves of dendritic reservoirs (Webb 1961; Lindeman 1999c). It is commonly reported from sloughs and lentic floodplain habitats (Moll 1977; Vogt 1980a; J. L. Carr 2001; Tollefson 2004; Dreslik and Phillips 2005; Dreslik et al. 2005).

Diet

Webb (1961) reported that Ouachita map turtles fed primarily on midge and caddis fly larvae and ants in an Oklahoma reservoir. One adult female had eaten a large number of grasshoppers, another had eaten mostly vegetation, and one contained pieces of a bryozoan colony in addition to other prey. In a lake in the floodplain of the Illinois River, heavy siltation caused a historical change from a diverse diet of insects and aquatic vegetation to one dominated by bottom-dwelling midge larvae (table 4.5), while a population in the river channel fed almost exclusively on caddis fly larvae (Moll 1977). Vogt (1981a) reported on the diet of specimens from the Mississippi and Wisconsin Rivers, Wisconsin, and the White River in Arkansas, concentrating on adult females. Vegetation—principally pondweed, duckweed, and algae—was the primary food category for adult females in Wisconsin, followed by mayfly nymphs and caddis fly larvae (table 4.8). In Arkansas, adult females consumed mayfly nymphs, caddis fly larvae, vegetation, mollusks, and fish. Males in both regions fed primarily on insects and ate little plant matter or algae.

Caddis fly larvae predominated in the diet in the lower Wabash in Illinois and Indiana (83 percent of total identified prey by volume), with snails (9 percent), vegetation (4 percent), and beetles (2 percent) constituting most of the rest (Pierce 1992). In a Kentucky reservoir, the most important foods of unsexed juveniles were snails, amphipod shrimp, water milfoil, filamentous algae, and dipteran flies and their larvae (table 4.6; Lindeman 1997c, 2000b). Adult males fed most heavily on spiral egg cases of chironomid midges, fly larvae, and algae. Females were divided between small and large individuals (roughly large juveniles vs. adults): small females fed most heavily on algae, bryozoan colonies, sponges, fly larvae, and mayfly nymphs, while large females fed most heavily on mayfly nymphs, algae, bryozoan colonies, beetles, adult flies, plant leaves, and fish.

Reproduction and Life History

Growth in Wisconsin and Kentucky follows a von Bertalanffy pattern of steadily declining growth rate with age (Vogt 1980a; Lindeman 1999b). Males matured in their third full growing season in Kentucky and their fourth in Wisconsin. The youngest mature females in Wisconsin were in their ninth growing season, while growth curves suggested maturation occurred during the twelfth growing season in Kentucky; however, this estimate may be too high because it was based on a small sample of gravid mature females.

Reproductive output varies by latitude, as evidenced by data from the Mississippi River in Wisconsin, a reservoir in Oklahoma, and the Red River in Louisiana (Webb 1961; Vogt 1980a; Janzen et al. 1995; Rosenzweig 2003). Wisconsin females had average clutch sizes of 10.5 and 11.3, while average clutch sizes farther south were 6.0 in Oklahoma and 6.4 in Louisiana. As many as three or four clutches may be laid by females annually, but two clutches appears typical; there is no evidence of latitudinal variation in clutch frequency. The latitudinal variation in clutch size coincides with variation in body size; females reach 220 mm midline plastron length in Wisconsin, 185 mm in Kentucky, and 182 mm in Louisiana (J. L. Carr 2001; Lindeman 2008a). Males ranged from 76 to 109 mm in Wisconsin, from 77 to 99 mm in Kentucky, and from 77 to 105 mm in Louisiana.

Natural History

Courtship has been observed in October, November, and April, with courting males drumming the backs of their forefeet claws against the ocular regions of females in repeated half-second bouts, averaging about five strokes per bout (Vogt 1980a, 1993). In the upper Mississippi River, nesting occurred primarily on open, sandy beaches of islands (Vogt and Bull 1984). The few records of parasites that can confidently be assigned to the Ouachita map turtle include turtle leeches, trematode flukes, tapeworms, and roundworms (Webb 1961; Vogt 1979). Emerging hatchlings in Wisconsin were preyed upon heavily by ring-billed gulls, American crows, grackles, and red-winged blackbirds (Vogt 1980a). Vogt (1979) described a symbiotic relationship in which basking turtles allowed grackles to approach and pull leeches from them.

Populations

Ouachita map turtles constituted 12 percent of the total catch of six species on the Tensas River in Louisiana, ranking third in relative abundance behind Mississippi map turtles (*G. pseudogeographica kohnii*) and sliders (*Trachemys scripta*; Tinkle 1958b). In the upper Mississippi River in Wisconsin, Vogt's (1980a) sample of three map turtle species in fyke, gill, and trammel nets was approximately 48 percent Ouachita map turtles, and 73 percent of captures at a hibernaculum site in the fall were Ouachita map turtles. The species was not taken in baited traps used to sample 283 turtles of seven species in the Mississippi in Illinois (R. A. Anderson et al. 2002). Undifferentiated mixes of Ouachita and false map turtles (*G. pseudogeographica*) constituted just 9 percent of a sample of seven species taken in fyke nets in the Mississippi in Iowa (Gritters and Mauldin 1994) but 82 percent of a sample of six species taken in hoop, fyke, and gill nets from the Mississippi above its confluence with the Ohio River (Barko et al. 2004).

Ouachita map turtles were 15 percent of the catch and were the third most frequently captured of six turtle species taken in fyke nets, baited hoop nets, and trammel nets in the lower Wabash River (Pierce 1992). They were 11 percent of the catch and were the third most frequently captured of ten species in baited hoop traps deployed in forty-one counties in Kansas (Fuselier and Edds 1994). In the Kansas River and its oxbow lakes, the species ranked third in abundance of six species, accounting for 12 percent of all turtles captured (Busby and Parmelee 1996). In the tributary creeks of the Arkansas River in Oklahoma's Sequoyah National Wildlife Refuge, Ouachita map turtles were the third most abundant of nine species in trapping conducted with baited hoop nets, although they accounted for just 3.7 percent of all captures (Riedle et al. 2008). In hoop net surveys in eastern Oklahoma, it was the second most abundant of 13 species, behind *T. scripta*, at 8 percent of all captures (Riedle et al. 2009). In floodplain habitats of the Mississippi River in southern Missouri, only a single Ouachita map turtle was captured, as compared with 723 specimens of seven other species (J. E. Wallace et al. 2007).

In Kentucky Lake and the Tennessee River below its dam, mean basking density of Ouachita map turtles was 25/km of shoreline. The species constituted 27 percent of the turtles seen in spotting scope surveys and 29 percent of the turtles captured, primarily in basking traps and fyke nets; in both cases the species was second most abundant of 11 species, after slider turtles (*Trachemys scripta*; Lindeman 1999c, unpublished data). In the upper reaches of the Tradewater River, a much smaller river in western Kentucky, only 2 Ouachita map turtles were seen among 344 total turtles recorded basking (<1 percent), making it the third rarest of eight species recorded (Lindeman 2001c). In basking surveys of seventeen aquatic habitats conducted by J. L. Carr (2001) in northern Louisiana in 1998–2000, Ouachita map turtles were seen at four sites, where they accounted for 14 percent of 552 emydids spotted, with their highest relative abundance being on two rivers, the Red (57 percent) and Ouachita (16 percent). Basking density ranged from 4 to 85/river km (J. L. Carr 2001).

Conservation Status and Actions

No federal or state listings. Listed as a species of least concern with stable populations by the IUCN.

The most extensive federal habitat preserve is the Upper Mississippi River National Wildlife and Fish Refuge along a 420 km reach of the river in Illinois, Iowa, Wisconsin, and Minnesota. The refuge's 97,000 ha include numerous forested islands and extensive backwaters. Other prominent national wildlife refuges (NWRs) on the Mississippi are Trempealeau NWR in Wisconsin (2,521 ha), Port Louisa NWR in Iowa and Illinois (4,364 ha), Great River NWR in Missouri and Illinois (6,073 ha), and Chickasaw NWR in Tennessee (10,124 ha). White River NWR (64,775 ha) protects 145 km of the lower White in Arkansas near its confluence with the Mississippi. Additional federal refuges of importance include Upper Ouachita NWR (17,611 ha) in northern Louisiana, Little River NWR (6,073 ha) in southeastern Oklahoma, Felsenthal NWR (26,316 ha) on the Ouachita and Saline Rivers in southern Arkansas, Sequoyah NWR (8,421 ha) on the Arkansas River in eastern Oklahoma, and Meredosia NWR (1,560 ha) on the Illinois River in Illinois. Land Between the Lakes National Recreation Area protects 480 km of forested reservoir shoreline along impoundments of the lower Tennessee and Cumberland Rivers in Kentucky and Tennessee. In Arkansas, a 1,175 ha bottomland forest along a 10 km reach in the Ouachita River Nature Preserve is managed under a conservation easement by the Nature Conservancy.

Needed Further Research

A status survey in the eastern (Ohio River drainage) portions of the range is needed to better understand the species' distribution and biogeographic history, particularly the nature of its supposed range disjuncture. In those northern portions of its range where it is sympatric with the false map turtle (*G. p. pseudogeographica*), historical confusion regarding the two taxa has left several aspects of their ranges uncertain; much careful work is needed to more accurately determine which of these species occupies various tributaries. Latitudinal trends in life history should be examined and compared with other widespread North American turtles that have been more extensively studied.

Ringed Sawback, *Graptemys oculifera*

Figure 8.23. Male, West Pearl River at Interstate 59 crossing, St. Tammany Parish, Louisiana.

> This is one of the most beautiful of the American tortoises.
>
> GEORG BAUR (1890, P. 262)

Alternate Common Names

Ocellated terrapin (Ditmars 1907), Baur's turtle (Strecker 1909), and ocellated map turtle (Strecker 1909), never widely adopted; ringed sawback turtle (A. Carr 1952; K. P. Schmidt 1953); ringed map turtle (Collins et al. 1978), see discussion in chapter introduction. First called ringed sawback by Cagle (1954).

Description and Diagnosis

Postorbital markings small ovals or rectangles, not usually connecting to neck stripes but often connecting to anterior lines over orbits. Interorbital marking a prominent stripe, typically flanked by one or more thinner stripes on each side. Supramandibular spots subsumed by prominent ventrolateral neck stripes. Chin with transverse bar subsuming elongated submandibular spots. Markings on head, neck, limbs, and tail yellow. Females and males with narrow heads and narrow alveolar surfaces. Second vertebral scute steeply sloped, distinctly higher than the third. Carapace highly peaked at posterior of second vertebral, nearly triangular in cross section. Carapacial keels producing

Figure 8.24. The four USNM adult female syntypes of *Graptemys oculifera* (Baur 1890). (a) Shell of USNM 15510 and (left to right) alcohol-preserved specimens USNM 15509, USNM 15508, and USNM 15511. (b) Close-up of the designated lectotype of *G. oculifera*, USNM 15511.

exceptionally high, sharply angled points in males and juveniles, reduced and blunted in females. Black pigment along midline of carapace, particularly at keel points. Carapace marked with prominent, black-bordered, pale orange rings on pleural scutes (one or more rings incomplete on some specimens; e.g., USNM 15511 in fig. 8.24) and irregular, thin, black-bordered, pale orange lines on vertebral scutes. Marginals with prominent, black-bordered, pale orange lines that curve inward and rearward. Pleural scutes with prominently raised central bosses, particularly in juveniles and males. Plastron pale yellow to light orange, marked with dark markings along seams. Male foreclaws conspicuously elongated. Irises yellow, bisected by black stripes. Differs from congeners by the wide, pale orange, ring-like markings of the carapace, the sharply angled keels of the carapace, and the transverse chin bar.

Taxonomic History and Phylogeny

The earliest specimens of this species were from the private collection of Gustave Kohn. Beyer (1900) stated Kohn had purchased them in the French Market in New Orleans. Kohn placed four preserved adult females in the U.S. National Museum (cataloged on 10 June 1889 as USNM 15508–11, ranging from 146 to 185 mm in midline plastron length; fig. 8.24). Two additional live specimens (USNM 15506–507) were discarded after they died, and one specimen laid an egg (preserved as USNM 15512) while in transport to the museum. Baur (1890) described the species as *Malacoclemmys oculifera* using the USNM series to augment his examination of specimens he borrowed from Kohn's private collection. The specific epithet is from the Latin terms *oculus* (eye) and *fero* (to bear), undoubtedly referring to the rings on the carapace.

Kohn later placed eight specimens in the Tulane University collections. Two, from 9 and 20 May 1888, can be considered part of Baur's hypodigm; the other six were collected in 1891 and thus are not. Kohn's labels indicate the 1888 specimens were from Mandeville, Louisiana (TU 20), and Pensacola, Florida (TU 7628). Baur (1890, p. 263) regarded the former as type locality while not designating type specimens per se, simply noting, "Such specimens are also in the collection of the Smithsonian Institution, Washington, D.C., sent by Mr. G. Kohn, No. 15,511, etc." Cagle (1953a, p. 138) restricted the type locality to "Pearl River, 26 miles east of Mandeville," which is consistent with the localities given for the specimens Kohn collected in 1891. Indeed, the most specific locality information for any of Kohn's Pearl River *Graptemys* specimens is written on the plastrons of TU 24 (a ringed sawback; fig. 2.9) and TU 15072 (a Pearl map turtle, *G. pearlensis*) and places their site of collection at the crossing of the Northeastern Railroad, about 1 km northeast of the town of Pearl River, Louisiana, and due east of Mandeville.

Two specimens have been referred to by different authors as "holotype" of the species: MCZ 6430 by Barbour and Loveridge (1929, p. 303) and USNM 15511 by D. M. Cochran (1961, p. 233), who also listed USNM 15508–10 as "paratypes." Cagle (1953a, p. 138) listed USNM 15508–11 as "cotypes" (= syntypes) but erroneously stated that USNM 15510 had been exchanged and recataloged as MCZ 6430; the Museum of Comparative Zoology (MCZ) specimen is an alcohol-preserved adult female measuring 186 mm in plastron length, according to A. Carr (1952, p. 201), while USNM 15510 is a dry, complete skeleton of an adult female (Reynolds et al. 2007). The MCZ specimen may or may not be part of Baur's hypodigm, as Carr stated the specimen was sent by Baur to the museum in 1895 and Baur's private correspondence lists three *G. oculifera* among specimens he received from Kohn in May 1891, six months after he published his description of the species. The syntype series thus numbers seven and possibly eight extant specimens (USNM 15508–12, TU 20, and TU 7628, with the status of MCZ 6430 uncertain). To avoid future confusion, I hereby designate USNM 15511 the lectotype of the species (fig. 8.24); the remaining six USNM and Tulane University (TU) specimens listed above should be regarded as paralectotypes, as should MCZ 6430 if information on an appropriate date of collection comes to light.

Stejneger and Barbour (1917, 1923, 1933) relegated *oculifera* to a subspecies of *pseudogeographica* but then recognized it as a full species in the fourth (1939) and fifth (1943) editions of their checklist. A. Carr (1949) initially dismissed the taxon as a variant of *pseudogeographica* resembling the Texas endemic, *versa* (then considered a subspecies of *pseudogeographica*), but later he recognized it (A. Carr 1952). Cagle (1953a) and his students rediscovered the species in the Pearl and its Bogue Chitto tributary. Following fruitless searches for the species in the field in 1947 and 1948, the "Pearl River" inscriptions were found on the plastra of specimens in the Kohn collection at Tulane University (see chapter 2). Trapping with hoop nets and trotlines was initially unsuccessful—only a single adult female was captured, while on land to nest—but development of a nighttime collecting technique in 1950 (Chaney and Smith 1950) resulted in a large series of specimens, which Cagle (1953a) used to redescribe the species. Despite relegation by Mertens and Wermuth (1955) to conspecific status with the other two sawback taxa, *flavimaculata* and *nigrinoda*, as well as earlier relegation to conspecific status with *G. pseudogeographica* in the Stejneger and Barbour checklists, the ringed sawback has generally been treated since Cagle's work as a separate, monotypic species (e.g., C. H. Ernst and Lovich 2009). Phylogenetic studies have placed it in a polytomy with *G. flavimaculata* and *G. sabinensis* (Lamb and Osentoski 1997) and as the sister species of either *G. flavimaculata* (Lamb et al. 1994; Stephens and Wiens 2003) or a clade composed of *G. flavimaculata* and *G. nigrinoda* (Wiens et al. 2010).

Range

The ringed sawback is restricted to the Pearl drainage in south-central Mississippi and two parishes in southeastern Louisiana. It ranges upstream in the Pearl to the Nanih Waiya Wildlife Management Area in Neshoba County

(Keiser 2000). It occurs in tributary streams, including the Yockanookany River, Copiah Creek, the Strong River, Mikes River, and the Bogue Chitto River (Kofron 1991; Keiser 1994; Lindeman 1998, 1999c; Shively 1999). The most significant occurrence in a tributary is in the Bogue Chitto, in which it ranges upstream of the east-west border of Louisiana and Mississippi (Shively 1999; pers. observations 2010). Downstream, it occurs to the vicinity of Interstate 10 in the interconnected channels of the East and West Pearl, the Pearl River Canal, and the Porter River (McCoy and Vogt 1979; Kofron 1991; Lindeman 1998, 1999c; Jones and Selman 2009).

Habitat

McCoy and Vogt (1979) suggested that the primary features promoting abundance of ringed sawbacks are fast to moderate currents, abundant deadwood and lack of shading to promote basking and provide substrates for prey organisms, and flat sandbars known to be used for nesting (see also Cagle 1953a; P. K. Anderson 1958). Lindeman (1999c) further stressed the importance of deadwood: maximum numbers were significantly positively correlated with the total number of potential basking substrates and nonsignificantly but positively correlated with an index of deadwood abundance, and mean numbers were nonsignificantly but positively correlated with both total substrates and the index. Of five deadwood categories, the crowns of fallen trees and large logs were especially important to ringed sawbacks and other turtles, with both the highest percentage occupancies and the highest mean numbers of turtles per occupied substrate.

The ringed sawback persists at low densities in the upper section of the only major impoundment of the Pearl, Ross Barnett Reservoir (McCoy and Vogt 1979; Lindeman 1998, 1999c). It has also been recorded in oxbow lakes (Peter's Lake, Kofron 1991; Mayes Lake, Lindeman 1998, 1999c) and may well be found in other oxbows along the main-stem Pearl. Keiser (1994) observed several ringed sawbacks in gravel pit ponds on the grounds of Stennis Space Center.

Diet

Cagle (1953a) reported that six juveniles, three males, and a female contained only insect remains, which Dundee and Rossman (1989) examined and found to include adult and larval beetles, mayfly nymphs, adult damselflies, dragonfly nymphs, midge larvae, and homopterans. Cagle suggested that two specimens observed grazing on the emergent portion of a log were attempting to ingest snails; Dundee and Rossman also found snails in the stomach contents of his specimens. The only detailed study of diet is Kofron's (1991) study of twenty-nine dissected museum specimens (although unfortunately he did not report their gender or status as adults or juveniles). Adult and larval caddis flies predominated in the diet, followed by flies, beetles, mayflies, plant material (chiefly aster flowers and seeds), dragonflies, and earthworms (table 4.4). No snails were found, but two-thirds of the specimens had wood fragments Kofron felt were consistent with grazing on the *aufwuchs* of submerged deadwood. Foraging on insects taken from a log has also been described (Lindeman 2009a).

Reproduction and Life History

Cagle (1953a), Kofron (1991), and Jones and Hartfield (1995) studied growth of ringed sawbacks. The comparatively rapid growth in juvenile females claimed by Cagle was disputed by Kofron, who re-examined Cagle's fastest-growing female specimen and found it had more growth annuli than Cagle had counted. Kofron found similar growth rates (mean 8.4 mm/year) in juvenile males and females. He used linear regression to relate size to age, with similar slopes for males and females but a slightly higher *y* intercept for females. Jones and Hartfield (1995) found that the von Bertalanffy model, with steadily decreasing growth rates, was a better fit to their large data set than the logistic model, with its increasing growth rates during the first half of growth followed by decreasing rates of growth thereafter. Females tended to be more variable in size at any given age.

Across five populations, adult males ranged from 53 to 98 mm in midline plastron length, while adult females ranged from about 115 to 180 mm (Jones and Hartfield 1995; Jones 2006). Maturity is reached in the fourth or fifth growing season in males and after about the tenth growing season in females (Cagle 1953a; Kofron 1991; Jones and Hartfield 1995).

In dissected specimens, clutch sizes ranged from 2 to 4 eggs, and the presence of smaller follicles suggested that annual production of two clutches was possible (Cagle 1953a; Kofron 1991). In a much larger sample from the Ratliff Ferry population, Jones (2006) recorded 1–10 eggs (mean 3.7). His data confirmed production of two clutches by females, with two females possibly having produced three clutches in a year. Oviposition occurred from mid-May through mid-July. The average time to pipping was sixty-four days, with emergence on average twelve days later.

Natural History

Courting males drum the backs of their forefeet claws against the ocular regions of females (Hertwig 2001). Nests are concentrated near sandbars' elevated vegetation lines on their inner edges (P. K. Anderson 1958). Habitat at nest sites averaged 37 percent cover, with 9.4 herbaceous stems and 3.8 woody stems within a half-meter radius (Jones 2006).

Populations

Both trapping and basking surveys indicate that in the Pearl drainage, the ringed sawback is a predominant species that may have increased in abundance relative to its sympatric congener, the Pearl map turtle (*G. pearlensis*), over recent decades. In nighttime hand captures in 1950–51, it was outnumbered by the

Pearl map turtle by a ratio of 2.1:1 but still made up 26 percent of the total catch of turtles; however, the technique is probably biased toward captures of map turtles and sawbacks (Cagle 1953a). In the mid-1950s, Tinkle (1958b) used a variety of capture techniques and took 1 ringed sawback for every 1.90 Pearl map turtles, with ringed sawbacks being the second most frequently captured turtle species overall, at 27 percent of all captures. In the late 1960s, hand collecting of juveniles and adult males on the main-stem Pearl produced a 2.1:1 ratio of the two *Graptemys*, with Pearl map turtles again predominant (Cliburn 1971). All subsequent studies have shown a preponderance of ringed sawbacks relative to Pearl map turtles. In fyke net captures in Copiah County, Mississippi, ringed sawbacks outnumbered Pearl map turtles 1.3:1 and constituted 15 percent of the total catch of nine species, second in abundance to the slider turtle, *Trachemys scripta* (Vogt 1980b). Ringed sawbacks predominated in drainage-wide basking surveys conducted in 1994 and 1995, outnumbering Pearl map turtles 9.6:1 and constituting 75 percent of all turtles identified (Lindeman 1999c). In basking surveys of the Bogue Chitto River in 1999, ringed sawbacks outnumbered Pearl map turtles 1.4:1 and constituted 32 percent of all turtles identified, second in abundance after river cooters, *Pseudemys concinna* (Shively 1999).

Two areas in Mississippi house the densest populations of ringed sawbacks, separated by the Ross Barnett Reservoir (which inundates 48 km of river habitat; Stewart 1986a, 1986b) and the city of Jackson. The first stretches between Carthage and the reservoir, while the second is located approximately between Georgetown and Columbia (McCoy and Vogt 1979; Jones and Hartfield 1995; Lindeman 1998, 1999c). The Ringed Sawback Turtle Refuge near the Ratliff Ferry and River Bend access points in Madison County has the densest populations; means of replicated basking surveys were 300 and 190 turtles/river km, respectively (Lindeman 1998, 1999c), and two independent mark-recapture estimates were 341 and 1,170 turtles/river km (Jones and Hartfield 1995; they also averaged seeing 98 turtles/river km along a 3.2 km reach at Ratliff Ferry). Mean basking densities on the main-stem Pearl were 103/river km at five sites above the reservoir and 41/river km at seven sites below the reservoir (including an oxbow lake), while the reservoir and tributaries had densities fewer than 25/river km (Lindeman 1998, 1999c).

Population estimates were published by Jones and Hartfield (1995) for five sections of the main-stem Pearl, based on mark-recapture techniques that assume open populations over the long term and closed populations over shorter time frames. For the latter estimates, they marked turtles with paint splotches and conducted basking surveys to generate "recaptures." They concluded that while estimates were similar for three sites, the mark-resight technique gave better estimates for the two remaining sites, where recapture frequency in traps was low. Estimates using this technique ranged from 85 to 341 turtles/km. The total estimated population seen basking was 7–14 percent for four study reaches but 29 percent for the Ratliff Ferry reach. Recent repetition of population estimates suggest population declines at two of the five sites (Jones 2009). Using the assumption that 10 percent of ringed sawbacks were basking during his surveys and that he recorded 75 percent of all basking individuals, Shively (1999) estimated total densities ranging from 22 to 220 turtles/km for five segments of the Bogue Chitto River, with 16,348 total turtles.

Conservation Status and Actions

Listed as threatened in 1986 under the U.S. Endangered Species Act. State-listed as endangered by Mississippi and as threatened by Louisiana. Listed as vulnerable with unknown population trends by the IUCN. Recovery efforts have focused on population estimation and studies of growth, maturation, reproductive ecology, and status surveys (Jones and Hartfield 1995; Lindeman 1998, 1999c; Shively 1999; Jones 2006, 2009).

Federally protected habitat occurs chiefly in 14,500 ha of the Bogue Chitto National Wildlife Refuge. A portion of the Natchez Trace Parkway (managed by the National Park Service) also provides a relatively undeveloped western shoreline along the upper Pearl, lower Yockanookany, and Ross Barnett Reservoir. Mississippi protects populations in two oxbows called Mayes Lake in LeFleur's Bluff State Park in Jackson, although these habitats will be lost if the proposed LeFleur Lakes project is completed (see chapter 5). The Pearl River Valley Water Supply District also is required to maintain 18 km on the upper Pearl in Madison and Rankin Counties as the Ringed Sawback Turtle Refuge, where channel management (particularly of deadwood) is to be carried out in a way that is not detrimental to the species and informational signs are to be maintained at a public boat ramp. However, a sign at Ratliff Ferry fell into disrepair (pers. observation 2007) and has been removed (Jones and Selman 2009).

Needed Further Research

More detailed reporting of diet would be of interest, particularly if paired with study of prey availability and foraging habits. There has been little focus on this species in the lower reaches of the Pearl. It would be interesting to know whether the populations located farthest downstream mirror those of the other two sawbacks in showing relatively large body size and reduced sympatry with a megacephalic congener (Lindeman 2000a; see chapter 4). From a conservation perspective, the populations in Mayes and Peter's Lakes raise the question of the extent of use of oxbow lakes range-wide, in particular whether they are used ephemerally by individuals or support relatively independent breeding populations. Lastly, further range-wide analyses examining the ecological correlates of population densities should assist recovery.

Yellow-Blotched Sawback, *Graptemys flavimaculata*

Figure 8.25. Adult male, Pascagoula River at Highway 596 in Merrill, George County, Mississippi.

> This very attractive turtle has been advertised for retail sale at $65 each. The turtle is very vulnerable to knowledgeable commercial collectors, who can easily decimate a local population in a short period of collecting.
>
> JIM STEWART (1989, P. 3)

Alternate Common Name

Yellow-blotched map turtle (Collins et al. 1978); see discussion in chapter introduction. First called yellow-blotched sawback by Cagle (1954).

Description and Diagnosis

Postorbital markings upright or slanted rectangular bars or ovals, usually joining neck stripes above, often connecting to anterior lines over orbits. Interorbital marking a prominent stripe, typically flanked by one or more thinner stripes on each side. Supramandibular spots subsumed by prominent ventrolateral neck stripes. Chin with transverse bar subsuming elongated submandibular spots. Markings on head, neck, limbs, and tail yellow. Females and males with narrow heads and narrow alveolar surfaces. Second vertebral scute steeply sloped, distinctly higher than the third. Carapace highly peaked at posterior of second vertebral, nearly triangular in cross section. Carapacial keels produc-

Figure 8.26. A female *Graptemys flavimaculata* (Leaf River at Wingate Road, Perry County, Mississippi) that exhibits ring-shaped blotches on both first pleural scutes.

ing exceptionally high, sharply angled points in males and juveniles, reduced and blunted in females. Black pigment along midline of carapace, particularly at keel points. Carapace marked with large, black-bordered, yellow-orange blotches of irregular shape on pleural and vertebral scutes; blotches on pleurals sometimes shaped as rings (e.g., fig. 8.26). Marginals with prominent, black-bordered, yellow-orange lines that curve inward and rearward. Pleural scutes with prominently raised central bosses, particularly in juveniles and males. Plastron pale yellow, marked with dark markings along seams. Male foreclaws conspicuously elongated. Irises yellow, bisected by black stripes. Differs from congeners by the large, brightly colored blotches of the carapace and the sharply angled keels of the carapace.

Taxonomic History and Phylogeny

The earliest known specimen of this species (AMNH 46774) was taken in 1930 from the Pascagoula River in Jackson County, Mississippi, but was identified as a *Graptemys pseudogeographica kohnii* (Allen 1932; Cagle 1954). The species was described by Cagle (1954, p. 167), who designated as holotype a juvenile male (TU 14798) collected on 18 August 1952 "from the Pascagoula River, 13 miles S.W. of Lucedale, George Co., Mississippi." He also designated eighty-three paratypes (FMNH 69806–808; TU 14752, 14754, 14756–66, 14768,

14772, 14774–76, 14778–85, 14788, 14790, 14795, 14799–802, 14804, 14806–809, 14811–12, 14815, 14818, 14821–22, 14825–27, 14829, 14832–33, 14842, 14845–46, 14850, 14852–54, 14857–58, 14862–63, 14865–71, 14873–75, 14920–21, 14935, 14938; UMMZ 108567–71) from the type locality and examined an additional forty preserved specimens and seventy-four discarded specimens from the type locality and a site on the lower Chickasawhay River. Cliburn (1971, p. 17) noted that the distance from Lucedale cited by Cagle is incorrect and restricted the type locality to "Pascagoula River at Old Benndale Crossing," which is consistent with labeling on Cagle's specimens that notes Mississippi State Highway 26 as the collecting locality. The specific epithet, from the Latin terms *flavus* (yellow) and *macula* (spot), refers to prominent yellow-orange markings on carapacial scutes that inspired the common name. Mertens and Wermuth (1955) regarded the taxon as a subspecies of *G. oculifera* without comment on their reason for the change, but otherwise the yellow-blotched sawback has been regarded as a separate species, a status recently confirmed by Ennen, Kreiser, et al. (2010). Phylogenetic studies have placed it in a polytomy with *G. oculifera* and *G. sabinensis* (Lamb and Osentoski 1997) and as the sister species of either *G. oculifera* (Lamb et al. 1994; Stephens and Wiens 2003) or *G. nigrinoda*, within a clade that also includes *G. oculifera* (Wiens et al. 2010).

Range

The yellow-blotched sawback is restricted to the Pascagoula River drainage in southeastern Mississippi. It ranges upstream in the two major tributaries that form the Pascagoula River, into Smith County in the Leaf River and into Clarke County in the Chickasawhay River (Selman and Qualls 2009a). It also inhabits the lower reaches of several smaller tributaries: the Bowie River, the Bogue Homa, and Bowie, Okatoma, Tallahala, Thompson, and Gaines Creeks on the Leaf; Buckatunna Creek on the Chickasawhay; Black, Red, and Bluff Creeks on the western side of the Pascagoula; and the Escatawpa River east of the lower Pascagoula (Cliburn 1971; Jones 1993b; Lindeman 1998, 1999c; Selman and Qualls 2009a). The downstream range limit in the main-stem Pascagoula occurs at approximately the Interstate 10 bridges over the East and West Pascagoula channels (McCoy and Vogt 1979; Jones 1993b; Selman and Qualls 2009a), although occasional reports have been made farther south, possibly concerning vagrants or individuals displaced by flooding (Selman and Jones 2011).

Habitat

Range-wide, the largest populations of yellow-blotched sawbacks are associated with wide, sandy rivers having medium to fast current, abundant deadwood, and large sandbars used for nesting (McCoy and Vogt 1979). Telemetered turtles on the lower Pascagoula were relocated mainly along the outer eroding banks of river bends, areas with strong currents, deep water, and abundant deadwood (Jones 1996). Average basking density of the species was significantly positively correlated with both an index of deadwood abundance and the total number of potential basking substrates, while maximum density was positively but nonsignificantly correlated with both variables (Lindeman 1999c).

There are no major impoundments within the range of the species. The turtles followed by Jones (1996) used an oxbow lake and bayous that connected it to the lower Pascagoula. He also documented seasonal overland movement between the river and a cypress pond used in late winter and spring.

Diet

Seigel and Brauman (1994) dissected museum specimens from localities throughout much of the species' range and examined the feces of specimens from the lower Pascagoula (table 4.2). In the museum specimens, sponges predominated in both sexes and juveniles. Other important prey were algae, insects, and bryozoan colonies in males; aquatic insects in females; and insects and algae in juveniles. In the lower Pascagoula, sponges predominated in males and juveniles, but females ate bivalve mollusks (a prey absent from females farther upstream) at about the same level of importance as sponges. The authors observed yellow-blotched sawbacks grazing on submerged deadwood in the field and suggested that much of the algae ingestion might be incidental to this style of feeding; however, Selman and Qualls (2008a) observed grazing on the algae of a log that appeared to have few associated animal prey.

Reproduction and Life History

Average clutch size on the lower Pascagoula is 4.7 eggs (range 3–9) and tends to increase by 1 egg for every 19 mm increase in carapace length of the female (Horne et al. 2003). Most females appear to lay only one clutch annually (Shelby et al. 2000; Horne et al. 2003). Emergence of hatchlings occurs in the fall (Jones 1993a; Horne et al. 2003).

Males mature in their second or third growing season (Jones 1993a), but age at maturity has not been reported for females. In upstream segments of the range, midline plastron length ranges from 61 to 102 mm in adult males and reaches 156 mm in adult females (Lindeman 2008a). In the lower Pascagoula, males have slightly larger midline plastron lengths (50–118 mm) but females are considerably larger (to ca. 200 mm; Jones 1993a; Seigel and Brauman 1995; Selman 2012). See figure 4.3 for additional body-size data.

Natural History

Courting males drum the backs of their forefeet claws against the ocular regions of females and bob their heads (Wahlquist 1970; Hertwig 2001; Selman and Jones 2011). Most nests on the lower Pascagoula were on sandbars, but sites with clay soils in small clearings on steep sections of riverbank were also used (Horne et al. 2003). The devastation wrought by Hurricane Katrina in August 2005 provided the opportunity to study impacts of tidal surge on yellow-blotched sawbacks (Selman and Qualls 2008e, 2009a). Following the saltwater surge measuring 3.7–4.6 m, 16 percent of marked turtles were seen at their pre-hurricane sites of capture in a single-pass basking survey. The population near Vancleave was approximately halved over the course of the next year, declining from an estimated 602 turtles/river km just after the hurricane in October 2005 to 321/km in October and November 2006, with no return of larger populations subsequently in 2007 (281/km) or 2008 (292/km). An impacted prey base was considered the most likely reason for migration from the area. Body condition (mass per unit shell length) did not differ before and after the hurricane, nor did it decline a year later, when population density had declined. Parasite loads were greater in females than males, with the main enteric parasite species being a spiny-headed worm, five trematode flukes, and two roundworms (Steinauer and Horne 2002). Fish crows were the main nest predator (Horne et al. 2003).

Populations

Both trapping and basking surveys indicate that the yellow-blotched sawback is the predominant species in the Pascagoula drainage. It may have increased in abundance relative to its sympatric congener, the Pascagoula map turtle (*G. gibbonsi*), over recent decades, although a strong upstream-downstream cline in abundance of the yellow-blotched sawback complicates interpretation of the data showing this trend. At the type locality on the middle Pascagoula in 1952, the Tulane field crews collected the two species at a ratio of 11.6 yellow-blotched sawbacks for every Pascagoula map turtle (chapter 2), but all other early references generally indicate ratios that approached unity. In the mid-1950s, Tinkle (1958b) used a variety of capture techniques at unnamed sites on the Pascagoula drainage and took 1.4 yellow-blotched sawbacks for every Pascagoula map turtle, with yellow-blotched sawbacks being the most frequently captured turtle species overall, at 46 percent of all captures. In the late 1960s, hand collecting of juveniles and adult males at various sites on the Pascagoula drainage produced a ratio of the two *Graptemys* of 1.03:1, with Pascagoula map turtles slightly predominant (Cliburn 1971). Nearly three decades later, the two species were the most abundant basking turtles in visual surveys, with 53 percent being yellow-blotched sawbacks and 22 percent being Pascagoula map turtles, a ratio of 2.4:1 (Lindeman 1999c). High numbers of the former at three sites on the main-stem Pascagoula produced an especially strong skew for this section of the drainage (80 percent yellow-blotched sawbacks and 11 percent Pascagoula map turtles, a ratio of 7.3:1; Lindeman 1998). Selman and Qualls (2009a) recorded 2.6 yellow-blotched sawbacks for every Pascagoula map turtle in range-wide basking surveys. Ratios of the two species were 0.7:1 in the Chickasawhay, 1.4:1 in the Leaf, and 7.5:1 in the Pascagoula. Annual ratios in basking trap captures ranged between 1.6 and 2.0 to 1 on the lower Chickasawhay, between 2.1 and 3.7 to 1 on the upper Leaf, and between 15 and 102 to 1 on the lower Pascagoula (Selman and Qualls 2009a).

Basking surveys over the past three decades have consistently revealed a gradient of population density, ranging from highest in the lower Pascagoula to very low in the upper Leaf and Chickasawhay Rivers (McCoy and Vogt 1979; Stewart 1990; Lindeman 1998, 1999c; Selman and Qualls 2009a). Basking densities in the lower Pascagoula were reported as 42–73 turtles/river km for the period 1989–1992 (Stewart 1990; Jones 1993a). The farthest downstream site surveyed in 1994 and 1995 averaged 138 turtles/river km, with two sites farther upstream on the main-stem Pascagoula averaging 28 and 56 turtles/river km (Lindeman 1998, 1999c). In two counts on the Chickasawhay in 1989, yellow-blotched sawback basking densities were 1.3 and 1.8 turtles/river km (Stewart 1990), whereas 1994–95 counts at eight sites along the length of the Chickasawhay yielded means of 0–29 turtles/river km, with an overall mean of 6.0 (Lindeman 1999c). McCoy and Vogt (1979) described abundant basking yellow-blotched sawbacks in the lower Leaf near its confluence with the Chickasawhay and fewer turtles farther upstream in 1979. These findings were mirrored by 1994–95 surveys in which the site farthest downstream averaged 20 turtles/river km but three upstream sites had few to no turtles, with an overall mean of 6.0 (Lindeman 1998, 1999c). Contrasting the general trend in falling densities at upstream sites, the mouth of the Bowie River on the upper Leaf River averaged 19 turtles/km—the third highest density for this species in the entire drainage (Lindeman 1998). Cliburn's (1971) most upstream sites for specimens on both the Leaf and Chickasawhay produced negative results in later basking surveys (McCoy and Vogt 1979; Lindeman 1998, 1999c), suggesting possible range reduction, but more recent surveys have found the species at these sites again (Selman and Qualls 2009a). In 2007–2008 surveys, Selman and Qualls (2009a) recorded basking densities averaging 87.8 yellow-blotched sawbacks/km on the Pascagoula, 22.0/km on the Leaf, 11.4/km on the Chickasawhay, and 1.4/km on the Escatawpa and ranging between 0 and 4.3/km on various smaller tributaries.

Jones (1993a) estimated population densities near Vancleave, Mississippi, just above the species' downstream range limit, at 370 turtles/km for October 1991, compared with an earlier estimate of 210 turtles/km in April 1989. These estimates, which relied on sightings of turtles trapped earlier and outfitted with plastic carapacial tags, are problematic due to low "recapture" rates, the possibility of tag loss, and the likelihood that the population was not closed during sampling. A large fluctuation in population size is nevertheless possible, given seasonal movements into and out of more lentic habitats that were not sampled for these estimates (Jones 1996). Selman and Qualls (2009a) reported capture-mark-resight population estimates for the same area as 602 turtles/km for 2005 and 281–321 for 2006-2008, the highest estimate coming immediately after Hurricane Katrina and the subsequent lower estimates suggesting movement out of the area. On the upper Leaf they estimated population density at 80–120 turtles/km in 2005-2008, and for the lower Chickasawhay they estimated 93 turtles/km in 2005. A recent estimate of a range-wide population size of 50,000 (Klinkenborg 2009) was roughed out based on the relationship between mark-resight estimates and basking density extrapolated over the extent of known inhabited river reaches (W. Selman, pers. comm. 2009).

Conservation Status and Actions

Listed as threatened in 1990 under the U.S. Endangered Species Act. State-listed as endangered in Mississippi. Listed as vulnerable with decreasing populations by the IUCN. Recovery efforts have focused on basic reproductive and dietary biology, status surveys, accumulation of contaminants and their effects on hormone levels, and the effects of human disturbance on nesting and basking behavior (Seigel and Braumann 1994; Lindeman 1998, 1999c; Kannan et al. 2000; Shelby and Mendonça 2001; Horne et al. 2003; Moore and Seigel 2006; Selman and Qualls 2009a; Selman et al. 2012, 2013).

Portions of some minor tributaries of the Pascagoula occur within DeSoto National Forest, including 34 km of Black Creek under federal designation as a National Wild and Scenic River, although populations of the yellow-blotched sawback are small to nonexistent in these tributaries. Extensive state protection of habitat occurs within the Pascagoula River and Ward Bayou game management areas (26,000 ha). The Nature Conservancy protects 1,988 ha near the confluence of the Leaf and Chickasawhay Rivers in the Charles M. Deaton and Herman Murrah Preserves (see chapter 6).

Needed Further Research

Given that recovery of populations under the federal plan is predicated on minimum basking densities (see chapter 5), a much-needed study would relate abundance to habitat features. In particular, the clinal variation in abundance should be more closely studied with an eye on possibly refining minimum densities in a revised recovery plan. The low clutch frequency suspected in recent studies requires confirmation, and its possible link to endocrine disruption should be further investigated. The species' small range and status as a federally listed species make it a good candidate for a concerted effort to standardize data collection in a long-term survey of population abundance at several localities within its range. Standardized data are critical to evaluating population trends, the degree of threat imposed by various factors, and the outcomes of conservation management.

Black-Knobbed Sawback, *Graptemys nigrinoda*

Figure 8.27. Adult male, Black Warrior River at Tuscaloosa, Tuscaloosa County, Alabama.

> The extent of the Alabama and Tombigbee systems as contrasted with the Pearl and Pascagoula is suggestive of greater age. Perhaps in this situation rests the explanation of the greater degree of divergence in *G. nigrinoda*, it being perhaps the older of the three [sawback species]. . . . Particularly striking is the contrast between the flattened carapace and knob-like projections in *G. nigrinoda* and the elevated carapace and laterally compressed spines of the other two species.
>
> FRED CAGLE (1954, P. 181)

Alternate Common Names

Black-knobbed map turtle (Collins et al. 1978; see discussion in chapter introduction); black knob map turtle (Burpo 2004); northern black-knobbed map turtle (Iverson 1985; Crother et al. 2012) for the nominate subspecies; southern black-knobbed sawback (Folkerts and Mount 1969) for the subspecies *delticola*; southern black-knobbed map turtle, delta map turtle (Iverson 1985; Crother et al. 2001; J. T. Collins and Taggart 2002), alternatives for *delticola*. First called black-knobbed sawback by Cagle (1954).

Description and Diagnosis

Postorbital bars continuous with neck stripes above and strongly recurved, extending from lower end rearward; posteriorly, their adjoining neck stripes either coalescing in V shape or approaching one another and then diverging. Interorbital marking a thin medial longitudinal stripe widening slightly posteriorly, typically flanked by one to three lines to each side. Supramandibular spots subsumed by prominent neck stripes. Chin with a prominent transverse bar anteriorly, subsuming submandibular spots. Markings on head yellow, those of neck, limbs, and tail yellow to light orange. Females and males with narrow heads and narrow alveolar surfaces. Second vertebral scute only slightly sloped, extending little higher than the third. Carapace not highly peaked, rounded in cross section. Carapace with high keels with knobby, black, rounded tips in juveniles and males, tips often hooking posteriorly; reduced in females. Carapace strongly serrated at rear edges of all marginal scutes. Carapace marked with large, thin, black-bordered, pale yellow rings on pleurals and irregular, thin, black-bordered, pale yellow lines on vertebrals. Females sometimes with large melanistic blotches on carapace. Marginal scutes with thin yellow lines that curve inward and rearward. Pleural scutes with prominently raised central bosses, particularly in juveniles and males. Plastron pale yellow to light orange, marked with dark markings along seams. Male foreclaws conspicuously elongated. Irises yellow, bisected by black stripes. Differs from congeners by the blunt-tipped black keels of the carapace; the low profile of the carapace; the jagged serrations of the anterior marginals; the narrow, pale, ring-like markings of the carapace; and the transverse chin bar.

The delta subspecies differs from the nominate subspecies by having little or no rearward curvature of the postorbital mark and higher incidence of separation of postorbitals from superior neck stripes; a more highly domed carapace; more extensive dark patterning, covering more than 60 percent of the plastron; and dark stripes on the limbs and neck that are black rather than dark brown and substantially wider than the light stripes. Purported intergrades from the Tombigbee drainage and the lower Alabama River have intermediate characteristics, particularly with respect to the plastral pattern (Folkerts and Mount 1969).

Taxonomic History and Phylogeny

Lahanas (1986) noted that lithographs of a hatchling black-knobbed sawback appeared in the turtle catalog of Agassiz (1857b) almost a century prior to the description of the species (fig. 8.28); Agassiz considered it to be a false map turtle and labeled it with the erroneous name *Graptemys LeSueurii.* Leonhard Stejneger of the U.S. National Museum noted the lithographs, and in 1925 he wrote to Thomas Barbour of the Museum of Comparative Zoology, Agassiz's home museum, to inquire about specimens, but none were found. The exact collection date and locality of the oldest extant specimen, USNM 17820, are unknown, as a letter from ichthyologist David Starr Jordan dated 1892 indicates it was thought to have been collected in 1884 with several other turtle specimens in Arkansas, where the species does not occur. Baur listed the speci-

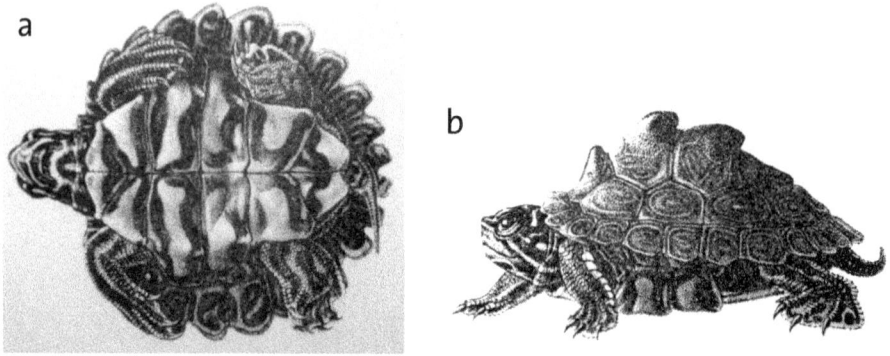

Figure 8.28. Lithographs (a and b) of a hatchling *Graptemys nigrinoda* from volume 2 of *Contributions to the Natural History of the United States of America* (Agassiz 1857b), published nearly a century before description of the species by Cagle (1954). Agassiz also depicted top and frontal views of the head. The specimen appears to be of the delta subspecies (Lahanas 1986).

men under the name *G. oculifera* in an unpublished manuscript from the 1890s. The species was described by Cagle (1954), who designated as holotype a small juvenile female (TU 14662) collected between 7 and 10 August 1952 from the "Black Warrior River, above Lock 9, 17.5 miles SSW of Tuscaloosa, Tuscaloosa County, Alabama" (p. 173). He also designated thirty-one paratypes (FMNH 69809–11; TU 14643, 14647–48, 14652–53, 14655–57, 14659, 14664–65, 14682, 14691, 14694, 14697, 14700, 14706, 14708–10, 14714, 14720, 14723, 14725, 14729; UMMZ 108572–74) from the type locality and examined eighty-five specimens not given type status from the type locality and a site on the Alabama River in Monroe County, Alabama. The specific epithet, from the Latin terms *niger* (black) and *nodus* (knot), refers to the black, knobby projections of the vertebral scutes that inspired the common name. Mertens and Wermuth (1955) regarded the taxon as a subspecies of *G. oculifera* without commenting on their reason for the change, but otherwise the black-knobbed sawback has been regarded as a separate species. Phylogenetic analyses based on mitochondrial DNA placed the species in relatively basal positions of unresolved polytomies involving six or seven other species (Lamb et al. 1994; Lamb and Osentoski 1997), while combined morphological and molecular analyses placed it as sister species to the clade composed of the other sawbacks, *G. flavimaculata* and *G. oculifera* (Stephens and Wiens 2003), and sister species to *G. flavimaculata* within a sawback clade (Wiens et al. 2010).

Folkerts and Mount (1969) described the southern black-knobbed sawback (*Graptemys nigrinoda delticola*) based on an adult male holotype (FLMNH 26238) collected from "Hubbard's Landing on Tensaw Lake, 2.6 air miles SW of Latham, Baldwin County, Alabama" (p. 677) on 4 May 1968. They also designated an adult female allotype (FLMNH 26239) and 24 paratypes (AUM 8749,

8968–70, 8979–81, 9228, 9230–38, 9334–36, 9366, 9399–401) and examined four specimens from other localities that were not given type status. All specimens assigned to this subspecies were from Baldwin and Mobile Counties in extreme southwestern Alabama; the subspecific epithet is from the Greek *delta* (triangle) and the Latin *colere* (inhabit), referencing the Mobile-Tensaw Delta. Folkerts and Mount recognized sites typified by intergrade specimens in the Tombigbee, lower Black Warrior, and lower Alabama Rivers, with sites typified by *G. n. nigrinoda* farther upstream on the two latter drainages. Freeman (1970) took issue with the designation. He suggested the authors had merely named either end of multiple clines in characteristics and criticized the small size of the range of the southern black-knobbed sawback relative to the zone of intergradation. Folkerts and Mount (1970) defended the intermediate status of characteristics in their proposed zone of intergradation and noted the relatively much more extensive habitat, with its "multiplicity of branching and anastomosing rivers" (p. 4) within the region occupied by the southern black-knobbed sawback. Lamb et al. (1994) did not find mitochondrial DNA variation between the two subspecies, and Stephens and Wiens (2003) did not differentiate subspecies of *G. nigrinoda* in their comprehensive phylogenetic study of emydid taxa.

Range

The black-knobbed sawback is restricted to the drainages that converge at the Mobile Delta. Its range encompasses most of southwestern and central Alabama and part of east-central Mississippi. For the most part the species occurs below the fall line (Folkerts and Mount 1970), although it occurs above the fall line in the Coosa (Godwin 2001, 2003) and Cahaba Rivers (pers. observation 2009). Within the Mobile Bay drainages, it inhabits the Mobile-Tensaw Delta; the Tombigbee River upstream to its tributary, Bull Mountain Creek, in Itawamba County, Mississippi, with an occurrence in lower Horse Creek; the Noxubee River into Noxubee County, Mississippi; the lower Sipsey River in Alabama (pers. observations 2007, 2010); the entire reach of the Black Warrior River; portions of the two streams that meet to form the Black Warrior, Mulberry and Locust Forks; the entire length of the Alabama River; the Cahaba River upstream into Bibb County, Alabama; the lower Tallapoosa River in Elmore County, Alabama, with an occurrence in its tributary, Line Creek; and the Coosa River upstream at least to Mitchell Dam in Chilton and Coosa Counties, Alabama (Shoop 1967; Folkerts and Mount 1969; Cliburn 1971; Mount 1975; McCoy and Vogt 1979; Godwin 2001, 2003; Blankenship et al. 2008; AUM 15829; AMNH 123809). McCoy and Vogt (1979) reported seeing three basking individuals about 110 km farther north on the Coosa River in Talladega County, Alabama, but their sighting has not been confirmed. At the southern periphery of the range, a population is centered on the nesting habitat of Gravine Island in the lower Tensaw River in Baldwin County, Alabama (Lahanas 1982). A specimen was taken in 1965 from the Dog River in Mobile

(AUM 3879), farther south, although its exclusion from consideration by Folkerts and Mount (1969) and Mount (1975) suggests they doubted the validity of the record.

Habitat

Upstream of the Mobile-Tensaw Delta, the eroding, deep, swiftly flowing outer bends of streams have abundant deadwood providing cover for high densities of black-knobbed sawbacks (Cagle 1954; Godwin 2001, 2003). Sites along the high sandstone bluffs of the northern bank of the Alabama and the high limestone chalk bluffs of the Tombigbee and the Black Warrior have less deadwood and fewer turtles (Godwin 2001, 2003). Substrates are a mix of sand and gravel, which accumulate as large sandbars on inner bends.

In the Mobile-Tensaw Delta, multiple wide rivers and their backwaters diverge and reconnect, creating an extensive aquatic labyrinth lined predominantly by two trees that define the great southern swamps: bald cypress (*Taxodium distichum*) and tupelo gum (*Nyssa sylvatica*; Folkerts and Mount 1970; Lahanas 1982). Current is strong, and the rivers are turbid and subject to a tidal influence that raises and lowers their level by as much as 30 cm (Lahanas 1982). Small juveniles occur in greatest numbers in quieter backwaters (Lahanas 1982). In the southern delta, Gravine Island is a major nesting site augmented for many years by dredge spoil (Lahanas 1982).

Tinkle's (1959) suggestion that shoals limit the upstream range to below the fall line is apparently incorrect. The species occurs above fall line shoals in the Black Warrior (Folkerts and Mount 1969), the Cahaba (pers. observation 2009), and the Coosa (Godwin 2001, 2003). It is more likely that prey communities' light limitation, caused by the narrowing river channel and closing of the riparian canopy, is the reason for the lack of records farther upstream, a situation similar to that described for Sabine map turtles (Shively and Jackson 1985).

Lentic backwater habitats are occupied in the Mobile-Tensaw Delta, including Tensaw Lake (Folkerts and Mount 1969), Coon Neck (INHS 13410), and David Lake (LSUMZ 74157), but occurrence in oxbow lakes farther up the drainage has not been reported. Much of the species' main-stem habitat in the Alabama, the Tombigbee, and the Black Warrior is impounded as long, riverlike reservoirs (see chapter 5) that house populations, while Jordan Lake, a dendritic, lake-like reservoir with several embayments on the lower Coosa River, is also inhabited (Godwin 2001, 2003).

Diet

Diet has been studied in detail only in the Gravine Island population of the southern black-knobbed sawback (table 4.3; Lahanas 1982). Three animal and three algal taxa constituted more than 95 percent of the food by volume, with very little intersexual difference in relative proportions. Sponges were the most important prey. For males, the second most important prey category was the green alga *Spirogyra*, followed by bryozoans, mussels, and the algae *Cladophora* and *Ulothrix*. For females, the second most important prey category was bryozoans, followed by mussels, *Spirogyra*, and *Cladophora*. Insects, crabs, and barnacles were the remaining prey. Feeding was apparently opportunistic, as most turtles contained only one type of prey.

Captives showed a predilection for grazing *aufwuchs* from submerged deadwood (Lahanas 1982), a behavior also observed in the wild (Waters 1974). In upstream populations, anecdotal reports indicate that aquatic insects are the most important prey, and snails, spiders, and fish may also be taken (Wahlquist 1970; Waters 1974; Mount 1975; Lahanas 1982, p. 225; C. H. Ernst and Lovich 2009). An adult female from the Cahaba contained only Asian clams (Lahanas 1982).

Reproduction and Life History

Lahanas (1982) studied growth at Gravine Island. Growth appears to follow the typical von Bertalanffy pattern of decelerating with age. Seasonally, growth seems to be confined to the months of May–September, although feeding begins in April and does not cease until November. Males mature in their third season of growth, but females do not mature until they are in their eighth or ninth growing season.

Average clutch size at Gravine Island was 5.6 eggs (range 3–7) with three or perhaps four clutches laid annually (Lahanas 1982), but females from the Alabama River lay only 2–5 eggs (P. Lindeman, unpublished data). Range-wide, for the nominate subspecies and intergrade specimens, midline plastron length ranges from 63 to 97 mm in adult males and reaches 180 mm in adult females; however, the delta subspecies reaches 102 mm in males and 202 mm in females (Lindeman 2008a and unpublished data).

Natural History

In the Cahaba, Waters (1974) observed basking in all months, with highest numbers on sunny days in March and April. Nesting on Gravine Island occurs at night (Lahanas 1982; Godwin 2002b). Major nest predators are armadillos, fish crows, and raccoons (Lahanas 1982; Godwin 2002b). Waters (1974) reported relatively high incidence of algal growth on black-knobbed sawbacks, particularly on adult males and particularly in the spring, after a winter of reduced basking opportunity. Both males and females had high incidence of infection by enteric spiny-headed worms and turtle leeches, and nematodes were also reported as parasites (Lahanas 1982).

Populations

The black-knobbed sawback was the predominant turtle species captured on the Black Warrior River below the fall line, at 68 percent of a large sample, and ranked third in a smaller sample from the Tombigbee River, at 19 percent of the catch (Tinkle 1959). Near Gravine Island, black-knobbed sawbacks were the most abundant of four emydids trapped in fyke nets by Lahanas (1982), at 41 percent of the catch, but dropped to fourth in abundance (of seven total turtle species), at just 6 percent of the total catch, in the fyke net captures of Turner (2001). Lahanas trapped in deep, moving water, whereas Turner trapped in coves, which may account for some of the difference in relative abundance, but the Tensaw River surrounding Gravine Island has also become increasingly dominated by an exotic plant, *Hydrilla verticillata* (Turner 2001).

McCoy and Vogt (1979) saw the greatest numbers of basking black-knobbed sawbacks in the Tensaw, lower Tombigbee, and lower Black Warrior Rivers, with fewer in the Cahaba, Alabama, Coosa, and Tallapoosa Rivers. In more recent basking surveys (Godwin 2001, 2003), black-knobbed sawbacks were the predominant turtle species in the Black Warrior (93 percent of all turtles seen, 4.2/river km), Tensaw (92 percent, 4.8), Tombigbee (88 percent, 4.9), Alabama (82 percent, 19.8), Cahaba (81 percent, 12.8), and Tallapoosa (45 percent, 3.4); only in the Coosa was it second most abundant (22 percent and 1.5, compared with 57 percent for the Alabama map turtle, *G. pulchra*, counting only surveys on the lower river, where both species were present).

Conservation Status and Actions

Briefly considered for listing under the Endangered Species Act in the late 1970s and early 1980s but dropped from further consideration; a new federal review was recently announced. State-listed as a protected nongame species in Alabama and as endangered in Mississippi. Listed as a species of least concern with stable populations by the IUCN.

Choctaw National Wildlife Refuge protects riparian habitat, including numerous backwaters and oxbows, on the west side of the lower Tombigbee. A stretch of the upper Noxubee is protected within Noxubee National Wildlife Refuge, and a section of the middle Cahaba occurs within or borders Talladega National Forest. The Cahaba River National Wildlife Refuge protects 5.6 km of the middle river. Alabama protects much of the Mobile-Tensaw Delta (37,850 ha) in two large wildlife management areas. In addition, Roland Cooper State Park occurs on the lower Alabama River, and Fort Toulouse-Jackson State Historical Park is located at the confluence of the Coosa and the Tallapoosa. The Nature Conservancy owns two small preserves along the lower Cahaba, one of which includes one of the river's largest nesting beaches, Barton Beach.

Needed Further Research

A detailed study of diet and life history of an upstream population is much needed for population management and identifying contrasts with the delta population Lahanas (1982) studied. A more detailed understanding of the importance of deadwood as a habitat feature for this species would also be rewarding.

9

Future Perspectives
Ecology

Following decades dominated by taxonomic studies, in the last half century or so ecology has become the best-studied aspect of map turtle and sawback biology. Yet research on the ecology of these animals is far from complete, as new discoveries beg new questions and more attention is paid to the multitude of ecological patterns that can be found across this diverse set of species in its diverse set of habitats. It is in fact possible to spend weeks or months reading vast stacks of articles, unpublished reports, and graduate theses and dissertations on map turtles and sawbacks and yet come away impressed by how little we really know about the ecology of these animals. The next half century should see exponential growth in our understanding of their ecology—albeit perhaps coupled with new insights into just how much we do *not* know. This chapter constitutes a sort of wish list for future studies, with suggestions on how to approach some of the research questions.

Habitat Use and Movements of *Graptemys*

Field study and observation of map turtles and sawbacks leaves one with the strong impression that habitat features strongly influence fine-scale distribution, particularly in rivers; to date, though, little study has focused on microhabitat preferences. It would be a relatively simple matter to choose a short but heterogeneous stretch of river and use trapping sites or telemetry locations to determine which habitat features—for instance, basking sites, deep pools, prey-rich patches, and nesting beaches—are associated with more or less use than would be expected based on their overall proportional availability. The generality of the identified habitat features in the ecology of the species in question could then be tested by combining habitat mapping with trapping or visual surveys upstream and downstream of the focal study reach. Of particular interest in this regard would be comparisons of sympatric species. Casual observations suggest nothing in the way of spatial segregation, but subtle differences may occur, perhaps relating to differences in prey distributions.

Only recently have any *Graptemys* species become the subject of long-term life-history studies. A longitudinal study of temporal variation in habitat quality in a river habitat would be rewarding, as observations suggest the importance of continuous replenishment of basking sites, fast-flowing deep pools, and nesting sandbars, all driven by bank erosion and river channel migration (chapter 4). Long-term studies should include regular data collection on rates of tree fall, deadwood availability, river depth, sandbar area, and population density to examine how the turtles' population density tracks changes in their habitat.

Two recent studies of intraspecific sexual differences in habitat use have considered the habitat use of males, small juvenile females that overlap males in body size, and larger females. Results of the two studies contrasted, suggesting a strong role of sexual dietary differences in structuring habitat use in a small riffle-and-run river habitat (Lindeman 2003b) but little role of diet in structuring habitat use in a lake (Bulté, Gravel, et al. 2008; chapter 4). In both cases, the authors strongly cautioned against perceiving a universal pattern for the genus. More studies are needed to examine the same three sex and size groups for a greater variety of species, in smaller versus larger rivers and in lakes or other lentic habitats such as sloughs and oxbows, in order to understand under what circumstances, and to what extent, habitat use depends on prey distribution. Only the lake study compared the prey base in different microhabitats (Bulté, Gravel, et al. 2008). Future studies should quantify available prey and focus on how the major dietary taxa taken by males and females differ in their distribution within the environment.

Long-distance movements by map turtles and sawbacks are apparently common in both rivers and lakes (chapter 4) and will be an important topic of future research. Telemetry of large samples of individuals and continued tracking of individuals over several years will allow researchers to determine the motivations behind such movements, potentially including mate searching, nesting habitat, food, and hibernation sites. Long-distance movements also have important implications for refining mark-recapture techniques for estimating population size. The Jolly-Seber estimator, for example, is the most popular model for estimating population size using recapture data taken over long periods because it allows for a population open to emigration, immigration, death, and births. One critical assumption is that individuals do not leave the population and then re-enter it at a later date (Seber 1982), although newer models purport to deal with temporary emigration (Kendall et al. 1997). Therefore, it will be important to follow some individuals over several years to determine the extent to which long-distance movements are tied to habitat fidelity—that is, how often individuals move long distances but then return to areas they have frequented in the past. Such fidelity would also have obvious relevance to habitat conservation initiatives.

Dietary Ecology of *Graptemys*

Strong patterns corresponding to ecomorphological groupings among adult females are apparent from studies summarizing map turtle and sawback diets (chapter 4). Nevertheless, dietary information for the genus remains incomplete and additional studies are warranted. Perhaps most needed are new detailed studies of diet in the five species of the megacephalic *pulchra* clade (*sensu* Lamb et al. 1994); three species have never been studied for diet, and a more detailed look at diet is desirable for the other two. Studies of how diet varies seasonally and with variation in habitat are also relatively few at this time and will be a valuable addition to the future literature on the genus.

It is surprising that no direct dietary comparison has been conducted of sympatric species in any of the isolated southern drainages, where trophic morphology is most variable. Of particular interest would be the presumed contrast in interspecific dietary overlap that may exist between sympatric males, which have similar trophic morphology, and females, with their highly divergent trophic morphology. If interspecific competition for prey has been important in structuring patterns of coexistence via character assortment, as suspected (chapter 4), one question that needs to be answered is the extent to which males of sympatric species compete for food and its impact upon species coexistence.

The map turtles and sawbacks provide excellent subjects for further biomechanical studies of feeding. Several correlational studies demonstrate increased capacity for mollusk crushing with increasing head or alveolar width and even link trophic morphology to individual fitness (see chapter 7). Further studies of the linkages of molluscan prey size, trophic morphology, and fitness in meso- and megacephalic females should be undertaken, together with detailed studies of the behaviors and muscular characteristics that enable crushing of hard-shelled prey. The adaptive nature of the narrower alveolar surfaces that typify males and microcephalic females has not been explored, but it may be important in their reported "grazing" behavior on submerged deadwood.

Recent studies documenting the extent to which meso- or megacephalic females depend on invasive exotic bivalves, such as Asian clams and zebra and quagga mussels (chapter 4), should stimulate interest in the ecological services map turtles and sawbacks provide. No study has assessed the degree of population regulation females may exert upon invasive bivalves. In the Great Lakes or Mississippi River, for example, it would be a relatively simple matter to place predator exclosures around artificial substrates, pair them with substrates to which turtles are allowed access, and then record differences in zebra and quagga mussel density, size, and biomass. While it is perhaps unlikely that map turtles will be found to exert a strong regulatory influence on these pest species, some effect on population dynamics is likely, given that turtle biomass and intake rates are both high (Lindeman 2006b; Bulté and Blouin-Demers 2008).

Life History of *Graptemys*

Map turtles and sawbacks have a life history that in most respects is fairly typical of other deirochelyine emydid turtles: growth follows a von Bertalanffy pattern of being most rapid in juveniles and progressively slowing; eggs and hatchlings have low survivorship due to predation, but adult survivorship is high and individuals live for decades; maturity is delayed, with females maturing later in life and at larger body sizes than males; typically two or more clutches composed of a few to more than a dozen and a half eggs are produced in midsummer, with both clutch size and egg size increasing in larger females; body size and clutch size increase with latitude, while clutch frequency is greater at southern localities; and hatchlings in most species emerge in the late summer or fall, but overwintering in the nest is also known (see chapter 4). The only truly distinctive life-history aspect relative to other deirochelyines is the exceptional intersexual differences in age at maturity and adult body size. The ultimate selective factors that have forged the exceptional sexual size dimorphism in these turtles has been and will continue to be a strong focus of ecological research on these turtles, as will the ecological consequences with regard to dietary and habitat differences.

Intragenerically, map turtles and sawbacks show little variation in egg size but a strong effect of body size on reproductive output, with body size influenced by both latitude and female diet. Two questions stand out from the existing data on interspecific variation in reproductive output, presented in chapter 4. First, why do the megacephalic females of the southeastern Gulf Coast lay smaller clutches than the three species of the Great Lakes and northern Mississippi River region, when their body sizes are so similar? Perhaps greater clutch frequency in the south allows for decreased per-clutch reproductive effort. It would be interesting to compare per-clutch and per-season reproductive effort (e.g., in grams of eggs per gram of female) for megacephalic southern species and northern populations of the three wide-ranging species. Second, if female body size constrains egg width in large-bodied populations in the north, how is it possible that smaller-bodied females of southern species or populations lay eggs that are similar in mean size? Answering this question will require comparisons of data from species or populations with dissimilar body sizes as well as increased attention to geographic and interspecific variation in the morphological aspects that constrain egg size (e.g., in diameter of the pelvic opening or size of the gap between plastron and carapace).

Studies of reproductive allometry using log-log correlation suggest that small female common and Texas map turtles lay smaller-than-optimum eggs.

As maternal body size then increases, there are competing increases in both the size and number of eggs, such that each increases at a rate slower than the rate at which body size increases (chapter 4). More studies from a variety of species in the genus will be necessary to test the generality of this pattern. Further studies of the effect of egg size on hatchling fitness are also much needed as a logical next step in these investigations, as the hypothesis that anatomy constrains egg size suggests that small eggs are laid by small females, in a trade-off favoring current reproductive investment at the expense of hatchling fitness.

A life table uses data on age at maturity and age-specific fecundity, reproductive frequency, and survival to calculate a population's rate of growth or decline and its mean generation length. Its calculations are thus invaluable for assessing conservation status (e.g., Congdon et al. 1993). Life tables used in elasticity analysis would be of particular value to the conservation of map turtles and sawbacks. In elasticity analysis, a researcher models the demographic response of hypothetical changes in factors such as fecundity or survivorship (i.e., changes that might be possible under revised management of populations and their habitats), with the aim of determining which parameters of the life history have the most impact on rate of population growth (Heppell 1998). Analyses of other freshwater turtle species suggest that changes in adult survival exert the greatest impact on population growth and so adult survival is the most appropriate target for conservation efforts (Heppell 1998); while there is presently nothing in our knowledge of map turtle and sawback demography to suggest that this pattern does not also hold true for them, confirmation is warranted.

Life tables have been constructed for several freshwater turtle species from a diverse set of taxa (Heppell 1998). None of the map turtles or sawbacks has ever been the subject of a life table, however, probably because long-term studies of freshwater turtle life history and population dynamics have generally been conducted on relatively discrete populations in small, lentic bodies of water. This should change as more long-term studies of map turtle and sawback species are undertaken. For a life table to accurately depict a population's rate of growth or decline for a species of *Graptemys*, however, two parameters are sure to prove to be most elusive, due to the highly continuous and linear nature of populations in rivers and along lake shorelines. First, accurately estimating survival rates is problematic because of the difficulty in separating mortality from long-distance dispersal out of a population (i.e., out of range of a researcher's traps and telemetry gear). Trapping and telemetry efforts will need to be widely dispersed over several years to obtain good estimates of survival. Second, clutch frequency is a very uncertain parameter (see chapter 4). While it is best determined via intensive recaptures during the nesting season, the linear dispersion of potential nesting habitats practically ensures that many cases of oviposition are missed. Certainly along rivers, no researcher will ever be able to surround entire *Graptemys* habitats with drift fences (as has been done in nesting studies of turtles in small lentic habitats; e.g., Tinkle et al. 1981; Burke et al. 1998). Perhaps intensive day-to-day radiotelemetric monitoring of the activities of small samples of females each year will allow confidence in clutch frequency data.

If life tables can be constructed for a variety of species, it will be possible to examine the interplay of diet, age at maturity, body size, and susceptibility to population declines. Species with microcephalic females are generally smaller in body size and thus are expected to mature relatively early in life, while species with mega- or mesocephalic females are larger and expected to mature later. The larger-bodied, later-maturing species should thus be more susceptible to anthropogenic impacts that increase adult mortality, and their populations may recover from such impacts more slowly. Contemporary data on abundance almost always show substantially greater numbers of microcephalic species relative to sympatric, broader-headed congeners, although historical data suggest there was far less of a discrepancy decades ago in some cases (see chapter 8). Although the only federally protected U.S. species of *Graptemys* are two sawbacks with small-bodied, microcephalic females, it is entirely possible that our attention to the conservation status of the five megacephalic species has been inadequate and that they are equally—if not more—worthy of protection and population management.

Temperature-Dependent Sex Determination in *Graptemys*

Map turtles and sawback nests produce males at cooler incubation temperatures and females at warmer incubation temperatures, with little interspecific or geographic variation in pivotal temperature (see chapter 4). Perhaps the most obvious direction for future studies of temperature-dependent sex determination (TSD) in map turtles and sawbacks is to determine whether females select nest sites depending on the size of the eggs that they will lay, as was reported for their sister species, the diamondback terrapin (Roosenburg 1996). Larger eggs laid in more open, sunnier areas would tend to develop more rapidly and hatch earlier, as females and at relatively large body sizes, giving them a head start on growing to their larger adult body size. The possibility of natal homing by nesting females should also be further investigated in a variety of species, as any tendency to use natal nest sites likely produces a female bias in sex ratio (Freedberg et al. 2005). Also, the role endocrine hormones play as temperature-mediated proximate triggers of embryonic sex determination (Crews 1996) may be a fertile field for future studies of TSD in map turtles and sawbacks.

Ultimately, the question regarding TSD that is arguably of greatest interest is the effect of hatchling sex ratios on functional sex ratios. In between the sex ratio at hatching and that at maturity, there is the possibility of different survival rates between the sexes as well as the effect of the much later maturation of females. To establish the connection between hatchling and functional sex ratio, it will first be necessary to examine the former over several seasons, with care given to sampling in proper proportion to the factors that influence sex ratio, including nest densities in shaded and more open areas and seasonal patterns of oviposition dates (Vogt and Bull 1984). Sampling to accurately determine the functional sex ratio is the bigger obstacle, however, requiring mark-recapture estimates of male and female segments of the population, with proper attention to factors such as use of backwater habitats and long-distance movements into and out of the sampling area. To date only Jones and Hartfield (1995) and Bulté and Blouin-Demers (2009) have reported functional sex ratios based on population estimates for each gender, rather than on numbers trapped or observed (see table 4.22).

Community Ecology and Interspecific Interactions of *Graptemys*

In the southeastern United States, map turtles and sawbacks occur in some of the world's richest turtle assemblages; thus studies of interspecific competition need not be limited to congeners. One of the more intriguing possibilities concerns the possible dietary overlap between the five species of map turtles that have megacephalic females and sympatric kinosternid mud and musk turtles (*Kinosternon* spp. and *Sternotherus* spp.; Lindeman 2000b). The latter are much smaller in body size but have large, powerful jaws that allow them to consume snails and bivalves (Tinkle 1958b). Studies of dietary competition among these species should be accompanied by examination of the degree of habitat overlap, however, as microhabitats used for foraging may differ more than the types of prey taken would. In addition, kinosternids may be more nocturnal in their activity than the diurnal map turtles and sawbacks, possibly relating to the question of resource competition between the two taxa (Ennen and Scott 2008).

While egg predation occurs at high rates with a diverse set of predators, predation on hatchlings, juveniles, and adults is known only anecdotally. Data on female carcasses found at nesting areas would be useful, but of greatest interest would be new dietary studies of potential major predators such as American alligators, alligator snapping turtles, largemouth bass, and nesting bald eagles—carried out within the habitats of dense map turtle and sawback populations, with careful identification of turtle remains. These and additional studies of nest depredation will be important to expanding our understanding of ecological factors that select for various life-history parameters and will also aid conservation.

Our current understanding of map turtle and sawback parasites is little more than a list of which parasites have been found in which species in the wild and how prevalent they are, with rudimentary knowledge of transmission routes. Future studies should emphasize the fitness effects parasites exert on map turtles and sawbacks. Body condition index makes a convenient surrogate for fitness (Bulté, Irschick, et al. 2008), but the possible influences of parasite load on fecundity, egg size, or juvenile growth rates could also be studied. More detailed understanding of transmission routes among hosts will require work with captives in the laboratory.

Ecophysiology of *Graptemys*

The ecophysiology of basking is one of the most obvious targets for increased research emphasis in the biology of the map turtles and sawbacks. Anyone familiar with these animals knows that basking is an activity of great importance. Nevertheless, the physiological literature on *Graptemys* has been dominated by studies of hibernation in one species, the common map turtle, which outnumber studies of basking for all fourteen species of the genus (chapter 7). In fact, detailed physiological study of basking is effectively limited to an unpublished master's thesis on black-knobbed sawbacks (Waters 1974) and recent studies of basking and locomotion in common map turtles (Ben-Ezra et al. 2008; Bulté and Blouin-Demers 2010a, 2010b); Boyer (1965) used other species as his subjects for the most part, and other observations regarding basking have generally been incidental to studies that were focused more on life history and general autecology (see chapters 4 and 7). While detailed studies similar to Waters's and Boyer's on the factors that influence basking incidence will be of use, especially as they are conducted for a variety of species, controlled laboratory studies will ultimately be the most efficient means of revealing the influences basking has on digestion, follicle development, sperm formation, and ectoparasite control.

Future investigations of hibernation needs to be expanded beyond the excellent work conducted to date on northern populations of the common map turtle, the outgroup species of the genus. It will be interesting to determine the degree to which the two other northerly distributed species are similar to common map turtles with regard to dependence on oxygen during hibernation, plasma sequestration, and ion balance. Likewise, the winter physiology of southern species should be investigated, given that their habitats cool considerably in winter but remain free of ice (e.g., Waters 1974; Shealy 1976; Craig 1992; Jones 1996).

Behavioral Ecology of *Graptemys*

The existing literature on map turtles and sawbacks is almost completely devoid of topics that fall under the heading of behavioral ecology. Potential avenues for fruitful behavioral research include studies of foraging behavior, basking aggression, and the social aspects of basking and reproduction.

Observations of foraging behavior are few and purely anecdotal (chapter 4). Intensive monitoring of individuals by telemetry might reveal movement patterns related to foraging. In clear, spring-fed rivers, it may also be possible for a snorkeler or scuba diver to discreetly make more detailed observations of forag-

Figure 9.1. An aggregation of nine Ouachita map turtles basking with their heads oriented in almost a 360° ring. Watching turtles in basking aggregations begs two questions: Do they engage in a sort of cooperative scanning for potential threats? Do single individuals and small groups shift the orientation of their heads more frequently than turtles in large groups to compensate for the relative lack of information available to them via the behaviors of other turtles? Photo from the marina at Paris Landing State Park, Henry County, Tennessee.

ing. Such observations could eventually guide the design of studies of foraging decisions, either in the wild or under controlled laboratory conditions.

Limited study has been made of basking aggression among map turtles and sawbacks and other turtle species. The suggestion that individuals assess and avoid one another based on body size (Lindeman 1999a; Selman and Qualls 2008b) is worthy of further investigation, as are the consequences of thermoregulatory opportunities lost to displacement from basking sites.

Flaherty (1982) and Flaherty and Bider (1984) suggested that aggregation occurs to enhance detection of threats by groups of basking turtles, presumably by increased scanning of the visual field. If so, one might expect that a group of basking turtles would orient their heads in directions overdispersed from a random distribution of orientations and that turtles basking singly or in small groups would change head orientation more frequently than turtles in larger groups (fig. 9.1). Either of these hypotheses would best be tested using digital video recordings of basking groups of various sizes. Also, the hypothesis that larger groups have increased detection of threats could be tested by recording flight distances of various-sized aggregations to a standard threat (e.g., a research assistant or model predator appearing in the open by a riverbank).

Based on our knowledge of map turtle and sawback life history, and by extrapolation from the behavioral ecology literature, one would expect that these species lack territoriality and male-male aggression during mating, given the males' exceptionally small body sizes. Instead, competition among males should occur solely by sperm competition. Females likely accept multiple copulations, and clutches would be expected to often have multiple sires. Confirmation is needed, however, and may be obtained with a combination of DNA paternity testing and observations of courtship and mating.

10

Future Perspectives
Evolutionary Biology and Phylogeny

To date, the study of the stunning evolutionary radiation of the genus *Graptemys* has been distinguished by two major attempts at deriving phylogeny from character data, some published thoughts on biogeography that essentially stand as yet-untested hypotheses, and comparative analyses that have been applied to questions regarding the evolution of trophic morphology and body size. Given the evolving nature of species concepts, recognition of new species of *Graptemys* is a distinct possibility in the near future. Much of the future study of the species diversity, phylogeny, biogeography, and evolutionary ecology of the genus will depend on identification of molecular markers that show sufficient variability to allow high confidence in studies of species limits, possible hybridization events, phylogeny, historical biogeography, and phylogeographic genetic structuring within species. These markers will need to come from as-yet unidentified nuclear genes, as mitochondrial DNA shows variation insufficient to allow confident answers to questions on these topics.

Species Diversity in *Graptemys*

At a species richness of fourteen, *Graptemys* falls short of only the African hinged terrapins (*Pelusios*, eighteen), American mud turtles (*Kinosternon*, eighteen), and slider turtles (*Trachemys*, fifteen) among the world's turtle genera (Fritz and Havaš 2007). Future taxonomic studies may yet increase the number of species recognized if the evolutionary or phylogenetic species concepts are rigorously applied. These species concepts emphasize the diagnosability of isolated groups that can be recognized as separate species lacking gene flow connecting them with other groups. A "unified species concept" (de Queiroz 2007) seeks to recognize species using *any* of the lines of evidence, which, under diverse concepts, support a finding of "separately evolving metapopulation lineages" (pp. 879–80). This concept could greatly increase the likelihood of future recognition of separate species among *Graptemys* populations currently recognized as conspecific but found in isolated river drainages. Molecular genetic studies combined with detailed morphological analyses may well one day demonstrate consistent, diagnosable differences among isolated populations of *G. geographica* (Mississippi, St. Lawrence, Susquehanna, Hudson, Delaware, and Mobile Bay drainages), *G. pseudogeographica* (San Bernard, Brazos, Trinity, Sabine-Neches, Calcasieu, Mermentau, and Mississippi), *G. sabinensis* (Sabine-Neches, Calcasieu, and Mermentau), *G. ernsti* (Escambia-Yellow and Choctaw-hatchee), and *G. barbouri* (Choctawhatchee, Apalachicola, Ochlockonee, and Aucilla).

Of course, some of the more recently discovered drainage occurrences of *Graptemys*—particularly those of the Hudson, Delaware, Ochlockonee, and Aucilla drainages—may merely represent human introductions, perhaps via released pets. In such cases, molecular genetic analyses may identify source populations. On the other hand, taxonomic splitting is a plausible outcome for isolated drainage populations of *G. pseudogeographica* and *G. sabinensis*. In particular, the differences in the iris and chin patterns noted for *G. pseudogeo-graphica* in the Calcasieu drainage (J. F. Jackson and Shively 1983; see chapter 7) make the status of its metapopulation worthy of investigation.

Any possible increase in the recognized diversity of the map turtles and saw-backs in the near future, as envisioned above, would be driven by the discovery of "cryptic" species: species already represented in collections but labeled with the names of other species. Interestingly, the taxonomic history of the genus blurs the distinction between what constitutes cryptic species and species that are truly newly discovered. Only the first two species of the genus to be described were recognized immediately as new and distinct species (LeSueur 1817, 1827; J. E. Gray 1831); all the others were initially labeled under other names by museum personnel or in the literature or, in the case of *sabinensis*, were neglected taxonomically for several decades (table 10.1). Only *ouachitensis*, *ernsti*, *gibbonsi*, and *pearlensis* were abundantly collected and discussed in the literature under other names before being described, however, so only they might be said to fit the conventional concept of cryptic species.

Discovery of new species never before collected is an intriguing possibility for *Graptemys*, albeit perhaps not a particularly likely one. For example, if any of the specimens from the Ochlockonee or Aucilla drainages should turn out to represent natural occurrences and also be diagnosable as distinct species (as unlikely as either event seems), they would be species discovered only very recently (1993 and 2002, respectively). More intriguingly, the *Graptemys* from the Choctawhatchee River (fig. 10.1) that are currently referred to as *G. barbouri*, *G. ernsti*, and hybrids of the two species (Godwin 2002a; Godwin et al. 2013) should be investigated as a possible new taxon or taxa. The Choctawhatchee

Table 10.1. Earliest-known specimens of the fourteen species of *Graptemys*

Species	Earliest specimen(s)	Year of collection	Year of taxon author-ship	Previously applied specific epithets
geographica	LeSueur (1817); illustrated specimen* lost	1816	1817	None
	ANSP 250	1820		
pseudogeographica	MNHN 9136–37*, 9139*, 9147*	1827–28	1831	None
ouachitensis	MNHN 9146	1827	1953	*pseudogeographica* (J. E. Gray 1831)
nigrinoda	Agassiz (1857b); illustrated specimen lost	by 1857	1954	*lesueurii* (Agassiz 1857b)
	USNM 17820	1884[a]		*oculifera* (USNM catalog)
pulchra	USNM 8808*, 318254*	1876	1893	*geographicus* (Yarrow 1883)
versa	USNM 13339; specimen lost	1883	1925	*lesueuri* (USNM catalog)
	USNM 27473–80*; MCZ 42346*	1900		*oculifera* (Strecker 1909)
oculifera	TU 20, 7628; USNM 15506–12*; USNM 15506–507 lost	1888–89	1890	*lesueurii* (USNM catalog)
pearlensis	TU 7657, 7680, 15070, 15072; USNM 220884*[b]; FMNH 22171*[b]	1891	1992	*pulchra* (Baur 1893; Cagle 1952b); *gibbonsi* (Lovich and McCoy 1992)
sabinensis	TU 7632, 7634, 7638, 7640, 7643, 7690	1893–94	1953	*intermedia* (Beyer 1900)
barbouri	Brimley (1910); specimen lost	1910	1942	*pulchra* (Brimley 1910)
	FLMNH 65918	1937		*pseudogeographica pseudo-geographica* (FLMNH catalog)
flavimaculata	AMNH 46774	1930	1954	*pseudogeographica kohnii* (Allen 1932)
ernsti	TU 13446–48, 13456–63, 13472	1950	1992	*pulchra* (Cagle 1952b)
gibbonsi	TU 14739.1–39.12, TU 14919.1–19.4	1952	2010	*pulchra* (Cagle 1954)
caglei	TNHC 23053; two additional specimens lost	1957–58	1974	*pseudogeographica* (Raun 1959); *versa* (Webb 1962)

Note: Data include dates of formal description and names of other taxa that were applied in museum catalog entries and literature accounts prior to formal description. Asterisks indicate specimens of type series. For species whose earliest specimens have been lost, the oldest-known surviving specimen are also given. See species accounts in chapter 8 for further discussion.

[a] Date of collection uncertain; catalogued in 1892.

[b] Two specimens now regarded as *pearlensis* were part of Baur's (1893) type series for *pulchra*.

Figure 10.1. Looking downstream from the point at the confluence of the Choc-
tawhatchee River (left) and Pea River (right) in Geneva County, Alabama. The
Choctawhatchee drainage was recently found to house populations of Barbour's and
Escambia map turtles and purported hybrids of the two species.

discoveries—made incidental to sampling of alligator snapping turtles in 1996
(Godwin 2002a)—beg the question of how many other small drainages along
the Gulf Coast may have had their *Graptemys* populations overlooked, populations that could possibly include undescribed species. Of course, even if the turtles of the Choctawhatchee drainage were to be recognized as an undescribed
species, the case may merely demonstrate again the lack of a fine distinction
between cryptic and new species: two hatchling *G. barbouri* in the Auburn University Museum (AUM 3880, 3881) were collected from the Choctawhatchee in
1965 but are labeled as having questionable locality information.

Godwin (2002a) and Godwin et al. (2013) presented plausible stream-capture
scenarios for both species found in the Choctawhatchee: upstream tributaries of
the river may have been captured by the Chattahoochee to the east, introducing
barbouri into the greater Apalachicola drainage, while the Pea may be a former
part of the Yellow drainage to the west that has been captured by the Choctawhatchee, introducing *ernsti* populations. While anthropogenic introduction
is a possibility, the abundance of the taxa in the Choctawhatchee (especially for
G. barbouri; G. E. Wallace 2000; Godwin 2002a; Enge and Wallace 2008) would
seem to require that introduction occurred many decades ago.

If the Choctawhatchee turtles are natural occurrences, then regardless of
whether or not they are too recently derived from their respective parent species

to warrant recognition as separate taxa, their apparent hybridization (Godwin 2002a; Godwin et al. 2013) is an intriguing situation worthy of further research. The fate of hybrids with regard to viability and fertility would be of interest, and given the similar trophic morphology of the two species, the situation would become an in-progress test of the hypothesis that character assortment occurs in southeastern drainages (see below and chapter 4).

The status of populations of the *pseudogeographica/kohnii* complex is also worthy of additional study. Vogt (1993) concluded that the taxa were conspecific with a broad band of intergradation in the middle latitudes of the range of the species, and indeed a close relationship has been supported (Lamb et al. 1994; Stephens and Wiens 2003), but the nature of the interaction of the two taxa remains poorly explored. Of particular interest is the situation in Reelfoot Lake and the nearby Mississippi River in western Tennessee, where large numbers of specimens have been taken that have markings of one taxon or the other (Vogt 1993) and *kohnii* females have significantly wider heads (Lindeman 2000a). In the lower Tennessee River in western Kentucky—essentially well upstream of Reelfoot Lake in the greater Mississippi drainage—specimens showed primarily *kohnii*-like characteristics with only rare morphological evidence of influence from the northern taxon, *pseudogeographica* (Lindeman 2003a). Perhaps the two taxa vary geographically in the degree to which they interbreed, mixing freely in some localities but hybridizing only rarely at others; indeed, Freedberg and Myers (2012) recently suggested the same scenario may hold for *pseudogeographica* (both subspecies) and *ouachitensis* in their broadly overlapping ranges. More morphological and especially molecular genetic studies from Vogt's (1993) putative area of intergradation are necessary to fully describe the interaction of *pseudogeographica* and *kohnii* populations. In particular, molecular genetic investigation of bimodal and unimodal hybrid zones (Jiggins and Mallet 2000; Fitzpatrick et al. 2008) may prove a fruitful avenue for research on this complex.

Phylogeny and Historical Biogeography of *Graptemys*

The three published attempts to discern the phylogeny of *Graptemys* have been based on mitochondrial DNA sequences (Lamb et al. 1994) and combined analyses of gene sequences and morphological data (Stephens and Wiens 2003; Wiens et al. 2010). All produced phylogenies with only a few highly supported clades. It appears there are three well-supported clades of *Graptemys*, with only one species that is of uncertain affinity to these clades. The basal split forms a monotypic outgroup, *G. geographica*; a second clade contains the five species with megacephalic females; and a third contains a mix of species with meso- or microcephalic females.

Placement of *G. caglei* is uncertain, it being placed in a basal position within either one of the two polytypic clades (Lamb et al. 1994; Stephens and Wiens

2003; most recently, it was placed as sister to *G. versa* in a clade also containing *G. sabinensis*; Wiens et al. 2010). The result in the two earlier analyses is thus a monotypic *caglei* clade located in the extreme western portion of the range of the genus, while the megacephalic clade is located in the east and the large remaining clade is more central (fig. 10.2). Uncertainty regarding how these three clades are related is suggestive of two cladogenetic events that may have occurred in rapid succession, obscuring efforts to determine which split occurred first (i.e., a minor version of a "starburst" phylogeny; Shaffer et al. 1997). Bootstrap percentages indicate high confidence for placement of *caglei* with the larger mixed clade using mitochondrial DNA sequences versus low confidence for its placement with the smaller megacephalic clade in the combined analysis. Association of the western *caglei* with the eastern megacephalic clade produces the biogeographic conundrum of a broad, intervening region that lacks surviving species. It would be worthwhile to identify which characters support a sister-clade association of *caglei* with one clade or the other (*caglei* could also conceivably be outgroup to a sister-clade arrangement of the other two clades; fig. 10.2) and to determine the degree to which various data sets support one arrangement over the other. Partitioned analyses in Wiens et al. (2010) suggest that the morphological data may drive association of *caglei* with the megacephalic clade. The most recent combined analyses supporting a clade combining *caglei* with *sabinensis* and *versa* is also of interest, given the three species' western range centers relative to all other Gulf Coast species. Hence future phylogenetic studies should emphasize the placement of *caglei* and should calculate and compare support for its placement in various alternative arrangements.

Additional points of interest in the phylogeny of *Graptemys* include the placement of *sabinensis*, relationships among the five members of the megacephalic clade, relationship among the three sawbacks, and the relationship of *ouachitensis* and *pseudogeographica*. Placement of *sabinensis* as sister to *versa* or to *caglei* + *versa* using the combined data sets was unexpected, given the association of the former as a subspecies of *ouachitensis*, but makes sense biogeographically, as the range of *sabinensis* lies near that of *versa*. If the relationship is confirmed with higher bootstrap percentages, recognition of *sabinensis* as a separate species would of course be supported, but, perhaps more importantly, the case would become an example of how the recognition of allopatrically distributed subspecies sometimes has the effect of obscuring alpha diversity.

Regarding the relationships within the megacephalic clade, one could seemingly hazard a pretty good guess at relationships even without character data: because four of the species were long considered conspecific, *barbouri* is the logical outgroup, while relationships among the other four species would seem to relate to occurrence in adjacent drainages. The outgroup within the clade was not *barbouri* in the analysis of mitochondrial DNA, but analysis of a larger data set did place *barbouri* in the outgroup position. On the other hand, whereas mitochondrial DNA placed the neighboring species *pulchra* and *ernsti* as sister

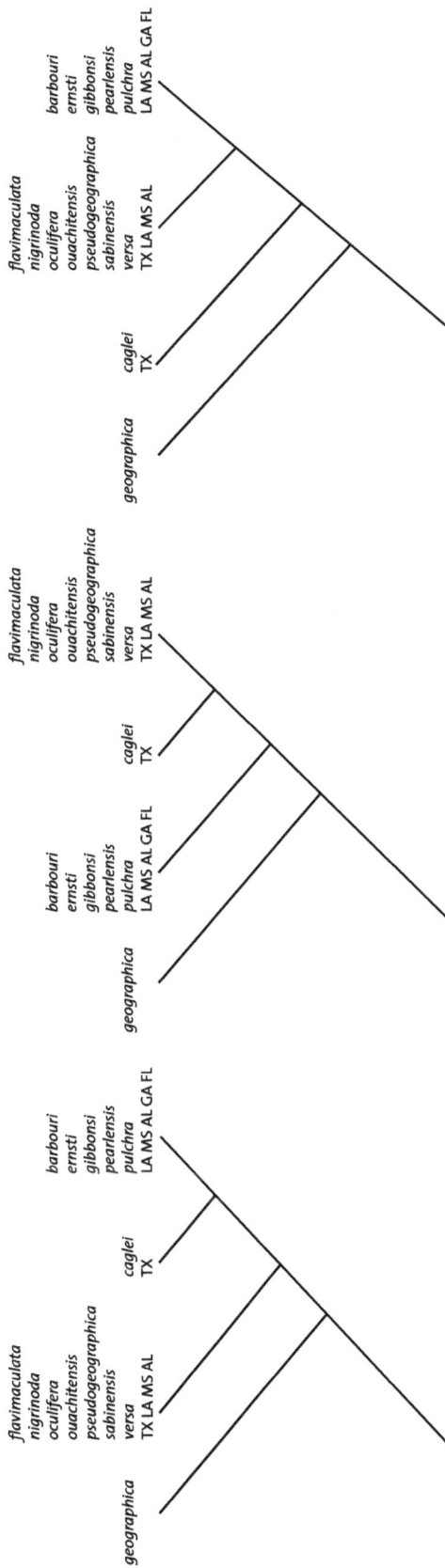

Figure 10.2. Three alternative phylogenetic arrangements for three well-supported clades of *Graptemys* with emphasis on their affinity to *caglei*. The arrangement on the left was weakly supported by a combined analysis of molecular and morphological data (Stephens and Wiens 2003); the middle arrangement was more strongly supported by mitochondrial DNA sequences (Lamb et al. 1994); and the arrangement on the right is also plausible. Gulf Coast states that are inhabited by species of each of the three southern clades are indicated with abbreviations under each list of species.

taxa, the larger data set placed *ernsti* and *gibbonsi sensu lato* as sister taxa, a scenario that is difficult to envision because the wider-ranging *pulchra* is located between them. None of these placements has been strongly supported by bootstrap percentages, however, so resolution awaits additional character data together with more detailed analysis of the support for alternative phylogenetic arrangements within the clade.

A "sawback clade" containing *nigrinoda*, *oculifera*, and *flavimaculata* was not recovered in the analysis of mitochondrial DNA, as *nigrinoda* was not placed as a close relative of the other two species, which were highly supported sister species. In the two combined analyses, however, the three formed a highly supported clade with either *nigrinoda* or *oculifera* as outgroup, the former outcome confirming the opinion Cagle (1954). Future analyses should determine which characters support a sawback clade—in particular whether there is incongruence between the molecular and morphological data, as is suggested in analyses of partitioned data sets in Stephens and Wiens (2003)—as well as relationships within the clade.

Given their broad sympatry and the long history of confusion between *ouachitensis* and *pseudogeographica*, their relationship is of particular interest. While molecular data show no particularly close relationship, partitioned morphological data not surprisingly suggest a sister relationship, and the combined analysis placed *pseudogeographica* as sister to a clade containing *ouachitensis* and the three sawbacks (Stephens and Wiens 2003; Wiens et al. 2010). Such a close relationship, if confirmed, begs the question of how the species became isolated from one another to achieve speciation. It has not generally been appreciated that within the Mississippi drainage, *pseudogeographica* is a more western species (with an extensive distribution in the Missouri River) while *ouachitensis* is a more eastern species (with a much more extensive distribution upstream into the Tennessee River and occurrences in the Muskingum, Little Kanawha, Great Kanawha, Scioto, White, and possibly the Whitewater Rivers of the upper Ohio drainage). Perhaps this difference is a clue to the isolation that allowed differentiation of these taxa, especially given the back-and-forth nature of the connection the Tennessee drainage has had to the lower Ohio and Mississippi Rivers and to the Mobile Bay drainages (Swift et al. 1986; Mayden 1988; Bănărescu 1992). A close relationship between *pseudogeographica* and *ouachitensis* also would make hybridization, potentially with multigenerational back-crossing, a real possibility, and recent evidence suggests hybridization has occurred (Freedberg and Myers 2012). Further studies of molecular genetic variation in these species, perhaps including analysis of whether any hybrid zones that may be found are bimodal or unimodal (Jiggins and Mallet 2000; Fitzpatrick et al. 2008), would shed light on the subject.

Fully understanding the biogeographic history of *Graptemys* will depend in large part on a better-resolved phylogeny, but comparisons with other animal taxa of the rivers of eastern North America will also be useful. Biogeographically, the crucial questions for *Graptemys* concern the geographic pathways by which species dispersed and established sympatry—the species *ouachitensis* with *pseudogeographica* in the Mississippi drainage, as noted above; *pseudogeographica* with *sabinensis* in three drainages to the west; and *pearlensis* with *oculifera*, *gibbonsi* with *flavimaculata*, and *nigrinoda* with *pulchra* in three drainages to the east. Examinations of the biogeography of fishes, snails, and crayfishes establish a Tennessee-Mobile Bay drainage connection and a connection from the Mobile Bay drainages to the Pearl and Pascagoula drainages (Swift et al. 1986; Mayden 1988; Bănărescu 1992). Regarding the hypothesis of character assortment related to trophic morphology (Lindeman 2000a; see chapter 4), it is difficult to say how often species of similar trophic morphology have come into sympatry, resulting in exclusion of one species or the other; a better biogeographic hypothesis may shed some light on the question.

Phylogeography of *Graptemys*

Phylogeography, the study of geographic patterns in intraspecific differentiation, has relied on molecular rather than morphological data, with most studies using mitochondrial DNA (Avise 2000). Published studies of intraspecific phylogeography have been conducted on numerous aquatic turtle species of the eastern United States, including the painted, chicken, and slider turtles, snapping turtles, mud and musk turtles, and softshells (Walker and Avise 1998; Walker et al. 1998; Roman et al. 1999; Weisrock and Janzen 2000; Serb et al. 2001; Starkey et al. 2003). Species of *Graptemys* were conspicuous by their absence from the list until A. D. Brown et al. (2012) published a phylogeography of *G. ouachitensis* (see chapter 7). Undoubtedly one reason for their relative neglect is the exceptionally low level of differentiation in mitochondrial sequences, which is likely related to how recently species within each of the major clades shared common ancestry (Lamb et al. 1994; Walker and Avise 1998). Nevertheless, given identification of the appropriate markers (i.e., those with a high enough degree of geographic structuring within species of *Graptemys*), phylogeographic methods have a great potential for yielding insights into the history of the map turtles and sawbacks and will undoubtedly be key in answering many of the questions posed above.

Phylogeographic studies should yield important insights into the status of species distributed over multiple drainages, helping to resolve the issues of species delimitation discussed previously. A phylogeographic study of the false map turtle/Mississippi map turtle complex in the greater Mississippi drainage should be helpful in delineating the nature of the interaction of the taxa *pseudogeographica* and *kohnii*, including intergradation or areas experiencing hybridization or lack thereof. At present our understanding of that interaction is clouded by the reliance on only morphological characters (Vogt 1993; Lindeman 2003a), as we have only preliminary information on geographic structure in genetic characters of the complex (samples from five northern populations; Myers 2008). Phylogeographic methods could further identify possible hybridization between *pseudogeographica* and *ouachitensis*. Similarly, a phylogeographic tree for the map turtles of the Escambia-Yellow, Choctawhatchee, and Apalachicola drainages would help to resolve the possibilities discussed above.

Fine-scale phylogeographies may help to determine whether closely related species have arisen via stream capture or the truncation of river drainages that occurred as sea levels rose following glaciation. If the sawbacks are a true clade, their derivation within a larger clade containing *ouachitensis* and *pseudogeographica* from the Mississippi drainage would likely be associated with capture of part of the Tennessee drainage in the Mobile Bay drainage and, possibly, from there into the Pearl and Pascagoula drainages to the west. Stream capture was also evoked by Godwin (2002a) and Godwin et al. (2013) as a potential explanation for the two species recently found in the Choctawhatchee drainage. On the other hand, separation of river drainages by rising sea levels has also been suggested as a major influence on speciation events in *Graptemys* of the Gulf Coast (Lovich and McCoy 1992; Lamb et al. 1994) and on genetic differentiation in other southeastern turtles (Walker and Avise 1998; Avise 2000).

Two hypothetical scenarios (fig. 10.3) demonstrate how phylogeography might distinguish between vicariant hypotheses involving stream capture versus rising sea levels in the case of two sister species, *oculifera* and *flavimaculata* (based on the phylogenies of Lamb et al. [1994] and Stephens and Wiens [2003]). Both scenarios assume that geographic structuring will be found to occur among populations inhabiting different sections of drainages (not a certainty, given the continuous nature of habitat and dispersal capabilities) and that one species may be paraphyletic with respect to another for two close relatives that diverged very recently. (For an argument that paraphyletic relationships occur commonly in allopatric speciation, see D. J. Funk and Omland [2003].) In the first hypothetical set of results, *oculifera* is paraphyletic with respect to *flavimaculata*, which shows evidence of having spread into the Pascagoula drainage via capture of a stream formerly associated with the Strong River, the major eastern tributary of the Pearl drainage. In the second hypothetical set of results, *oculifera* is again paraphyletic, but both species have differentiated from downstream toward upstream, suggesting a former connection of the lower reaches of their respective drainages, as may have occurred at low sea levels. Scenarios with reciprocal monophyly also may distinguish between hypotheses, depending on which populations are basal within each species (upstream vs. downstream or eastern vs. western tributaries).

Mitochondrial DNA may not provide sufficient resolution to answer the questions posed in this chapter because of low levels of sequence variation (Lamb et al. 1994; Walker and Avise 1998; Wiens et al. 2010; A. D. Brown et al. 2012). Increasingly, phylogeographic analyses employ nuclear genes. While they generally evolve more slowly, nuclear genes are advantageous in that sev-

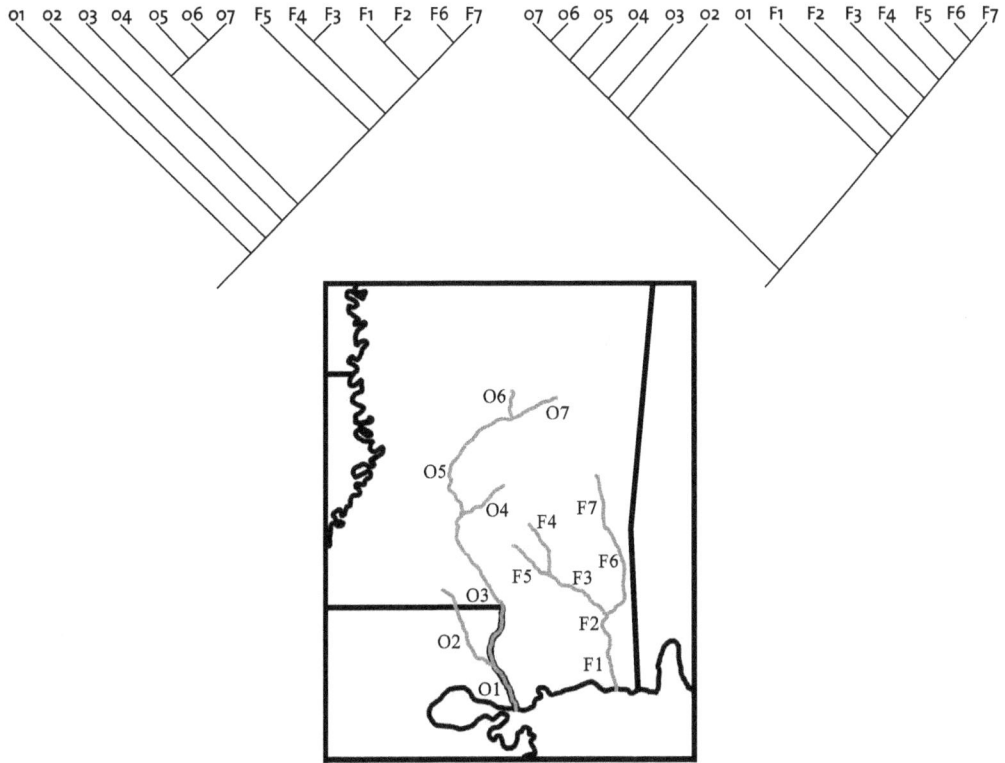

Figure 10.3. Two hypothetical phylogeographic scenarios for the relationships of popula-tions of G. oculifera in the Pearl drainage (O1–O7) and G. flavimaculata in the Pascagoula drainage (F1–F7). The scenario on the left would support the hypothesis of an origin of G. flavimaculata populations by stream capture from the Pearl drainage, while the scenario on the right would support the hypothesis of a common drainage that was separated by rising sea levels, leading to speciation.

eral genes can be analyzed and their results compared; in addition, phylogeo-graphic studies increasingly employ sequences from intron regions of genes, which have high mutation rates (Avise 2000; Hare 2001). In *Graptemys*, mic-rosatellite DNA sequences (short repeats occurring several times in a row) can generally be targeted with primers developed for other emydid turtle species (Y. L. King and Julian 2004; Freedberg et al. 2005; Myers 2008; Selman, Ennen, et al. 2009) and will likely be helpful in future efforts at describing the phylo-geography of species in the genus.

Identification of new molecular markers will undoubtedly also improve con-fidence in future phylogenies and determination of species limits. For example, single nucleotide polymorphisms have been suggested as an important new tool for studying species limits in turtles and other taxa (Shaffer and Thomson

2007), but they may also be important in phylogeographic and phylogenetic studies, particularly for the radiation of *Graptemys*.

Evolutionary Biology of *Graptemys*

Application of comparative methods to map turtle and sawback evolution has successfully addressed questions regarding trophic morphology, character displacement versus character assortment, and body size (see chapters 4 and 7). Life history is a fruitful area for future comparative research. Within the genus, there is a considerable range in female body size associated with a similarly wide range in average clutch size, both of which appear to covary with age at maturity. Body size in both sexes and clutch frequency also show latitudinal variation that is likely tied to climatic selective factors, while diet is an additional influence on the body size of females. Unfortunately, life-history data are scanty for Alabama, Cagle's, Pascagoula, Pearl, and Sabine map turtles; hence further robust comparative analyses await the collection of a more complete set of input data.

There are also undoubtedly many more morphological analyses that might be conducted and would be of interest; while these, too, lack complete sets of input data at present, collecting the data from multiple species using museum specimens or live captives should be less arduous than it will be to complete the collection of life-history data. Do leg and toe lengths vary with habitat, especially with strength of current, or with tendency to occur in lakes (common map turtles) or lentic floodplain habitats (false and Ouachita map turtles)? Do height of the shell and vertebral keels vary according to overlap with predator species—that is, are these characteristics that have been selected for because of predation? Is shell shape affected by current speed, river size, or predators that are present? What features of the skull, jaw musculature, and neck vertebrae covary with the evolution of micro-, meso-, and megacephaly? A genus strongly characterized by fascinating and diverse morphologies awaits much more detailed analyses of its functional morphology than have been carried out to date.

11

Future Perspectives
Conservation Biology

To better conserve populations of map turtles and sawbacks, improved understanding of their present status is foremost among the objectives that must be met in the coming decades. While identifying potential *Graptemys* habitat is simple enough with reference to any map that shows rivers and their tributaries, it is clear that many species occurrences remain to be identified, particularly in upstream tributaries and floodplain lakes. There is also a need to expand our knowledge from simply knowing where species are present to quantifying their abundance in different localities, which will require increased attention to population estimation and the correlation between basking counts and total population density. Beyond the need for a more complete inventory of what we have on hand, five major areas have received relatively little attention but are worthy of increased future attention: management of deadwood and its supply; a reservoir-ecology perspective on how river impoundment affects habitat and populations; research on the effects of contaminants on reproductive physiology; determination of the effects of global climate change on populations; and strategies for combating recent expansion of the pet trade.

Distribution and Abundance of *Graptemys*

Effective conservation requires a good understanding of where a species occurs and how abundant it is in different habitats, both under anthropogenic threat and in relatively more-pristine (or less-impacted) conditions. The range maps presented in this volume appear to be far from complete. Perusal of the maps shows counties for some species that would be expected to have populations, based on upstream and downstream occurrences, but which have no specimens or sight records. There is also a need to standardize basking survey techniques and to learn more about the relationship of sighted abundance to actual abundance.

Upstream Range Limits

Many of the upstream limits depicted in range maps are tenuous, given researchers' apparent preferences for trapping or sighting specimens farther downstream, in areas known to house larger populations. Selman and Qualls (2009a) discussed extensions of known geographic ranges for yellow-blotched sawbacks and Pascagoula map turtles they observed far upstream of historical localities (compare their maps with the maps of McCoy and Vogt 1987 and Lovich and McCoy 1994b). In March 2012, I photographed basking Escambia map turtles at each of four bridges I visited on the Conecuh River above Gantt Reservoir, their previous known upstream range limit (Godwin 2000). The site farthest upstream extends the known range by 75 river km.

As a further example, consider the range extensions that resulted since 1994 from spotting-scope observations of the ringed sawback and the Pearl map turtle on the Bogue Chitto, Strong, and Yockanookany Rivers and Lobutcha and Topisaw Creeks (tributaries of the Pearl drainage; see range maps). In the Bogue Chitto, mapped records had previously been confined to St. Tammany and southern Washington Parishes, Louisiana (McCoy and Vogt 1988; Dundee and Rossman 1989; Kofron 1991; Iverson 1992), although Cliburn (1971) had also reported the Pearl map turtle farther upstream in Walthall County, Mississippi, and McCoy and Vogt (1988) had questioned whether ringed sawbacks might also range farther upstream. McCoy and Vogt (1979) had reported seeing ringed sawbacks in the lower Strong near its confluence with the Pearl, although Stewart (1986a, 1986b) deemed their record doubtful based on his surveys; the only two specimens from the Strong are Pearl map turtles. As of the early 1990s, there were no specimens or literature reports of *Graptemys* from the Yockanookany or from Lobutcha or Topisaw Creeks.

Shively (1999) observed both species at five sites extending to the Louisiana-Mississippi state line in the Bogue Chitto. In 2008, I saw three female Pearl map turtles at a new locality in southern Walthall County and a juvenile in Pike County, the latter record extending the known range 10 river km (Lindeman 2010). In 2010, I recorded ringed sawbacks at two Walthall County sites (fig. 11.1), 74 river km above a specimen record in Washington Parish, Louisiana, and Pearl map turtles at four sites up to 27 river km above the previous record in Pike County, including the lower reaches of Topisaw Creek.

I observed both species in 1994–95 in the Strong River's downstream section in Simpson County: the ringed sawback 7 km from its confluence with the Pearl, where McCoy and Vogt (1979) had reported the species, and the Pearl map turtle at the same site and farther upstream, 9 river km above the higher of two specimen localities (Lindeman 1998, 1999c). In 2008 I saw a ringed sawback 16 river km upstream of my previous reported site. I also saw Pearl map

Figure 11.1. A female ringed sawback photographed in 2010 on the Bogue Chitto River at Walkers Bridge Water Park in Walthall County, Mississippi, the first record of the species upstream of the Louisiana-Mississippi border in the Bogue Chitto.

turtles at the two previously reported sites and at two additional sites 19 and 37 river km farther upstream, the latter site in Rankin County.

I saw both species in the lower Yockanookany 7 river km from its confluence with the Pearl in 1994 and 1995. I saw four female ringed sawbacks 32 river km farther upstream in 2008 and two female Pearl map turtles 18 river km upstream of the original record in 2010. A recent Pearl map turtle specimen, the first *Graptemys* record for Lobutcha Creek, was taken 7 river km from the creek's confluence with the Pearl (Ennen, Lovich, et al. 2010), and in 2010 I saw the species 21 river km upstream of that site.

In summary, since federal listing of the ringed sawback in 1986, our understanding of its range has expanded by 74 river km in the Bogue Chitto, 23 river km in the Strong, and 39 river km in the Yockanookany; corresponding expansions for the Pearl map turtle are 37, 46, and 25 river km, plus 28 river km in Lobutcha Creek and 1 river km in Topisaw Creek. While extensions of known range are obviously good news in a conservation context, the size and viability of these populations are unknown. Selman and Qualls (2009a) suggested yellow-blotched sawbacks were making year-round use of upstream sites in the Pascagoula drainage. More detailed survey work (not limited to counts from bridge crossings, which may fail to detect presence; Selman and Qualls

2009a) will be necessary to determine true upstream range limits and population densities at these sites.

New Tributary and Drainage Records

The importance of smaller tributaries is an open question for many species of map turtles and sawbacks. Table 11.1 lists small tributaries that were not published localities prior to 1995. Such sites should be an important consideration in determining the legal status of imperiled species under federal and state or provincial laws as well as status on the IUCN Red List. Also of critical importance is the degree of connection between tributary and main-stem river populations: Is use of these smaller streams ephemeral for individual turtles? If so, are there biases in tributary use with regard to sex or age class? Do tributaries provide important resources, perhaps seasonally, that river main stems do not? For example, the only use common map turtles made of a tributary creek of the lower Susquehanna River in Maryland was for nesting forays (Richards-Dimitrie 2011).

Table 11.1. Recent range extensions for *Graptemys* species

Species	State	Tributaries or drainages with first specimens or reported sightings coming after 1995
barbouri	GA	Buck Creek, Line Creek
	AL/FL	Choctawhatchee/Pea River*
	FL	Aucilla River,* Ochlockonee River*
caglei	TX	None
ernsti	AL	Sepulga River, Murder Creek, Persimmon Creek
	AL/FL	Choctawhatchee/Pea River*
flavimaculata	MS	Escatawpa River,* Bogue Homa, Bluff Creek,* Bowie Creek, Buckatunna Creek, Gaines Creek, Okatoma Creek, Thompson Creek
geographica	AL	Paint Rock River
	AR	Twin Creek
	GA	Little Chickamauga Creek
	KY	Tradewater River
	MD	Deer Creek
	MI	Flat River
	MO	James River, Little Black River, Little Sac River, Moreau River, Big Creek, Crane Creek, Young's Creek
	OH	Little Auglaize River
	PA	Schuylkill River
	TN	Hiwassee River, Stones River, East Fork Stones River, West Fork Stones River, Bear Creek
	WI	Milwaukee River, Namekagon River, Sugar River

Table 11.1. Recent range extensions for *Graptemys* species *(continued)*

Species	State	Tributaries or drainages with first specimens or reported sightings coming after 1995
gibbonsi	MS	Escatawpa River,* Chunky River, Bogue Homa, Buckatunna Creek, Gaines Creek, Long Creek, Oakohay Creek, Souinlovey Creek, Thompson Creek
nigrinoda	AL	Sipsey River, Line Creek
oculifera	MS	Yockanookany River
ouachitensis	IL	Sangamon River
	IA	Cedar River, Skunk River
	KS	Big Blue River, Big Caney River, Cottonwood River, Fall River, Little Arkansas River, Little Caney River, Marais des Cygnes River, Pawnee River, Stranger Creek
	KY	Tradewater River
	LA	Comite River
	MO	Big River, Current River, Gasconade River
	OH	Muskingum River
	OK	Blue River, North Canadian River, Spring River, Big Cabin Creek, Grave Creek
	WV	Great Kanawha River
pearlensis	LA	Pushepatapa Creek, Topisaw Creek
	MS	Lobutcha Creek
pseudogeographica	AR	Salado Creek
	KS	Saline River, Salt River, Mill Creek
	LA	Comite River, Houston River, West Fork Calcasieu River, Bundick Creek, Bayou Castor, Bayou Nezpique, Bayou Plaquemine Brule, Bayou Toro, Indian Bayou
	MO	Current River, Salt River, Bryant Creek
	OK	Blue River, Caney River, Spring River, Big Cabin Creek, Grave Creek
	TX	Bois d'Arc Creek, San Bernard River*
pulchra	MS	Noxubee River, Tibbee Creek
sabinensis	LA	Houston River, Little River, West Fork Calcasieu River, Bundick Creek, Anacoco Bayou, Bayou des Cannes, Bayou Nezpique, Bayou Plaquemine Brule, Bayou Queue de Tortue, Bayou Toro, English Bayou, Indian Bayou, Lacassine Bayou
versa	TX	Clear Creek, East Johnson Fork, Live Oak Creek, White Oak Creek

Note: For sources, see species accounts in chapter 8. Asterisks denote isolated drainages.

Unrecognized diversity may exist in some smaller, poorly sampled drainages, as new occurrences in isolated drainages have also been reported in recent years (marked with asterisks in table 11.1; see also chapter 10). Selman et al. (2008a) described an unsuccessful attempt to verify a report of Pascagoula map turtles—based on a live specimen they had been shown—in the Biloxi River. It is important from a conservation perspective to distinguish populations introduced by humans from populations isolated by natural processes centuries and millennia ago. It is also important to fully recognize, with taxonomic names where appropriate, the full extent of existing diversity. Continued searches outside the Gulf Coast drainages marked in the range maps in this book combined with molecular genetic studies and careful morphological studies of isolated drainage populations will accomplish these objectives.

Use of Floodplain Habitats

There is increasing appreciation of the fact that map turtles and sawbacks make significant use of oxbows and other river backwaters (see chapter 4 and species accounts in chapter 8). To better understand the ecology and conservation status of any species, its use of such habitats must be more fully explored. Turtles in oxbows and backwaters are excluded from basking counts or mark-recapture population estimates on river main stems, potentially skewing our view of population status. Use of lentic habitats may be transient and seasonal (e.g., Jones 1996) but may nevertheless be pivotal for productivity of populations. Naturally, use of lentic habitat means conservation planning must also prioritize preservation of these habitats.

Basking Surveys

While trapping map turtles and sawbacks is necessary for a host of ecological studies, including those that better inform conservation efforts, visual surveys of basking turtles are also an important conservation tool. Response to trap bait is weak at best in these species, and it is difficult to tend traps in rivers that rise and drop with changes in discharge, so it is perhaps fortunate that map turtles and sawbacks show themselves with such regularity, allowing ease of observation. For example, consider the false map turtle, listed as threatened by South Dakota, where it occurs primarily in the upper Missouri. In a two-year, statewide trapping survey, Bandas (2003) captured 683 turtles, with 95 percent taken in baited hoop nets. Of 45 turtles she captured on the Missouri and in two floodplain lakes, only 5 were false map turtles. In contrast, at about noon on a late August day on LaFramboise Island in Pierre, I used my spotting scope to "capture" 7—on a single log!—of the 18 I recorded in two hours that day (fig. 11.2). The objectives of Bandas's study were more generalized than my own, and no criticism is meant of her work, but the example makes it clear that any assessment of *Graptemys* population status that is not augmented by visual surveys would have to be considered deficient, or at least inefficient, in

Figure 11.2. Seven false map turtles basking on a log installed for basking turtles on LaFramboise Island in Pierre, Hughes County, South Dakota. The species is state-listed as threatened, but statewide visual surveys have not been conducted.

its methodology. Basking surveys can cover considerably more habitat area per unit of researcher time than trapping and, with the advent of digiscoping, are also useful for documenting previously overlooked localities (e.g., Lindeman 2008b, 2010).

There is a need to standardize basking surveys to enable comparison among studies, particularly for population studies carried out in different years or decades. Surveys have been conducted from boats (e.g., Jones and Hartfield 1995), on foot after stopping a boat at a sandbar (e.g., Selman and Qualls 2008e, 2009a), and from fixed vantage points accessed at highway crossings (e.g., Lindeman 1999c). More attention to the relationship between basking densities and mark-recapture population estimates will also become increasingly important as the database on population status of map turtles and sawbacks grows. Although researchers sometimes use basking densities and trapping results to compare sites, species, or time periods, only population estimates actually quantify abundance (and then only if proper attention is paid to the assumptions of the estimators).

Four studies have examined the relationship between basking density and population density in map turtles and sawbacks, and their widely divergent results probably relate at least in part to differences in methodology and time of year, as well as to possible interspecific differences. Shively and Jackson (1985) saw 50 percent of the estimated population of Sabine map turtles in ten-hour observations. Babitzke (1992) and Killebrew et al. (2002) averaged sighting 2.8 and 3.0 percent, respectively, of their population estimates for Cagle's map turtle in counts taken throughout the year. Jones and Hartfield (1995) saw 8–10 percent of estimated population sizes of ringed sawbacks in July surveys at four sites but 30 percent at a fifth site, where turtles appeared habituated to human recreation. Selman and Qualls (2007) and Selman, Qualls, and Mendonça (2007, 2008) conducted mark-resight surveys from sandbars and riverbanks. They saw 44 percent of paint-marked Pascagoula map turtles in April surveys and 14–33 percent of paint-marked yellow-blotched sawbacks in October surveys. Future emphasis should be placed on comparing the relationship of basking density to population density among species, especially those that are sympatric, and on ascertaining general patterns in the relationship as it is influenced by latitude, habitat, season, climatic conditions, or population sex ratio.

Deadwood Management and Restoration

While a burgeoning literature on the important ecological role of deadwood in rivers has developed over the last four decades (Gregory 2003), little of that attention has focused on freshwater turtles (Wondzell and Bisson 2003), as compared with aquatic invertebrates and fish (J. B. Wallace and Benke 1984; Benke and Wallace 2003; Dolloff and Warren 2003). Given their habitat preferences and thermoregulatory behavior, the map turtles and sawbacks are prime subjects for developing a much larger literature on how deadwood and its management affect turtles. Such a literature will need to adopt a long-term perspective to properly evaluate the retention of wood in rivers inhabited by *Graptemys* populations and the source of that wood in surrounding landscapes that are subjected to various types of use and management (Sterrett et al. 2011; Carrière and Blouin-Demers 2010). Of particular interest regarding *Graptemys* may be the fact that emergent wood exposed to air decays much more rapidly than submerged wood (disappearing in years vs. decades or centuries; Abbe et al. 2003; Benke and Wallace 2003), as a steady supply of both the submerged and emergent sections is vital to *Graptemys* populations.

Restoration of river habitat must include some combination of engineered wood reintroduction in the short term and restoration of riparian sources of wood for the long term (Abbe et al. 2003; Reich et al. 2003; Zalewski et al. 2003; Angradi et al. 2009). Placement of large, anchored pieces of deadwood in rivers has been mentioned as a possible management strategy for *Graptemys* (Lindeman 1999c). Although not related to aquatic turtle conservation per se, an increasing research focus has developed on producing "engineered log jams" for the benefit of river ecosystems and placing them in such a way as to avoid negatively impacting infrastructure, navigation, and recreation (Abbe et al. 2003; Piégay 2003). Overreliance on this strategy, however, without also restoring riparian forests and the erosional forces that introduce large snags, would be treating the symptoms (historical reduction of deadwood densities in rivers) without correcting the problem (bank stabilization and loss of riparian forests). This would produce a sort of halfway technology *sensu* Frazer (1992), who first used the term in turtle conservation for head-starting programs that incubate eggs and raise juveniles in captivity before release without addressing the causes of elevated adult mortality. Nevertheless, just as head-starting may be an accessory tool used to increase a population while mortality issues are addressed, engineered log jams can be used as a short-term fix during the decades that may be required to restore riparian functioning (Reich et al. 2003; Zalewski et al. 2003). Efforts at habitat restoration via deadwood placement or riparian forest restoration will provide a fertile new field for research on the ecology and conservation management of *Graptemys*.

Reservoirs and *Graptemys*

Although it has generally been accepted that reservoir construction is harmful to *Graptemys* populations (e.g., Gibbons 1997; Bodie 2001), the reaction of map turtles and sawbacks to river impoundment has in fact been little studied. While Cagle's map turtle appears particularly sensitive to impoundment and loss of lotic habitat (Babitzke 1992; Killebrew et al. 2002), many other species have been found to persist, sometimes abundantly, in impoundments (e.g., fig. 11.3; see also species accounts in chapter 8). Further, while the quality of aquatic habitat may decrease with impoundment, habitat quantity is greatly increased. For example, in western Kentucky, impoundment of the lower Tennessee River created 313 km of reservoir shoreline from what had been 86 km of riverbank because of the many embayments formed in small tributary creek beds of the original river valley (Lindeman 1997a). The observed density of a rare emydid, the river cooter *Pseudemys concinna* (2.8 percent of all basking turtles recorded), was estimated to be 16 turtles/km of reservoir shoreline, based on the assumption that on average 20 percent of turtles were seen during surveys conducted during optimal basking weather. This equates to a density of 58 turtles/river km, had the same number of turtles occurred in the original river—a number so high as to be unlikely for *P. concinna* at its northern range periphery. Similar calculations for two map turtles in the same surveys (Lindeman 1998, 1999c) are 162 Ouachita map turtles/km of reservoir shoreline (or 590/km in the original river) and 93 Mississippi map turtles/km of reservoir shoreline (337/km in the original river). The two higher figures are likely unattainable based on population estimates for *Graptemys* (see species accounts in chapter 8). Hence the question is how reservoir construction changes map turtle and sawback populations, both in absolute abundance (likely increased in some cases, as habitat area is greatly increased) and in relative abundance

Figure 11.3. Four Barbour's map turtles basking on the supports of a private dock adjacent to Smoak Bridge public boat ramp on Lake Blackshear, an impoundment of the Flint River, in Crisp County, Georgia.

within the assemblage (likely decreased in most cases, due to decreased habitat quality with the loss of current and increases in generalist species such as the slider turtle, *Trachemys scripta*, the river cooter, and the painted turtle, *Chrysemys picta*).

Reservoir Productivity and *Graptemys* Abundance

The river-lake hybrid model postulates that a reservoir's ecological dynamics are best understood by examination of axes of longitudinal continua, with rough division of the reservoir into a riverine zone upstream, a lacustrine zone abutting the dam, and a middle transitional zone (Kimmel et al. 1990). In the riverine zone, current is slowing but strong enough to carry suspended sediments and particulate organic matter, causing turbidity and consequent light limitation of primary productivity, thereby limiting community production. In the lacustrine zone, light penetration is maximal but, with lack of current, low organic nutrients limit primary and thus community production. The transitional zone may thus be most productive, having the best combination of light penetration and current-mediated allochthonous organic inputs. An investigation of reservoir dynamics as they relate to longitudinal patterns in primary productivity, the productivity of prey organisms, and the productivity of map

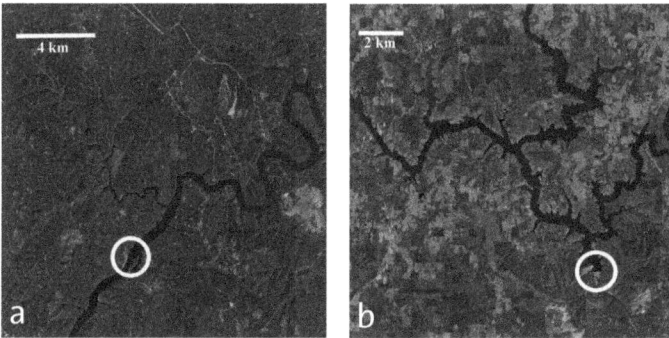

Figure 11.4. Satellite images of two dams (circled) and their contrasting reservoirs in Alabama. (a) Bankhead Lock and Dam on the Black Warrior River, forming Bankhead Lake to the northeast, a run-of-the-river reservoir with few lateral embayments. (b) Lewis Smith Dam on Sipsey Fork, forming Lewis Smith Lake to the north, a dendritic, lake-like reservoir with numerous lateral embayments and a water residence time of 435 days. Base map data for both © 2012 Google, Landsat.

turtles and sawbacks would be a valuable contribution to the conservation literature for the genus.

Positioning of the three zones of a reservoir depends upon dam placement, discharge, and water residence time (Kimmel et al. 1990). River-like reservoirs ("high-flow" or "run-of-the-river" reservoirs) are linear, with few small embayments, short residence times (measured in weeks or months), and transitional zones positioned near their dams. Lake-like ("low-flow") reservoirs have more irregular shapes, larger embayments, long residence times (a year or more), and transitional zones positioned far upstream. The two types of reservoirs (fig. 11.4) may differ dramatically with regard to their ability to support map turtles and sawbacks. I predict that high-flow, river-like reservoirs will be found to provide more favorable habitat, supporting greater densities and higher abundance relative to syntopic turtle species.

Temporal changes in the productivity of a reservoir also occur as a vegetated landscape is initially inundated and the biomass of its forest slowly decays (Wetzel 1990). Such changes may impact the productivity of turtle populations and could also be the subject of long-term studies, particularly for new dams.

Reservoirs and Deadwood Supply

Another factor vitally important to map turtle and sawback populations is the management of reservoir shorelines. On the east shore of Kentucky Lake and the west shore of Lake Barkley, designation of 69,000 ha as Land Between the Lakes National Recreation Area preserves 480 km of forested shoreline that provides ample basking structure (Lindeman 1999c). Shore erosion is medi-ated not by current but, rather, by pounding wave action on windy days (pers. observations). Residential development along the shorelines of other reservoirs, such as the Ross Barnett Reservoir (Pearl River), Jordan Lake (Coosa River), and Lake McQueeny (Guadalupe River), replaces forests with grassy lawns and steep-sided structures designed to eliminate erosion, such that any basking sites must wash in from far upstream (Lindeman 1999c; Godwin 2001, 2003; Kille-brew et al. 2002). On the upper Mississippi and upper Missouri Rivers, riprap-ping eliminates shoreline erosion and snags are few and scattered.

Increased recognition of the importance of shoreline deadwood is evident in a recent story involving the state-listed false map turtle in South Dakota. In the Missouri River in Pierre, revegetation is being carried out on LaFram-boise Island (Ode 2004). In response to riprapping of the shoreline by the U.S. Army Corps of Engineers, biologists of the U.S. Fish and Wildlife Service and South Dakota Department of Game, Fish, and Parks lobbied successfully for the installation of large logs for the benefit of basking turtles and other wild-life (fig. 11.2; D. Backlund and D. Ode, pers. comm. 2008). How populations respond to management of reservoir shoreline forests will be a key subject of future investigations.

Contaminant Effects on *Graptemys*

To better understand the effects of toxic substances on map turtles and saw-backs, an increasingly interdisciplinary approach will be necessary. Techniques to quantify contaminants in the environment and in turtle tissues are labori-ous, expensive, and beyond the capabilities of most field biologists. In addition, testing of hypotheses that concern the identification and magnitude of adverse effects of toxic chemicals will best be carried out in the laboratory. Potentially, these could employ expendable, commercially produced subjects, such as Mis-sissippi or Ouachita map turtles raised on turtle farms.

The most important experimental research objectives concerning contami-nants will include assessing their effects on reproduction. The hypothesized depressive effects on clutch frequency (Shelby et al. 2000; Horne et al. 2003; Shelby-Walker et al. 2009) and hatchling survival (Neuman-Lee and Janzen 2011) would be a good place to start. In addition, the effects of contaminants on clutch size, sperm motility, and egg fertility should all be investigated.

At the same time, an inventory of contaminant exposures is necessary to fully comprehend the effects that they may be having in the wild. Rivers habi-tats should be assayed to determine which species are being exposed to what toxic chemicals. A broad database of toxins in their environments would surely assist conservation efforts. (Naturally, the utility of this objective is not limited to species of *Graptemys* or to aquatic turtles in general.) Extensive testing of tissue samples and eggs from wild specimens will allow determination of which contaminants are being sequestered and at what concentrations, guiding experimental investigation of adverse effects.

Potential Impacts of Global Climate Change

Global climate change mediated by anthropogenic greenhouse gases has already begun to exhibit a multitude of effects on native species around the world (e.g., McCarty 2001; Root et al. 2003). While no effects on map turtles and sawbacks have been identified, changes in climate potentially could affect sex determination (and thus sex ratios), river hydrology, and downstream range limits. Investigation of these effects will be an important focus of *Graptemys* research in future decades.

Changes in Sex Determination

Janzen (1994) studied the impact of annual temperature fluctuations on hatchling sex ratios in painted turtles (*Chrysemys picta*). Over five years, July temperatures varied enough to produce years in which hatchlings in natural nests were either all males or all females. Over several decades in the mid-1900s, he modeled fluctuating hatchling sex ratios that produced an approximate balance over time. With a warming of 4°C, however, hatchlings would cease to develop as males, assuming no change in nesting behavior or phenology. Species of *Graptemys* may be inherently susceptible to dramatic shifts in sex ratio, as they have a relatively narrow range over which clutches with mixed sexes are produced, suggesting that few mixed-sex clutches should occur in the wild (Hulin et al. 2009).

Tucker et al. (2008) found that nesting phenology of slider turtles (*Trachemys scripta*) might be more important than realized in a steadily warming climate. While average annual temperature increased by 0.16°C annually over fourteen years, both hatchling and secondary sex ratios became progressively biased toward males, which are produced at cooler incubation temperatures. The most plausible explanation for the contrary bias in sex ratio was their observation that with warming, nesting season began earlier (by about two weeks by the end of the study) and lasted longer, allowing production of more clutches early in the season when the soil was cooler.

The findings of Janzen (1994) and Tucker et al. (2008) have particular relevance for map turtles and sawbacks, given their similarity to painted turtles and slider turtles in their pattern of temperature-dependent sex determination. Studies of nesting phenology and sex ratios in areas studied decades ago (e.g., the Mississippi River in Wisconsin; Vogt and Bull 1984) would be of particular value, allowing for historical comparisons.

Changes in Hydrology

Climate-change models predict increased precipitation in some areas and decreased precipitation in others. Models generally do not allow for accurate predictions for specific localities, so which catchment basins within the ranges of map turtle and sawback species should increase or decrease in precipitation in the coming decades is not known. Long-term declines in precipitation would decrease the flow of a river, potentially impacting bank erosion and thus food sources, basking site availability, and maintenance of sandbars. To a researcher, a long-term drought provides an opportunity to assess the degree to which reduced flow impacts populations as well as how populations respond with movements upstream or downstream. With a better understanding of shifting precipitation patterns, it would then be possible to predict population changes.

Rising Sea Levels

The most recent prediction of the Intergovernmental Panel on Climate Change regarding sea level change is for a rise of 18–59 cm during the remainder of the twenty-first century (Hansen 2007), a change that would only slightly alter the downstream range limits of *Graptemys* species. However, recent concerns over the accelerating rate of ice melt in Greenland and West Antarctica are causing climatologists to consider more drastic scenarios, with sea level rises of several meters conceivable over the next few centuries (Hansen 2007).

Consider how a 5 m rise in sea level by 2100 would alter three Gulf Coast river drainages (fig. 11.5; the scenario can be envisaged by moving upstream along rivers in Google Earth and noting elevation). The effect would be only a 2 percent loss of land worldwide (Hansen 2007), but of course the first part of a continent to feel the effects of rising saltwater would be its coastal freshwater rivers. In the Mobile Bay drainages, a 5 m rise would inundate the rivers and connected sloughs of the entire Mobile-Tensaw Delta, with saltwater extending up the Alabama River to about Claiborne in Monroe County and up the Tombigbee River to the Coffeeville Lock and Dam on the border of Choctaw and Clarke Counties. Vast reaches of productive habitat for the black-knobbed sawback and Alabama map turtle would be eliminated. In the Pascagoula drainage, a 5 m inundation would impact all of the Pascagoula River's reach in Jackson County and several kilometers in southern George County, as well as lower portions of the Escatawpa River and Black and Red Creeks. The largest populations of yellow-blotched sawbacks would thus be eliminated (at 7 m elevation, the species' type locality at Benndale would be spared); smaller populations of Pascagoula map turtles would also be affected. In the West and East Pearl, ringed sawbacks and Pearl map turtles would disappear from a short section of their range (~15–20 river km) south of Holmes Island in St. Tammany Parish, Louisiana, and Hancock County, Mississippi. Abundance at downstream range

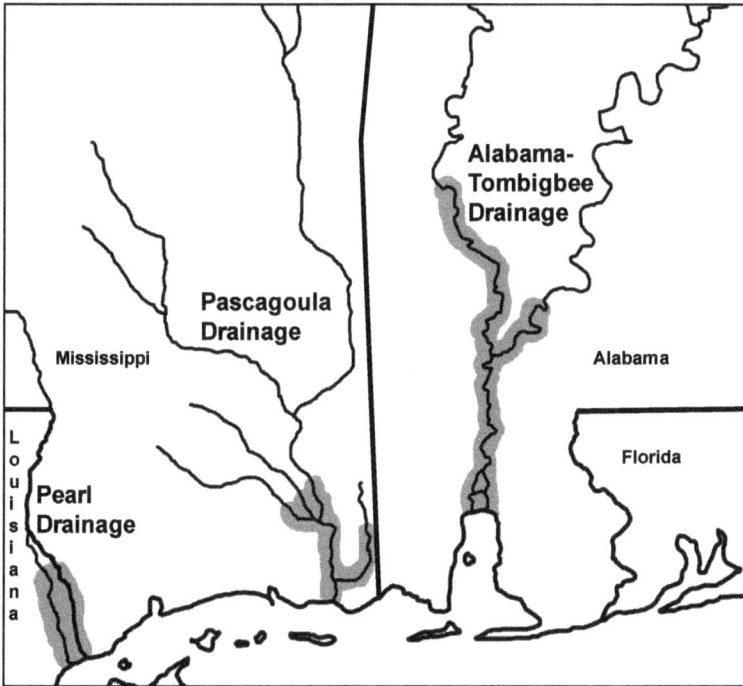

Figure 11.5. Potential loss of freshwater habitat due to global climate change in the Pearl, Pascagoula, and the Mobile Bay drainages. River reaches that would be inundated by a 5 m rise in sea level are marked in gray.

limits will need to be monitored over decades to determine the impact of rising sea levels.

Regulation of Trade in *Graptemys*

Trade data demonstrate a surging mania among hobbyists who desire to own map turtles and sawbacks (see chapter 5). The Appendix III listing of the genus under CITES is a promising development that will enable more detailed tracking, at least of trade that is both legal and international. Likewise, the proliferation of state listings of species is promising in that it legislates against overexploitation. The key, however, will lie in effective enforcement, which will require training at two levels. State wildlife field officers must be knowledgeable about these species and the threats they face if they are to recognize and seriously confront organized poaching efforts (e.g., Muir 1984). Likewise, customs officials will need good identification guides to recognize falsely labeled live specimens. At both levels, undercover sting operations will be helpful in interrupting organized poaching efforts. Because the geographic area in which vigilance is needed is extensive, there is strong potential for citizen involvement through education programs that alert the public to the threat of poaching and encourage reporting of suspicious trapping activity, particularly in the ranges of species that may be most vulnerable to the pet trade (e.g., the sawbacks, Barbour's map turtle, and Cagle's map turtle). Letting people know about the resource and how they are being robbed of it should be an effective strategy in combating illegal take, given the public's sympathy for endangered species and their generally favorable attitude toward turtles.

Recent debates among turtle biologists have concerned the advisability and workability of allowing commercial trade in endangered species, provided they are produced in captivity by private hobbyists. In 2008, for example, an extensive list server debate concerned whether captive-bred individuals of the angonoka (*Geochelone yniphora*), a Madagascan tortoise critically endangered in large part by illegal trade, should be made available in legal markets in order to ease extreme poaching pressures on the species in the wild (and perhaps even to help fund conservation research). In theory, allowing hobbyists to buy and sell ringed or yellow-blotched sawbacks they have hatched in captivity has no bearing on conservation of wild populations of these federally listed species in Mississippi and Louisiana; the problem lays in how captive-bred status would be confidently certified and thereby eliminate the incentive for illegal poaching. While such allowances may be part of the future strategy for conserving map turtles and sawbacks, as with other turtle species, no such allowances should be given without a well-thought-out and field-tested plan for eliminating faked "captive-bred" labels. Until that is possible, "no" simply must mean no and enforcement efforts and penalties will need to be of a magnitude sufficient to deter illegal take from the wild.

Ultimately and idealistically, however, discussions over how to regulate the trade in *Graptemys* offer answers for the wrong question. In a hypothetical world in which every individual human is blessed with at least a modicum of common sense, none of us would seek to have map turtles and sawbacks as pets in our homes. Instead, we would seek them out on their own terms, in their natural habitats. More of the people who have an appreciation for turtles would spend their money and time on equipment such as spotting scopes and learning to canoe or kayak instead of on tanks, bulbs, pelleted food, books on turtle culture, and the interminable changing of fetid water. I have often spoken with both those who keep map turtles and sawbacks as pets and those who have taken trips with the express purpose of seeing the species in the wild. Watch their eyes—people in the latter group are more vibrant and come much more fully alive in discussing their interactions with these species. Perhaps one day a market could even develop to make map turtles, sawbacks, and other sympatric turtles the focus of eco-tour operations that draw in fanciers from around the United States, as well as from Europe and anywhere else that turtle diversity is

Figure 11.6. An adult female Texas map turtle basks in the setting sun on the Peder-nales River in Lyndon B. Johnson State Historical Park, Gillespie County, Texas—a natural experience and a visual tableau that cannot be replicated with a hobbyist's captive kept in an aquarium.

celebrated and appreciated (e.g., see Schulz 2001, 2004 for descriptions of trans-Atlantic trips made to view *Graptemys* in the wild). Tour buses could swing across southern river drainages, stopping at bridge crossings or for short float trips so that the customers could see these highly visible animals in real rivers, with their flat, white inner-bend sandbars, slumping outer banks, and accumulations of tangled tree limbs, root masses, and logs—features no basement tank can ever replicate.

On more than one occasion I have sat at my office desk lost in remembrances of work I did years ago on the South Llano River and two more-recent trips all around the Colorado drainage to take photographs for this book (fig. 11.6). The Texas map turtle intrigues me even more than do most members of the genus, for reasons both scientific and aesthetic: it occurs in unique habitats, including some very narrow, shallow, clear streams; it has really small males compared with most species; it has beautiful markings and coloration; and the country-side where it occurs is breathtaking. In short, I frequently get lost in thought, wishing that I could get to the Hill Country of Texas on a more regular basis to feed my *versa* jones. Lost in this recurring reverie, I sometimes remember that right behind me on a counter, in a twenty-gallon tank, sits a male Texas map turtle that hatched from an egg laid after an administration of oxytocin during my studies. He is not nearly enough, though—and so I need to make it back to Texas, just as I perpetually crave returning to every other river drainage that holds these fascinating and beautiful creatures. Take my word, if you are drawn to these animals, you will make yourself much happier visiting them out there than they could ever make you by adorning your living space.

References

Abbe, T. B., A. P. Brooks, and D. R. Montgomery. 2003. Wood in river rehabilitation and management. Pages 367–389 in S. V. Gregory, K. L. Boyer, and A. M. Gurnell (Eds.), *The Ecology and Management of Wood in World Rivers*. American Fisheries Society, Bethesda, Md.

Acholonu, A. D. 1969a. Acanthocephala of Louisiana turtles with a redescription of *Neoechinorhynchus stunkardi* Cable and Fisher 1961. *Proceedings of the Helmintho-logical Society of Washington* 36:177–183.

———. 1969b. Some monogenetic trematodes from Louisiana turtles. *Proceedings of the Louisiana Academy of Sciences* 32:20–25.

Acholonu, A. D., and K. Arny. 1970. Incidence of nematode parasites in Louisiana turtles. *Proceedings of the Louisiana Academy of Sciences* 33:25–34.

Adams, M. S., and H. F. Clark. 1958. A herpetofaunal survey of Long Point, Ontario, Canada. *Herpetologica* 14:8–10.

Adler, K. 1968. Turtles from archaeological sites in the Great Lakes states. *Michigan Archaeologist* 14:147–163.

———. 2007. The development of systematic reviews of the turtles of the world. *Verte-brate Zoology* 57:139–148.

Agassiz, L. 1857a. *Contributions to the Natural History of the United States of America*, vol. 1. Little, Brown and Company, Boston.

———. 1857b. *Contributions to the Natural History of the United States of America*, vol. 2. Little, Brown and Company, Boston.

Akers, T. K., and M. G. Damm. 1963. The effect of temperature on the electrocardio-grams of two species of turtles. *Copeia* 1963:629–634.

Allen, M. J. 1932. A survey of the amphibians and reptiles of Harrison County, Missis-sippi. *American Museum Novitates* 542:1–20.

Anderson, P. K. 1958. The photic responses and water-approach behavior of hatchling turtles. *Copeia* 1958:211–215.

Anderson, R. A., M. L. Gutierrez, and M. A. Romano. 2002. Turtle habitat use in a reach of the upper Mississippi River. *Journal of Freshwater Ecology* 17:171–177.

Angradi, T. R., D. W. Bolgrien, T. M. Jicha, M. S. Pearson, D. L. Taylor, and B. H. Hill. 2009. Multispatial-scale variation in benthic and snag-surface macroinvertebrate assemblages in mid-continent US great rivers. *Journal of the North American Ben-thological Society* 28:122–141.

Angradi, T. R., E. W. Schweiger, D. W. Bolgrien, P. Ismert, and T. Selle. 2004. Bank stabilization, riparian land use and the distribution of large woody debris in a regulated
reach of the upper Missouri River, North Dakota, USA. *River Research and Applications* 20:829–846.

Anonymous. 1973. Barbour's map turtle nearing extinction. *Florida Conservation News*
8(5):2.

———. 1976. Barbour's map turtle said increasing. *Florida Conservation News* 11(10):8.

———. 2000. Untitled. *International Zoo News* 47:61.

———. 2003. Untitled. *International Zoo News* 50:178.

Aresco, M. J., and R. M. Shealy. 2006. *Graptemys ernsti*—Escambia map turtle. *Chelonian Research Monographs* 3:273–278.

Arndt, R. G., and W. A. Potter. 1973. A population of the map turtle, *Graptemys geographica*, in the Delaware River, Pennsylvania. *Journal of Herpetology* 7:375–377.

Artner, H. 2001a. Bemerkungen zu den Höckerschildkröten des *Graptemys
pseudogeographica*-Formenkreises mit Berichten über Haltung and Nachzucht der
falschen Landkarten-Höckerschildkröte *Graptemys pseudogeographica pseudogeographica* Gray, 1831 und der Sabine-Höckerschildkröte *Graptemys ouachitensis
sabinensis* Cagle, 1953. *Emys* 8(3):4–17.

———. 2001b. Die Pascagoula-Höckerschildkröte *Graptemys gibbonsi* Lovich & McCoy,
1992: Erstbericht über eine erfolgreiche Vermehrung in menschlicher Obhut. *Emys*
8(2):4–19.

———. 2003. Die rezenten Schildkrötenarten der Erde. *Emys* 10(6):4–38.

Ash, L. R. 1962. Development of *Gnathostoma procyonis* Chandler, 1942, in the first and
second intermediate hosts. *Journal of Parasitology* 48:298–305.

Ashley, E. P., and J. T. Robinson. 1996. Road mortality of amphibians, reptiles and other
wildlife on the Long Point Causeway, Lake Erie, Ontario. *Canadian Field-Naturalist*
110:403–412.

Atkinson, D. A. 1901. The reptiles of Allegheny County, Pennsylvania. *Annals of the
Carnegie Museum* 1:145–157.

Auger, P. J., and P. Giovannone. 1979. On the fringe of existence: Diamondback terrapins at Sandy Neck. *Cape Naturalist* 89:44–58.

Avise, J. C. 2000. *Phylogeography: The History and Formation of Species*. Harvard University Press, Cambridge, Mass.

Babitzke, J. B. 1992. An Analysis of Population Size of *Graptemys caglei*. Unpublished
M.S. thesis, West Texas State University, Canyon.

Bailey, V., F. M. Bailey, and L. Giovannoli. 1933. Cave life of Kentucky mainly in the
Mammoth Cave region. *American Midland Naturalist* 14:385–635.

Baker, P. J., J. P. Costanzo, R. Herlands, R. C. Wood, and R. E. Lee, Jr. 2006. Inoculative
freezing promotes winter survival in hatchling diamondback terrapin, *Malaclemys
terrapin*. *Canadian Journal of Zoology* 84:116–124.

———, J. P. Costanzo, J. B. Iverson, and R. E. Lee, Jr. 2003. Adaptations to terrestrial
overwintering of hatchling northern map turtles, *Graptemys geographica*. *Journal of
Comparative Physiology B* 173:643–651.

———, J. B. Iverson, R. E. Lee, Jr., and J. P. Costanzo. 2010. Winter severity and phenology of spring emergence from the nest in freshwater turtles. *Naturwissenschaften*
97:607–615.

Ballinger, R. E., J. D. Lynch, and G. R. Smith. 2010. *Amphibians and Reptiles of Nebraska*. Rusty Lizard Press, Oro Valley, Ariz.

Bănărescu, P. 1992. *Zoogeography of Fresh Waters, vol. 2: Distribution and Dispersal
of Freshwater Animals in North America and Eurasia*. AULA-Verlag, Wiesbaden,
Germany.

Bandas, S. J. 2003. Geographical Distribution and Morphometrics of South Dakota Turtles. Unpublished M.S. thesis, South Dakota State University, Brookings.

Barbour, T., and A. Loveridge. 1929. Typical reptiles and amphibians. *Bulletin of the Museum of Comparative Zoology, Harvard* 69:205–360.

Barger, M. A. 2004. The *Neoechinorhynchus* of turtles: Specimen base, distribution, and host use. *Comparative Parasitology* 71:118–129.

Barker, F. D., and M. Parsons. 1914. A new aspidobothrid trematode from Lesseur's terrapin. *Transactions of the American Microscopical Society* 33:261–262.

Barko, V. A., J. T. Briggler, and D. E. Ostendorf. 2004. Passive fishing techniques: A cause of turtle mortality in the Mississippi River. *Journal of Wildlife Management* 68:1145–1150.

Barr, B. 1997. Food Habits of the American Alligator, *Alligator mississippiensis*, in the Southern Everglades. Unpublished Ph.D. dissertation, University of Miami, Coral Gables, Fla.

Barrett Beehler, K. M. 2007. An Investigation of the Abundance and Key Habitat Parameters of the Northern Map Turtle (*Graptemys geographica*) in an Eastern Ontario Bay: A Baseline Study. Unpublished M.E.S. thesis, University of Waterloo, Waterloo, Ontario, Canada.

Baur, G. 1890. Two new species of tortoises from the south. *Science* 16:262–263.

———. 1893. Two new species of North American Testudinata. *American Naturalist* 27:675–677.

Beane, J. C., L. T. Pusser, and R. A. Davis. 2009. *Graptemys geographica* (northern map turtle): Geographic distribution. *Herpetological Review* 40:449.

Behler, J. L., C. M. Castellano, and T. J. Crockett. 2004. *Graptemys geographica* (northern map turtle): Geographic distribution. *Herpetological Review* 35:186.

Belusz, L. C., and R. J. Reed. 1969. Some epizoophytes on six turtle species collected in Massachusetts and Michigan. *American Midland Naturalist* 81:598–601.

Ben-Ezra, E., G. Bulté, and G. Blouin-Demers. 2008. Are locomotor performances coadapted to preferred basking temperature in the northern map turtle (*Graptemys geographica*)? *Journal of Herpetology* 42:322–331.

Benke, A. C., and J. B. Wallace. 2003. Influence of wood on invertebrate communities in streams and rivers. Pages 149–177 in S. V. Gregory, K. L. Boyer, and A. M. Gurnell (Eds.), *The Ecology and Management of Wood in World Rivers*. American Fisheries Society, Bethesda, Md.

Bennett, A. M. 2009. Effects of Habitat Fragmentation on the Spatial Ecology and Genetics of Northern Map Turtles (*Graptemys geographica*). Unpublished M.S. thesis, Laurentian University, Sudbury, Ontario, Canada.

———, M. Keevil, and J. D. Litzgus. 2009. Demographic differences among populations of northern map turtles (*Graptemys geographica*) in intact and fragmented sites. *Canadian Journal of Zoology* 87:1147–1157.

———, M. Keevil, and J. D. Litzgus. 2010. Spatial ecology and population genetics of northern map turtles (*Graptemys geographica*) in fragmented and continuous habitats in Canada. *Chelonian Conservation and Biology* 9:185–195.

Benson, K. R. 1978. Herpetology on the Lewis and Clark expedition 1804–1806. *Herpetological Review* 9:87–91.

Bertl, J., and F. C. Killebrew. 1983. An osteological comparison of *Graptemys caglei* Haynes and McKown and *Graptemys versa* Stejneger (Testudines: Emydidae). *Herpetologica* 39:375–382.

Beyer, G. E. 1900. Louisiana herpetology. *Proceedings of the Louisiana Society of Naturalists* 1897–99:25–46.

Bickham, J. W., T. Lamb, P. Minx, and J. C. Patton. 1996. Molecular systematics of the genus *Clemmys* and the intergeneric relationships of emydid turtles. *Herpetologica* 52:89–97.

Bishop, S. C. 1921. The map turtle, *Graptemys geographica* (Le Suer) in New York. *Copeia* 1921:80–81.

Bjork, P. R. 1967. Latest Eocene vertebrates from northwestern South Dakota. *Journal of Paleontology* 41:227–236.

Bjorndal, K. A., A. B. Bolten, C. J. Lagueux, and D. R. Jackson. 1997. Dietary overlap in three sympatric congeneric freshwater turtles (*Pseudemys*) in Florida. *Chelonian Conservation and Biology* 2:430–433.

Black, J. H., W. A. Carter, and J. Pigg. 1993. County records, voucher specimens, and notes on turtles (Reptilia: Testudines) in Oklahoma. *Proceedings of the Oklahoma Academy of Science* 73:19–22.

———, J. Pigg, and R. L. Lardie. 1987. Distribution records of *Graptemys* in Oklahoma. *Bulletin of the Maryland Herpetological Society* 23:65–68.

Blair, W. F., A. P. Blair, P. Brodkorb, F. R. Cagle, and G. A. Moore. 1957. *Vertebrates of the United States*. McGraw-Hill, New York.

Blankenship, E. L., B. P. Butterfield, and J. C. Godwin. 2008. *Graptemys nigrinoda* Cagle 1954—black-knobbed map turtle, black-knobbed sawback. *Chelonian Research Monographs* 5:5.1–5.6.

Boarman, W. I. 1997. Predation by turtles and tortoises by a "subsidized" predator. Pages 103–104 in J. Van Abbama (Ed.), *Proceedings: Conservation, Restoration, and Management of Turtles and Tortoises: An International Conference*. New York Turtle and Tortoise Society, New York.

Bochkov, A. V., and B. M. OConnor. 2008. A new mite superfamily Cloacaroidea and its position within the Prostigmata (Acariformes). *Journal of Parasitology* 94:335–344.

Bodie, J. R. 2001. Stream and riparian management for freshwater turtles. *Journal of Environmental Management* 62:443–455.

———, and R. D. Semlitsch. 2000a. Size-specific mortality and natural selection in freshwater turtles. *Copeia* 2000:732–739.

———, and R. D. Semlitsch. 2000b. Spatial and temporal use of floodplain habitats by lentic and lotic species of aquatic turtles. *Oecologia* 122:138–146.

———, R. D. Semlitsch, and R. B. Renken. 2000. Diversity and structure of turtle assemblages: Associations with wetland characters across a floodplain landscape. *Ecography* 23:444–456.

Bonnemains, J., and R. Bour. 1996. Les chéloniens de la collection Lesueur du Muséum d'Histoire Naturelle du Havre. *Bulletin Trimestriel de la Société Géologique de Normandie et des Amis du Muséum du Havre* 83:5–45.

Bour, R., and A. Dubois. 1983. Statut nomenclatural et specimens-types d'*Emys pseudogeographica* Gray, 1831 et d'*Emys lesueuri* Gray, 1831 (Reptilia, Chelonii, Emydidae). *Bulletin de la Societe Linneenne de Lyon* 52:42–46.

Boyd, C. E., and D. H. Vickers. 1963. Distribution of some Mississippi amphibians and reptiles. *Herpetologica* 19:202–205.

Boyer, D. R. 1965. Ecology of the basking habit in turtles. *Ecology* 46:99–118.

Brauman, R. J., and R. A. Fiorillo. 1995. *Lampropeltis getula holbrooki* (speckled kingsnake): Oophagy. *Herpetological Review* 26:101.

Brimley, C. S. 1910. Records of some reptiles and batrachians from the southeastern United States. *Proceedings of the Biological Society of Washington* 23:9–18.

Bringsøe, H. 1998. Quo vadis? Three American CITES proposals for American reptiles. *Herpetological Review* 29:70–71.

Britson, C. A. 1998. Predatory responses of largemouth bass (*Micropterus salmoides*) to conspicuous and cryptic hatchling turtles: A comparative experiment. *Copeia* 1998:383–390.

——, and W. H. N. Gutzke. 1993. Antipredator mechanisms of hatchling freshwater turtles. *Copeia* 1993:435–440.

Brodman, R., S. Cortwright, and A. Resetar. 2002. Historical changes of reptiles and amphibians of northwest Indiana Fish and Wildlife properties. *American Midland Naturalist* 147:135–144.

Brooks, D. R. 1975. A review of the genus *Allassostomoides* Stunkard 1924 (Trematoda: Paramphistomidae) with a redescription of *A. chelydrae* (MacCallum 1919) Yamaguti 1958. *Journal of Parasitology* 61:882–885.

——, and M. A. Mayes. 1975. Platyhelminths of Nebraska turtles with descriptions of two new species of spirorchiids (Trematoda: Spirorchiidae). *Journal of Parasitology* 61:403–406.

——, and M. A. Mayes. 1976. *Telorchis gutturosi* sp. n. (Trematoda: Telorchiidae) from *Graptemys pseudogeographica* Gray in Nebraska, with reports of additional species of trematodes from Nebraska turtles. *Journal of Parasitology* 62:901–905.

Brown, A. D., K. Temple-Miller, W. M. Roosenberg, and M. M. White. 2012. Mitochondrial DNA variation in the Ouachita map turtle. *Copeia* 2012:301–306.

Brown, A. E. 1908. Generic types of Nearctic Reptilia and Amphibia. *Proceedings of the Academy of Natural Sciences of Philadelphia* 60:112–127.

Brown, B. C. 1950. *An Annotated Check List of the Reptiles and Amphibians of Texas.* Baylor University Press, Waco, Tex.

Brown, D. J., M. R. J. Forstner, and J. R. Dixon. 2008. *Graptemys pseudogeographica kohnii* (Mississippi map turtle): Geographic distribution. *Herpetological Review* 39:481.

Brown, G., J. B. Jensen, and S. P. Graham. 2011. *Graptemys pulchra* (Alabama map turtle): Geographic distribution. *Herpetological Review* 42:389.

——, and O. M. Kinney. 2011. *Graptemys geographica* (northern map turtle): Geographic distribution. *Herpetological Review* 42:565.

Browne, C. L., and S. J. Hecnar. 2005. Capture success of northern map turtles (*Graptemys geographica*) and other turtle species in basking vs. baited hoop traps. *Herpetological Review* 36:145–147.

——, and S. J. Hecnar. 2007. Species loss and shifting population structure of freshwater turtles despite habitat protection. *Biological Conservation* 138:421–429.

Buhlmann, K. A. 2006. Map turtles: A complete overview. *Reptiles Magazine* November 2006:48–61.

——, and J. W. Gibbons. 2006. Habitat Management Recommendations for Turtles of Conservation Concern on National Wildlife Refuges. Unpublished report to the National Fish and Wildlife Foundation, Washington, D.C.

——, T. Tuberville, and W. Gibbons. 2008. *Turtles of the Southeast.* University of Georgia Press, Athens.

Bull, J. J. 1985. Sex ratio and nest temperature in turtles: Comparing field and laboratory data. *Ecology* 66:1115–1122.

——, and R. C. Vogt. 1979. Temperature-dependent sex determination in turtles. *Science* 206:1186–1188.

——, and R. C. Vogt. 1981. Temperature-sensitive periods of sex determination in emydid turtles. *Journal of Experimental Zoology* 218:435–440.

——, R. C. Vogt, and M. G. Bulmer. 1982. Heritability of sex ratio in turtles with environmental sex determination. *Evolution* 36:333–341.

——, R. C. Vogt, and C. J. McCoy. 1982. Sex determining temperatures in turtles: A geographic comparison. *Evolution* 36:326–332.

———, T. Wibbels, and D. Crews. 1990. Sex-determining potencies vary among female incubation temperatures in a turtle. *Journal of Experimental Biology* 256:339–341.

Bulté, G., and G. Blouin-Demers. 2008. Northern map turtles (*Graptemys geographica*) derive energy from the pelagic pathway through predation on zebra mussels (*Dreissena polymorpha*). *Freshwater Biology* 53:497–508.

———, and G. Blouin-Demers. 2009. Does sexual bimaturation affect the cost of growth and the operational sex ratio in an extremely size-dimorphic reptile? *Écoscience* 16:175–182.

———, and G. Blouin-Demers. 2010a. Estimating the energetic significance of basking behaviour in a temperate-zone turtle. *Écoscience* 17:387–393.

———, and G. Blouin-Demers. 2010b. Implications of extreme sexual size dimorphism for thermoregulation in a freshwater turtle. *Oecologia* 162:313–322.

———, M.-A. Carrière, and G. Blouin-Demers. 2010. Impact of recreational power boating on two populations of northern map turtles (*Graptemys geographica*). *Aquatic Conservation: Marine and Freshwater Ecosystems* 19:31–38.

———, M.-A. Gravel, and G. Blouin-Demers. 2008. Intersexual niche divergence in northern map turtles (*Graptemys geographica*): The roles of diet and habitat. *Canadian Journal of Zoology* 86:1235–1243.

———, D. J. Irschick, and G. Blouin-Demers. 2008. The reproductive role hypothesis explains trophic morphology dimorphism in the northern map turtle. *Functional Ecology* 22:824–830.

Burger, J., and S. D. Garber. 1995. Risk assessment, life history strategies, and turtles: Could declines be prevented or predicted? *Journal of Toxicology and Environmental Health* 46:483–500.

Burke, V. J., S. L. Rathbun, J. R. Bodie, and J. W. Gibbons. 1998. Effect of density on predation rate for turtle nests in a complex landscape. *Oikos* 83:3–11.

Burling, J. W. 1972. *Graptemys pseudogeographica ouachitensis*: Geographic distribution. *Herpetological Review* 4:170.

Burpo, N. 2004. Consequences of Variation in Dietary Protein on Captive-Raised Black Knob Map Turtles (*Graptemys nigrinoda*, Emydidae). Unpublished M.S. thesis, Texas State University-San Marcos.

Busby, W. H., and J. R. Parmelee. 1996. Historical changes in a herpetofaunal assemblage in the Flint Hills of Kansas. *American Midland Naturalist* 135:81–91.

Byrd, E. E. 1939. Studies on the blood flukes of the family Spirorchidae: Part II. Revision of the family and description of new species. *Journal of the Tennessee Academy of Science* 14:116–161.

Cable, R. M., and F. M. Fisher, Jr. 1961. A fifth species of *Neoechinorhynchus* (Acanthocephala) in turtles. *Journal of Parasitology* 47:666–668.

———, and W. B. Hopp. 1954. Acanthocephalan parasites of the genus *Neoechinorhynchus* in North American turtles with the descriptions of two new species. *Journal of Parasitology* 40:674–680.

Cagle, F. R. 1952a. *A Key to the Amphibians and Reptiles of Louisiana*. Tulane Book Store, Tulane University, New Orleans, Louisiana.

———. 1952b. The status of the turtles *Graptemys pulchra* Baur and *Graptemys barbouri* Carr and Marchand, with notes on their natural history. *Copeia* 1952:223–234.

———. 1953a. The status of the turtle *Graptemys oculifera* (Baur). *Zoologica* 38:137–144.

———. 1953b. Two new subspecies of *Graptemys pseudogeographica*. *Occasional Papers of the Museum of Zoology, University of Michigan* 546:1–17.

———. 1954. Two new species of the genus *Graptemys*. *Tulane Studies in Zoology* 11:167–186.

———. 1955. Courtship behavior in juvenile turtles. *Copeia* 1955:307.

———, and A. H. Chaney. 1950. Turtles populations in Louisiana. *American Midland Naturalist* 43:383–388.

Cahn, A. R. 1937. The turtles of Illinois. *University of Illinois Bulletin* 35:5–218.

Carpenter, C. C. 1958. An unusual Ouachita map turtle. *Herpetologica* 14:116.

Carr, A. 1949. The identity of *Malacoclemmys kohnii* Baur. *Copeia* 1949:9–10.

———. 1952. *Handbook of Turtles: The Turtles of the United States, Canada, and Baja California.* Cornell University Press, Ithaca, New York.

———, and L. J. Marchand. 1942. A new turtle from the Chipola River, Florida. *Proceedings of the New England Zoölogical Club* 20:95–100.

Carr, J. L. 2001. Louisiana Map Turtle Survey: Map Turtles in Northern Louisiana. Unpublished report to the U.S. Geological Survey-Biological Resources Division, Cooperative Agreement no. 99CRAG0017, Washington, D.C.

———. 2008. Terrestrial foraging by two species of semiaquatic turtles (Testudines: Emydidae). *Southeastern Naturalist* 7:748–752.

———, and M. A. Messinger. 2002. *Graptemys gibbonsi* (Pascagoula map turtle): Predation. *Herpetological Review* 33:201–202.

Carrière, M.-A. 2007. Movement Patterns and Habitat Selection of Common Map Turtles (*Graptemys geographica*) in St. Lawrence Islands National Park, Ontario, Canada. Unpublished M.S. thesis, University of Ottawa, Ottawa, Ontario, Canada.

———, and G. Blouin-Demers. 2010. Habitat selection at multiple spatial scales in northern map turtles (*Graptemys geographica*). *Canadian Journal of Zoology* 88:846–854.

———, G. Bulté, and G. Blouin-Demers. 2009. Spatial ecology of northern map turtles (*Graptemys geographica*) in a lotic and a lentic habitat. *Journal of Herpetology* 43:597–604.

Casley, S. W., and G. Sievert. 2006. *Graptemys ouachtensis ouachitensis* (Ouachita map turtle): Geographic distribution. *Herpetological Review* 37:491.

Casper, G. S. 1996. *Geographic Distributions of the Amphibians and Reptiles of Wisconsin.* Milwaukee Public Museum, Milwaukee, Wis.

———. 1999. New distribution records from the Wisconsin Herpetology Atlas Project, 1998. *Herpetological Review* 30:181–182.

Cecala, K. K., J. W. Gibbons, and M. E. Dorcas. 2009. Ecological effects of major injuries in diamondback terrapins: Implications for conservation and management. *Aquatic Conservation: Marine and Freshwater Ecosystems* 19:421–427.

Chaney, A., and C. L. Smith. 1950. Methods for collecting mapturtles. *Copeia* 1950:323–324.

Charnov, E. L., and J. Bull. 1977. When is sex environmentally determined? *Nature* 266:828–830.

Christensen, D. L., B. R. Herwig, D. E. Schindler, and S. R. Carpenter. 1996. Impacts of lakeshore residential development on coarse woody debris in north temperate lakes. *Ecological Applications* 6:1143–1149.

Christiansen, J. L., and B. J. Gallaway. 1984. Raccoon removal, nesting success, and hatchling emergence in Iowa turtles with special reference to *Kinosternon flavescens* (Kinosternidae). *Southwestern Naturalist* 29:343–348.

Clark, H. W., and J. B. Southall. 1920. Fresh-water turtles: A source of meat supply. *U.S. Bureau of Fisheries Document* 889:1–20.

Clark, J. 1937. The stratigraphy and paleontology of the Chadron formation in the Big Badlands of South Dakota. *Annals of the Carnegie Museum* 25:261–350.

Clarke, R. F. 1953. Additional turtle records for Kansas. *Transactions of the Kansas Academy of Science* 56:438–439.

————. 1956. Distributional notes on some amphibians and reptiles of Kansas. *Transactions of the Kansas Academy of Science* 59:213–219.

Claude, J., P. C. H. Pritchard, H. Tong, E. Paradis, and J.-C. Auffray. 2004. Ecological correlates and evolutionary divergence in the skull of turtles: A geometric morphometric assessment. *Systematic Biology* 53:933–948.

Cliburn, J. W. 1971. The ranges of four species of *Graptemys* in Mississippi. *Journal of the Mississippi Academy of Sciences* 16:16–19.

Cochran, D. M. 1961. Type specimens of reptiles and amphibians in the U.S. National Museum. *U.S. National Museum Bulletin* 220:xv, 1–291.

Cochran, P. A. 1987. *Graptemys geographica* (map turtle): Adult mortality. *Herpetological Review* 18:37.

Coleman, J. L. 2006. Basking Habits of Map Turtles (Emydidae: *Graptemys*) along the Sabine River in East Texas. Unpublished M.S. thesis, University of Texas at Tyler.

————, and R. L. Gutberlet, Jr. 2008. Seasonal variation in basking in two syntopic species of map turtles (Emydidae: *Graptemys*). *Chelonian Conservation and Biology* 7:276–281.

Collins, D., and P. V. Lindeman. 2006. The influence of body size and trophic morphology on the size of molluscan prey of female Texas map turtles (*Graptemys versa*). *Herpetological Review* 37:416–418.

Collins, J. T. 1990. Standard common and current scientific names for North American amphibians and reptiles, 3rd ed. *Society for the Study of Amphibians and Reptiles Herpetological Circular* 19:1–41.

————. 1991. Viewpoint: A new taxonomic arrangement for some North American amphibians and reptiles. *Herpetological Review* 22:42–43.

————. 1993. *Amphibians and Reptiles in Kansas*, 3rd ed. University of Kansas Museum of Natural History, Lawrence.

————. 1997. Standard common and current scientific names for North American amphibians and reptiles, 4th ed. *Society for the Study of Amphibians and Reptiles Herpetological Circular* 25:1–40.

————, R. Conant, J. E. Huheey, J. L. Knight, E. M. Rundquist, and H. M. Smith. 1982. Standard common and current scientific names for North American amphibians and reptiles, 2nd ed. *Society for the Study of Amphibians and Reptiles Herpetological Circular* 12:1–28.

————, J. E. Huheey, J. L. Knight, and H. M. Smith. 1978. Standard common and current scientific names for North American amphibians and reptiles. *Society for the Study of Amphibians and Reptiles Herpetological Circular* 7:1–36.

————, and T. W. Taggart. 2002. *Standard Common and Current Scientific Names for North American Amphibians, Turtles, Reptiles & Crocodilians*, 5th ed. Center for North American Herpetology, Lawrence, Kans.

————, and T. W. Taggart. 2009. *Standard Common and Current Scientific Names for North American Amphibians, Turtles, Reptiles & Crocodilians*, 6th ed. Center for North American Herpetology, Lawrence, Kans.

Conant, R. 1951. *The Reptiles of Ohio*, 2nd ed. American Midland Naturalist, Notre Dame, Ind.

————. 1958. *A Field Guide to Reptiles and Amphibians of the United States and Canada East of the 100th Meridian*. Houghton Mifflin, Boston.

————. 1975. *A Field Guide to Reptiles and Amphibians: Eastern and Central North America*, 2nd ed. Houghton Mifflin, Boston.

————, F. R. Cagle, C. J. Goin, C. H. Lowe, Jr., W. T. Neill, M. G. Netting, K. P. Schmidt, C. E. Shaw, R. C. Stebbins, and C. M. Bogert. 1956. Common names for North American amphibians and reptiles. *Copeia* 1956:172–185.

———, and J. T. Collins. 1991. *A Field Guide to Reptiles and Amphibians: Eastern and Central North America*, 3rd ed. Houghton Mifflin, Boston.

———, M. B. Trautman, and E. B. McLean. 1964. The false map turtle, *Graptemys pseudogeographica* (Gray), in Ohio. *Copeia* 1964:212–213.

Congdon, J. D., A. E. Dunham, and R. C. van Loben Sels. 1993. Delayed sexual maturity and demographics of Blanding's turtles (*Emydoidea blandingii*): Implications for conservation and management of long-lived organisms. *Conservation Biology* 7:826–833.

———, and J. W. Gibbons. 1987. Morphological constraints on egg size: A challenge to optimal egg size theory? *Proceedings of the National Academy of Science USA* 84:4145–4147.

Conner, C. A., B. A. Douthitt, and T. J. Ryan. 2005. Descriptive ecology of a turtle assemblage in an urban landscape. *American Midland Naturalist* 153:428–435.

Connor, P. 1993. Notice of 12-month finding on petition to list Cagle's map turtle. *Federal Register* 58:5701–5704.

Cooper, J. E., and R. E. Sharp. 1982. Suctoria from ponds and streams in Rutherford County, Tennessee. *Journal of the Tennessee Academy of Science* 57:73–75.

Cope, E. D. 1875. A check-list of North American Reptilia and Batrachia. *Bulletin of the U.S. National Museum* 32:1–98.

COSEWIC (Committee on the Status of Endangered Wildlife in Canada). 2002. *COSEWIC Assessment and Status Report on the Northern Map Turtle* Graptemys geographica *in Canada.* COSEWIC, Ottawa, Ontario, Canada.

Costanzo, J. P., R. E. Lee, Jr., and G. R. Ultsch. 2008. Physiological ecology of overwintering in hatchling turtles. *Journal of Experimental Zoology* 309A:297–379.

———, J. D. Litzgus, J. B. Iverson, and R. E. Lee, Jr. 2001. Cold-hardiness and evaporative water loss in hatchling turtles. *Physiological and Biochemical Zoology* 74:510–519.

Craig, M. J. 1992. Radio-Telemetry and Tagging Study of Movement Patterns, Activity Cycles, and Habitat Utilization in Cagle's Map Turtle, *Graptemys caglei.* Unpublished M.S. thesis, West Texas State University, Canyon.

Crenshaw, J. W., Jr., and G. B. Rabb. 1949. Occurrence of the turtle *Graptemys barbouri* in Georgia. *Copeia* 1949:226.

Crews, D. 1996. Temperature-dependent sex determination: The interplay of steroid hormones and temperature. *Zoological Science* 13:1–13.

Crocker, C. E., T. E. Graham, G. R. Ultsch, and D. C. Jackson. 2000. Physiology of common map turtles (*Graptemys geographica*) hibernating in the Lamoille River, Vermont. *Journal of Experimental Zoology* 286:143–148.

Crother, B. I., J. Boundy, F. T. Burbink, J. A. Campbell, K. de Queiroz, D. R. Frost, R. Highton, et al. 2008. Scientific and standard English names of amphibians and reptiles of North America north of Mexico, with comments regarding confidence in our understanding, 6th ed. *Society for the Study of Amphibians and Reptiles Herpetological Circular* 37:1–84.

———, J. Boundy, J. A. Campbell, K. de Queiroz, D. R. Frost, R. Highton, D. M. Green, et al. 2012. Scientific and standard English names of amphibians and reptiles of North America north of Mexico, with comments regarding confidence in our understanding, 7th ed. *Society for the Study of Amphibians and Reptiles Herpetological Circular* 39:1–92.

———, J. Boundy, J. A. Campbell, K. de Queiroz, D. R. Frost, R. Highton, J. B. Iverson, et al. 2001. Scientific and standard English names of amphibians and reptiles of North America north of Mexico, with comments regarding confidence in our understanding, 5th ed. *Society for the Study of Amphibians and Reptiles Herpetological Circular* 29:1–82.

Daigle, C., A. Desroisiers, and J. Bonin. 1994. Distribution and abundance of common map turtles, *Graptemys geographica*, in the Ottawa River, Québec. *Canadian Field-Naturalist* 108:84–86.

———, and M. Lepage. 1997. Tortues du fleuve Saint-Laurent et de ses affluents États des connaissances sur la distribution 1980–1994. Unpublished report to the Ministère de l'Environnement et de la Faune, Quebec City, Quebec, Canada.

———, and D. St-Hilaire. 2000. Inventaire de la tortue-molle à épines (*Apalone spinifera*) dans la rivière des Outaouais, secteur Montebello–Hull. Unpublished report to the Société de la faune et des parcs du Québec, Quebec City, Quebec, Canada.

Daniel, R. E., and B. S. Edmond. 2012. Atlas of Missouri Amphibians and Reptiles for 2011, http://atlas.moherp.org/pubs/atlas11.pdf.

Daniels, S. D., S. A. Dykes, and R. L. P. Wyatt. 2012. New amphibian and reptile county records for eight counties in east Tennessee, USA. *Herpetological Review* 43:313–315.

Daugherty, A. E. 1942. A record of *Graptemys pseudogeographica versa*. *Copeia* 1942:51.

Davis, J. G., and P. J. Krusling. 2010. New distribution records for six species of turtles in Ohio. *Herpetological Review* 41:391–392.

———, and L. Barnhart. 2012. *Graptemys ouachitensis*: Geographic distribution. *Herpetological Review* 43:303.

Dayan, T., and D. Simberloff. 2005. Ecological and community-wide character displacement: The next generation. *Ecology Letters* 8:875–894.

DeCatanzaro, R., and P. Chow-Fraser. 2010. Relationship of road density and marsh condition to turtle assemblage characteristics in the Laurentian Great Lakes. *Journal of Great Lakes Research* 36:357–365.

DeGiusti, D. L., and P. J. Batten, Jr. 1951. Notes on *Haemoproteus metchinikovi* in turtles from Wisconsin, Michigan, and Louisiana. *Journal of Parasitology* 37:12.

———, C. R. Sterling, and D. Dobrzechowski. 1973. Transmission of the chelonian haemoproteid *Haemoproteus metchnikovi* by a tabanid fly *Chrysops callidus*. *Nature* 242:50–51.

DeKay, J. E. 1842. *Natural History of New York: Zoology of New York, vol. 3; Reptiles and Amphibia*. White and Visscher, Albany, N.Y.

Delany, M. F., and C. L. Abercrombie. 1986. American alligator food habits in north-central Florida. *Journal of Wildlife Management* 50:348–353.

de Queiroz, K. 2007. Species concepts and species delimitation. *Systematic Biology* 56:879–886.

Dexter, R. W. 1948. More records of reptiles from Portage County, Ohio. *Copeia* 1948:139.

Dinkelacker, S. A., J. P. Costanzo, and R. E. Lee, Jr. 2005. Anoxia and freeze tolerance in hatchling turtles. *Journal of Comparative Physiology B* 175:209–217.

Ditmars, R. L. 1907. *The Reptiles of North America*. Doubleday & Company, N.Y.

Dixon, J. R. 1960. Epizoophytic algae on some turtles of Texas and Mexico. *Texas Journal of Science* 12:36–38.

———. 2000. *Amphibians and Reptiles of Texas*, 2nd ed. Texas A&M University Press, College Station.

Dixon, L. A. 2009. False Map, Spiny Softshell and Smooth Softshell Turtle Nest and Nest-Site Habitat Characteristics along the Lower Stretch of the Missouri National Recreation River in South Dakota. Unpublished M.S. thesis, South Dakota State University, Brookings.

Dobie, J. L. 1972. Correction of distributional records for *Graptemys barbouri* and *Graptemys pulchra*. *Herpetological Review* 4:23.

———. 1981. The taxonomic relationship between *Malaclemys* Gray, 1844 and *Grapte-*

mys Agassiz, 1857 (Testudines: Emydidae). *Tulane Studies in Zoology and Botany* 23:85–102.

Dodd, C. K., Jr. 1977a. Amphibians & reptiles: The declining species. *Water Spectrum* 10(1):24–32.

———. 1977b. Endangered and threatened wildlife and plants; review of status of twelve species of turtles. *Federal Register* 42:28903–28904.

———. 1988a. Disease and population declines in the flattened musk turtle *Sternotherus depressus. American Midland Naturalist* 119:394–401.

———. 1988b. Patterns of distribution and seasonal use of the turtle *Sternotherus depressus* by the leech *Placobdella parasitica. Journal of Herpetology* 22:74–81.

Dolan, C., V. Polton, J. Tucker, and J. Parmalee. 2011. *Graptemys ouachitensis* (Ouachita map turtle): Geographic distribution. *Herpetological Review* 42:388–389.

Dolley, J. S. 1933. Preliminary notes on the biology of the St. Joseph River. *American Midland Naturalist* 14:193–227.

Dolloff, C. A., and M. L. Warren, Jr. 2003. Fish relationships with large wood in small streams. Pages 179–193 in S. V. Gregory, K. L. Boyer, and A. M. Gurnell (Eds.), *The Ecology and Management of Wood in World Rivers.* American Fisheries Society, Bethesda, Md.

DonnerWright, D. M., M. A. Bozek, J. R. Probst, and E. M. Anderson. 1999. Responses of turtle assemblage to environmental gradients in the St. Croix River in Minnesota and Wisconsin, U.S.A. *Canadian Journal of Zoology* 77:989–1000.

Douglas, N. H. 1972. *Graptemys geographica:* Geographic distribution. *Herpetological Review* 4:93.

Dowling, H. G. 1993. Viewpoint: A reply to Collins (1991, 1992). *Herpetological Review* 24:11–13.

Dreslik, M. J., A. R. Kuhns, and C. A. Phillips. 2005. Structure and composition of a southern Illinois freshwater turtle assemblage. *Northeastern Naturalist* 12:173–186.

———, and C. A. Phillips. 2005. Turtles communities in the upper Midwest, USA. *Journal of Freshwater Ecology* 20:149–164.

Du, W.-G., B. Zhao, and R. Shine. 2010. Embryos in the fast lane: High-temperature heart rates of turtles decline after hatching. *PLOS ONE* 5:e9557.

Dundee, H. A. 1974. Evidence for specific status of *Graptemys kohnii* and *Graptemys pseudogeographica. Copeia* 1974:540–542.

———, and D. A. Rossman. 1989. *The Amphibians and Reptiles of Louisiana.* Louisiana State University Press, Baton Rouge.

Dyer, W. G., and A. K. Wilson. 1997. *Falcaustra wardi* (Nematoda: Kathlaniidae) in the false map turtle (*Graptemys pseudogeographica*) (Testudines: Emydidae) from southern Illinois. *Transactions of the Illinois State Academy of Science* 90:135–138.

Ecksdine, V. 1985. Meteorological influence on fall basking of the map turtle, *Graptemys geographica,* in a north temperate stream. *Bulletin of the Chicago Herpetological Society* 20:82–85.

Edds, D. R., P. A. Shipman, and L. E. Shipman. 1991. *Graptemys pseudogeographica ouachitensis* (Ouachita map turtle): Geographic distribution. *Herpetological Review* 22:134.

Edgren, R. A., M. K. Edgren, and L. H. Tiffany. 1953. Some North American turtles and their epizoophytic algae. *Ecology* 34:733–740.

Ehret, D. J., and J. R. Borque. 2011. An extinct map turtle *Graptemys* (Testudines, Emydidae) from the late Pleistocene of Florida. *Journal of Vertebrate Paleontology* 31:575–587.

Elosegi, A., and L. B. Johnson. 2003. Wood in streams and rivers in developed landscapes. Pages 337–353 in S. V. Gregory, K. L. Boyer, and A. M. Gurnell (Eds.), *The Ecology and Management of Wood in World Rivers*. American Fisheries Society, Bethesda, Md.

Elsey, R. M. 2006. Food habits of *Macrochelys temminckii* (alligator snapping turtle) from Arkansas and Louisiana. *Southeastern Naturalist* 5:443–452.

Emerson, D. N. 1967. Preliminary study on seasonal liver lipids and glycogen, and blood sugar levels in the turtle *Graptemys pseudogeographica* (Gray) from South Dakota. *Herpetologica* 23:68–70.

Enge, K. M., R. L. Cailteux, and J. J. Nordhaus. 1996. *Graptemys barbouri* (Barbour's map turtle): Geographic distribution. *Herpetological Review* 27:150–151.

———, and G. E. Wallace. 2008. Basking survey of map turtles (*Graptemys*) in the Choctawhatchee and Ochlockonee Rivers, Florida and Alabama. *Florida Scientist* 71:310–322.

Ennen, J. R., B. R. Kreiser, C. P. Qualls, and J. E. Lovich. 2010. Morphological and molecular reassessment of *Graptemys oculifera* and *Graptemys flavimaculata* (Testudines: Emydidae). *Journal of Herpetology* 44:544–554.

———, J. E. Lovich, B. R. Kreiser, W. Selman, and C. P. Qualls. 2010. Genetic and morphological variation between populations of the Pascagoula map turtle (*Graptemys gibbonsi*) in the Pearl and Pascagoula rivers with description of a new species. *Chelonian Conservation and Biology* 9:98–113.

———, and A. F. Scott. 2008. Diel movement behavior of the stripe-necked musk turtle (*Sternotherus minor peltifer*) in middle Tennessee. *American Midland Naturalist* 160:278–288.

———, W. Selman, and B. R. Kreiser. 2007. *Graptemys gibbonsi* (Pascagoula map turtle): Diet. *Herpetological Review* 38:200.

Ernst, C. H. 1973. The distribution of the turtles of Minnesota. *Journal of Herpetology* 7:42–47.

———. 1974. Observations on the courtship of male *Graptemys pseudogeographica*. *Journal of Herpetology* 8:377–378.

———, and R. W. Barbour. 1972. *Turtles of the United States*. University Press of Kentucky, Lexington.

———, and R. W. Barbour. 1989. *Turtles of the World*. Smithsonian Institution Press, Washington, D.C.

———, and J. E. Lovich. 2009. *Turtles of the United States and Canada*, 2nd ed. Johns Hopkins University Press, Baltimore, Md.

———, J. E. Lovich, and R. W. Barbour. 1994. *Turtles of the United States and Canada*. Smithsonian Institution Press, Washington, D.C.

Ernst, E. M., and C. H. Ernst. 1977. Synopsis of helminths endoparasitic in native turtles of the United States. *Bulletin of the Maryland Herpetological Society* 13:1–75.

Esch, G. W., J. W. Gibbons, and J. E. Borque. 1979. The distribution and abundance of enteric helminths in *Chrysemys s. scripta* from various habitats on the Savannah River Plant in South Carolina. *Journal of Parasitology* 65:624–632.

Etchberger, C. R., M. A. Ewert, J. B. Phillips, and C. E. Nelson. 2002. Carbon dioxide influences environmental sex determination in two species of turtles. *Amphibia-Reptilia* 23:169–175.

Evermann, B. W., and H. W. Clark. 1916. The turtles and batrachians of the Lake Maxinkuckee region. *Proceedings of the Indiana Academy of Science* 1916:472–518.

Ewert, M. A. 1979. *Graptemys pseudogeographica ouachitensis* (Ouachita map turtle): Geographic distribution. *Herpetological Review* 10:102.

———. 1989. Field Survey of the False Map Turtle Group and the Hieroglyphic Cooter in Southwestern Indiana. Unpublished report to the Indiana Department of Natural Resources Division of Fish and Wildlife Non-game and Endangered Wildlife Program, Indianapolis.

——, J. S. Doody, and J. L. Carr. 2004. *Graptemys ouachitensis sabinensis* (Sabine map turtle). Reproduction. *Herpetological Review* 35:382–383.

——, C. R. Etchberger, and C. E. Nelson. 2004. Turtle sex-determining modes and TSD patterns, and some TSD pattern correlates. Pages 21–32 in N. Valenzuela and V. Lance (Eds.), *Temperature-Dependent Sex Determination in Vertebrates*. Smithsonian Books, Washington, D.C.

——, and D. R. Jackson. 1994. Nesting Ecology of the Alligator Snapping Turtle, *Macroclemys temminckii*, along the Lower Apalachicola River, Florida. Unpublished report to the Nongame Wildlife Program, Florida Game and Fresh Water Fish Commission, Tallahassee.

——, D. R. Jackson, and C. E. Nelson. 1994. Patterns of temperature-dependent sex determination in turtles. *Journal of Experimental Zoology* 270:3–15.

——, and C. E. Nelson. 1991. Sex determination in turtles: Diverse patterns and some possible adaptive values. *Copeia* 1991:50–69.

——, P. C. H. Pritchard, and G. E. Wallace. 2006. *Graptemys barbouri*—Barbour's map turtle. *Chelonian Research Monographs* 3:260–272.

Feldman, C. R., and J. F. Parham. 2002. Molecular phylogenetics of emydine turtles: Taxonomic revision and the evolution of shell kinesis. *Molecular Phylogenetics and Evolution* 22:388–398.

Felsenstein, J. 1985. Phylogenies and the comparative method. *American Naturalist* 125:1–15.

Fisher, F. M., Jr. 1960. On acanthocephala of turtles, with the description of *Neoechinorhynchus emyditoides* n. sp. *Journal of Parasitology* 46:257–266.

Fitzpatrick, B. M., J. S. Placyk, Jr., M. L. Niemiller, G. S. Casper, and G. M. Burghardt. 2008. Distinctiveness in the face of gene flow: Hybridization between specialist and generalist garter snakes. *Molecular Ecology* 17:4107–4117.

Flaherty, N. C. 1982. Home Range, Movement, and Habitat Selection in a Population of Map Turtle, *Graptemys geographica* (Le Sueur), in Southwestern Quebec. Unpublished M.S. thesis, McGill University, Montreal, Quebec, Canada.

——, and J. R. Bider. 1984. Physical structures and the social factor as determinants of habitat use by *Graptemys geographica* in southwestern Quebec. *American Midland Naturalist* 111:259–266.

Folkerts, G. W., and R. H. Mount. 1969. New subspecies of the turtle *Graptemys nigrinoda* Cagle. *Copeia* 1969:677–682.

——, and R. H. Mount. 1970. Reply to H. L. Freeman's (Herpetol. Rev. 2[1]:3) comments on: A new subspecies of the turtle *Graptemys nigrinoda* Cagle. *Herpetological Review* 2(3):3–4.

Fontenot, C. L., Jr. 2004. Cajun-French common names for Louisiana amphibians and reptiles. *Herpetological Review* 35:337–338.

Fratto, Z. W., V. A. Barko, and J. S. Scheibe. 2008. Development and efficacy of a bycatch reduction device for Wisconsin-type fyke nets deployed in freshwater systems. *Chelonian Conservation and Biology* 7:205–212.

——, and B. K. Swallow. 2007. Range extensions for eight species of turtles in Gasconade, Osage, and New Madrid Counties, Missouri. *Herpetological Review* 38:359.

Frazer, N. B. 1992. Sea turtle conservation and halfway technology. *Conservation Biology* 6:179–184.

Freedberg, S., M. A. Ewert, and C. E. Nelson. 2001. Environmental effects on fitness and consequences for sex allocation in a reptile with environmental sex determination. *Evolutionary Ecology Research* 3:953–967.

———, M. A. Ewert , B. J. Ridenhour, M. Neiman, and C. E. Nelson. 2005. Nesting fidelity and molecular evidence for natal homing in the freshwater turtle, *Graptemys kohnii*. *Proceedings of the Royal Society of London B* 272:1345–1350.

———, and E. M. Myers. 2012. Cytonuclear equilibrium following interspecific introgression in a turtle lacking sex chromosomes. *Biological Journal of the Linnean Society* 106:405–417.

———, A. L. Stumpf, M. A. Ewert, and C. E. Nelson. 2004. Developmental environment has long-lasting effects on behavioural performance in two turtle with environmental sex determination. *Evolutionary Ecology Research* 6:739–747.

Freeman, H. L. 1970. A comment on: A new subspecies of the turtle *Graptemys nigrinoda* Cagle. *Herpetological Review* 2(1):3.

Fritz, U. 1991. Balzverhalten und Systematik in der Subtribus Nectemydina 2: Vergleich oberhalb des Artniveaus und Anmerkungen zur Evolution. *Salamandra* 27:129–142.

———, and P. Havaš. 2007. Checklist of chelonians of the world. *Vertebrate Zoology* 57:149–368.

Frum, W. G. 1947. *Graptemys geographica* in West Virginia. *Copeia* 1947:211.

Funk, D. J., and K. E. Omland. 2003. Species-level paraphyly and polyphyly: Frequency, causes, and consequences, with insights from animal mitochondrial DNA. *Annual Review of Ecology and Systematics* 34:397–423.

Funk, R. E. 1976. Recent contributions to Hudson Valley prehistory. *New York State Museum Memoir* 22:1–325.

Fuselier, L, and D. Edds. 1994. Habitat partitioning among three sympatric species of map turtles, genus *Graptemys*. *Journal of Herpetology* 28:154–158.

Gaffney, E. S., and P. A. Meylan. 1988. A phylogeny of turtles. Pages 157–219 in M. J. Benton (Ed.), *The Phylogeny and Classification of the Tetrapods, vol. 1: Amphibians, Reptiles, Birds*. Clarendon Press, Oxford, United Kingdom.

Galat, D. L., L. H. Fredrickson, D. D. Humburg, K. J. Bataille, J. R. Bodie, J. Dohrenwend, G. T. Gelwicks, et al. 1998. Flooding to restore connectivity of regulated, large-river wetlands. *BioScience* 48:721–733.

Galois, P. 2005. Plan de rétablissement de cinq espèces de tortues au Québec pour les années 2005 à 2010: la tortue des bois (*Glyptemys insculpta*), la tortue géographique (*Graptemys geographica*), la tortue mouchetée (*Emydoidea blandingii*), la tortue musquée (*Sternotherus odoratus*) et la tortue ponctuée (*Chelydra serpentina*). Unpublished report to the Ministère des Ressources Naturelles et de la Faune, Quebec City, Quebec, Canada.

———, and M. Ouellet. 2007. Traumatic injuries in eastern spiny softshell turtles (*Apalone spinifera*) due to recreational activities in the northern Lake Champlain basin. *Chelonian Conservation and Biology* 6:288–293.

Gamble, T., and J. J. Moriarty. 2006. New county records for amphibians and reptiles in Minnesota. *Herpetological Review* 37:114–116.

Garman, H. 1890. The differences between the geographic turtles (*Malacoclemmys geographicus* and *M. lesueuri*). *Essex Institute Bulletin* 22:70–83.

Garrott, R. A., P. J. White, and C. A. Vanderbilt White. 1993. Overabundance: An issue for conservation biologists? *Conservation Biology* 7:946–949.

Geller, G. A. 2012a. Notes on the nesting ecology of Ouachita map turtles (*Graptemys ouachitensis*) at two Wisconsin sites using trail camera monitoring. *Chelonian Conservation and Biology* 11:206–213.

———. 2012b. Notes on the nest predation dynamics of *Graptemys* at two Wisconsin sites using trail camera monitoring. *Chelonian Conservation and Biology* 11:197–205.

Gentry, G. 1956. An annotated check list of the amphibians and reptiles of Tennessee. *Journal of the Tennessee Academy of Science* 31:242–251.

George, S. G., J. Killgore, and S. L. Harrel. 1995. *Graptemys pseudogeographica ouachitensis* (Ouachita map turtle): Geographic distribution. *Herpetological Review* 26:43.

Georges, A. 1989. Female turtles from hot nests: Is it duration of incubation or proportion of development at high temperatures that matters? *Oecologia* 81:323–328.

Gibbons, J. W. 1990. Sex ratios and their significance among turtle populations. Pages 171–182 in J. W. Gibbons (Ed.), *Life History and Ecology of the Slider Turtle*. Smithsonian Institution Press, Washington, D.C.

––––––. 1997. Measuring declines and natural variation in turtle populations: Spatial lessons from long-term studies. Pages 243–246 in J. van Abbema (Ed.), *Proceedings: Conservation, Restoration, and Management of Tortoises and Turtles—An International Conference*. New York Turtle and Tortoise Society, New York.

––––––, and J. E. Lovich. 1990. Sexual dimorphism in turtles with emphasis on the slider turtle (*Trachemys scripta*). *Herpetological Monographs* 4:1–29.

––––––, and D. H. Nelson. 1978. The evolutionary significance of delayed emergence from the nest by hatchling turtles. *Evolution* 32:297–303.

Gibbs, J. P., A. R. Breisch, P. K. Ducey, G. Johnson, J. L. Behler, and R. C. Bothner. 2007. *The Amphibians and Reptiles of New York State: Identification, Natural History, and Conservation*. Oxford University Press, New York.

Giles, L. W., and V. L. Childs. 1949. Alligator management of the Sabine National Wildlife Refuge. *Journal of Wildlife Management* 13:16–28.

Gillingwater, S. D. 2001. A Selective Herpetofaunal Survey Inventory and Biological Research Study of Rondeau Provincial Park. Unpublished report to Rondeau Provincial Park, Ontario, Canada.

Gist, D. H., and J. M. Jones. 1989. Sperm storage within the oviducts of turtles. *Journal of Morphology* 199:379–384.

––––––, and J. D. Congdon. 1998. Oviductal sperm storage as a reproductive tactic of turtles. *Journal of Experimental Biology* 282:526–534.

Godwin, J. 2000. Escambia Map Turtle (*Graptemys ernsti*) Status Survey. Unpublished report to the Alabama Department of Conservation and Natural Resources, Montgomery.

––––––. 2001. Black-Knobbed Sawback (*Graptemys nigrinoda*) Status Survey. Unpublished report to the Alabama Department of Conservation and Natural Resources, Montgomery.

––––––. 2002a. Distribution and Status of Barbour's Map Turtle (*Graptemys barbouri*) in the Choctawhatchee River System, Alabama. Unpublished report to the Alabama Department of Conservation and Natural Resources, Montgomery.

––––––. 2002b. Turtle Nest Success on Gravine Island with Emphasis on the Alabama Red-Bellied Turtle (*Pseudemys alabamensis*) and Delta Map Turtle (*Graptemys nigrinoda delticola*). Unpublished report to the Alabama Department of Conservation and Natural Resources, Montgomery.

––––––. 2003. Alabama Map Turtle (*Graptemys pulchra*) Status Survey. Unpublished report to the Alabama Department of Conservation and Natural Resources, Montgomery.

––––––, J. E. Lovich, J. R. Ennen, B. R. Kreiser, B. P. Folt, and C. Lechowicz. 2013. Taxonomic Assessment of Map Turtles (*Graptemys*) of the Choctawhatchee and Pea Rivers, Alabama and Florida. Unpublished report to the Alabama Department of Conservation and Natural Resources, Montgomery.

Goode, M. 1997. Eggs and clutch production of captive *Graptemys*. Page 478 in J. van Abbema (Ed.), *Proceedings: Conservation, Restoration, and Management of Tortoises and Turtles—an International Conference*. New York Turtle and Tortoise Society, New York.

Goodpaster, W. W., and D. F. Hoffmeister. 1952. Notes on the mammals of western Tennessee. *Journal of Mammalogy* 33:362–371.

Goodrich, J. M., and S. W. Buskirk. 1995. Control of abundant native vertebrates for conservation of endangered species. *Conservation Biology* 9:1357–1364.

Gooley, A. C., H. J. Stanton, C. J. Bartkus, and T. K. Pauley. 2011. The distribution of aquatic turtles along the Ohio, Great Kanawha, and Little Kanawha rivers, West Virginia, with emphasis on *Graptemys ouachitensis* and *G. geographica*. *Ohio Biological Survey Notes* 3:21–28.

Gordon, D. M., and R. D. MacCulloch. 1980. An investigation of the ecology of the map turtle, *Graptemys geographica* (Le Sueur), in the northern part of its range. *Canadian Journal of Zoology* 58:2210–2219.

Graham, T. E. 1984. *Pseudemys rubriventris* (Red-bellied turtle): Predation. *Herpetological Review* 15:19–20.

———, and A. A. Graham. 1992. Metabolism and behavior of wintering common map turtles, *Graptemys geographica*, in Vermont. *Canadian Field-Naturalist* 106:517–519.

———, C. B. Graham, C. E. Crocker, and G. R. Ultsch. 2000. Dispersal from and fidelity to a hibernaculum in a northern Vermont population of common map turtles, *Graptemys geographica*. *Canadian Field-Naturalist* 114:405–408.

———, R. A. Saumure, and B. Ericson. 1997. Map turtle winter leech loads. *Journal of Parasitology* 83:1185–1186.

Gray, B. S., and M. Lethaby. 2008. The amphibians and reptiles of Erie County, Pennsylvania. *Bulletin of the Maryland Herpetological Society* 44:49–69.

Gray, J. E. 1831. *Synopsis Reptilium, or Short Descriptions of the Species of Reptiles, Part 1: Cataphracta; Tortoises, Crocodilians, and Enaliosaurians*. Treuttel, Wurtz, and Co., London.

Green, D. M., G. Blouin-Demers, Y. Dubois, C. Fontenot, P. Galois, J. Lefebvre, D. Lesbarrères, et al. 2012. Noms Français standardisés des amphibiens et des reptiles d'Amérique du Nord au nord du Méxique. *Society for the Study of Amphibians and Reptiles Herpetological Circular* 40:1–63.

Green, N. B., and T. K. Pauley. 1987. *Amphibians and Reptiles in West Virginia*. University of Pittsburgh Press, Pa.

Gregory, K. J. 2003. The limits of wood in world rivers: Present, past, and future. Pages 1–19 in S. V. Gregory, K. L. Boyer, and A. M. Gurnell (Eds.), *The Ecology and Management of Wood in World Rivers*. American Fisheries Society, Bethesda, Md.

Gritters, S. A., and L. M. Mauldin. 1994. Four Years of Turtle Collections on Navigation Pool 13 of the Upper Mississippi River. Unpublished report LTRMP 94-S010 of the Iowa Department of Natural Resources, Bellevue, Iowa, for the National Biological Survey, Environmental Management Technical Center, Onalaska, Wisc.

Guilday, J. E., H. W. Hamilton, E. Anderson, and P. W. Parmalee. 1978. The Baker Bluff cave deposit, Tennessee, and the late Pleistocene faunal gradient. *Bulletin of the Carnegie Museum of Natural History* 11:1–67.

Haislip, N. 2008. *Graptemys geographica* (northern map turtle): Habitat. *Herpetological Review* 39:82–83.

Hall, C. B., and C. J. Parmenter. 2008. Necrotic egg and hatchling remains are key factors attracting dipterans to sea turtle (*Caretta caretta*, *Chelonia mydas*, and *Natator depressus*) nests in central Queensland, Australia. *Copeia* 2008:75–81.

Hansen, J. 2007. Climate catastrophe. *New Scientist* 195(2614):30–34.

Hare, M. P. 2001. Prospects for nuclear gene phylogeography. *Trends in Ecology and Evolution* 16:700–706.

Harrah, E. C. 1922. North American monostomes primarily from fresh water hosts. *Illinois Biological Monographs* 7(3):1–107.

Harris, H. S., Jr. 1969. Distributional survey: Maryland and the District of Columbia. *Bulletin of the Maryland Herpetological Society* 5:97–161.

Harris, J. L., J. Laerm, and L. J. Vitt. 1982. *Graptemys pulchra* (Alabama map turtle). *Herpetological Review* 13:24.

Harvey, M. B. 1992. The distribution of *Graptemys pseudogeographica* on the upper Sabine River. *Texas Journal of Science* 44:257–258.

Hay, O. P. 1892a. The batrachians and reptiles of the state of Indiana. *Annual Report of the Indiana Department of Natural Resources* 17:412–602.

———. 1892b. Some observations on the turtles of the genus *Malaclemys*. *Proceedings of the National Museum* 15:379–383.

———. 1908. The fossil turtles of North America. *Carnegie Institution of Washington Publication* 75:1–568.

———. 1923. Characteristics of sundry fossil vertebrates. *Pan-American Geologist* 9:101–120.

Haynes, D. 1976. *Graptemys caglei* Haynes and McKown: Cagle's map turtle. *Catalogue of American Amphibians and Reptiles* 184:1–2.

———, and R. R. McKown. 1974. A new species of map turtle (genus *Graptemys*) from the Guadalupe River system in Texas. *Tulane Studies in Zoology and Botany* 18:143–152.

Hebert, P. D. N., B. W. Muncaster, and G. L. Mackie. 1989. Ecological and genetic studies on *Dreissena polymorpha* (Pallas): A new mollusk in the Great Lakes. *Canadian Journal of Fisheries and Aquatic Sciences* 46:1587–1591.

Heppell, S. 1998. Application of life-history theory and population model analysis to turtle conservation. *Copeia* 1998:367–375.

Hernandez-Divers, S. J., P. Hensel, J. Gladden, S. M. Hernandez-Divers, K. A. Buhlmann, C. Hagen, S. Sanchez, K. S. Latimer, M. Ard, and A. C. Camus. 2009. Investigation of shell disease in map turtles (*Graptemys* spp.). *Journal of Wildlife Diseases* 45:637–652.

Herrel, A., and J. C. O'Reilly. 2006. Ontogenetic scaling of bite force in lizards and turtles. *Physiological and Biochemical Zoology* 79:31–42.

———, J. C. O'Reilly, and A. M. Richmond. 2002. Evolution of bite performance in turtles. *Journal of Evolutionary Biology* 15:1083–1094.

Hertwig, S. 2001. Ökologie, Haltung, und Fortpflanzung im Terrarium von *Graptemys caglei*, *G. flavimaculata*, *G. nigrinoda nigrinoda* und *G. oculifera*. *Salamandra* 37:21–48.

———, and C. Hohl. 2000a. Sawbacks: Ernährung und Zucht von Sägerückenschildkröten. *Schildkröte* 2(2):54–59.

———, and C. Hohl. 2000b. Sawbacks: Haltung und Zucht kleiner Graptemysarten. *Schildkröte* 2(1):4–9.

Hill, S. K. 2008. The Influence of Urbanization on the Basking Behavior of a Central Texas Freshwater Turtle Community. Unpublished Ph.D. dissertation, Baylor University, Waco, Tex.

Hively, C. L. 2009. A Comparative Analysis of Two Turtle Assemblages in an Altered Floodplain. Unpublished M.S. thesis, University of Texas at Tyler.

Holbrook, J. E. 1842. *North American Herpetology: Or, a Description of the Reptiles Inhabiting the United States*. J. Dobson, Philadelphia.

Holman, J. A. 1972. Herpetofauna of the Kanapolis local fauna (Pleistocene: Yarmouth) of Kansas. *Michigan Academician* 5:87–98.

———. 1988. The status of Michigan's Pleistocene herpetofauna. *Michigan Academician* 20:125–132.

———, G. Bell, and J. Lamb. 1990. A late Pleistocene herpetofauna from Bell Cave, Alabama. *Herpetological Journal* 1:521–529.

———, and R. L. Richards. 1993. Herpetofauna of the Prairie Creek site, Daviess County, Indiana. *Proceedings of the Indiana Academy of Science* 102:115–131.

Hopp, W. B. 1946. Notes on the life history of *Neoechinorhynchus emydis* (Leidy), an acanthocephalan parasite of turtles. *Proceedings of the Indiana Academy of Science* 55:183.

———. 1954. Studies on the morphology and life cycle of *Neoechinorhynchus emydis* (Leidy), an acanthocephalan parasite of the map turtle, *Graptemys geographica* (Le Sueur). *Journal of Parasitology* 40:284–299.

Horne, B. D., R. J. Brauman, M. J. C. Moore, and R. A. Seigel. 2003. Reproductive and nesting ecology of the yellow-blotched map turtle, *Graptemys flavimaculata*: Implications for conservation and management. *Copeia* 2003:729–738.

Horsfall, M. W. 1935. Observations on the life history of *Macravestibulum obtusicaudum* Mackin, 1930 (Trematoda: Pronocephalidae). *Proceedings of the Helminthological Society of Washington* 2:78–79.

Howe, M. A., and B. MacBryde. 1996. Species being considered for amendments to the appendices to the Convention on International Trade in Endangered Species of Wild Flora and Fauna; Request for information. *Federal Register* 61:44324–44332.

———, and B. MacBryde. 1997. Changes in list of species in appendices to the Convention on International Trade in Endangered Species of Wild Flora and Fauna. *Federal Register* 62:44627–44634.

Hsü, D.Y.-M. 1937. Life history and morphology of *Macravestibulum eversum* sp. nov. (Pronocephalidae, Trematoda). *Transactions of the American Microscopical Society* 56:478–504.

Hudson, G. E. 1942. The amphibians and reptiles of Nebraska. *Nebraska Conservation Bulletin* 24:1–146.

Hulin, V., V. Delmas, M. Girondot, M. H. Godfrey, and J.-M. Guillon. 2009. Temperature-dependent sex determination and global change: Are some species at greater risk? *Oecologia* 160:493–506.

Huey, R. B., and A. F. Bennett. 1987. Phylogenetic studies of coadaptation: Preferred temperatures versus optimal performance temperatures of lizards. *Evolution* 41:1098–1115.

Hughes, R. C., J. W. Higginbotham, and J. W. Clary. 1942. The trematodes of reptiles, part 1: Systematic section. *American Midland Naturalist* 27:109–134.

Hulse, A. C., C. J. McCoy, and E. J. Censky. 2001. *Amphibians and Reptiles of Pennsylvania and the Northeast*. Cornell University Press, Ithaca, N.Y.

Hutchison, J. H. 1996. Testudines. Pages 337–353 in D. R. Prothero and R. J. Emry (Eds.), *The Terrestrial Eocene-Oligocene Transition in North America*. Cambridge University Press, Cambridge, United Kingdom.

Irwin, K. 2007a. Commercial Aquatic Turtle Harvest 2004–2006. Unpublished report to the Arkansas Game & Fish Commission, Little Rock.

———. 2007b. Commercial harvest of aquatic turtles in Arkansas. *Life in the Rocks* 10(2):3–5.

———, and J. T. Collins. 2005. A survey for selected species of herpetofauna in the lower Marais des Cygnes River valley, Linn and Miami Counties, Kansas. *Journal of Kansas Herpetology* 14:12–13.

Iverson, J. B. 1985. Checklist of the turtles of the world with English common names. *Society for the Study of Amphibians and Reptiles Herpetological Circular* 14:1–14.

———. 1986. *A Checklist with Distribution Maps of the Turtles of the World.* Privately published, Richmond, Ind.

———. 1988. Growth in the common map turtle, *Graptemys geographica. Transactions of the Kansas Academy of Science* 91:153–157.

———. 1992. *A Revised Checklist with Distribution Maps of the Turtles of the World.* Privately published, Richmond, Ind.

———, and C. R. Etchberger. 1989. The distribution of the turtles of Florida. *Florida Scientist* 52:119–144.

———, and R. Hudson. 2005. *Macrochelys temminckii* (alligator snapping turtle): Diet. *Herpetological Review* 36:312–313.

———, and G. R. Smith. 1993. Reproductive ecology of the painted turtle (*Chrysemys picta*) in the Nebraska Sandhills and across its range. *Copeia* 1993:1–21.

Jackson, D. C., C. E. Crocker, and G. R. Ultsch. 2001. Mechanisms of homeostasis during long-term diving and anoxia in turtles. *Zoology* 103:150–158.

———, S. E. Taylor, V. S. Asare, D. Villarnovo, J. M. Gall, and S. A. Reese. 2007. Comparative shell buffering properties correlate with anoxia tolerance in freshwater turtles. *American Journal of Physiology—Regulatory, Integrative and Comparative Physiology* 292:R1008-R1015.

Jackson, D. R. 1975. A Pleistocene *Graptemys* (Reptilia: Testudines) from the Santa Fe River of Florida. *Herpetologica* 31:213–219.

———. 2003. *Graptemys barbouri* (Barbour's map turtle): Geographic distribution. *Herpetological Review* 34:164.

———, and R. N. Walker. 1997. Reproduction in the Suwannee cooter, *Pseudemys concinna suwanniensis. Bulletin of the Florida Museum of Natural History* 41:69–167.

Jackson, J. F., and S. H. Shively. 1983. A distinctive population of *Graptemys pseudogeographica* from the Calcasieu River system. *Association of Southeastern Biologists Bulletin* 30:64.

Jacobson, E. R., J. M. Gaskin, and H. Wahlquist. 1982. Herpesvirus-like infection in map turtles. *Journal of the American Veterinary Medicine Association* 181:1322–1324.

———, C. H. Gardiner, S. L. Barten, D. H. Burr, and A. L. Bourgeois. 1989. *Flavobacterium meningosepticum* infection of a Barbour's map turtle (*Graptemys barbouri*). *Journal of Zoo and Wildlife Medicine* 20:474–477.

Janzen, F. J. 1994. Climate change and temperature-dependent sex determination in reptiles. *Proceedings of the National Academy of Sciences USA* 91:7487–7490.

———, J. C. Ast, and G. L. Paukstis. 1995. Influence of the hydric environment and clutch on eggs and embryos of two sympatric map turtles. *Functional Ecology* 9:913–922.

———, and J. G. Krenz. 2004. Phylogenetics: Which was first, TSD or GSD? Pages 121–130 in N. Valenzuela and V. Lance (Eds.), *Temperature-Dependent Sex Determination in Vertebrates.* Smithsonian Books, Washington, D.C.

———, G. L. Paukstis, and E. D. Brodie, III. 1992. Observations on the basking behavior of hatchling turtles in the wild. *Journal of Herpetology* 26:217–219.

Jenkins, J. D. 1978. Notes on the courtship of the map turtle *Graptemys pseudogeographica* (Gray) (Reptilia, Testudines, Emydidae). *Journal of Herpetology* 13:129–131.

Jensen, J. B. 2011. *Graptemys barbouri* (Barbour's map turtle): Geographic distribution. *Herpetological Review* 42:565.

———, C. D. Camp, W. Gibbons, and M. J. Elliott. 2008. *Amphibians and Reptiles of Georgia.* University of Georgia Press, Athens.

Jiggins, C. D., and J. Mallet. 2000. Bimodal hybrid zones and speciation. *Trends in Ecology and Evolution* 15:250–255.

Johansen, E. P. 2011. A Survey of the Freshwater Turtles of Eastern Oklahoma. Unpublished M.S. thesis, Oklahoma State University, Stillwater.

Jones, R. L. 1993a. Density, population structure, and movements of the yellow-blotched map turtle (*Graptemys flavimaculata*): Year one report. *Mississippi Museum of Natural Science Technical Report* 23:1–20.

———. 1993b. *Yellow-Blotched Map Turtle (*Graptemys flavimaculata*) Recovery Plan.* U.S. Fish and Wildlife Service, Atlanta, Ga.

———. 1996. Home range and seasonal movements of the turtle *Graptemys flavimaculata*. *Journal of Herpetology* 30:376–385.

———. 2006. Reproduction and nesting of the endangered ringed map turtle, *Graptemys oculifera*, in Mississippi. *Chelonian Conservation and Biology* 5:195–209.

———. 2009. Population Estimates of the Ringed Sawback Turtle, *Graptemys oculifera*. Unpublished Museum Technical Report no. 158, Mississippi Department of Wildlife, Fisheries, and Parks, Museum of Natural Science, Jackson.

———, and P. D. Hartfield. 1995. Population size and growth in the turtle *Graptemys oculifera*. *Journal of Herpetology* 29:426–436.

———, T. C. Majure, and C. L. Knight. 1996. *Graptemys pulchra* (Alabama map turtle): Geographic distribution. *Herpetological Review* 27:151.

———, and W. Selman. 2009. *Graptemys oculifera* (Baur 1890)—Ringed map turtle, ringed sawback. *Chelonian Research Monographs* 5:33.1–33.8.

Jordan, D. S., and A. W. Brayton. 1878. Contributions to North American ichthyology, 3: On the distribution of the fishes of the Allegheny region of South Carolina, Georgia, and Tennessee, with descriptions of new and little known species. *Bulletin of the U.S. National Museum* 12:7–95.

Jordan, F., and D. A. Arrington. 2001. Weak trophic interactions between large predatory fishes and herpetofauna in the channelized Kissimmee River, Florida, USA. *Wetlands* 21:155–159.

Jundt, J. A. 2000. Distributions of Amphibians and Reptiles in North Dakota. Unpublished M.S. thesis, North Dakota State University, Fargo.

Kannan, K., M. Ueda, J. A. Shelby, M. T. Mendonça, M. Kawano, M. Matsuda, T. Wakimoto, and J. P. Giesy. 2000. Polychlorinated dibenzo-*p*-dioxins (PCDDs), dibenzofurans (PCDFs), biphenyls (PCBs), and organochlorine pesticides in yellow-blotched map turtle from the Pascagoula River basin, Mississippi, USA. *Archives of Environmental Contamination and Toxicology* 38:362–370.

Keiser, E. D. 1994. A Summer Survey of Amphibians, Reptiles, and Mammals within Designated Western Acreage of the Stennis Space Center, Hancock County, Mississippi. Unpublished report to Lockheed Stennis Operations, Stennis Space Center, Miss.

———. 2000. A Survey on the Nanih Waiya Wildlife Management Area to Locate Turtles and Turtle Nesting Sites, with Emphasis on the Ringed Map Turtle (*Graptemys oculifera*). Unpublished report to the Mississippi Museum of Natural Science, Jackson.

Kendall, W. L., J. D. Nichols, and J. E. Hines. 1997. Estimating temporary emigration using capture-recapture data with Pollock's robust design. *Ecology* 78:563–578.

Killebrew, F. C. 1977. Mitotic chromosomes of turtles, 4: The Emydidae. *Texas Journal of Science* 29:245–253.

———. 1979. Osteological variation between *Graptemys flavimaculata* and *Graptemys nigrinoda* (Testudines: Emydidae). *Herpetologica* 35:146–153.

———, and J. B. Babitzke. 1996. Population Analysis and Nesting Study of Cagle's Map Turtle. Unpublished report to the U.S. Fish and Wildlife Service, Washington, D.C.

———, K. B. Blair, D. Chiszar, and H. M. Smith. 1996. New records for amphibians and reptiles from Texas. *Herpetological Review* 27:90–91.

———, and D. A. Porter. 1989. *Graptemys caglei* (Cagle's map turtle): Size maxima. *Herpetological Review* 20:70.

———, and D. A. Porter. 1991. *Graptemys caglei* (Cagle's map turtle): Distribution. *Herpetological Review* 22:24.

———, W. J. Rogers, and J. B. Babitzke. 2002. Assessment of Instream Flow and Habitat Requirements for Cagle's Map Turtle (*Graptemys caglei*). Unpublished report to the Edwards Aquifer Authority, San Antonio, Tex.

Kimmel, B. L., O. T. Lind, and L. J. Paulson. 1990. Reservoir primary production. Pages 133–193 in K. W. Thornton, B. L. Kimmel, and F. E . Payne (Eds.), *Reservoir Limnology: Ecological Perspectives*. John Wiley and Sons, New York.

King, F. W., and R. L. Burke. 1989. *Crocodilian, Tuatara, and Turtle Species of the World: A Taxonomic and Geographic Reference*. Association of Systematics Collections, Washington, D.C.

King, R. B., M. J. Oldham, and W. F. Weller. 1997. Historic and current amphibian and reptile distributions in the island region of western Lake Erie. *American Midland Naturalist* 138:153–173.

King, T. L., and S. E. Julian. 2004. Conservation of microsatellite DNA flanking sequence across 13 emydid genera assayed with novel bog turtle (*Glyptemys muhlenbergii*) loci. *Conservation Genetics* 5:719–725.

Kiviat, E., and D. C. Buso. 1977. *Graptemys geographica* (map turtle): Geographic distribution. *Herpetological Review* 8:84.

Kizirian, D. A., W. K. King, and J. R. Dixon. 1990. *Graptemys versa* (Texas map turtle): Size maximum and diet. *Herpetological Review* 21:60.

Klinkenborg, V. 2009. Last one. *National Geographic* 215(1):82–107.

Kofron, C. P. 1991. Aspects of the ecology of the threatened ringed sawback turtle, *Graptemys oculifera*. *Amphibia-Reptilia* 12:161–168.

Koper, N., and R. J. Brooks. 1998. Population-size estimators and unequal catchability in painted turtles. *Canadian Journal of Zoology* 76:458–465.

Krusling, P. J., J. G. Davis, and R. Lisi. 2010a. *Graptemys ouachitensis ouachitensis* (Ouachita map turtle): Geographic distribution. *Herpetological Review* 41:509–510.

———, J. G. Davis, and R. Lisi. 2010b. *Graptemys pseudogeographica pseudogeographica* (false map turtle): Geographic distribution. *Herpetological Review* 41:510.

Lagler, K. F. 1943. Food habits and economic relations of the turtles of Michigan with special reference to fish management. *American Midland Naturalist* 29:257–312.

Lahanas, P. N. 1982. Aspects of the Life History of the Southern Black-Knobbed Sawback, *Graptemys nigrinoda delticola* Folkerts and Mount. Unpublished M.S. thesis, Auburn University, Auburn, Ala.

———. 1986. *Graptemys nigrinoda* Cagle: Black-knobbed sawback. *Catalogue of American Amphibians and Reptiles* 396:1–2.

Lamb, T., C. Lydeard, R. B. Walker, and J. W. Gibbons. 1994. Molecular systematics of map turtles (*Graptemys*): A comparison of mitochondrial restriction site versus sequence data. *Systematic Biology* 43:543–559.

———, and M. F. Osentoski. 1997. On the paraphyly of *Malaclemys*: A molecular genetic assessment. *Journal of Herpetology* 31:258–265.

Lamer, J. T., J. K. Tucker, C. R. Dolan, K. Irons, M. Smith, and N. Michaels. 2008. *Graptemys pseudogeographica* (false map turtle): Geographic distribution. *Herpetological Review* 39:236.

Lane, T. J., and D. R. Mader. 1996. Parasitology. Pages 185–203 in D. R. Mader (Ed.), *Reptile Medicine and Surgery*. W. B. Saunders, Philadelphia.

Lardie, R. L. 1999. *Graptemys pseudogeographica ouachitensis* (Ouachita map turtle): Geographic distribution. *Herpetological Review* 30:108.

Lechowitz, C. J., and J. Archer 2007. *Graptemys ernsti* (Escambia map turtle): Geographic distribution. *Herpetological Review* 38:479.

LeClere, J. B., C. E. Smith, and R. E. Blasus. 2009. New and updated county records for amphibians and reptiles in North Dakota, USA. *Herpetological Review* 40:246–247.

Lee, D. S. 1969. Save Barbour's map turtle. *Florida Naturalist* 42:38.

———, R. Franz, and R. A. Sanderson. 1975. A note on the feeding habits of male Barbour's map turtles. *Florida Field Naturalist* 3:45–46.

Leidy, J. 1851. Contributions to helminthology. *Proceedings of the Academy of Natural Science of Philadelphia* 5:205–209.

LeSueur, C. A. 1817. An account of an American species of tortoise, not noticed in the systems. *Journal of the Academy of Natural Sciences of Philadelphia* 1:86–88.

———. 1827. Note sur deux espèces de tortues, du genre *Trionyx* de M. Geoffroy Saint-Hilaire. *Mémoires du Muséum d'Histoire Naturelle* 15:257–268.

Lieberman, S., and T. VanNorman. 2000. Proposed rule: Notice of intent to include several native U.S. species in Appendix III to the Convention on International Trade in Endangered Species of Wild Fauna and Flora. *Federal Register* 65:4217–4221.

Lincicome, D. R. 1948. Observations on *Neoechinorhynchus emydis* (Leidy), an acanthocephalan parasite of turtles. *Journal of Parasitology* 34:51–54.

———, and A. Whitt, Jr. 1947. On the occurrence of *Neoechinorhynchus emydis* (Acanthocephala) in snails. *Transactions of the Kentucky Academy of Science* 12:19.

Lindeman, P. V. 1990. Closed and open model estimates of abundance with tests of model assumptions for two populations of the turtle, *Chrysemys picta*. *Journal of Herpetology* 24:78–81.

———. 1993. Aerial basking by hatchling freshwater turtles. *Herpetological Review* 24:84–87.

———. 1996. Distribution, relative abundance, and basking ecology of the razorback musk turtle, *Kinosternon carinatum*, in the Pearl and Pascagoula River drainages. *Herpetological Natural History* 4:23–34.

———. 1997a. A comparative spotting-scope study of the distribution and relative abundance of river cooters (*Pseudemys concinna*) in western Kentucky and southern Mississippi. *Chelonian Conservation and Biology* 2:378–383.

———. 1997b. Contributions toward improvement of model fit in nonlinear regression modelling of turtle growth. *Herpetologica* 53:179–191.

———. 1997c. Effects of Competition, Phylogeny, Ontogeny, and Morphology on Structuring the Resource Use of Freshwater Turtles. Unpublished Ph.D. dissertation, University of Louisville, Louisville, Ky.

———. 1998. Of deadwood and map turtles (*Graptemys*): An analysis of species status for five species in three river drainages using replicated spotting-scope counts of basking turtles; Linnaeus Fund research report. *Chelonian Conservation and Biology* 3:137–141.

———. 1999a. Aggressive interactions during basking among four species of emydid turtles. *Journal of Herpetology* 33:214–219.

———. 1999b. Growth curves for *Graptemys*, with a comparison to other emydid turtles. *American Midland Naturalist* 142:141–151.

———. 1999c. Surveys of basking map turtles *Graptemys* spp. in three river drainages and the importance of deadwood abundance. *Biological Conservation* 88:33–42.

———. 2000a. The evolution of relative width of the head and alveolar surfaces in map turtles (Testudines: Emydidae: *Graptemys*). *Biological Journal of the Linnean Society* 69:549–576.

———. 2000b. Resource use of five sympatric turtle species: Effects of competition, phylogeny, and morphology. *Canadian Journal of Zoology* 78:992–1008.

———. 2001a. Investigations of the ecology of *Graptemys versa* and *Pseudemys texana* in southcentral Texas; Linnaeus Fund research report. *Chelonian Conservation and Biology* 4:223–224.

———. 2001b. Notes on nesting of the smooth softshell turtle (*Apalone mutica*) in a river impoundment in western Kentucky. *Journal of the Kentucky Academy of Science* 62:117–120.

———. 2001c. Turtle fauna of the upper Tradewater River near Dawson Springs, Kentucky. *Journal of the Kentucky Academy of Science* 62:121–124.

———. 2003a. Diagnostic characteristics in lower Tennessee River populations of the map turtles *Graptemys pseudogeographica* and *Graptemys ouachitensis*. *Chelonian Conservation and Biology* 4:564–568.

———. 2003b. Sexual difference in habitat use of Texas map turtles (Emydidae: *Graptemys versa*) and its relationship to size dimorphism and dietary differences. *Canadian Journal of Zoology* 81:1185–1191.

———. 2004. Private lands, public access, and the conservation status of *Graptemys versa*. *Turtle and Tortoise Newsletter* 7:6–8.

———. 2005. Aspects of the life history of the Texas map turtle (*Graptemys versa*). *American Midland Naturalist* 153:378–388.

———. 2006a. Diet of the Texas map turtle (*Graptemys versa*): relationship to sexually-dimorphic trophic morphology and changes over five decades as influenced by an invasive mollusk. *Chelonian Conservation and Biology* 5:25–31.

———. 2006b. Zebra and quagga mussels (*Dreissena* spp.) and other prey of a Lake Erie population of common map turtles (Emydidae: *Graptemys geographica*). *Copeia* 2006:268–273.

———. 2007. Diet, growth, body size, and reproductive potential of the Texas river cooter (*Pseudemys texana*). *Southwestern Naturalist* 52:586–594.

———. 2008a. Evolution of body size in the map turtles and sawbacks (Emydidae: Deirochelyinae: *Graptemys*). *Herpetologica* 64:32–46.

———. 2008b. Geographic distribution: *Graptemys pulchra*. *Herpetological Review* 39:107.

———. 2009a. *Graptemys oculifera* (ringed map turtle): Foraging behavior. *Herpetological Review* 40:215.

———. 2009b. On the type locality of *Testudo geographica* LeSueur 1817. *Chelonian Conservation and Biology* 8:95–98.

———. 2010. Geographic distribution: *Graptemys gibbonsi*. *Herpetological Review* 40:105.

———, and M. A. Barger. 2005. Acanthocephalan (*Neoechinorhynchus emydis*) infections in Texas map turtles (*Graptemys versa*). *Southwestern Naturalist* 50:12–16.

———, and M. J. Sharkey. 2001. Comparative analyses of functional relationships in the evolution of trophic morphology in the map turtles (Emydidae: *Graptemys*). *Herpetologica* 57:313–318.

Little, R. B. 1973. Variation in the plastral scutellation of *Graptemys pulchra* (Reptilia, Chelonia, Emydidae). *Association of Southeastern Biologists Bulletin* 20:65–66.

Logier, E. B. S. 1925. Notes on the herpetology of Point Pelee, Ontario. *Canadian Field-Naturalist* 29:91–95.

———, and G. C. Toner. 1961. *Check List of the Amphibians and Reptiles of Canada and Alaska*. Contribution no. 53, Life Sciences Division, Royal Ontario Museum, Toronto, Canada.

Lohoefener, R. 1991. Endangered and threatened wildlife and plants; threatened status for the yellow-blotched map turtle, *Graptemys flavimaculata*. *Federal Register* 56:1459–1463.

Loomis, F. B. 1904. Two new river reptiles from the titanothere beds. *American Journal of Science* 18:427–432.

Losos, J. B. 1990. A phylogenetic analysis of character displacement in Caribbean *Anolis* lizards. *Evolution* 44:558–569.

Loveridge, A., and E. E. Williams. 1957. Revision of the African tortoises and turtles of the suborder Cryptodira. *Bulletin of the Museum of Comparative Zoology, Harvard* 115:163–557.

Lovich, J. E. 1985. *Graptemys pulchra* Baur: Alabama map turtle. *Catalogue of American Amphibians and Reptiles* 360:1–2.

———, and C. H. Ernst. 1989. Variation in the plastral formulae of selected turtles with comments on taxonomic utility. *Copeia* 1989:304–318.

———, and J. W. Gibbons. 1990. Age at maturity influences adult sex ratio in the turtle *Malaclemys terrapin*. *Oikos* 59:126–134.

———, J. C. Godwin, and C. J. McCoy. 2011. *Graptemys ernsti* Lovich and McCoy 1992—Escambia map turtle. *Chelonian Research Monographs* 5:51.1–51.6.

———, S. W. Gotte, C. H. Ernst, J. C. Harshbarger, A. F. Laemmerzahl, and J. W. Gibbons. 1996. Prevalence and histopathology of shell disease in turtles from Lake Blackshear, Georgia. *Journal of Wildlife Diseases* 32:259–265.

———, and C. J. McCoy. 1992. Review of the *Graptemys pulchra* group (Reptilia: Testudines: Emydidae), with descriptions of two new species. *Annals of the Carnegie Museum* 61:293–315.

———, and C. J. McCoy. 1994a. *Graptemys ernsti* Lovich and McCoy: Escambia map turtle. *Catalogue of American Amphibians and Reptiles* 585:1–2.

———, and C. J. McCoy. 1994b. *Graptemys gibbonsi* Lovich and McCoy: Pascagoula map turtle. *Catalogue of American Amphibians and Reptiles* 586:1–2.

———, W. Selman, and C. J. McCoy. 2009. *Graptemys gibbonsi* Lovich and McCoy 1992—Pascagoula map turtle, Gibbons' map turtle, Pearl River map turtle. *Chelonian Research Monographs* 5:29.1–29.8.

———, A. D. Tucker, D. E. Kling, J. W. Gibbons, and T. D. Zimmerman. 1991. Behavior of hatchling diamondback terrapins (*Malaclemys terrapin*) released in a South Carolina marsh. *Herpetological Review* 22:81–83.

Lynch, J. D. 1985. Annotated checklist of the amphibians and reptiles of Nebraska. *Transactions of the Nebraska Academy of Sciences* 13:33–57.

Mabie, D. W., M. T. Merendino, and D. H. Reid. 1995. Prey of nesting bald eagles in Texas. *Journal of Raptor Research* 29:10–14.

MacCulloch, R. D. 1981. Leech parasitism on the western painted turtle, *Chrysemys picta belli*, in Saskatchewan. *Journal of Parasitology* 67:128–129.

Magath, T. B. 1919. *Camallanus americanus*, nov. spec., a monograph on a nematode species. *Transactions of the American Microscopical Society* 38:49–170.

Maginniss, L. A., S. A. Ekelund, and G. R. Ultsch. 2004. Blood oxygen transport in common map turtles during simulated hibernation. *Physiological and Biochemical Zoology* 77:232–241.

Maltese, M. T. 2005. Inclusion of alligator snapping turtle (*Macroclemys* [=*Macrochelys*] *temminckii*) and all species of map turtles (*Graptemys* spp.) in Appendix III to the Convention on International Trade in Endangered Species of Wild Fauna and Flora. *Federal Register* 70:74700–74712.

Marr, J. C. 1944. Notes on amphibians and reptiles from the central United States. *American Midland Naturalist* 32:478–490.

Martins, E. P., and T. Garland, Jr. 1991. Phylogenetic analyses of the correlated evolution of continuous characters: A simulation study. *Evolution* 45:534–557.

Martof, B. S. 1956. *Amphibians and Reptiles of Georgia*. University of Georgia Press, Athens.

Marx, H. 1958. Catalogue of type specimens of reptiles and amphibians in Chicago Natural History Museum. *Fieldiana Zoology* 36:409–496.

May, B., and J. E. Marsden. 1992. Genetic identification and implications of another invasive species of dreissenid mussel in the Great Lakes. *Canadian Journal of Fisheries and Aquatic Sciences* 49:1501–1506.

Mayden, R. L. 1988. Vicariance biogeography, parsimony, and evolution in North American freshwater fishes. *Systematic Zoology* 37:329–355.

McAllister, C. T., and S. J. Upton. 1989. The coccidian (Apicomplexa: Eimeriidae) of Testudines, with descriptions of three new species. *Canadian Journal of Zoology* 67:2459–2467.

———, S. J. Upton, and F. C. Killebrew. 1991. Coccidian parasites (Apicomplexa: Eimeriidae) of *Graptemys caglei* and *Graptemys versa* (Testudines: Emydidae) from Texas. *Journal of Parasitology* 77:500–502.

———, R. Ward, and J. R. Glidewell. 1983. New distributional records for selected amphibians and reptiles of Texas. *Herpetological Review* 14:52–53.

McAuliffe, J. R. 1977. An hypothesis explaining variations of hemogregarine parasitemia in different aquatic turtle species. *Journal of Parasitology* 63:580–581.

McCallum, M. L. 2003. *Graptemys geographica* (northern map turtle): Nest overwintering. *Herpetological Review* 34:241.

McCarthy, R. 2010. Geographic distribution: *Graptemys ouachitensis*. *Herpetological Review* 40:105.

McCarty, J. P. 2001. Ecological consequences of recent climate change. *Conservation Biology* 15:320–331.

McCloud, K. 1985. Letter to the editor. *Herpetological Review* 16:33.

McCord, J. S., Jr., and M. E. Dorcas. 1989. New Texas herpetological records from the University of Texas at Arlington collection of vertebrates. *Herpetological Review* 20:94, 96.

McCord, W. P., J. B. Iverson, P. Q. Spinks, and H. B. Shaffer. 2000. A new genus of geoemydid turtle from Asia. *Hamadryad* 25:86–90.

McCoy, C. J., and R. C. Vogt. 1979. Distribution and Population Status of the Ringed Sawback (*Graptemys oculifera*), Blotched Sawback (*Graptemys flavimaculata*), and Black-Knobbed Sawback (*Graptemys nigrinoda*) in Alabama and Mississippi. Unpublished report to the U.S. Fish and Wildlife Service, Jackson, Mississippi.

———, and R. C. Vogt. 1987. *Graptemys flavimaculata* Cagle: Yellow-blotched sawback. *Catalogue of American Amphibians and Reptiles* 403:1–2.

———, and R. C. Vogt. 1988. *Graptemys oculifera* (Baur): Ringed sawback. *Catalogue of American Amphibians and Reptiles* 422:1–2.

———, and R. C. Vogt. 1990. *Graptemys geographica* (Le Sueur): Map turtle. *Catalogue of American Amphibians and Reptiles* 484:1–4.

———, and R. C. Vogt. 1994. *Graptemys* Agassiz: Map turtles. *Catalogue of American Amphibians and Reptiles* 584:1–3.

McDowell, S. B. 1964. Partition of the genus *Clemmys* and related problems in the taxonomy of the aquatic Testudinidae. *Proceedings of the Zoological Society of London* 143:239–279.

McGowan, K. J. 2001. Fish crow. *Birds of North America* 589:1–27.

McIlroy, S. K., C. Montagne, C. A. Jones, and B. L. McGlynn. 2008. Identifying linkages between land use, geomorphology, and aquatic habitat in a mixed-use watershed. *Environmental Management* 42:867–876.

McKinney, J. M. 1987. *Graptemys versa* (Texas map turtle): Maximum size. *Herpetological Review* 18:17.

McKnight, T. J. 1959. A Taxonomic Study of the Helminth Parasites of the Turtles of Lake Texoma. Unpublished Ph.D. dissertation, University of Oklahoma, Norman.

McKown, R. R. 1972. Phylogenetic Relationships within the Turtle Genera *Graptemys* and *Malaclemys*. Unpublished Ph.D. dissertation, University of Texas, Austin.

McMahon, R. F. 1982. The occurrence and spread of the introduced Asiatic freshwater bivalve, *Corbicula fluminea* (Mueller) in North America: 1924–1981. *Nautilus* 96:134–141.

McMullen, D. B. 1934. The life cycle of the turtle trematode, *Cercorchis medius*. *Journal of Parasitology* 20:248–250.

McNair, D. B. 2000. Fish crow predation on eggs being laid by a Florida softshell turtle. *Oriole* 65:12–13.

Means, D. B., and A. Harvey. 1999. Barbour's map turtle in the diet of nesting bald eagles. *Florida Field Naturalist* 27:14–16.

Mertens, R., and H. Wermuth. 1955. Die rezenten Schildkröten, Krokodile und Brückenechsen. *Zoologisches Jahrbuch Abteilung Systematik* 83:323–440.

Miller, B. T., J. W. Lamb, and J. L. Miller. 2005. The herpetofauna of Arnold Air Force Base in the Barrens of Tennessee. *Southeastern Naturalist* 4:51–62.

———, and J. L. Miller. 2011. *Graptemys geographica* (northern map turtle): Geographic distribution. *Herpetological Review* 42:388.

Minton, S. A., Jr. 2001. *Amphibians & Reptiles of Indiana*, 2nd ed. Indiana Academy of Science, Indianapolis.

Mitchell, J. C. 1994. *The Reptiles of Virginia*. Smithsonian Institution Press, Washington, D.C.

Mittleman, M. B. 1947. Miscellaneous notes on Indiana amphibians and reptiles. *American Midland Naturalist* 38:466–484.

Moler, P. E. 1986. Barbour's Map Turtle Census. Unpublished report E-1-10 III-A to the Florida Game and Freshwater Fish Commission, Tallahassee.

Moll, D. 1976a. Environmental influence on growth rate in the Ouachita map turtle, *Graptemys pseudogeographica ouachitensis*. *Herpetologica* 32:439–443.

———. 1976b. Food and feeding strategies of the Ouachita map turtle (*Graptemys pseudogeographica ouachitensis*). *American Midland Naturalist* 96:478–482.

———. 1976c. A review of supposed insect catching by basking *Graptemys geographica*. *Transactions of the Illinois Academy of Science* 69:302–303.

———. 1977. Ecological Investigations of Turtles in a Polluted Ecosystem: The Central Illinois River and Adjacent Flood Plain Lakes. Unpublished Ph.D. dissertation, Illinois State University, Normal.

———, and E. O. Moll. 2004. *The Ecology, Exploitation, and Conservation of River Turtles*. Oxford University Press, New York.

Moore, M. J. C., and R. A. Seigel. 2006. No place to nest or bask: Effects of human disturbance on the nesting and basking habits of yellow-blotched map turtles (*Graptemys flavimaculata*). *Biological Conservation* 130:386–393.

Moriarty, J. J. 2004. *Turtles and Turtle Watching in the North Central States*. Minnesota Department of Natural Resources, St. Paul.

Mosimann, J. E. 1958. An analysis of allometry in the chelonian shell. *Revue Canadienne de Biologie* 17:137–228.

Mossa, J., and D. Coley. 2006. River Corridor Sand and Gravel Mining, Louisiana and Mississippi: A Database and Comparison of Different Data Sources. Unpublished report to the Minerals Resources Program, U.S. Geological Survey, Reston, Va.

Moulis, R. A. 1997. Status Survey of Barbour's Map Turtle (*Graptemys barbouri*) in Georgia. Unpublished report to the Georgia Department of Natural Resources, Atlanta.

Mount, R. H. 1975. *The Reptiles and Amphibians of Alabama*. Alabama Agricultural Experimental Station, Auburn.

———. 1981. The red imported fire ant, *Solenopsis invicta* (Hymenoptera: Formicidae), as a possible serious predator on some native southeastern vertebrates: Direct observations and subjective impressions. *Journal of the Alabama Academy of Science* 52:71–78.

Muir, J. H. 1984. Commercial exploitation of *Graptemys* and *Sternotherus* turtles. *Bulletin of the Chicago Herpetological Society* 19:98–100.

Myers, E. M. 2008. Post-orbital Color Pattern Variation and the Evolution of a Radiation of Turtles (*Graptemys*). Unpublished Ph.D. dissertation, Iowa State University, Ames.

Nagle, R. D., C. L. Lutz, and A. L. Pyle. 2004. Overwintering in the nest by hatchling map turtles (*Graptemys geographica*). *Canadian Journal of Zoology* 82:1211–1218.

National Science Foundation. 1952. *The Second Annual Report of the National Science Foundation: Fiscal Year 1952*. U.S. Government Printing Office, Washington, D.C.

Near, T. J., P. A. Meylan, and H. B. Shaffer. 2005. Assessing concordance of fossil calibration points in molecular clock studies: An example using turtles. *American Naturalist* 165:137–146.

Neill, W. T. 1951. Notes on the natural history of certain North American snakes. *Publications of the Research Division, Ross Allen's Reptile Institute* 1:47–60.

———. 1971. *The Last of the Ruling Reptiles: Alligators, Crocodiles, and Their Kin*. Columbia University Press, New York.

———, and E. R. Allen. 1954. Algae on turtles: Some additional considerations. *Ecology* 35:581–584.

Nelson, S. K., B. Fitzpatrick, and P. Colclough. 2012. *Graptemys geographica* (northern map turtle): Geographic distribution. *Herpetological Review* 43:615–616.

Neuman-Lee, L. A., and F. J. Janzen. 2011. Atrazine exposure impacts behavior and survivorship of neonatal turtles. *Herpetologica* 67:23–31.

Newman, H. H. 1906a. The habits of certain tortoises. *Journal of Comparative Neurology and Psychology* 16:126–152.

———. 1906b. The significance of scute and plate "abnormalities" in Chelonia. *Biological Bulletin* 10:68–114.

Nickerson, M. A., and A. L. Pitt. 2012. Historical turtle population decline and community changes in an Ozark river. *Bulletin of the Florida Museum of Natural History* 51:257–267.

Niemiller, M. L., R. G. Reynolds, B. M. Glorioso, J. Spiess, and B. T. Miller. 2011. Herpetofauna of the cedar glades and associated habitats of the inner central basin of middle Tennessee. *Herpetological Conservation and Biology* 6:127–141.

Nijs, J. 1999. Verzorging en kweek van de zwartknobbelzaagrugschildpad, *Graptemys nigrinoda nigrinoda*, in gevangenschap. *Lacerta* 57:91–96.

Norton, V. M., and M. J. Harvey. 1975. Herpetofauna of Hardeman County, Tennessee. *Journal of the Tennessee Academy of Science* 50:131–136.

Nutting, W. L., and T. E. Graham. 1993. Preferred body temperatures in five Nearctic freshwater turtles: A preliminary study. *Comparative Biochemistry and Physiology* 104A:243–246.

Ode, D. J. 2004. Wildlife Habitats of LaFramboise Island: Vegetational Change and Management of a Missouri River Island. Unpublished report no. 2004–14 to the South Dakota Game, Fish and Parks Department, Pierre.

Oldfield, B., and J. J. Moriarty. 1994. *Amphibians and Reptiles Native to Minnesota.* University of Minnesota Press, Minneapolis.

Olson, R. E. 1959. Notes on some Texas herptiles. *Herpetologica* 15:48.

———. 1967. Peripheral range extensions and some new records of Texas amphibians and reptiles. *Texas Journal of Science* 19:99–106.

Ortenburger, A. I. 1929. Reptiles and amphibians from northeastern Oklahoma. *Copeia* 1929(170):26–28.

Ouellette, M., and J. A. Cardille. 2011. The complex linear home range estimator: Representing the home range of river turtles in multiple channels. *Chelonian Conservation and Biology* 10:259–265.

Pappas, M. J., and B. J. Brecke. 1992. Habitat selection of juvenile Blanding's turtles, *Emydoidea blandingii. Journal of Herpetology* 26:233–234.

———, B. J. Brecke, and J. D. Congdon. 2000. The Blanding's turtles (*Emydoidea blandingii*) of Weaver Dunes, Minnesota. *Chelonian Conservation and Biology* 3:557–568.

Parren, S. G., and M. A. Rice. 2004. Terrestrial overwintering of hatchling turtles in Vermont nests. *Northeastern Naturalist* 11:229–233.

Paulmier, F. C. 1902. Lizards, tortoises and batrachians of New York. *Bulletin of the New York State Museum* 51:389–409.

Pauly, G. B. 2010. *Graptemys geographica* (northern map turtle): Geographic distribution. *Herpetological Review* 41:509.

Pearse, D. E., and J. C. Avise. 2001. Turtle mating systems: Behavior, sperm storage, and genetic paternity. *Journal of Heredity* 92:206–211.

———, F. J. Janzen, and J. C. Avise. 2001. Genetic markers substantiate long-term storage and utilization of sperm by female painted turtles. *Heredity* 86:378–384.

Perkins, K., III, and D. C. Backlund. 2000. Freshwater Mussels of the Missouri National Recreational River below Gavins Point Dam, South Dakota and Nebraska. Unpublished report no. 2000–1 to the South Dakota Department of Game, Fish, and Parks, Pierre.

Peterman, W. E., and T. J. Ryan. 2009. Basking behavior of emydid turtles (*Chrysemys picta, Graptemys geographica,* and *Trachemys scripta*) in an urban landscape. *Northeastern Naturalist* 16:629–636.

Peters, J. A. 1942. Reptiles and amphibians of Cumberland County, Illinois. *Copeia* 1942:182–183.

Peterson, R. L. 1950. Amphibians and reptiles of Brazos County, Texas. *American Midland Naturalist* 43:157–164.

Phelps, J. R. 2004. Aquatic turtles of diversely managed watersheds in the Ouachita Mountains, Arkansas. Pages 183–186 in J. M. Guldin (Ed.), *Ouachita and Ozark Mountain Symposium: Ecosystem Management Research.* General Technical Report SRS-74, U.S. Department of Agriculture, Forest Service, Southern Research Station, Asheville, N.C.

Phillips, C. A., R. A. Brandon, and E. O. Moll. 1999. *Field Guide to Amphibians and Reptiles of Illinois.* Illinois Natural History Survey, Champaign.

Phillips, G. E. 2006. Paleofaunistics of Nonmammalian Vertebrates from the Late Pleistocene of the Mississippi-Alabama Black Prairie. Unpublished M.S. thesis, North Carolina State University, Raleigh.

Pianka, E. R. 1992. The state of the art in community ecology. Pages 141–162 in K. Adler (Ed.), *Herpetology: Current Research on the Biology of Amphibians and Reptiles; Proceedings of the First World Congress of Herpetology, September 1989.* Society for the Study of Amphibians and Reptiles, Oxford, Ohio.

Piégay, H. 2003. Dynamics of wood in large rivers. Pages 109–133 in S. V. Gregory, K. L.

Boyer, and A. M. Gurnell (Eds.), *The Ecology and Management of Wood in World Rivers*. American Fisheries Society, Bethesda, Md.

Pierce, L. 1992. Diet Content and Overlap of Six Species of Turtle among the Wabash River. Unpublished M.S. thesis, Eastern Illinois University, Charleston.

Pitt, A. L., and M. A. Nickerson. 2012. Reassessment of the turtle community in the North Fork of the White River, Ozark County, Missouri. *Copeia* 2012:367–374.

Pluto, T. G., and E. D. Bellis. 1986. Habitat utilization by the turtle, *Graptemys geographica*, along a river. *Journal of Herpetology* 20:22–31.

———, and E. D. Bellis. 1988. Seasonal and annual movements of riverine map turtles, *Graptemys geographica*. *Journal of Herpetology* 22:152–158.

———, and H. Rothenbacher. 1976. *Eimeria juniataensis* sp. n. (Protozoa: Eimeriidae) from the map turtle, *Graptemys geographica*, in Pennsylvania. *Journal of Parasitology* 62:207–208.

Poly, W. J. 1999. Herpetofauna of the Little South Fork basin (Cumberland River drainage), Wayne and McCreary Counties, Kentucky. *Ohio Journal of Science* 99:26–29.

Pope, C. H. 1939. *Turtles of the United States & Canada*. Alfred A. Knopf, New York.

Porter, D. A. 1990. Feeding Ecology of *Graptemys caglei* Haynes and McKown in the Guadalupe River, Dewitt County, Texas. Unpublished M.S. thesis, West Texas State University, Canyon.

Powell, M. A., and R. Powell. 2011. Aquatic turtles feasting on periodical cicadas. *Missouri Herpetological Association Newsletter* 24:21.

Preston, R. E. 1979. Late Pleistocene cold-blooded vertebrate faunas from the mid-continental United States, 1: Reptilia; Testudines, Crocodilia. *University of Michigan Museum of Paleontology Papers on Paleontology* 19:1–53.

Pritchard, P. C. H., and W. F. Greenhood. 1968. The sun and the turtle. *Journal of the International Turtle and Tortoise Society* 2:20–25, 34.

Proctor, V. W. 1958. The growth of *Basicladia* on turtles. *Ecology* 39:634–645.

Rakowitz, V. A., R. R. Fleet, and F. L. Rainwater. 1983. New distributional records of Texas amphibians and reptiles. *Herpetological Review* 14:85–89.

Raun, G. G. 1959. Terrestrial and aquatic vertebrates of a moist, relic area in central Texas. *Texas Journal of Science* 11:158–171.

———, and F. R. Gelbach. 1972. Amphibians and reptiles in Texas. *Dallas Museum of Natural History Bulletin* 2:1–61.

Rausch, R. 1946. New host records for *Microphallus ovatus* Osborn, 1919. *Journal of Parasitology* 32:93–94.

———. 1947. Observations on some helminths parasitic in Ohio turtles. *American Midland Naturalist* 38:434–442.

Ream, C., and R. Ream. 1966. The influence of sampling methods on the estimation of population structure in painted turtles. *American Midland Naturalist* 75:325–338.

Reed, R. N., and J. W. Gibbons. 2004. Conservation Status of Live U.S. Nonmarine Turtles in Domestic and International Trade. Unpublished report to the U.S. Fish and Wildlife Service, Washington, D.C.

Reese, S. A., C. E. Crocker, M. E. Carwile, D. C. Jackson, and G. R. Ultsch. 2001. The physiology of hibernation in common map turtles (*Graptemys geographica*). *Comparative Biochemistry and Physiology* A 130:331–340.

———, G. R. Ultsch, and D. C. Jackson. 2004. Lactate accumulation, glycogen depletion, and shell composition of hatchling turtles during simulated aquatic hibernation. *Journal of Experimental Biology* 207:2889–2895.

Reich, M., J. L. Kershner, and R. C. Wildman. 2003. Restoring streams with large wood: A synthesis. Pages 355–366 in S. V. Gregory, K. L. Boyer, and A. M. Gurnell (Eds.),

The Ecology and Management of Wood in World Rivers. American Fisheries Society, Bethesda, Md.

Reynolds, R. P., S. W. Gotte, and C. H. Ernst. 2007. Catalog of type specimens of recent Crocodilia and Testudines in the National Museum of Natural History, Smithsonian Institution. *Smithsonian Contributions to Zoology* 626:1–49.

Richards-Dimitrie, T. M. 2011. Spatial Ecology and Diet of Maryland Endangered Northern Map Turtles (*Graptemys geographica*) in an Altered River System: Implications for Conservation and Management. Unpublished M.S. thesis, Towson University, Towson, Md.

——, S. E. Gresens, S. A. Smith, and R. A. Seigel, 2013. Diet of northern map turtles (*Graptemys geographica*): Sexual differences and potential impacts of an altered river system. *Copeia* 2013: 477–484.

Richmond, N. D. 1953. The false map turtle: A new addition to the fauna of West Virginia. *Proceedings of the West Virginia Academy of Science* 25:35.

Riedle, J. D., and A. Hynek. 2002. Amphibian and reptile inventory of the Kansas Army Ammunition Plant, Labette County, Kansas. *Journal of Kansas Herpetology* 2:18–20.

——, P. A. Shipman, S. F. Fox, J. C. Hackler, and D. M. Leslie, Jr. 2008. Population structure of the alligator snapping turtle, *Macrochelys temminckii*, on the western edge of its distribution. *Chelonian Conservation and Biology* 7:100–104.

——, P. A. Shipman, S. F. Fox, and D. M. Leslie, Jr. 2009. Habitat associations of aquatic turtle communities in eastern Oklahoma. *Proceedings of the Oklahoma Academy of Science* 89:19–30.

Rizkalla, C. E., and R. K. Swihart. 2006. Community structure and differential responses of aquatic turtles to agriculturally induced habitat fragmentation. *Landscape Ecology* 21:1361–1375.

Roberts, L. S., and J. Janovy, Jr. 2005. *Foundations of Parasitology*, 7th ed. McGraw-Hill, Boston.

Rodriguez, D., J. Duvall, and M. R. J. Forstner. 2006. *Graptemys pseudogeographica kohnii* (Mississippi map turtle): Geographic distribution. *Herpetological Review* 37:492.

Roman, J., S. D. Santhuff, P. E. Moler, and B. W. Bowen. 1999. Population structure and cryptic evolutionary units in the alligator snapping turtle. *Conservation Biology* 13:135–142.

Roosenburg, W. M. 1996. Maternal condition and nest site choice: An alternative for the maintenance of environmental sex determination? *American Zoologist* 36:157–168.

Root, T. L., J. T. Price, K. R. Hall, S. H. Schneider, C. Rosenzweig, and J. A. Pounds. 2003. Fingerprints of global warming on wild animals and plants. *Nature* 421:57–60.

Rosenzweig, A. 2003. The Reproductive Ecology and Life History of *Graptemys ouachitensis* (Testudines: Emydidae) in the Red River of Louisiana. Unpublished M.S. thesis, University of Louisiana at Monroe.

Roughgarden, J. 1995. Anolis *Lizards of the Caribbean: Ecology, Evolution, and Plate Tectonics.* Oxford University Press, New York.

Rudolph, D. C. 1983. Vertebrate Survey (Exclusive of Fishes) of the Roy E. Larsen Sandyland Sanctuary, Hardin County, Texas. Unpublished report to the Texas Chapter of the Nature Conservancy, Austin.

Ruhl, J. D. 1991. Barbour's Map Turtle Survey: Apalachicola River Wildlife and Environmental Area. Unpublished report to the Florida Game and Freshwater Fish Commission, Tallahassee.

Ryan, K. M., and P. V. Lindeman. 2007. Reproductive allometry in the common map turtle, *Graptemys geographica*. *American Midland Naturalist* 158:49–59.

Ryan, T. J., C. A. Conner, B. A. Douthitt, S. C. Sterrett, and C. M. Salsbury. 2008. Movement and habitat use of two aquatic turtles (*Graptemys geographica* and *Trachemys scripta*) in an urban landscape. *Urban Ecosystems* 11:213–225.

———, and A. Lambert. 2005. Prevalence and colonization of *Placobdella* on two species of freshwater turtles (*Graptemys geographica* and *Sternotherus odoratus*). *Journal of Herpetology* 39:284–287.

Sanderson, R. A. 1974. Sexual Dimorphism in the Barbour's Map Turtle, *Malaclemys barbouri* (Carr and Marchand). Unpublished M.A. thesis, University of South Florida, Tampa.

———, and J. E. Lovich. 1988. *Graptemys barbouri* Carr and Marchand. Barbour's map turtle. *Catalogue of American Amphibians and Reptiles* 421:1–2.

Santhuff, S. D., and L. A. Wilson. 1990. *Graptemys pulchra* (Alabama map turtle): Geographic distribution. *Herpetological Review* 21:39.

Saumure, R. A., and P. J. Livingston. 1994. *Graptemys geographica* (common map turtle): Parasites. *Herpetological Review* 25:121.

———, A. D. Wade, and T. A. Wheeler. 2006. Non-predatory fly larvae (*Delia platura*: Anthomyiidae) in a nest of a northern map turtle (*Graptemys geographica*). *Chelonian Conservation and Biology* 5:274–275.

Say, T. 1825. On the fresh water and land tortoises of the United States. *Journal of the Academy of Natural Sciences of Philadelphia* 1:203–219.

Schmidt, K. P. 1953. *A Check List of North American Amphibians and Reptiles*. University of Chicago Press, Chicago.

Schmidt, R. E., T. W. Hunsinger, T. Coote, E. Griffin-Noyes, and E. Kiviat. 2004. The mudpuppy (*Necturus maculosus*) in the tidal Hudson River, with comments on its status as native. *Northeastern Naturalist* 11:179–188.

Schoener, T. W. 1974. Resource partitioning in ecological communities. *Science* 185:27–39.

Schulz, S. 2001. Erfarhrungsbericht über die Haltung und Zucht der Höckerschildkröten der Gattung *Graptemys*, im Besonderen der Arten *pseudogeographica pseudogeographica, ouachitensis ouachitensis, nigrinoda nigrinoda*, und *barbouri*. *Radiata* 10:3–14.

———. 2004. *Graptemys barbouri* "Barbour-Höckershildkröte" Carr und Marchand 1942. *Emys* 11:22–31.

Seber, G. A. F. 1982. *The Estimation of Animal Abundance and Related Parameters*. Macmillan, New York.

Seidel, M. E., and D. R. Jackson. 1990. Evolution and fossil relationships of slider turtles. Pages 68–73 in J. W. Gibbons (Ed.), *Life History and Ecology of the Slider Turtle*. Smithsonian Institution Press, Washington, D.C.

Seigel, R. A., and R. J. Brauman. 1994. *Food Habits of the Yellow-Blotched Map Turtle* (Graptemys flavimaculata*). Museum Technical Report 28, Mississippi Museum of Natural Science, Jackson.

———, and R. J. Brauman. 1995. Reproduction and Nesting of the Yellow-Blotched Map Turtle, *Graptemys flavimaculata*. Unpublished report to the U.S. Fish and Wildlife Service, Jackson, Miss.

Selman, W. 2010. Conservation and Ecology of the Yellow-Blotched Sawback (*Graptemys flavimaculata*). Unpublished Ph.D. dissertation, University of Southern Mississippi, Hattiesburg.

———. 2011. *Graptemys flavimaculata* (yellow-blotched map turtle): Unique aerial basking behaviors. *Herpetological Review* 42:593–595.

———. 2012. Intradrainage variation in population structure, shape morphology, and sexual size dimorphism in the yellow-blotched sawback, *Graptemys flavimaculata*. *Herpetological Conservation and Biology* 7:427–436.

———, J. R. Ennen, B. R. Kreiser, and C. P. Qualls. 2009. Cross-species amplification of emydid microsatellite loci in three *Graptemys* species and their utility as a conservation tool. *Herpetological Review* 40:32–37.

———, and A. Holbrook. 2010. *Graptemys gibbonsi* (Pascagoula map turtle): Geographic distribution. *Herpetological Review* 41:509.

———, and R. L. Jones. 2011. *Graptemys flavimaculata* Cagle 1954—Yellow-blotched sawback, yellow-blotched map turtle. *Chelonian Research Monographs* 5:52.1–52.11.

———, and R. L. Jones. 2012. Growth in kyphotic ringed sawbacks, *Graptemys oculifera* (Testudines: Emydidae). *Chelonian Conservation and Biology* 11:259–261.

———, B. Kreiser, and C. Qualls. 2007. Conservation Genetics of the Yellow-Blotched Sawback (Emydidae: *Graptemys flavimaculata*). Unpublished report to the U.S. Fish and Wildlife Service, Mississippi Department of Wildlife, Fisheries, and Parks, Jackson.

———, B. Kreiser, and C. Qualls. 2013. Conservation genetics of the yellow-blotched sawback *Graptemys flavimaculata* (Testudines: Emydidae). *Conservation Genetics* 14: in press.

———, J. M. Lawlor, and C. P. Qualls. 2012. Seasonal variation of corticosterone levels in *Graptemys flavimaculata*, an imperiled freshwater turtle. *Copeia* 2012:698–705.

———, and C. Qualls. 2006. Steroid Hormone Levels and Current Population Status of the Yellow-Blotched Sawback Turtle (*Graptemys flavimaculata*). Unpublished report to the Mississippi Department of Wildlife, Fisheries, and Parks, Jackson.

———, and C. Qualls. 2007. Distribution, Status, and Conservation of the Pascagoula Map Turtle (*Graptemys gibbonsi*) in the Pascagoula River System. Unpublished report to the U.S. Fish and Wildlife Service, Mississippi Department of Wildlife, Fisheries, and Parks, and Louisiana Department of Wildlife and Fisheries, Jackson.

———, and C. Qualls. 2008a. *Graptemys flavimaculata* (yellow-blotched map turtle): Foraging behavior. *Herpetological Review* 39:215.

———, and C. Qualls. 2008b. *Graptemys flavimaculata* (yellow-blotched map turtle): Interspecific basking site competition. *Herpetological Review* 39:214–215.

———, and C. Qualls. 2008c. *Graptemys gibbonsi* (Pascagoula map turtle): Interactions with ducks. *Herpetological Review* 39:216–217.

———, and C. Qualls. 2008d. *Graptemys gibbonsi* (Pascagoula map turtle): Interspecific competition for basking sites. *Herpetological Review* 39:216.

———, and C. Qualls. 2008e. The impacts of Hurricane Katrina on a population of yellow-blotched sawbacks (*Graptemys flavimaculata*) in the lower Pascagoula River. *Herpetological Conservation and Biology* 3:224–230.

———, and C. Qualls. 2009a. Distribution and abundance of two imperiled *Graptemys* species of the Pascagoula River system. *Herpetological Conservation and Biology* 4:171–184.

———, and C. Qualls. 2009b. *Graptemys flavimaculata* (yellow-blotched map turtle): Basking and parasite removal. *Herpetological Review* 40:78–79.

———, and C. Qualls. 2011. Basking ecology of the yellow-blotched sawback (*Graptemys flavimaculata*), an imperiled turtle species of the Pascagoula River system, Mississippi, United States. *Chelonian Conservation and Biology* 10:188–197.

———, C. Qualls, and M. Mendonça. 2007. Assessment of the Impact of Hurricane Katrina on the Yellow-Blotched Sawback (*Graptemys flavimaculata*). Unpublished report to the Mississippi Department of Wildlife, Fisheries, and Parks and the U.S. Fish and Wildlife Service, Jackson.

———, C. Qualls, and M. Mendonça. 2008. Assessment of the Impact of Hurricane Katrina on the Yellow-Blotched Sawback (*Graptemys flavimaculata*): Year 2. Unpublished report to the Mississippi Department of Wildlife, Fisheries, and Parks and the U.S. Fish and Wildlife Service, Jackson.

———, C. Qualls, and J. Owen. 2013. Effects of human disturbance on the behavior and physiology of an imperiled freshwater turtle. *Journal of Wildlife Management* 77:877–885.

———, D. Strong, and C. Qualls. 2008. *Graptemys gibbonsi* (Pascagoula map turtle): Basking and parasite removal. *Herpetological Review* 39:216.

———, D. Strong, and C. Qualls. 2009. *Graptemys flavimaculata* (Yellow-blotched map turtle): Unusual basking disturbance. *Herpetological Review* 40:78.

Semlitsch, R. D., and J. W. Gibbons. 1989. Lack of largemouth bass predation on hatchling turtles (*Trachemys scripta*). *Copeia* 1989:1030–1031.

Semple, R. E. 1964. Effects of temperature on blood and plasma volumes in the turtle. *Physiologist* 7:251.

———, D. Sigsworth, and J. T. Stitt. 1969. Composition and volume of fluids in winter and summer of turtles native to Ontario, Canada. *Journal of Physiology* 204:39P-40P.

———, D. Sigsworth, and J. T. Stitt. 1970. Seasonal observations on the plasma, red cell, and blood volumes of two turtle species native to Ontario. *Canadian Journal of Physiology and Pharmacology* 48:282–290.

Senneke, D. 2006. Declared Turtle Trade from the United States. Unpublished report to the World Chelonian Trust, Vacaville, Calif.

Serb, J. M., C. A. Phillips, and J. B. Iverson. 2001. Molecular phylogeny and biogeography of *Kinosternon flavescens* based on complete mitochondrial control region sequences. *Molecular Phylogenetics and Evolution* 18:149–162.

Sergeev, A. 1937. Some materials to the problem of the reptilian post-embryonic growth. *Zoological Journal of Moscow* 16:723–735.

Serrouya, R., A. Ricciardi, and F. G. Whoriskey. 1995. Predation on zebra mussels (*Dreissena polymorpha*) by captive-reared map turtles (*Graptemys geographica*). *Canadian Journal of Zoology* 73:2238–2243.

Shaffer, H. B., P. Meylan, and M. L. McKnight. 1997. Tests of turtle phylogeny: Molecular, morphological, and paleontological approaches. *Systematic Biology* 46:235–268.

———, and R. C. Thomson. 2007. Delimiting species in recent radiations. *Systematic Biology* 56:896–906.

Shankman, D. 1993. Channel migration and vegetation patterns in the southeastern coastal plain. *Conservation Biology* 7:176–183.

Shealy, R. M. 1976. The natural history of the Alabama map turtle, *Graptemys pulchra* Baur, in Alabama. *Bulletin of the Florida State Museum, Biological Sciences* 21:47–111.

Shelby, J. A., and M. T. Mendonça. 2001. Comparison of reproductive parameters in male yellow-blotched map turtles (*Graptemys flavimaculata*) from a historically contaminated site and a reference site. *Comparative Biochemistry and Physiology C* 129:233–242.

———, M. T. Mendonça, B. D. Horne, and R. A. Seigel. 2000. Seasonal variation in reproductive steroids of male and female yellow-blotched map turtles, *Graptemys flavimaculata*. *General and Comparative Endocrinology* 119:43–51.

Shelby-Walker, J. A., C. K. Ward, and M. T. Mendonça. 2009. Reproductive parameters in female yellow-blotched map turtles (*Graptemys flavimaculata*) from a historically contaminated site vs. a reference site. *Comparative Biochemistry and Physiology A* 154:401–408.

Shine, R. 1999. Why is sex determined by nest temperature in many reptiles? *Trends in Ecology and Evolution* 14:186–189.

Shipman, P. A., and D. Riedle. 1994. Alligator Snapping Turtle, *Macroclemys temminckii*, Trap, Mark, and Release Project. Unpublished report to the Missouri Department of Conservation, Columbia.

Shively, S. H. 1982. Factors Limiting the Upstream Distribution of the Sabine Map Turtle, *Graptemys ouachitensis sabinensis*, in Whisky Chitto Creek, Louisiana. Unpublished M.S. thesis, University of Southwestern Louisiana, Lafayette.

———. 1999. 1999 Survey for the Ringed Map Turtle (*Graptemys oculifera*) in the Bogue Chitto River, Louisiana. Unpublished report to the Louisiana Natural Heritage Program, Louisiana Department of Wildlife and Fisheries, Baton Rouge.

———. 2001. Sabine Map Turtle (*Graptemys ouachitensis sabinensis*) Survey in Western Louisiana. Unpublished report to the Louisiana Natural Heritage Program, Louisiana Department of Wildlife and Fisheries, Baton Rouge.

———, and J. F. Jackson. 1985. Factors limiting the upstream distribution of the Sabine map turtle. *American Midland Naturalist* 114:292–303.

———, and M. F. Vidrine. 1984. Fresh-water mollusks in the alimentary tract of a Mississippi map turtle. *Proceedings of the Louisiana Academy of Sciences* 47:27–29.

Shoop, C. R. 1967. *Graptemys nigrinoda* in Mississippi. *Herpetologica* 23:56.

Siddall, M. E., and S. S. Dresser. 2001. Transmission of *Haemogregarina balli* from painted turtles to snapping turtles through the leech *Placobdella ornata*. *Journal of Parasitology* 87:1217–1218.

Sievert, G., and J. T. Collins. 1998. *Graptemys ouachitensis* (Ouachita map turtle): Geographic distribution. *Herpetological Review* 29:109.

Simpson, G. G. 1946. The Duchesnean fauna and the Eocene-Oligocene boundary. *American Journal of Science* 244:52–57.

Simpson, T. R., and F. L Rose. 2007. Distribution of Cagle's map turtle (*Graptemys caglei*) in the Blanco and San Marcos Rivers. *Texas Journal of Science* 59:201–208.

Slaughter, B. H., W. W. Crook, Jr., R. K. Harris, D. C. Allen, and M. Seifert. 1962. The Hill-Shuler local faunas of the upper Trinity River, Dallas and Denton Counties, Texas. *University of Texas Bureau of Economic Geology Report of Investigations* 48:1–75.

Sloan, K. N., K. A. Buhlmann, and J. E. Lovich. 1996. Stomach contents of commercially harvested adult alligator snapping turtles, *Macroclemys temminckii*. *Chelonian Conservation and Biology* 2:96–99.

Smith, G. R., and J. B. Iverson. 2004. Diel activity patterns of the turtle assemblage of a northern Indiana lake. *American Midland Naturalist* 152:156–164.

———, J. B. Iverson, and J. E. Rettig. 2006. Changes in a turtle community from a northern Indiana lake: A long-term study. *Journal of Herpetology* 40:180–185.

Smith, H. M. 1948. The map turtles of Texas. *Proceedings and Transactions of the Texas Academy of Science* 30:60.

———, and H. K. Buechner. 1947. The influence of the Balcones Escarpment on the distribution of amphibians and reptiles in Texas. *Bulletin of the Chicago Academy of Sciences* 8:1–16.

———, and O. Sanders. 1952. Distributional data on Texan amphibians and reptiles. *Texas Journal of Science* 4:204–219.

Smith, L. L., D. A. Steen, J. M. Stober, M. C. Freeman, S. W. Golladay, L. M. Conner, and J. Cochrane. 2006. The vertebrate fauna of Ichauway, Baker County, Georgia. *Southeastern Naturalist* 5:599–620.

Snider, A. T., and J. K. Bowler. 1992. Longevity of reptiles and amphibians in North

American collections, 2nd ed. *Society for the Study of Amphibians and Reptiles Herpetological Circular* 21:1–40.

Spinks, P. Q., H. B. Shaffer, J. B. Iverson, and W. P. McCord. 2004. Phylogenetic hypotheses for the turtle family Geoemydidae. *Molecular Phylogenetics and Evolution* 32:164–182.

———, R. C. Thomson, G. A. Lovely, and H. B. Shaffer. 2009. Assessing what is needed to resolve a molecular phylogeny: Simulations and empirical data from emydid turtles. *BioMed Central Evolutionary Biology* 9:1–17.

Starkey, D. E., H. B. Shaffer, R. L. Burke, M. R. J. Forstner, J. B. Iverson, F. J. Janzen, A. G. J. Rhodin, and G. R. Ultsch. 2003. Molecular systematics, phylogeography, and the effects of Pleistocene glaciation in the painted turtle (*Chrysemys picta*) complex. *Evolution* 57:119–128.

Steen, D. A., M. J. Aresco, S. G. Beilke, B. W. Compton, E. P. Condon, C. K. Dodd, Jr., H. Forrester, et al. 2006. Relative vulnerability of female turtles to road mortality. *Animal Conservation* 9:269–273.

Steinauer, M. L., and B. D. Horne. 2002. Enteric helminths of *Graptemys flavimaculata* Cagle, 1954, a threatened chelonian species from the Pascagoula River in Mississippi, U.S.A. *Comparative Parasitology* 69:219–222.

Stejneger, L. 1925. New species and subspecies of American turtles. *Journal of the Washington Academy of Sciences* 15:462–463.

———, and T. Barbour. 1917. *A Check List of North American Amphibians and Reptiles*, 1st ed.. Harvard University Press, Cambridge, Mass.

———, and T. Barbour. 1923. *A Check List of North American Amphibians and Reptiles*, 2nd ed. Harvard University Press, Cambridge, Mass.

———, and T. Barbour. 1933. *A Check List of North American Amphibians and Reptiles*, 3rd ed. Harvard University Press, Cambridge, Mass.

———, and T. Barbour. 1939. *A Check List of North American Amphibians and Reptiles*, 4th ed. Harvard University Press, Cambridge, Mass.

———, and T. Barbour. 1943. A check list of North American amphibians and reptiles, 5th ed. *Bulletin of the Museum of Comparative Zoology* 93:1–260.

Stephens, P. R. 1998. Variation in the Cranial Osteological Morphology of Turtles in the Genus *Graptemys* (Reptilia; Anapsida; Testudines; Cryptodira; Emydidae; Deirochelyinae). Unpublished M.S. thesis, University of South Alabama, Mobile.

———, and J. J. Wiens. 2003. Ecological diversification and phylogeny of emydid turtles. *Biological Journal of the Linnean Society* 79:577–610.

———, and J. J. Wiens. 2008. Testing for evolutionary trade-offs in a phylogenetic context: Ecological diversification and evolution of locomotor performance in emydid turtles. *Journal of Evolutionary Biology* 21:77–87.

———, and J. J. Wiens. 2009. Evolution of sexual size dimorphisms in emydid turtles: Ecological dimorphism, Rensch's rule, and sympatric divergence. *Evolution* 63:910–925.

Sterrett, S. C. 2009. The Ecology and Influence of Land Use on River Turtles In Southwest Georgia. Unpublished M.S. thesis, University of Georgia, Athens.

———, and A. M. Grosse. 2009. *Graptemys barbouri* (Barbour's map turtle): Geographic distribution. *Herpetological Review* 40:110.

———, L. L. Smith, S. W. Golladay, S. H. Schweitzer, and J. C. Maerz. 2011. The conservation implications of riparian land use on river turtles. *Animal Conservation* 14:38–46.

———, L. L. Smith, S. H. Schweitzer, and J. C. Maerz. 2010. An assessment of two methods for sampling river turtle assemblages. *Herpetological Conservation and Biology* 5:490–497.

Stettner, A. 2005. Haltung und Naczucht von *Graptemys nigrinoda nigrinoda* Cagle 1954, der Schwarzknopf-Höckershildkröte. *Sacalia* 6(3):9–20.

Stewart, J. H. 1985. Status Review of Ringed Sawback Turtle *Graptemys oculifera*. Unpublished report to U.S. Fish and Wildlife Service, Jackson, Miss.

———. 1986a. Endangered and threatened wildlife and plants; determination of threatened status for the ringed sawback turtle (*Graptemys oculifera*). *Federal Register* 51:45907–45910.

———. 1986b. Endangered and threatened wildlife and plants; proposed threatened status for the ringed sawback turtle. *Federal Register* 51:2741–2744.

———. 1988. *A Recovery Plan for the Ringed Sawback Turtle* Graptemys oculifera. U.S. Fish and Wildlife Service, Atlanta, Ga.

———. 1989. Status Review of Yellow-Blotched Map Turtle, *Graptemys flavimaculata*. Unpublished report to U.S. Fish and Wildlife Service, Jackson, Miss.

———. 1990. Endangered and threatened wildlife and plants; proposed threatened status for the yellow-blotched map turtle, *Graptemys flavimaculata*. *Federal Register* 55:28570–28573.

———. 1992. Status Review: Barbour's Map Turtle, *Graptemys barbouri*. Unpublished report to U.S. Fish and Wildlife Service, Jackson, Miss.

Stitt, J. T., and R. E. Semple. 1971. Site of plasma sequestration induced by body cooling in turtles. *American Journal of Physiology* 221:1189–1191.

———, R. E. Semple, and D.W. Sigsworth. 1970. Effect of changes in body temperature on circulating plasma volume of turtles. *American Journal of Physiology* 219:683–686.

———, R. E. Semple, and D. W. Sigsworth. 1971. Plasma sequestration produced by acute changes in body temperature in turtles. *American Journal of Physiology* 221:1185–1188.

Stovall, J. W., and W. N. McAnulty. 1950. The vertebrate fauna and geologic age of Trinity River terraces in Henderson County, Texas. *American Midland Naturalist* 44:211–250.

Strecker, J. K., Jr. 1908. The reptiles and amphibians of McLennan County, Texas. *Proceedings of the Biological Society of Washington* 21:69–84.

———. 1909. Notes on the herpetology of Burnet County, Texas. *Baylor University Bulletin* 12:1–9.

———. 1930. A catalogue of the amphibians and reptiles of Travis County, Texas. *Contributions from the Baylor University Museum* 23:1–16.

———. 1922. An annotated catalogue of the amphibians and reptiles of Bexar County, Texas. *Bulletin of the Scientific Society of San Antonio* 4:1–31.

———, and W. J. Williams. 1927. Herpetological records from the vicinity of San Marcos, Texas, with distributional data on the amphibians and reptiles of the Edwards Plateau region and central Texas. *Contributions from the Baylor University Museum* 12:3–16.

Stuart, M. 1974. Reptile breeding data at Louisiana Purchase Gardens and Zoo. Pages 275–277 in *AAZPA Regional Conferences Proceedings*. American Association of Zoological Parks and Aquaria, Arlington, Tex.

Stunkard, H. W. 1915. Notes on the trematode genus *Telorchis* with descriptions of new species. *Journal of Parasitology* 2:57–66.

———. 1916. On the anatomy and relationships of some North American trematodes. *Journal of Parasitology* 3:21–27.

———. 1917. Studies on North American Polystomidae, Aspidogastridae, and Paramphistomidae. *Illinois Biological Monographs* 3(3):1–114.

———. 1919. On the specific identity of *Heronimus chelydrae* MacCallum and *Aorchis extensus* Barker and Parsons. *Journal of Parasitology* 6:11–18.

———. 1923. Studies on North American blood flukes. *Bulletin of the American Museum of Natural History* 48:165–221.

Surface, H. A. 1908. First report on the economic features of the turtles of Pennsylvania. *Zoological Bulletin of the Division of Zoology, Pennsylvania Department of Agriculture,* 6(4–5):106–196.

Swift, C. C., C. R. Gilbert, S. A. Bortone, G. H. Burgess, and R. W. Yerger. 1986. Zoogeography of the freshwater fishes of the southeastern United States: Savannah River to Lake Ponchartrain. Pages 213–265 in C. H. Hocutt and E. O. Wiley (Eds.), *The Zoogeography of North American Freshwater Fishes.* John Wiley and Sons, New York.

Taggart, T. W. 1992. *Graptemys pseudogeographica* (false map turtle): Geographic distribution. *Herpetological Review* 23:88.

Taulman, J. F., and L. W. Robbins. 1996. Recent range expansion and distributional limits of the nine-banded armadillo (*Dasypus novemcinctus*) in the United States. *Journal of Biogeography* 23:635–648.

Taylor, D. 1986. Fall foods of adult alligators from cypress lake habitat, Louisiana. *Proceedings of the Annual Conference of Southeastern Association of Fish and Wildlife Agencies* 40:338–341.

Taylor, S. J., C. A. Phillips, and M. L. DeNight. 2003. *Graptemys pseudogeographica kohnii* (Mississippi map turtle): Geographic distribution. *Herpetological Review* 34:261.

Telecky, T. M. 2001. United States import and export of live turtles and tortoises. *Turtle and Tortoise Newsletter* 4:8–13.

Telford, S. R., Jr. 2009. *Hemoparasites of the Reptilia: Color Atlas and Text.* CRC Press, Boca Raton, Fla.

Temple-Miller, K. G. 2008. Use of Radiotelemetry and GIS to Distinguish Habitat Use between *Graptemys ouachitensis* and *G. geographica* in the Scioto River. Unpublished M.S. thesis, Ohio University, Athens.

Thomas, R. B. 2002. Conditional mating strategy in a long-lived vertebrate: Ontogenetic shifts in the mating tactics of male slider turtles (*Trachemys scripta*). *Copeia* 2002:456–461.

———, and R. Altig. 2006. Characteristics of the foreclaw display behaviors of female *Trachemys scripta* (slider turtles). *Southeastern Naturalist* 5:227–234.

———, and J. Bradford. 1997. *Graptemys pulchra* (Alabama map turtle): Geographic distribution. *Herpetological Review* 28:209.

Tiebout, H. M., III. 2003. *An Inventory of the Herpetofauna of Valley Forge National Historical Park.* Technical Report NPS/PHSO/NRTR-03/088 to the National Park Service, Philadelphia.

Timken, R. L. 1968a. The Distribution and Ecology of Turtles in South Dakota. Unpublished Ph.D. dissertation, University of South Dakota, Vermillion.

———. 1968b. *Graptemys pseudogeographica* in the upper Missouri River of the north-central United States. *Journal of Herpetology* 1:76–82.

Tinkle, D. W. 1957. Tulane University field expedition. *Lacerta* 15:73–75.

———. 1958a. Experiments with censusing of southern turtle populations. *Herpetologica* 14:172–175.

———. 1958b. The systematics and ecology of the *Sternothaerus carinatus* complex (Testudinata, Chelydridae). *Tulane Studies in Zoology* 6:3–56.

———. 1959. The relation of the fall line to the distribution and abundance of turtles. *Copeia* 1959:167–170.

———. 1962. Variation in shell morphology of North American turtles 1: The carapacial seam arrangements. *Tulane Studies in Zoology* 9:331–349.

———, J. D. Congdon, and P. C. Rosen. 1981. Nesting frequency and success: Implications for the demography of painted turtles. *Ecology* 62:1426–1432.

———, and G. N. Knopf. 1964. Biologically significant distribution records for amphibians and reptiles in northwest Texas. *Herpetologica* 20:42–47.

Tollefson, J. F. 2004. Spatial Variation and Environmental Correlates of Turtle Assemblage Structure in a Great Plains River Ecosystem. Unpublished M.S. thesis, Emporia State University, Emporia, Kans.

Toner, G. C. 1936. Notes on the turtles of Leeds and Frontenac Counties, Ontario. *Copeia* 1936:236–237.

Tran, S. L., D. L. Moorhead, and K. C. McKenna. 2007. Habitat selection by native turtles in a Lake Erie wetland, USA. *American Midland Naturalist* 158:16–28.

Trauth, S. E., H. W. Robison, and M. V. Plummer. 2004. *The Amphibians and Reptiles of Arkansas*. University of Arkansas Press, Fayetteville.

Tryon, B. W. 1978. Some aspects of breeding and raising aquatic chelonians: Part 2. *Herpetological Review* 9:58–61.

Tucker, J. K. 2001. Clutch frequency in the red-eared slider (*Trachemys scripta elegans*). *Journal of Herpetology* 35:664–668.

———, C. R. Dolan, J. T. Lamer, and E. A. Dustman. 2008. Climatic warming, sex ratios, and red-eared sliders (*Trachemys scripta elegans*) in Illinois. *Chelonian Conservation and Biology* 7:60–69.

———, R. S. Funk, and G. L. Paukstis. 1978. The adaptive significance of egg morphology in two turtles (*Chrysemys picta* and *Terrapene carolina*). *Bulletin of the Maryland Herpetological Society* 14:10–22.

Turner, W. M., Jr. 2001. The Composition of the Diet of the Adult Alabama Redbelly Turtle (*Pseudemys alabamensis*). Unpublished M.S. thesis, University of South Alabama, Mobile.

Tyler, J. D. 2000. Noteworthy herpetological records for southwestern Oklahoma. *Proceedings of the Oklahoma Academy of Science* 80:119–122.

Ultsch, G. R., T. E. Graham, and C. E. Crocker. 2000. An aggregation of overwintering leopard frogs, *Rana pipiens*, and common map turtles, *Graptemys geographica*, in northern Vermont. *Canadian Field-Naturalist* 114:314–315.

———, and D. C. Jackson. 1995. Acid-base status and ion balance during simulated hibernation in freshwater turtles from the northern portions of their ranges. *Journal of Experimental Zoology* 273:482–493.

U.S. Fish and Wildlife Service. 1982. Endangered and threatened wildlife and plants; review of vertebrate wildlife for listing as endangered or threatened species. *Federal Register* 47:58454–58460.

———. 1985. Endangered and threatened wildlife and plants; review of vertebrate wildlife. *Federal Register* 50:37958–37967.

———. 1989. Endangered and threatened wildlife and plants; animal notice of review. *Federal Register* 54:554–579.

———. 1990. *Report to Congress: Endangered and Threatened Species Recovery Program*. U.S. Department of the Interior, U.S. Fish and Wildlife Service, Washington, D.C.

———. 1991. Endangered and threatened wildlife and plants; animal candidate review for listing as endangered or threatened species. *Federal Register* 56:58804–58836.

———. 1992. *Report to Congress: Endangered and Threatened Species Recovery Program*. U.S. Department of the Interior, U.S. Fish and Wildlife Service, Washington, D.C.

———. U.S. Fish and Wildlife Service. 1993. Endangered and threatened wildlife and plants; notice of 12-month finding on petition to list Cagle's map turtle. *Federal Register* 58:5701–5704.

———. 1994. Endangered and threatened wildlife and plants; animal candidate review for listing as endangered or threatened species. *Federal Register* 59:58892–59028.

———. 1996. Endangered and threatened wildlife and plants; review of plant and animal taxa that are candidates for listing as endangered or threatened species. *Federal Register* 61:7596–7613.

———. 2006. Endangered and threatened wildlife and plants; review of native species that are candidates or proposed for listing as endangered or threatened; annual notice of findings on resubmitted petitions; annual description of progress on listing actions. *Federal Register* 71:53756–53835.

———. 2011. Endangered and threatened wildlife and plants; partial 90-day finding on a petition to list 404 species in the southeastern United States as endangered or threatened with critical habitat. *Federal Register* 76:59836–59862.

Valentine, J. M., Jr., J. R. Walther, K. M. McCartney, and L. M. Ivy. 1972. Alligator diets on the Sabine National Wildlife Refuge, Louisiana. *Journal of Wildlife Management* 36:809–815.

Vance, T. 1986. Two unreported turtle records for Love County, Oklahoma. *Bulletin of the Oklahoma Herpetological Society* 11:13.

Van Cleave, H. J. 1913. The genus *Neorhynchus* in North America. *Zoologischer Anzeiger* 43:177–190.

———. 1919. Acanthocephala from the Illinois River, with descriptions of species and a synopsis of the family Neoechinorhynchidae. *Bulletin of the Illinois Natural History Survey* 13:225–257.

Vandewalle, T. J., and J. L. Christiansen. 1996. A relationship between river modification and species richness of freshwater turtles in Iowa. *Journal of the Iowa Academy of Science* 103:1–8.

Vermersch, T. G. 1992. *Lizards and Turtles of South-Central Texas.* Eakin Press, Austin, Tex.

Vetter, H. 2004. *Turtles of the World, vol. 2: North America.* Chimaira Buchhandelgesellschaft, Frankfurt am Main, Germany.

Vinyard, W.C. 1953. Epizoophytic algae from mollusks, turtles, and fish in Oklahoma. *Proceedings of the Oklahoma Academy of Science* 34:63–65.

Vogt, R. C. 1978. Systematics and Ecology of the False Map Turtle Complex *Graptemys pseudogeographica.* Unpublished Ph.D. dissertation, University of Wisconsin, Madison.

———. 1979. Cleaning/feeding symbiosis between grackles (*Quiscalus*: Icteridae) and map turtles (*Graptemys*: Emydidae). *Auk* 96:608–609.

———. 1980a. Natural history of the map turtles *Graptemys pseudogeographica* and *G. ouachitensis* in Wisconsin. *Tulane Studies in Zoology and Botany* 22:17–48.

———. 1980b. New methods for trapping aquatic turtles. *Copeia* 1980:368–371.

———. 1981a. Food partitioning in three sympatric species of map turtle, genus *Graptemys* (Testudinata, Emydidae). *American Midland Naturalist* 105:102–111.

———. 1981b. *Graptemys versa* Stejneger: Texas map turtle. *Catalogue of American Amphibians and Reptiles* 280:1–2.

———. 1981c. *Natural History of Amphibians and Reptiles in Wisconsin.* Milwaukee Public Museum, Milwaukee, Wisconsin.

———. 1981d. Turtle egg (*Graptemys*: Emydidae) infestation by fly larvae. *Copeia* 1981:457–459.

———. 1993. Systematics of the false map turtles (*Graptemys pseudogeographica* complex: Reptilia, Testudines, Emydidae). *Annals of the Carnegie Museum* 62:1–46.

———. 1995a. *Graptemys ouachitensis* Cagle: Ouachita map turtle. *Catalogue of American Amphibians and Reptiles* 603:1–4.

———. 1995b. *Graptemys pseudogeographica* (Gray): False map turtle. *Catalogue of American Amphibians and Reptiles* 604:1–6.

———, and J. J. Bull. 1982. Temperature controlled sex-determination in turtles: Ecological and behavioral aspects. *Herpetologica* 38:156–164.

———, and J. J. Bull. 1984. Ecology of hatchling sex ratio in map turtles. *Ecology* 65:582–587.

Wacha, R. S., and J. L. Christiansen. 1976. Coccidian parasites from Iowa turtles: Systematics and prevalence. *Journal of Protozoology* 23:57–63.

Wade, S. E., and C. E. Gifford. 1964. A preliminary study of the turtle population of a northern Indiana lake. *Proceedings of the Indiana Academy of Science* 74:371–374.

Wahlquist, H. 1970. Sawbacks of the Gulf Coast. *International Turtle & Tortoise Society Journal* 4(40):10–13, 28.

———, and G. W. Folkerts. 1973. Eggs and hatchlings of Barbour's map turtle, *Graptemys barbouri* Carr and Marchand. *Herpetologica* 29:236–237.

Walker, D., and J. C. Avise. 1998. Principles of phylogeography as illustrated by freshwater and terrestrial turtles in the southeastern United States. *Annual Review of Ecology and Systematics* 29:23–58.

———, P. E. Moler, K. A. Buhlmann, and J. C. Avise. 1998. Phylogeographic uniformity in mitochondrial DNA of the snapping turtle (*Chelydra serpentina*). *Animal Conservation* 1:55–60.

Wallace, G. E. 2000. Distribution and relative abundance of *Graptemys* c.f. *barbouri* on the Choctawhatchee River, Florida. *Florida Scientist* 63(suppl. 1):43.

Wallace, J. B., and A. C. Benke. 1984. Quantification of wood habitat in subtropical Coastal Plain streams. *Canadian Journal of Fisheries and Aquatic Sciences* 41:1643–1652.

Wallace, J. E., Z. W. Fratto, and V. A. Barko. 2007. A comparison of three sampling gears for capturing aquatic turtles in Missouri: The environmental variables related to species richness and diversity. *Transactions of the Missouri Academy of Science* 41:7–13.

Ward, H. B. 1921. A new blood fluke from turtles. *Journal of Parasitology* 7:114–128.

Ward, J. P. 1980. Comparative Cranial Morphology of the Freshwater Turtle Subfamily Emydinae: An Analysis of the Feeding Mechanisms and the Systematics. Unpublished Ph.D. dissertation, North Carolina State University, Raleigh.

Warwick, C., and C. Steedman. 1988. Report on the Use of Red-Eared Sliders (*Trachemys scripta elegans*) as a Food Source Utilized by Man. Unpublished report to the People's Trust for Endangered Species, London, UK.

Waters, J. C. 1974. The Biological Significance of the Basking Habit in the Black-Knobbed Sawback, *Graptemys nigrinoda* Cagle. Unpublished M.S. thesis, Auburn University, Auburn, Ala.

Watson, M. B., and T. K. Pauley. 2006. Spatial distribution of turtles along the Great Kanawha River, West Virginia. *Proceedings of the West Virginia Academy of Science* 78:14–25.

Webb, R. G. 1961. Observations on the life histories of turtles (genus *Pseudemys* and *Graptemys*) in Lake Texoma, Oklahoma. *American Midland Naturalist* 65:193–214.

———. 1962. North American Recent soft-shelled turtles (family Trionychidae). *University of Kansas Publications, Museum of Natural History* 13:429–611.

Weber, A. S., and J. B. Layzer. 2011. A comparison of turtle sampling methods in a small lake in Standing Stone State Park, Overton County, Tennessee. *Journal of the Tennessee Academy of Science* 86:45–52.

Weisrock, D. W., and F. J. Janzen. 2000. Comparative molecular phylogeography of North American softshell turtles (*Apalone*): Implications for regional and wide-scale historical evolutionary forces. *Molecular Phylogenetics and Evolution* 14:152–164.

Welter, W. A., and K. Carr. 1939. Amphibians and reptiles of northeastern Kentucky. *Copeia* 1939:128–130.

Wetzel, R. G. 1990. Reservoir ecosystems: Conclusions and speculations. Pages 227–238 in K. W. Thornton, B. L. Kimmel, and F. E. Payne (Eds.), *Reservoir Limnology: Ecological Perspectives*. John Wiley and Sons, New York.

Wharton, C. H., T. French, and C. Ruckdeschel. 1973. Recent range extensions for Georgia amphibians and reptiles. *Hiss News-Journal* 1:22.

Wheeler, W. M. 1899. George Baur's life and writings. *American Naturalist* 33(385):15–30.

White, D., Jr., and D. Moll. 1991. Clutch size and annual reproductive potential of the turtle *Graptemys geographica* in a Missouri stream. *Journal of Herpetology* 25:493–494.

———, and D. Moll. 1992. Restricted diet of the common map turtle *Graptemys geographica* in a Missouri stream. *Southwestern Naturalist* 37:317–318.

Whitlock, S. C. 1939. Snails as intermediate hosts of Acanthocephala. *Journal of Parasitology* 25:443.

Wibbels, T., F. C. Killebrew, and D. Crews. 1991. Sex determination in Cagle's map turtle: Implications for evolution, development, and conservation. *Canadian Journal of Zoology* 69:2693–2696.

Wiens, J. J., C. A. Kuczynski, and P. R. Stephens. 2010. Discordant mitochondrial and nuclear gene phylogenies in emydid turtles: Implications for speciation and conservation. *Biological Journal of the Linnean Society* 99:445–461.

Wilson, D. S., C. R. Tracy, and C. R. Tracy. 2003. Estimating age of turtles from growth rings: A critical evaluation of the technique. *Herpetologica* 59:178–194.

Wilson, R. L., and G. R. Zug. 1966. A fossil map turtle (*Graptemys pseudogeographica*) from central Michigan. *Copeia* 1966:368–369.

Winemiller, K. O., and E. R. Pianka. 1990. Organization in natural assemblages of desert lizards and tropical fishes. *Ecological Monographs* 60:27–55.

Wolfe, J. L., D. K. Bradshaw, and R. H. Chabreck. 1987. Alligator feeding habits: New data and a review. *Northeast Gulf Science* 9:1–8.

Wondzell, S. M., and P. A. Bisson. 2003. Influence of wood on aquatic biodiversity. Pages 249–263 in S. V. Gregory, K. L. Boyer, and A. M. Gurnell (Eds.), *The Ecology and Management of Wood in World Rivers*. American Fisheries Society, Bethesda, Md.

Woo, P. T. K. 1969a. The life cycle of *Trypanosoma chrysemydis*. *Canadian Journal of Zoology* 47:1139–1151.

———. 1969b. Trypanosomes in amphibians and reptiles in southern Ontario. *Canadian Journal of Zoology* 47:981–988.

Wood, J. T. 1946. *Graptemys geographica* (Le Sueur) added to herpetofaunal list of Great Smoky Mountains National Park. *Copeia* 1946:168.

Wood, R. C. 1977. Evolution of the emydine turtles *Graptemys* and *Malaclemys* (Reptilia, Testudines, Emydidae). *Journal of Herpetology* 11:415–421.

Wynn, D. E., and S. M. Moody. 2006. *Ohio Turtle, Lizard, and Snake Atlas*. Ohio Biological Survey, Columbus.

Yarrow, H. C. 1883. Check list of North American Reptilia and Batrachia, with catalogue of specimens in the U.S. National Museum. *Bulletin of the U.S. National Museum* 24:1–249.

Yntema, C. L. 1976. Effects of incubation temperatures on sexual differentiation in the turtle, *Chelydra serpentina*. *Journal of Morphology* 150:453–462.

Young, E. A., and M. C. Thompson. 1995. New county records for two reptiles in Kansas. *Transactions of the Kansas Academy of Science* 98:80–81.

Zalewski, M., M. Lapinska, and P. B. Bayley. 2003. Fish relationships with wood in large rivers. Pages 195–211 in S. V. Gregory, K. L. Boyer, and A. M. Gurnell (Eds.), *The Ecology and Management of Wood in World Rivers*. American Fisheries Society, Bethesda, Md.

Ziglar, C. L., and R. V. Anderson. 2002. Epizoic organisms on turtles in Pool 20 of the upper Mississippi River. *Journal of Freshwater Ecology* 17:389–396.

Index

P. iii: sexual dimorphism in head width in Cagle's map turtle (see p. 242).
P. 2: adult male Texas map turtles (see p. 139).

Copyedited by Darcy Wilson
Design and composition by Chris Crochetière, BW&A Books, Inc.
Set in Minion and The Sans
Jacket design by Tony Roberts
Image prepress by University of Oklahoma Printing Services
Range maps by Peter V. Lindeman and Gerry Krieg

www.ingramcontent.com/pod-product-compliance
Lightning Source LLC
Chambersburg PA
CBHW040253290326
41929CB00051B/3353